# Sources in the History of Mathematics and Physical Sciences

3

# Sources in the History of Mathematics and Physical Sciences

## Editor: G.J. Toomer

VOLUME 1

**Diocles on Burning Mirrors**
The Arabic Translation of the Lost Greek Original
Edited, with English Translation and Commentary by G.J. Toomer
1976. ix, 249 pages. With 37 Figures and 24 Plates.
ISBN 0-387-07478-3

VOLUME 2

**Wolfgang Pauli**
Scientific Correspondence with Bohr, Einstein, Heisenberg, A.O.
Volume I: 1919–1929
Edited by A. Hermann, K.v. Meyenn, V.F. Weisskopf
1979. xlvii, 577 pages.
ISBN 0-387-08962-4

VOLUME 3

**The Arabic Text of Books IV to VII of Diophantus' *Arithmetica***
in the Translation of Qusṭā ibn Lūqā
By Jacques Sesiano
1982. vii, 502 pages. With 4 Figures.
ISBN 0-387-90690-8

VOLUME 4

**Descartes on Polyhedra**
A Study of the *De Solidorum Elementis*
By P.J. Federico
1982. x, 144 pages. With 36 Figures.
ISBN 0-387-90760-2

Jacques Sesiano

# Books IV to VII of Diophantus' *Arithmetica*

in the Arabic Translation Attributed to Qusṭā ibn Lūqā

Springer-Verlag
New York Heidelberg Berlin

Jacques Sesiano
4, avenue du Mail
1205 Genève
Switzerland

AMS Subject Classifications: 01A20, 01A30

Library of Congress Cataloging in Publication Data

Sesiano, Jacques.
Books IV to VII of Diophantus' Arithmetica in the Arabic translation attributed to Qusṭā ibn Lūqā.

(Sources in the history of mathematics and physical sciences; 3)
Diophantus text in English and Arabic.
Originally presented as the author's thesis (doctoral—Brown University, 1975)
Bibliography: p.
Includes indexes.
1. Diophantus, of Alexandria. Arithmetica. 2. Mathematics, Greek. 3. Mathematics, Arabic. I. Diophantus, of Alexandria. Arithmetica. Book 4–7. Arabic & English. 1982. II. Title. III. Title: Books 4–7 of Diophantus' Arithmetica in the Arabic translation attributed to Qusṭā ibn Lūqā. IV. Title: Books four to seven of Diophantus' Arithmetica in the Arabic translation attributed to Qusṭā ibn Lūqā. V. Series.
QA22.D5623S47 1982 512 82-19077

With 4 Illustrations



Softcover reprint of the hardcover 1st edition 1982



Typeset by Composition House Ltd., Salisbury, England.

9 8 7 6 5 4 3 2 1

ISBN-13: 978-1-4613-8176-1 e-ISBN-13: 978-1-4613-8174-7
DOI: 10.1007/978-1-4613-8174-7

To my wife, Karin

# Preface

This edition of Books IV to VII of Diophantus' *Arithmetica*, which are extant only in a recently discovered Arabic translation, is the outgrowth of a doctoral dissertation submitted to the Brown University Department of the History of Mathematics in May 1975. Early in 1973, my thesis adviser, Gerald Toomer, learned of the existence of this manuscript in A. Gulchīn-i Maʿānī's just-published catalogue of the mathematical manuscripts in the Mashhad Shrine Library, and secured a photographic copy of it. In September 1973, he proposed that the study of it be the subject of my dissertation. Since limitations of time compelled us to decide on priorities, the first objective was to establish a critical text and to translate it. For this reason, the Arabic text and the English translation appear here virtually as they did in my thesis. Major changes, however, are found in the mathematical commentary and, even more so, in the Arabic index. The discussion of Greek and Arabic interpolations is entirely new, as is the reconstruction of the history of the *Arithmetica* from Diophantine to Arabic times.

It is with the deepest gratitude that I acknowledge my great debt to Gerald Toomer for his constant encouragement and invaluable assistance. It was under his guidance that I learned how to read mediaeval mathematical manuscripts and how to establish a critical text. He spared neither his time nor his energy, abandoning his own scholarly pursuits in order to facilitate my study of the Diophantus manuscript. This generous help also continued after the completion of the thesis; virtually all new ideas or interpretations have undergone his scrutiny.

I should also like to thank my former professor at the Swiss Federal Institute of Technology in Zurich (ETHZ), Dr. Peter Huber, who first encouraged me to study the History of Mathematics and who later helped procure a grant for me from ETHZ. This, together with Brown University's waiving of tuition fees, enabled me to pursue my studies at Brown University for three years.

During my stay in Providence and since then, I have enjoyed the continuing help and encouragement of Professor emer. O. Neugebauer, Professor and Mrs. A. Sachs, and Professor D. Pingree. All contributed to my formation and offered many valuable suggestions for my work.

Special thanks are due to the curator of the Shrine Library in Mashhad, Aḥmad Gulchīn-i Maᶜānī, for kindly having made available the necessary photographic reproductions to Gerald Toomer.

While preparing my thesis and this edition, I has occasion to consult a considerable amount of literature, and I should like to express my sincere thanks to the personnel of the Brown University Library and, more particularly, to the personnel of the Bibliothèque Publique et Universitaire of Geneva for their generous assistance.

Finally, I must express my gratitude to my wife, Karin, on whom devolved the delicate task of reading, polishing, and often reformulating the English text and many arguments in it. Despite family obligations, she found time to read the entire text several times and to rewrite the unsatisfactory parts of it until a coherent whole emerged. It is to her that I dedicate this, my first book.

*Geneva,*
*Switzerland*
September, 1982.

JACQUES SESIANO

# Table of Contents


Part One: Introduction 1

Chapter I. The Four Arabic Books and the *Arithmetica* 3

§1. Authenticity of the Arabic Books 3
1. That the Arabic Books Belong to the *Arithmetica* 4
2. Concerning Their Place Within the *Arithmetica* 4
a. Placement of the Arabic Book IV 4
b. Basic Methods Used in the Arabic Books 5
c. Placement of the Four Arabic Books 7

§2. Diophantus in Islamic, and Byzantine, Times 8
1. Qusṭā ibn Lūqā and the *Arithmetica* 8
2. Islamic Mathematicians and the *Arithmetica* 9
a. Abū Kāmil 9
b. Al-Ḫazin 10
c. Abū'l-Wafā᾿ 10
d. Al-Karaji 10
e. Ibn al-Haiṯam 11
f. Samaw᾿al ibn Yaḥyā 11
Appendix. Designation of the *Arithmetica* in Arabic 13
3. Mathematicians and the *Arithmetica* in Byzantium 14
a. The Time of Leon the Mathematician 14
α. The Seventh and Eighth Centuries 14
β. The Century of Leon 14
γ. From Leon to Planudes 16
b. The Time of Maximus Planudes 16
c. Oldest Greek Manuscripts Still Extant 18
α. Non-Planudean Class 18
β. Planudean Class 20

Chapter II. The Extant Arabic Text 21

§3. Description of the Manuscript 21

§4. Orthographical Remarks 23
1. Writing of the *hamza*[h] 23
2. Particular Endings 28
3. Numerals 28
4. Repeated, Erroneous Spellings 29

§5. Additions by Earlier Readers (or Copyists) 29

§6. On the Progenitor of Our Manuscript 36

§7. Grammatical and Lexicological Remarks 37
1. Numbers and Powers 37
A. Integers 37
a. Grammatical Peculiarities 37
b. Determination 38
B. Fractions 39
a. General Fractions 39
α. Expression 39
β. Determination 40
b. Aliquot Fractions and Related Cases 40
α. Expression 40
β. Decomposition of Some Fractions 41
γ. Grammatical Peculiarities Connected with Aliquot Fractions 41
C. Grammatical Number of a Mathematical Expression 42
a. Units 42
b. Multiple of a Power 42
c. Algebraic Polynomial Expression 43
D. Powers 43
a. The Greek Power-system 43
b. The Arabic Power-system 44
c. The Power-system in Our Text 44
α. $x^5$ 45
β. $x^8$ 45
d. Grammatical Determination of the Powers 46
α. Two Elements 46
β. Three Elements 46
2. Some Grammatical Remarks on Verbs 46
a. Verbal Persons Used 46
b. Jussives of Weak Verbs 46
c. The Verb ʿ*adala* 47
α. Agreement of ʿ*adala* 47
β. Agreement of the Auxiliary of ʿ*adala* 47

Chapter III. Tentative Reconstruction of the History of the *Arithmetica* 48

§8. Formal Subdivisions of a Problem 48
1. Analysis and Synthesis 48
2. Subdivisions of a Problem 49

§9. Major, Unsystematic Supplements in the *Arithmetica* 50
1. Interpolated Problems in the *Arithmetica* 51
2. Alternative Resolutions (ἄλλως) 54
a. In the Greek Books 54
b. In the Arabic Books 54
3. Other Supplements 55
a. Corollaries 55
b. Remarks 56
c. Additional Computations 56
Appendix. A Comparison Between al-Karajī's Version and the Extant *Arithmetica* 57

§10. Errors in the Problems of the Arabic Books 60

§11. Quality of the Translation 65
1. Imperfections in the Translation 65
2. General Character of the Translation 67

§12. Genealogy of the Mashhad Manuscript 68
1. Earliest Additions 68
2. The Major Commentary 68
a. Additions Originating with the Major Commentary 68
α. Additions in the Analysis 69
β. Additions in the Synthesis 69
b. Value of This Commentary 70
c. Possible Authorship of the Major Commentary 71
3. The Addition of the Final Statements 72
4. The Arabic Diophantus 73
5. Genealogical Tree of the Mashhad Manuscript 74

§13. On the Missing Part of the *Arithmetica* 76
1. New Aspects of the Problem 76
2. The Announcement in the Greek Introduction 77
3. Diophantus and the Equation $Ax^2 + Bx + C = \square$ 78
4. On Some Problems of a Diophantine Nature Found in Islamic Mathematics but Not in the Extant *Arithmetica* 81
a. Problems of Abū Kāmil 81
b. Problems of al-Karajī 82
5. Conclusion 83

## Part Two: Translation 85

Book IV 86
Book V 126
Book VI 139
Book VII 156

## Part Three: Mathematical Commentary 173

Book IV 175
Book V 223
Book VI 244
Book VII 261

## Part Four: Text 281

Book IV 283
Book V 352
Book VI 374
Book VII 403

## Part Five: Arabic Index 431

Appendix: Conspectus of the Problems of the *Arithmetica* 461

Bibliography 485

General Index 493

# Part One

# Introduction

# Chapter I

# The Four Arabic Books and the *Arithmetica*

## §1. Authenticity of the Arabic Books

The Greek mathematician Diophantus of Alexandria is known with certainty to have lived between 150 B.C. and A.D. 350, as we infer from his having mentioned Hypsicles and from his having been mentioned by Theon of Alexandria; it seems fairly probable, though, that he flourished about A.D. 250.[1] We can be sure that he wrote at least two treatises: one dealing with problems in indeterminate analysis, the *Arithmetica*, and another, smaller, tract on polygonal numbers, both of which are only partially extant today.

We gather from the *Arithmetica*'s introduction that it originally consisted of thirteen Books.[2] Of these, only six have survived until now in Greek, and they have been edited and translated several times.[3] The remaining seven were considered irretrievably lost until the recent discovery of four other, hitherto unknown Books in an Arabic translation which, since it is attributed to Quṣṭā ibn Lūqā, must have been made around or after the middle of the ninth century.

---

[1] On the (conjectural) basis for this date, see, e.g., Heath, *Diophantus*, p. 2. The only information (perhaps invented) we have relating to Diophantus' life is contained in a brief arithmetical epigram-problem (cf. *ibid.*, p. 3, or D.G. (= Tannery's two-volume edition of the Greek Diophantus), II, pp. 60–61).

[2] This is also said in a later source (D.G., II, p. 73,26 and *ibid.*, p. xiii).

[3] See the list given by K. Vogel in his article on Diophantus, *DSB*, IV, pp. 118–19, supplemented by my article, *ibid.*, XV, p. 122.

(N.B. I. Veselovsky and I. Bashmakova's *Arifmetika Diofanta*, which is a commented Russian translation of the Greek *Arithmetica*, is wrongly said in my article to have been translated into German. In fact, the work by I. Bashmakova translated into German has nothing to do with the above, and the error originates with the editors of the *DSB*, who added this reference without my knowledge. Incidentally, a reference to an article by E. S. Stamatis, which I have not seen, and therefore cannot judge, was also added.)

## 1. That the Arabic Books Belong to the *Arithmetica*

There can be little doubt that the four Arabic Books which we have were once part of Diophantus' *Arithmetica.* The first indication is in the text itself, for it is explicitly stated at the beginning and end of each Book that the author is Diophantus (*Diyūfanṭus*[4]). It is conceivable, at least for the sake of argument, that these assertions could be questioned by supposing that the Books are some pseudepigraphic tract. But this argument is considerably weakened if not refuted when one takes into account the existence of two problems, in one of the Greek Books, which are unquestionably interpolations stemming from problems original to the Arabic Books.[5] And subsequent considerations will show that no serious doubt can be raised about the Arabic Books' having once belonged to the *Arithmetica.*

## 2. Concerning Their Place Within the *Arithmetica*

The four parts of our Arabic version are labelled Books IV, V, VI and VII. But since three Books in Greek have been considered to be the fourth, fifth, and sixth Books since (at least) the thirteenth century, the next question is: which of the two sets actually follows the first Greek Books labelled I, II, and III? We conclude from several observations that the four Arabic Books must follow the Greek Books I–III, and must precede the three later Greek Books.[6]

### *a. Placement of the Arabic Book IV*

(α) In the introduction to Book IV, it is asserted that none of the problems found *before* involved any unknown raised to a degree higher than the second—whereas the coming problems will involve cubes also (as well as higher powers composed of squares and cubes; see §7,1,D). Now, the first part of this statement is accurate only in reference to Books I–III,[7] for in all the other Books,

---

[4] Remarks on the transliteration: (α) The *ḏāl* has been employed in part of our manuscript as the initial letter in Diophantus' name (e.g., lines 1 and 7 in our edition)—as well as in some other manuscripts wherein the name of Diophantus is mentioned—which is significant because the *ḏāl* represents a *scriptura difficilior.* Indeed, the common transliteration of δ is, in Arabic, *ḏāl* and not *dāl*, a use reflecting the transformation of the (originally) explosive *delta* into a spirant letter, which transformation occurred in the first centuries of the Christian Era. (β) As is usual in the transliteration of Greek names, *tau* becomes *ṭā*. (γ) The final vocalization *-us* is arbitrary: in the absence of any particular indication, one might just as easily write *Diyūfanṭis*. There are also instances of an *-ūs* ending (Abū'l-Faraj, *Hist. orient.* (ed. Pococke), p. 338) and of an *-īs* ending (cf. Ibn al-Nadīm's *Fihrist* (ed. Flügel), II, p. 125).

[5] See the mathematical commentary to V,7–8.

[6] We shall henceforth designate as Books "IV", "V", "VI" the last three Greek Books.

[7] On the only exception, see p. 178, n. 11.

Greek or Arabic, we find problems involving powers $x^n$ with $n \geq 3$. Thus the Arabic Book IV must in any event precede both set V–VII and set "IV"–"VI".

($\beta$) That Book IV must have come immediately after Book III in the Arabic Diophantus is confirmed twice by Arabic sources independent of our manuscript.

The first confirmation is an indirect one gleaned from the mathematician al-Karajī's *Faḫrī*, which contains a considerable number of problems taken from Books I–IV of the *Arithmetica* (cf. §2,2,d). All the problems taken from Book IV appear directly after all those taken from Book III, and undoubtedly al-Karajī was simply following the order of his version of Diophantus.

The second confirmation is found in a marginal gloss in the *Faḫrī* manuscript which Woepcke studied.[8] In his *Extrait du Fakhrî* (pp. 22–23), he translates it as follows: "J'ai vu[9] en cet endroit une glose de l'écriture d'Ibn Alsirâdj conçue en ces termes: Je dis, les problèmes de cette section et une partie de ceux de la section précédente, sont pris dans les livres de Diophante, suivant l'ordre. Ceci fut écrit par Ahmed Ben Abi Bekr Ben Ali Ben Alsirâdj Alkelânecî. Fin (de la glose)". This gloss is written in the margin of fol. 98$^r$, where the fourth section of the *Faḫrī* ends, and Woepcke took it to refer to the *third* and (just-ending) *fourth* sections of the *Faḫrī*. Now, the third section does indeed give many problems taken from the *Arithmetica*, but the fourth section includes twenty-five problems not at all Diophantine in type inserted between fourteen problems of Book II and the problems of Book III. This led Woepcke, who was of course acquainted only with the extant Greek Books of Diophantus, to assume that the assertion made in the gloss was inaccurate. Now that we know that the problems in the fifth section of the *Faḫrī* are all taken from Book IV, it is clear that the gloss must refer to sections *four* and *five*, in which case its assertion is not only perfectly correct but also confirms the sequence (Greek) Book III and (Arabic) Book IV.

### *b. Basic Methods Used in the Arabic Books*

Except for allusions to immediately preceding problems, the references to earlier propositions found in the set IV–VII concern only Books II–III, and, more generally, the intermediate problems at which one arrives in numerous propositions of the Arabic Books are all soluble by methods taught in Book II.[10] These basic methods are the following.

---

[8] Paris, B.N. arabe 2459 (*olim* Suppl. ar. 952).

[9] The author of the gloss is speaking.

[10] Only a numerical result is taken from Book III (see the allusion to III,6 in VII,7,1° and 2°).

N.B. There is not always an explicit reference when methods from Book II are used. Such references, when made, may be to Book II (in IV,26,b; IV,35; VI,12), or to "the treatise" (VI,1; VII,12–14), or be even more vague (VI,2; VII,4).

**II,8**: To divide a given square into two squares.[11]

$k^2$ being the given square and $a$, $b$ the required parts, one puts $a = x$, $b = hx - k$, with $h$ parameter; hence

$$k^2 = x^2 + (hx - k)^2,$$

whence, after performing the algebraical transformations explained in the introduction to Book I[12] and resolving for $x$,

$$x = \frac{2hk}{1 + h^2}.$$

**II,9**: To divide a given number, which is the sum of two squares, into two other squares.[13]

$k = k_1^2 + k_2^2$ being the given number and $a$, $b$ the required parts, one puts $a = x + k_1$, $b = hx - k_2$; hence

$$k = (x + k_1)^2 + (hx - k_2)^2,$$

and

$$x = \frac{2(hk_2 - k_1)}{h^2 + 1}.$$

**II,10**: To find two square numbers of given difference.[14]

$k$ being known, with $k = b^2 - a^2$, one will put $a = x$, $b = x + h$; hence

$$x = \frac{k - h^2}{2h}.$$

**II,11**: To add to two given numbers a (required) number making each of them a square.

$k$, $l$ being given, and $a$ being the required number, one must solve

$$\begin{cases} a + k = \square, \\ a + l = \square'. \end{cases}$$

---

[11] Required in problems IV,26,b; VI,2 (interpolated); VI,13; VII,12 and 13. Many problems of Book VI use the elementary relation $9 + 16 = 25$, and do not therefore require a real application of II,8 (or II,10); the same holds for VII,6 (interp.) and VII,16–18.

[12] Namely: elimination of the negative terms by the addition of their (absolute) value to both sides, and then elimination of the equal quantities from both sides (the *al-jabr* and the *al-muqābala*ʰ of the Arabians). See D.G., I, p. 14,11–20.

[13] Required in problems IV,35; 40; 42,b and VII,14.

[14] Required in numerous problems of Book IV (25; 26,a; 27; 34,2°; 36–39; 41; 42,a,2°; 43; 44,a–c), in VI,1 and 3 (interpolated), and in VI,12. In IV,44,a–c, the intermediate problems are stated in the form of Diophantus' II,11–13, which are all reducible to II,10. See also above note 11.

This may be done in two ways; these are explained again in the Arabic Book IV (see problems 34 and 42,a).

On the basis of what has been taught in these problems, we also know how to solve an equation of the type

$$Ax^2 + Bx + C = \square$$

where one of the extreme coefficients $A$, $C$ is either zero or a square. This is done by putting $\square = h^2$ or $\square = h^2x^2$ if $A = 0$ and $C = 0$, respectively, and $\square = (\sqrt{A}x + h)^2$ or $\square = (hx + \sqrt{C})^2$ for $A$ or $C$ square. The parameter $h$ of the resulting linear equation is chosen so as to give a positive solution.[15]

**II,19**: To find three squares such that the difference between the largest and the middle has to the difference between the middle and the smallest a given ratio.[16]

$p$, $q$ being given, to find $a$, $b$, $c$ such that

$$\frac{a^2 - b^2}{b^2 - c^2} = \frac{p}{q}.$$

Putting $a = x + h$, $b = x + 1$, $c = x$,[17] we have

$$(x + h)^2 - (x + 1)^2 = \frac{p}{q}[(x + 1)^2 - x^2],$$

whence
$$x = \frac{h^2 - \left(\frac{p}{q} + 1\right)}{2\left[\left(\frac{p}{q} + 1\right) - h\right]}.$$

We shall have $x > 0$ by choosing any $h$ fulfilling

$$\sqrt{\frac{p}{q} + 1} < h < \frac{p}{q} + 1.$$

### *c. Placement of the Four Arabic Books*

We have seen that the Arabic Book IV must have followed the Greek Book III. Since the problems at the beginning of Book V are perfectly similar in type

[15] See for such equations: ($\alpha$) problems IV,28–31 and the interpolated ones VI,4–10 and VII,5 (case $A$ or $C = 0$); ($\beta$) problems VI,13 (second part)—21 and 23, and VII,8–10 and 16–18 (case $A$ or $C$ square).

[16] Required in problems V,1–3 and 6. The identities given in V,4–5 can be obtained by the same method.

[17] Since any multiple of a solution is also a solution, taking the additive quantity in the middle term equal to 1 is not restrictive.

to those at the end of Book IV, Book V must likewise have followed Book IV in the original *Arithmetica*.

There are no such conclusive arguments for the placement of Books VI and VII. Neither the existence in Book VI of interpolations stemming from Book IV and in Book VII of an interpolation stemming from Book VI, nor the allusion in the preface to Book VII to the similarity of its problems to those in "Books IV and V" (VI might have been originally alluded to also, see p. 263) can be considered decisive proof for the sequence of the Books. Certain arguments of a more general nature are, however, convincing enough. First, knowledge of the methods listed under (*b*) is the only prerequisite for undertaking the study of the Arabic Books, so that the group IV–VII is the natural continuation of Books I–III.[18] Further, as we shall see later, the goal of the four Arabic Books seems to have been to train the reader in the use of the basic methods found in Book II; no fundamentally new procedures appear in these Books whereas some do in the later three Greek Books. Finally, that all the Arabic Books must be placed before Books "IV"–"VI" is suggested by the notably greater difficulty of the later three Greek Books.

## §2. Diophantus in Islamic, and Byzantine, Times

### 1. Qusṭā ibn Lūqā and the *Arithmetica*

The information we have concerning the life and works of Qusṭā ibn Lūqā (*fl. ca.* 860) comes essentially from three Arabic sources:

(a) the *Fihrist* of Ibn al-Nadīm (*ca.* 987);
(b) the *Dictionary of Philosophers and Scientists* by Ibn al-Qifṭī (d. 1248/49);
(c) the *History of the Physicians* by Ibn abī Uṣaibiᶜaʰ (d. 1269/70).

The following biographical sketch emerges from these sources.[19]

> Qusṭā, the son of Lūqā, was born in Baalbek (Heliopolis) and was a Christian of Greek origin.[20] He was a physician, philosopher, astrologer, mathematician, and musician, and was proficient in Greek, Arabic, and Syriac. He travelled in the Byzantine Empire and returned to Syria with

[18] In particular, Books I to VII all share the characteristic of leading to an equation with only one term on each side (cf. §13).

[19] See G. Gabrieli's *Nota biobibl.*, pp. 361–62. For other sources, see Daiber, *Placita*, p. 3.

[20] The Arabic codex Mus. brit. 407,10 also gives him the epithet *al-yūnānī* (cf. *Cat. codd. mss. orient. Mus. brit.*, pp. 193–94).

Gabrieli thought that the name Qusṭā could be a Syriac abbreviation of Constans or Constantinus. And in fact, in the Cairo manuscript containing the (Arabic version of the) *Mechanics* of Heron, the name of the translator is given as Qusṭanṭīn (*Heronis opera*, II, p. 3, app. crit.); of interest also is G. Toomer's observation that Κωστᾶ, in modern Greek, is a familiar form of Κωνσταντῖνος.

many manuscripts of Greek works. He was summoned to Baghdad to work as a translator, and he took with him many manuscripts which he himself translated or had others translate. He also revised many translations. Some time later he was invited to Armenia where he spent the rest of his life and wrote a number of books.

Among the many works attributed to him, two are of interest to us:

1°. A *translation* of Diophantus' treatise on algebra (*kitāb fī tarjama^h Diyūfanṭus fī'l-jabr wa'l-muqābala^h*) mentioned by Ibn abī Uṣaibiʿa^h (art. *Qusṭā*).

2°. A *commentary* on three and a half Books of the treatise by Diophantus on arithmetical problems (*tafsīr (li-) ṯalāṯ maqālāt wa-niṣf min kitāb Diyūfanṭus fī'l-masāʾil al-ʿadadīya^h*) mentioned by Ibn al-Nadīm and Ibn abī Uṣaibiʿa^h (both: art. *Qusṭā*).

The incipit of the manuscript containing Books IV to VII leaves little doubt about the origin of our text: it belongs to the *translation* made by Qusṭā; there is no allusion to any reworking of the text by the translator. Further, there are several indications that the translator did not follow the reasonings with great care—in any event not always—and some mistakes even point to a rapid translation of the text (see §11).

No sources make any mention of Qusṭā's translation having been limited to a certain number of Books for they all, as well as our manuscript's own incipit, credit him with "the translation of Diophantus' treatise". It is unlikely, however, that Qusṭā's translation was really complete, that is, that it included all thirteen Books. What we do know is that the first seven Books existed in Arabic translation—Books I–III (and IV) appearing in large part in al-Karajī's *Faḫrī*, and Books IV to VII in our manuscript.[21]

## 2. Islamic Mathematicians and the *Arithmetica*

### a. *Abū Kāmil* (*ca.* 880[22])

There is nothing to suggest that the Egyptian Abū Kāmil had any direct (or even indirect) knowledge of Diophantus' *Arithmetica*, although the problems in his *Algebra* dealing with indeterminate analysis are perfectly Diophantine in form and the basic methods are attested to in the *Arithmetica* (see my *Méthodes* (...) *chez Abū Kāmil*). I have strong suspicions that Abū

[21] We may not conclude with certainty from the explicit of our manuscript (*tamma al-kitāb*) that Qusṭā translated seven Books, and no more; for this "end of the treatise" may simply indicate the end of our manuscript (or of its progenitor). My assertion in the *Méthodes* (...) *chez Abū Kāmil*, p. 90, that Qusṭā's translation "n'a très certainement jamais contenu plus que les livres I à VII" may thus be too absolute. (N.B. "livres *IV* à VII" in the article is of course a mistake.)

[22] On the deduction of this date, see Anbouba, *Un Algébriste arabe*, pp. 7–9.

Kāmil had some (originally) Greek material at his disposal, a thesis which I shall examine in my proposed edition of his *Algebra.*

### b. *Al-Ḫazin* (*ca.* 940)

Abū Jaᶜfar al-Ḫazin wrote a short treatise, the core of which is the resolution of the indeterminate system $x^2 \pm k = \square$, and which ends with several propositions relating to the representation of numbers as a sum of two squares. This latter part will, he says, help "clarify the lemma (*muqaddimaʰ*) put by Diophantus as a preliminary to the nineteenth proposition of the third Book of his treatise on algebra (*fi'l-jabr*)" (Anbouba, *Traité d'Abū Jaᶜfar*, p. 161). Note that the proposition is given the same number here as in Tannery's edition.

### c. *Abū'l-Wafāʾ* (940–997/8)

That the Persian Abū'l-Wafāʾ al-Būzjānī commented on (part of) the *Arithmetica* is attested to by several, more or less interdependent, sources. Thus, Ibn al-Nadīm says that he wrote a "commentary (*tafsīr*) on the treatise of Diophantus on algebra (*fi'l-jabr*)", and on "proofs of the theorems (*al-qaḍāyā*) used by Diophantus in his treatise and of what he (Abū'l-Wafāʾ) used in his commentary" (*Fihrist*, art. Abū'l-Wafāʾ). These two statements are repeated by Ibn al-Qifṭī, who merely drops the second part of the latter (Casiri, *Bibliotheca*, I, pp. 433–34). Finally, Abū'l-Faraj, who notes that Abū'l-Wafāʾ arrived in Iraq in the year 348 (959/60) observes about him, among other things, that he "commented (*fassara*) on the treatise of Diophantus on algebra (*fi'l-jabr wa'l-muqābalaʰ*)" (cf. *Hist. orient.* (ed. Pococke), p. 338).

### d. *Al-Karajī* (*ca.* 1010)

In our discussion relative to the authenticity of the Arabic Books, we mentioned al-Karajī's *Faḫrī* and a reader's gloss found in one of the *Faḫrī* manuscripts at the beginning of the fifth section. It is from the *Faḫrī* that we can best infer which problems of the now missing first part of the Arabic version, covering Books I to III, were definitely known in Arabic times.[23] For al-Karajī reproduces in the *Faḫrī* with little if any change—except for the wording—nearly half of the first Book of the *Arithmetica*, most of Book II (absent are II,1–7 and 17), Book III except for one problem (Diophantus' III,4); next, almost all of Book IV appears.[24] Al-Karajī has not taken a single

[23] This is not to say that those problems not found in the *Faḫrī* were missing in the Arabic Diophantus; see p. 58, Remark.

[24] For a more detailed discussion, see Woepcke's *Extrait*, pp. 18–21 (Books I–III) and below, pp. 57–60. Note that Woepcke's numbering is not the same as Tannery's, but corresponds to the one used in Bachet's edition of the *Arithmetica*.

problem from Book V onwards; nor does his later *Badīʿ* contain any trace of problems from Books V–VII or "IV"–"VI".

In the said *Badīʿ*, the third part of which is devoted to indeterminate analysis, al-Karajī no longer slavishly reproduces Diophantus' problems but instead presents, for the benefit of the general reader, the methods of the first Books of the *Arithmetica*. Thus he provides an excellent introduction to the study of elementary Diophantine analysis as developed in Books II and III (see our *Traitement des éq. ind.*, pp. 305–6).

Though al-Karajī incorporated many of Diophantus' problems into his *Faḫrī*, never once is Diophantus' name mentioned in connection with these. Instead, his name appears in association with a "method" (*ṭarīq*, *maḏhab*) of solving determinate equations of the types $x^2 + px = q$ and $x^2 + q = px$ ($p, q > 0$). Since this "method" comes after the explanation of the formula and, particularly, of the "Euclidean" demonstration (see *Extrait*, pp. 65–68), it would seem that al-Karajī simply wished to contrast the geometrical demonstration with an algebraical one. It is unlikely that al-Karajī knew of any Diophantine method for solving complete quadratic equations, for not only do these equations occur in the *later* three Greek Books, with which al-Karajī was apparently not acquainted,[25] but also the approach in their resolutions in the *Arithmetica* is not that used by al-Karajī and explicitly attributed by him to Diophantus. One might hypothesize that al-Karajī knew of some other treatise by Diophantus or even a pseudepigraph on this subject, but we have no source which associates the name of Diophantus with any such work.

### e. *Ibn al-Haiṯam* (965–*ca.* 1040)

Among the writings of Ibn al-Haiṯam, Ibn abī Uṣaibiʿah cites the following: "Remarks made by Isḥāq ibn Yūnus the physician in Cairo on the authority of (*ʿan*) Ibn al-Haiṯam on the treatise of Diophantus on problems of algebra" (ed. Müller, II,98, bottom).[26] This is the only information we have relative to Ibn al-Haiṯam (and Isḥāq ibn Yūnus) with regard to the *Arithmetica*.

### *f. Samawʾal ibn Yaḥyā* (*ca.* 1180)

Samawʾal ibn Yaḥyā is the author of an extensive work on algebra entitled *al-Bāhir*, which is largely a commentary on materials gathered from other authors. At the very end of his treatise, after having spoken about the indeterminateness of problems, he refers the reader desiring some practice in the solving of nondeterminate problems to his *commentary* (šarḥ) *on the treatise by Diophantus the Alexandrian*, a work apparently lost now (if ever written).

---

[25] On his ignorance of Book "VI" in particular, see the *Traitement des éq. ind.*, pp. 317–18.

[26] Isḥāq ibn Yūnus was a pupil of Ibn al-Haiṯam; see Suter, *Math. und Astron.*, no. 248.

There are four other passages of his *Bāhir* relevant to the *Arithmetica*, and they are the following.

1°. (ed., p. 112). After giving the relation

$$\frac{1}{2}\left\{\frac{u^2 - v^2}{d} \pm d\right\} = u \text{ and } v, \quad \text{respectively } (u > v),$$

Samawʾal proceeds: "We have found that Diophantus used this relation (*maʿnaⁿ*) in his treatise and that the later algebraists (*jabriyūn*) used it in that science, though none of them gave a proof of it in those of their writings which have come down to us".

The formula, which he then goes on to prove (*d* represents the assumed numerical difference of the two magnitudes $u$, $v$), is the basis for the resolution of systems of two equations such as the one found in *Arithmetica* II,11.

2°. (ed., p. 150). "Diophantus said (*qāla*): The double of any number or quantity (*miqdār*) made up (*muʾallif*) of two square numbers is itself made up of two square numbers, and the half is (also) made up of two square numbers".

Indeed, if $k = p^2 + q^2$, then

$$2k = (p + q)^2 + (p - q)^2,$$

and $$\frac{k}{2} = \left(\frac{p+q}{2}\right)^2 + \left(\frac{p-q}{2}\right)^2 \quad \text{(cf. Euclid, \textit{Elements}, II,9–10).}$$

This proposition, stated in al-Ḫāzin's treatise (ed. Anbouba, pp. 147–48/167), is not found in the extant *Arithmetica*. Since Samawʾal apparently quotes Diophantus, he may have seen the proposition in some commented version of the *Arithmetica*, perhaps as an addition to II,9.

3°. (ed., p. 230). As an example of a problem with only one solution, Samawʾal gives the following: "We wish to find a number such that when we multiply it by two given numbers, the result of the multiplication by the one is a square number, and of the multiplication by the other, the side of that square"; the given numbers are then set as 200 and 5, and the problem is solved. Although no attribution is given, it is clearly an Arabic version of I,26.

4°. (ed., pp. 231–32). Samawʾal reproduces proposition I,16 of Diophantus, and adds two resolutions of his own. The problem consists in finding three numbers $a$, $b$, $c$ such that the sum of any two is given:

$$\begin{array}{ll} (1) & \\ (2) & \\ (3) & \end{array}\begin{cases} a + b = k, \\ b + c = l, \\ c + a = j. \end{cases}$$

Samawʾal gives Diophantus' diorism, i.e., $\frac{1}{2}(k + l + j) > k, l, j$, and chooses the values $k = 25, l = 35, j = 30$ (Diophantus: 20, 30, 40). The problem is thus determinate.

Samawʾal's first resolution: We put $a = x$. Introducing (1), or $b = 25 - x$, into (2), gives $c = x + 10$. This into (3) gives $x$. i.e., $a$.

Samawʾal's second resolution: We put $a = x$, $b = n$ (*ʿadad*), and $c = y$ (*majhūl*);[27] adding the three equations results in $2x + 2n + 2y = 90$, or $x + n + y = 45$, hence $x = 10$ by subtracting the second equation.

Samawʾal's third resolution is explicitly attributed to Diophantus, whose treatment consists in putting $a + b + c = x$, then expressing each required number in terms of $x$, and finally adding the three results, which yields the equation.

As indicated above, we do not possess Samawʾal's commentary to the *Arithmetica*. But such various approaches to a problem (whether indeterminate or not) taken from Diophantus are precisely what one would expect in a commentary intended to provide practice for the reader.

## Appendix. Designation of the *Arithmetica* in Arabic

Diophantus' work bears in Arabic several different appellations, even within single bibliographical works. Thus it is called:[28]

(a) "Treatise on algebra", i.e., *kitāb fīʾl-jabr waʾl-muqābalaʰ*, by Abūʾl-Faraj (*Hist. orient.*, pp. 141, 338)[29] and by Ibn abī Uṣaibiʿaʰ (Müller, I,245); *kitāb fīʾl-jabr*, by Ibn al-Nadīm (Flügel, I,283), by Ibn al-Qifṭī (Casiri, I,434), and by al-Ḫazīn (cf. *supra*, p. 10).

(b) "Treatise on arithmetical (*ʿadadīyaʰ*) problems", by Ibn al-Nadīm (I,295) and Ibn abī Uṣaibiʿaʰ (I,245).

(c) "Treatise on problems of algebra (*masāʾil al-jabr*)" by Ibn abī Uṣaibiʿaʰ (II,98).

(d) "Science of the algebra (*ṣināʿat al-jabr*)" by Ibn al-Nadīm (I,269); cf. Ibn al-Qifṭī (I,371).

Our manuscript itself refers to the *Arithmetica* in various ways. At the beginning and end of Book IV, it is called a "treatise on squares and cubes"; at the beginning and end of Book V, a "treatise on arithmetical problems";[30] at the beginning and end of Book VI, a "treatise", without further qualification; likewise at the beginning of Book VII, though at the end of the same Book it becomes a "treatise on algebra".

The vagueness of the word ʼΑριθμητικά may have provoked this inconsistency, although such variety is less easily explained when it occurs in a single manuscript.

---

[27] Observe the denomination of the various unknowns, different from that of Abū Kāmil (*Algebra*, e.g., fol. 95ʳ) or that of al-Karajī (cf. *Extrait*, pp. 11 and 139–42; *Badīʿ*, fol. 113ᵛ); see also al-Bīrūnī's *Elements of Astrology* (Wright), §114.

[28] We assume that titles such as (*kitāb*) *al-ariṯmāṭīqī* refer to Nicomachus' Εἰσαγωγή.

[29] Abūʾl-Faraj (p. 141) says Diophantus lived under the reign of Emperor Julian (361–363). Regarding this assertion (apparently a confusion), see Tannery, *A quelle époque vivait Diophante?*, p. 264 = *Mém. sc.*, I, pp. 65–66.

[30] See also the beginning of the introduction to Book IV (line 8 of the text).

## 3. Mathematicians and the *Arithmetica* in Byzantium

We have practically no information about the Greek text of Diophantus in Byzantium before the end of the thirteenth century, at which time it had the same form as it now has, with slight variations, in the twenty-seven manuscripts extant today.[31] These manuscripts can be divided into two classes, the first representing, except for isolated later glosses, a Diophantus-text in an early Greek tradition and the second a text established by the monk Maximus Planudes, with a partial commentary, around 1293 in Byzantium. At that time, Maximus Planudes was able to assemble a few manuscripts of Diophantus—perhaps all derived from a single copy, which happened to survive (as well as the version which went to the Arabs) one of the most unfavourable times for the preservation of science, that following the century of Justinian.[32]

### *a. The Time of Leon the Mathematician*

(α) *The Seventh and Eighth Centuries*

The demise of Greek mathematics came with the last commentators of classical works. For, however limited the contribution to mathematics of such fifth- and sixth-century writers as Proclus, Marinus, Simplicius, Anthemius, and Eutocius may seem in comparison with that of the classical authors from whom they drew their inspiration, the level of science in their centuries was overwhelmingly superior to that of seventh- and eighth-century Byzantium: in this period, Byzantine learning was scarcely more advanced than that of the Latin West, apparently being confined to the most rudimentary subjects. After all, the Empire had more serious problems with which to cope than education: war with its neighbours, the loss of provinces (particularly the rich Oriental ones), and, finally, the civil disorder prevailing during the iconoclast period.

(β) *The Century of Leon*

Ineluctable as it seemed, the disintegration of the Eastern Empire did not come to pass. Rather, it was in this turbulent period that the Byzantine Empire discovered its Greek essence and severed its bonds with the Occident; it also found its geographical equilibrium, the seemingly disastrous territorial losses turning out to be a necessary amputation for a territory lacking balance between its inherited and real power. Improvement in both the internal situation (re-establishment of orthodoxy in the first half of the ninth century)

[31] Twenty-six listed as extant in the edition of Tannery (see D.G., II, pp. xxii–xxxiii), to which may be added the manuscript (Tannery's 27 *deperditus*) described later by E. Gollob and M. Curtze independently (*Ein wiedergefundener Diophantuscodex* and *Eine Studienreise*, pp. 258 and 295).

[32] Vogel gives a general survey of the whole history of science in Byzantium in his *Byzantine Science*; and so does Hunger in the second volume of his *Hochsprachl. prof. Lit. der Byzant.* On the historical context, see, e.g., Ostrogorsky's classical account.

and in the external situation (relative *modus vivendi* with neighbouring countries) brought more peaceful times, and a renewed interest in ancient culture arose in literate circles.

This occurred just in time: many of the old codices were in very poor condition (cf. Impellizeri, *Lett. biz.*, pp. 322–23), and thus urgently needed to be copied. The moment was propitious, since copying activity had greatly increased with the introduction of the minuscule script, which led to the transcription of many uncial manuscripts.

One of the most prominent figures in the cultural revival of the ninth century was Leon the Mathematician. Having acquired what little knowledge he could from various teachers, he resolved to continue his studies by searching out old manuscripts in monastic libraries. After much patient study, he returned to the capital where he assumed an obscure position as a private teacher. An extraordinary sequence of events completely altered his humble situation. One of his students was taken captive by the Arabs and made a slave. Upon learning that the caliph al-Maʾmūn (813–833) was fervently interested in geometry, Leon's former student made himself known and was confronted with Arab geometers. His profound knowledge of Euclid so impressed al-Maʾmūn and his circle that they ardently desired to know the person from whom he had learned so much, with the result that the caliph invited Leon to Baghdad to teach. Eventually Emperor Theophilus heard of this and decided to offer his hitherto undistinguished subject an official teaching position.[33] This had considerable impact on the realization of Leon's desire to assemble and copy ancient works, thus allowing him to play the very important rôle for which he is celebrated in the preservation and transmission of early scientific works.

The titles of certain books which we know that Leon had acquired or had had copied give us an idea of the composition of his library:[34]

(1) A treatise on mechanics by Cyrinus and Marcellus (lost).
(2) The *Conica* of Apollonius.
(3) Works of Euclid.
(4) Works of Archimedes, in the manuscript which was the progenitor of today's main Greek Archimedean tradition.
(5) Ptolemy's *Almagest*.[35]
(6) A treatise on geometry by Proclus.
(7) A treatise on astronomy by Theon of Alexandria.
(8) An astrological treatise of Paulus of Alexandria (no doubt the Εἰσαγωγικά).

---

[33] Another, apparently less reliable source, has al-Muʿtaṣim (833–842) instead of al-Maʾmūn (see Lemerle, *Premier humanisme*, pp. 152–54).

[34] See, e.g., Lemerle, pp. 169–71.

[35] That Leon possessed a manuscript of the *Almagest* is highly probable. But it has been asserted recently that the *ex libris* of the manuscript considered until now to have been Leon's own copy (Vat. gr. 1594) was in fact written by a late Byzantine hand (cf. Wilson, *Three Byz. Scribes*, p. 223).

Certainly, Leon alone was not responsible for the preservation of Greek science in early Byzantine times. But he remains the symbol of an epoch in which "most of the manuscripts forming the vital link in the line of descent from antiquity were written" (Vogel, *Byz. Sc.*, p. 270); for he was the most prominent figure associated with the rescue of ancient science during this first Byzantine Renaissance.[36]

(γ) *From Leon to Planudes*

Thus, in Leon's lifetime a peak in scholarly activity was reached, the impetus of which was not lost afterwards, as may be inferred from the existence of several excellent manuscripts copied in the following period which are either extant today or of which we have copies.[37] It must not be understood from this, however, that the works copied were fully understood: judging from original works dating from the time of Leon to 1200, mathematics did not attain a high level. Still, there must have been a living mathematical tradition since around 1200 there were some Byzantine scholars who were capable of favourably impressing Leonard of Pisa.[38]

Mention of Diophantus is first made in the eleventh century. The polymath Michael Psellus (*ca.* 1018–*ca.* 1078) apparently saw a manuscript of the *Arithmetica*, for he wrote a letter concerned in part with algebraical terms used by Diophantus.[39] In addition to some extracts taken from the introduction to Book I, we find in this letter some very interesting information about two sets of denominations for the powers, different from that used by Diophantus (see pp. 43–44).

### b. *The Time of Maximus Planudes*

Ignorant as we are of the rôle played by Leon and his contemporaries in the transmission of the *Arithmetica*, we are fairly well informed as regards the fate of the Greek Books around 1300, the peak of the second Renaissance of Byzantine letters. This revival began in the first part of the thirteenth century, principally in the Empire of Nicea during the Latin rule in Byzantium (1204–1261), and continued in Byzantium after cessation of Latin rule.

Mention of Diophantus is found in the autobiography of Nicephorus Blemmydes (*ca.* 1197–*ca.* 1272) who learned arithmetic from the works of

---

[36] Further evidence of Leon's interest is perhaps seen in the fourth problem of the Byzantine collection edited by Hoche together with Nicomachus' *Arithmetic* (pp. 148–54). The attribution of this problem to Leon (VI) the Wise, who became emperor in 886, may have resulted from a confusion—all but rare in later literature—between the emperor and the mathematician.

[37] Best known are: Arethas' Euclid, copied in 888 and the oldest dated profane manuscript in minuscules (Bodl. d'Orville 301); the manuscript used by Peyrard for his edition of Euclid (Vat. gr. 190); the palimpsest manuscript containing Archimedes' *Method* (formerly in Constantinople, but since stolen); the Constantinopolitan manuscript of Heron's *Metrica*.

[38] Such as the "peritissimus Magister Muscus" whom he mentions in his *Liber abaci*; cf. *Scritti*, I, p. 249.

[39] See Tannery, *Psellus sur Diophante* = *Mém. sc.*, IV, pp. 275–82. The letter is also printed in D.G., II, pp. 37–42.

Nicomachus and of Diophantus (not the whole of the latter, he says, but what his teacher understood of it; cf. p. 5,1–4 in Heisenberg's edition).

The first real use of the *Arithmetica* was made by one of Blemmydes' pupils' pupils, Georgius Pachymeres (*ca.* 1240–*ca.* 1310). He is the author of a voluminous *Quadrivium*, the level of which contrasts very favourably with that of another quadrivium composed at the beginning of the eleventh century.[40] Pachymeres' *Quadrivium* gives a prolix paraphrase of Diophantus' definitions of powers and of the first problems of Book I (up to I,11).

This paraphrase, though, cannot compare with the methodical explanation of the introductory definitions and of the problems of both Books I and II written by the monk Maximus Planudes (*ca.* 1260–*ca.* 1310). His work represents the farthest-reaching commentary on the *Arithmetica* made in Byzantine times, and, though limited in length and content, it is particularly noteworthy coming from a man renowned as one of the foremost Byzantine humanists.

In order to establish a reliable text of the *Arithmetica*, Planudes endeavoured to assemble manuscripts of Diophantus. We gather from his correspondence that in 1293 there were at least *three* copies of the six Books.

1°. Planudes requested that the protobestiarios Theodorus Muzalon lend him a copy of the *Arithmetica.* When asked later to return it, he excused his delay by explaining that he had been obliged to repair the manuscript which was in poor condition: see his letter 67 (ed. Treu, p. 82; cf. p. 84).

2°. From letter 33, addressed to the mathematician Manuel Bryennius, we perceive that Planudes himself possessed a copy.[41]

3°. In this same letter, Planudes asked Bryennius to lend him his Diophantus so as to collate it with his own copy.

The form in which we know the Greek *Arithmetica* (with the fragment of the *De polygonis numeris* appended to it) was thus definitively established by the end of the thirteenth century. No attempt to comment on the *Arithmetica* was made after Planudes in Byzantine times—probably no one was capable enough to do so. Twice mention of Diophantus was made in the fourteenth century, which indicates that mathematicians still knew of the *Arithmetica.*

The first trace is in the hand of Nicolaus Rhabdas (*ca.* 1340). He seems to have found the most appealing part of the *Arithmetica* to have been the nonmathematical section of its introduction, for he reproduced (with no allusion to Diophantus) a lengthy passage from it, *ad verbum*, at the beginning of a letter (comp. Tannery, *Lettres de Rhabdas*, p. 142,7–16 = *Mém. sc.*, IV, p. 86,7–15 with D.G., I, p. 2,4–17); a subsequent letter reiterated, with minor

---

[40] The latter has been edited by Heiberg, the former by Tannery (the part relevant to Diophantus is also printed in D.G., II, pp. 78–122).

[41] Whether it was copied from the text of Muzalon or not, we do not know: whereas letter 67 is supposed to have been written at the beginning of 1293 (see Turyn, *Dated Gr. Mss.*, p. 80), no precise date can be attributed to letter 33.

alterations, part of this passage (comp. *ibid.*, p. 174,4–11 = p. 118,4–10 with D.G., I, p. 2,4–13). The remainder of his two letters does not suggest that he read more than the first few pages of the treatise of Diophantus, whom he recognizes, none the less, as "ὁ μέγιστος ἐν ἀριθμητικοῖς" (*ibid.*, p. 174,16 = p. 118,14–15).

The second mention of Diophantus appears in a letter written by Demetrius Cydones (*ca.* 1325–*ca.* 1400) to a friend to whom he sent an excerpt of Diophantus which he happened to find. He indicates that he has added numerical proofs (ἀποδείξεις) such as those he had already made for the arithmetical Books of Euclid (Epist. 347, ed. Loenertz).[42]

The rôle of Byzantium was essential in the preservation of the Greek *Arithmetica*, despite the fact that Byzantine scholars in general understood little more than the rudiments of Diophantus' indeterminate analysis of the second degree; and the remark of an irate reader in reference to II,8, reported by Tannery (D.G., II, p. 260,24–26), clearly delimits the point at which real difficulties began for the average Byzantine mathematician. No doubt most scholars never advanced beyond the first Book, if they even got that far.

### c. *Oldest Greek Manuscripts Still Extant*

(α) *Non-Planudean Class*

There are two extant copies of the six Books of Diophantus dating from the time of Planudes and belonging to the non-Planudean class. One of these is the Matritensis gr. 4678 (*olim* 48), to which we shall refer as M. Because of its great age and purity, Tannery chose to base his edition largely upon it. First dated by Iriarte in his *Catalogue* (p. 163), "quantùm suspicari licet", to the fourteenth century, this manuscript was later examined by Heiberg who attributed it to the thirteenth century (Tannery, *Rapport*, p. 413 = *Mém. sc.*, II, p. 274). The second extant manuscript is the Vaticanus gr. 191, which we shall designate as V; supposed by Cossali to be of the thirteenth century, it was, however, considered by Tannery to have been written in the fifteenth century (Tannery, *ibid.*). Reliable indications, though, set the date of its writing at about 1296, in any case no later than 1303 (cf. Turyn, *Codd. gr. vat.*, p. 94).

Because of his mistaken assumption relative to the age of manuscript V, and because of V's close relationship to M, Tannery considered V to have been copied from M (cf. D.G., I, p. iii). There is certainly no doubt that M and

[42] Concerning Cydones' Euclid-glosses, see *Euclidis opera* (ed. Heiberg), V, p. xxxiii. A further indication of Cydones' mathematical interests is seen in an (elementary) problem on summation of the natural numbers, which is the first of the six problems of Byzantine origin mentioned above (p. 16, n. 36). The attribution found in Hoche's edition, ΤΟΥ (misprinted as ΤΟΤ) ΚΥΝΟΣ (!), was emended to τοῦ κυδώνου by Tannery (*Lettres de Rhabdas*, p. 133, n. 2 = *Mém. sc.*, IV, p. 75(–76), n. 2).

V belong to the same family. A number of indications makes this indisputable, from the almost complete agreement which exists between the two manuscripts to the occurrence of similar characteristic signs or errors slavishly reproduced by both copyists, such as those found in Tannery's critical apparatus to pp. 296,17; 368,15; 382,23 and 438,5; in 180,20 (app.) we even run across the same dittography. This literal copying must be taken into account in any comparison of the two manuscripts; for, from this similarity one may conclude either that the later one was copied from the earlier one, as did Tannery, or, just as plausibly, that both are slavish copies with a common near ancestor.

We are inclined to choose the second possibility on the basis of the following considerations.[43]

1°. V *sometimes shows the same reading as Tannery's* B,[44] *and not that of* M. This reading may be a faulty one, as is the case in those places corresponding to D.G., I, pp. 6,25; 68,15 (*sine suppl.*); 160,1 (ποιῶν); 168,14; 182,5; 384,7; 408,12 (followed by τὸν). It may also be the correct reading, as in 12,21 and 26,27 (both of which look like corrections of a previous misreading); 78,12 (both); 90,14–15 (τοσούτων). Moreover, V sometimes omits, as does B, words found in M, as in the set 4,16; 4,18; 4,25; 6,2.

2°. Furthermore, V *shows readings which are different from those of both* M *and* B (where these differ), as in pp. 62,7 (ἤτοι τὸ ἥμισυ, ἤτοι not from corr.); 80,7 ($\bar{\beta}$ καὶ ἕβδομον, καὶ *supra lin.*; but 1ª m. in M not legible); 82,10 (τρίτον μέρος); 84,21–22 (noted by Tannery); 104,11 (also noted; ὅμοια *supra lin.*); 164,14 (also noted; but the $\mathring{\mu}$ι of Tannery is really the copyist's writing of Tannery's $\mathring{M}$); 326,17 (like M, but omitting the συγκείμενα); 328,23 (μερίσι). The reading found in V can even be the better or the correct one, as in 4,19; 30,9–10 (noted by Tannery); 54,16–17 (= text of Tannery).[45]

That V is not simply a copy of M seems evident from the above (assuming Tannery's apparatus is accurate!). But the strongest proof of this is the presence in V of words omitted in M, such as the εἶναι in 86,8 or the entire phrase in 8,21–23 (of which line 23 was added in the margin by the revisor of the manuscript).

Since M is the only extant manuscript antedating Planudes' revision, one may reasonably ask whether it could be one of the copies that Planudes himself possessed or used. C. Wendel has asserted that it is, and identifies M with the Muzalon-copy (*Planudea*, pp. 414–17), giving some arguments in favour of this. But it is surprising that a codex described by Planudes as being in poor condition whilst not particularly old could have survived many more centuries without being in worse condition than it now is.

---

43 Note that, for the text of M, we must rely entirely on Tannery's critical apparatus, as our letter to the Biblioteca Nacional requesting a microfilm of M was never answered.

44 By B, Tannery designates the Marcianus 308, possibly the oldest (complete) manuscript of the Planudean class (and also Bachet's reading, when the same: see D.G., I, p. iii).

45 Tannery's observation "ὁ suppl. V" to 270,12 is, however, wrong.

(*β*) *Planudean Class*

The progenitor of the Marcianus 308 is extant only in part, i.e., ten leaves of it are found in the Ambrosianus Et 157 sup. As this manuscript is supposed to have been written by Planudes himself, and was perhaps his final copy (see Turyn, *Dated Gr. Mss.*, pp. 78 *seqq.*), we know thus of a third extant manuscript dating from the time of Planudes.

# Chapter II

# The Extant Arabic Text

## §3. Description of the Manuscript

Books IV to VII of Diophantus' *Arithmetica* are found in a codex, apparently a unicum, which is described under the number 295 in the eighth volume of the catalogue of the manuscripts kept in the library attached to the shrine of Imam Rezā at Mashhad (cf. Gulchīn-i Maʿānī, *Fihrist*, pp. 235–36). This codex is said to have come to the Shrine Library as the result of an endowment (*waqf*) made in 1932 by a certain Mīrzā Rezā Khān from Nāʾīn (*Mīrzā Riḍāʾ Ḫān Nāʾīnī*).[1] The manuscript is protected by a card-board cover bound with and reinforced on the corners by leather. In recent times its eighty reddish-brown leaves (175 × 130 mm) have been numbered as pages.[2] On each of these—except for the title-page and the last page—figure twenty lines of text (128 × 92 mm).[3]

In certain portions of the text, vermilion ink was used. This is the case for the numbering of the problems,[4] for the titles of Books V, VI, and VII, and for some subtitles marking off alternative resolutions (see notes 90, 142, 331 of the critical apparatus). Signs of strong punctuation,[5] commonly used to separate problems, or corollaries and remarks, from other problems, may be filled in with red. It appears that the rubrication of the manuscript was not done simultaneously with the writing in dark ink, as in one place the space left blank for the red-ink inscription was insufficient whilst in another one it was unwittingly omitted (see notes 331 and 934).

---

[1] Nāʾīn is a small town on the road from Teheran to Yazd (32.52 N, 53.05 E).

[2] The title-page is not numbered. The leaf numbered 140–141 is out of place and should precede the leaf numbered 138–139.

[3] See plates II–IV. The pagination, written at the very top of the pages, is not visible on these reproductions.

[4] In *jum(m)al*-notation, i.e., with Arabic letters representing numerals.

[5] A point within a circle (see plate III, lines 5 and 16).

On the unnumbered title-page of the manuscript appear the signatures of (some of) its owners and a library's seal (see plate I). On this same page, we also see, written in a modern hand, the words "Traitant des carrés et des carrés cubiques" and "Ecrit en 595 de l'hégire". Since the same hand performed the subtraction 1343 − 595 = 748 on the facing page, no doubt to find the age of the manuscript in years of the Muslim era, the annotations must go back to the Christian year 1924/25 (and must thus antedate its endowment to the Shrine Library).

The year 595 of the hegira referred to by the French annotations is indicated in the Arabic text of the title-page, and the manuscript's concluding words specify the day of completion as Friday, the third of Ṣafar. This corresponds to a Friday, 4 December 1198 (Julian day 2,158,965), when one uses for the conversion the *astronomical epoch* beginning on 15 July 622, instead of the usual one beginning one day later (cf. Ginzel, *Hdbuch der Chronol.*, I, p. 259).

The scribe who copied, or, rather (see below), who began to copy the manuscript, a certain Muḥammad ibn abī Bakr ibn Ḥākīr,[6] declares himself to be an astronomer. This otherwise unknown person copied the initial pages of the manuscript in a very readable oriental *nasẖī*,[7] adding most diacritical points, and even vocalization signs (see plates I–II).[8] From page fourteen onwards, the writing changes abruptly to a beautiful calligraphic script (see plates III–IV). Unfortunately, although the text becomes aesthetically more pleasing, its legibility suffers somewhat, since, with very few exceptions, the second scribe chose not to add diacritical marks.[9] By and large, however, this omission of the diacritical points is of little importance since the manuscript remains quite legible. Only a few places presented any problem (see, e.g., notes 450, 621) or serious difficulty (notes 172, 751).

The two scribes barely if ever understood what they were writing. There are many blunders, meaningless interpolations, repetitions or omissions of words or sentences due to homoeoteleuton. This is true for the first hand (see, e.g., notes 32, 35, 75, 100, 124, 134) and even more so for the second one (characteristic examples in notes 164, 389, 575, 708, 788, 795).

---

[6] Gulchīn-i Maʿānī, in his description, reads *Jāgīr* (with the Persian *gāf*) meaning the land obtained as a reward for services (its possessor being the *jāgīr-dār*).

[7] Except for the first five words (in the title), which are written in the so-called qarmatian Kufic (plate I). This type of writing was indeed in use in Persia at the time our manuscript was copied (cf. Kühnel, *Islam. Schriftkunst*, p. 16). A further example of this writing is found on the title-page of the Ms Bodl. Marsh 667 (Apollonius' *Conics*), copied at Marāġa[h] in A.H. 472 = A.D. 1079/80 (concerning this date, see Beeston, *Marsh Ms of Apollonius*).

[8] Note, however, that these vocalization signs tend to be of little help since they are, as a rule, superfluous (sometimes wrong: see note 88), and absent when they would be truly useful (see note 6).

[9] The diacritical marks were added to only a few words, which are easy to read anyway, and are conspicuously absent when genuinely necessary. In some cases, they are even wrong and thus misleading for the reader (see notes 218, 382, 686). Sometimes the addition of diacritical points seems to coincide with words about which the scribe had some doubt (see notes 167, 172, 173, 452); points were added once to a badly written word (note 639).

There is no recognizable trace left by any reader of the manuscript, and its owners seem to have contended themselves with writing their names on the title-page. The few emendations found in the text were made by the copyists themselves, principally if not exclusively to correct their own scribal mistakes. They are:

(a) one marginal addition, by the second copyist, occasioned by an omission (note 782);
(b) one "mark" made by the first scribe, presumably to cancel a word (see note 54[10]), perhaps another by the second scribe (note 432; see also note 253).
(c) some supralinear additions made by the second copyist (notes 396, 532, 685, 739, 820, 898, 928, 953, 957; see also below, p. 37).

Finally, there are a number of places in which the second scribe miscopied and then corrected a word, or hesitated in the copying of a word (notes 198, 219, 241, 248, 660, 777, 971).[11]

## §4. Orthographical Remarks

The orthographical peculiarities listed below either occur inconsistently or are limited to one of the two scribes. Therefore they characterize our manuscript, and not the original copy written by the translator.

### 1. Writing of the *hamza*ʰ

As far as we can judge from the few relevant words—for the vocabulary of the manuscript is limited—, the writing of the *hamza*ʰ in our text appears to differ little from the normal use in other manuscripts. This is true for both hands, except that the second scribe, not surprisingly in view of his constant omission of diacritical marks, writes, if anything, only the *hamza*ʰ's support. Thus, only the first hand—and not always—writes the final *hamza*ʰ in *šaiʾ* or in *ašyāʾ*.[12] Both unify the radical *yā* with the support of the *hamza*ʰ when *šaiʾ* is in the indeterminate accusative or in the oblique dual (see, e.g.,

[10] I have not seen the scribe's abbreviation, formed by a *ṭā* and, apparently, a *hamza*ʰ, used elsewhere. The siglum *ṭā* (*ẓā*) was sometimes used in manuscripts to draw attention to an error (see, e.g., B.N. arabe 2459 (copy of the *Faḫrī*), fol. 102ᵛ et 105ʳ; Aya Sofya 4830, fol. 218ʳ), but it was written alone. If, in our case, the two letters are intended to abbreviate a single word, this word might be *ḫaṭaʾ*: it is suitable, and we know of several cases of abbreviations formed by the last letter(s) of a word (see Caspari–Wright, I, pp. 25–26; Flügel, *Wortabkürzungen*).

[11] The error indicated in note 965 might have arisen from the scribe's uncertainty about the reading of the word, thus causing him to write two similar words consecutively (so as to leave the choice to the reader?).

[12] The latter word often with *madda*ʰ on the final *alif*.

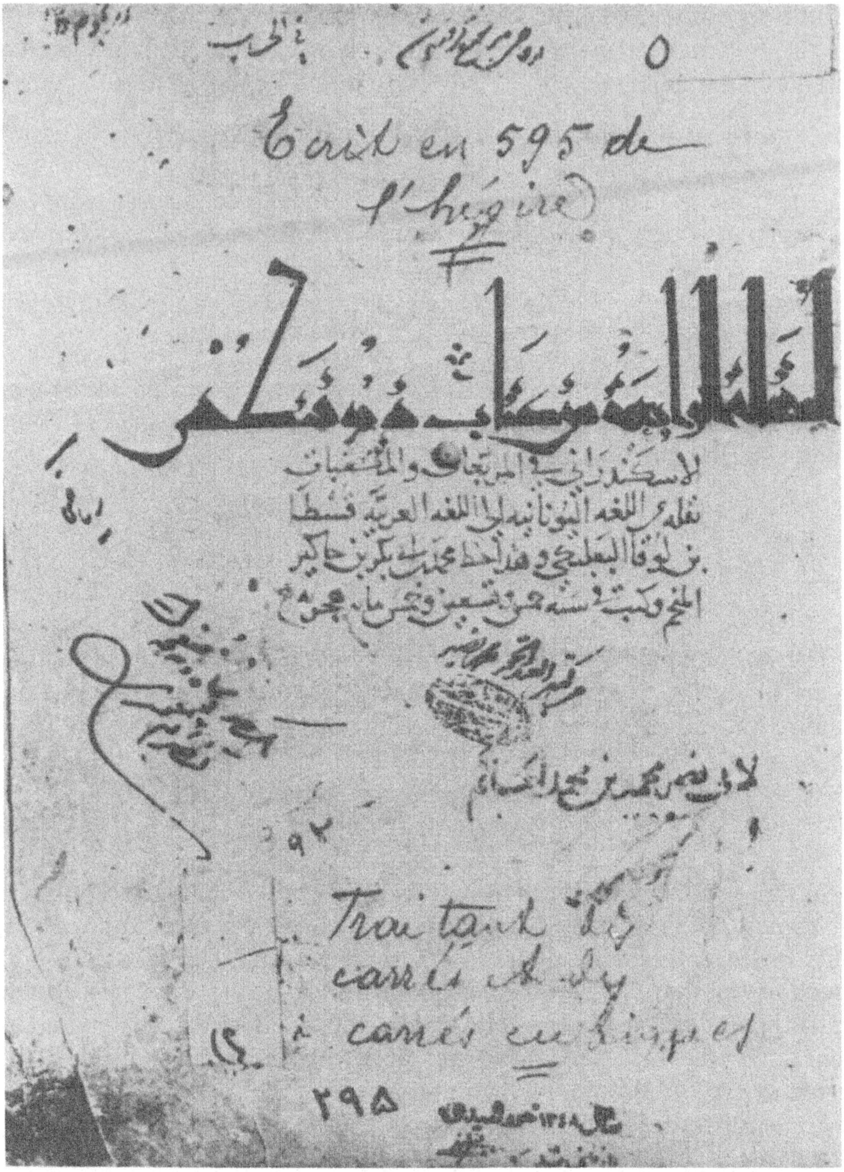


Plate I. Manuscript, title-page.

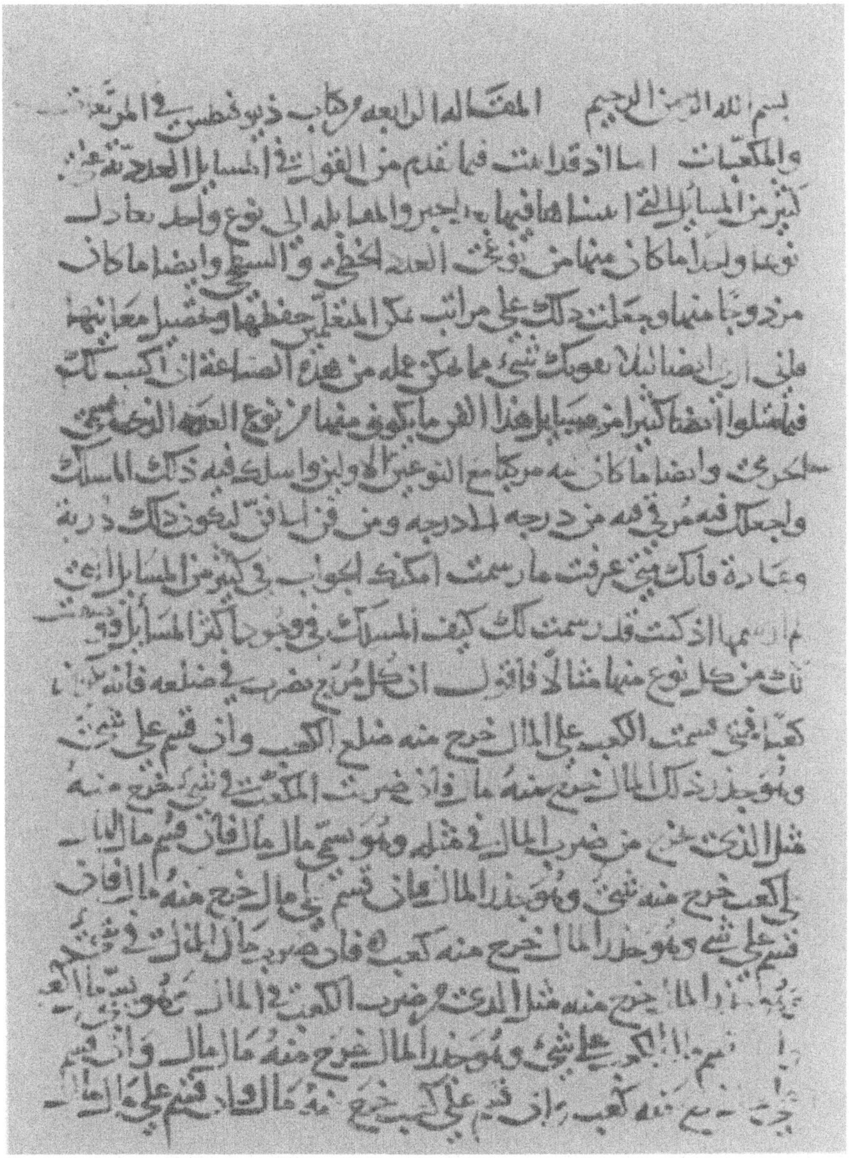

بسم الله الرحمن الرحيم المقالة الرابعة من كتاب ذيوفنطس في المربعات والمكعبات

Plate II. Manuscript, page 1.

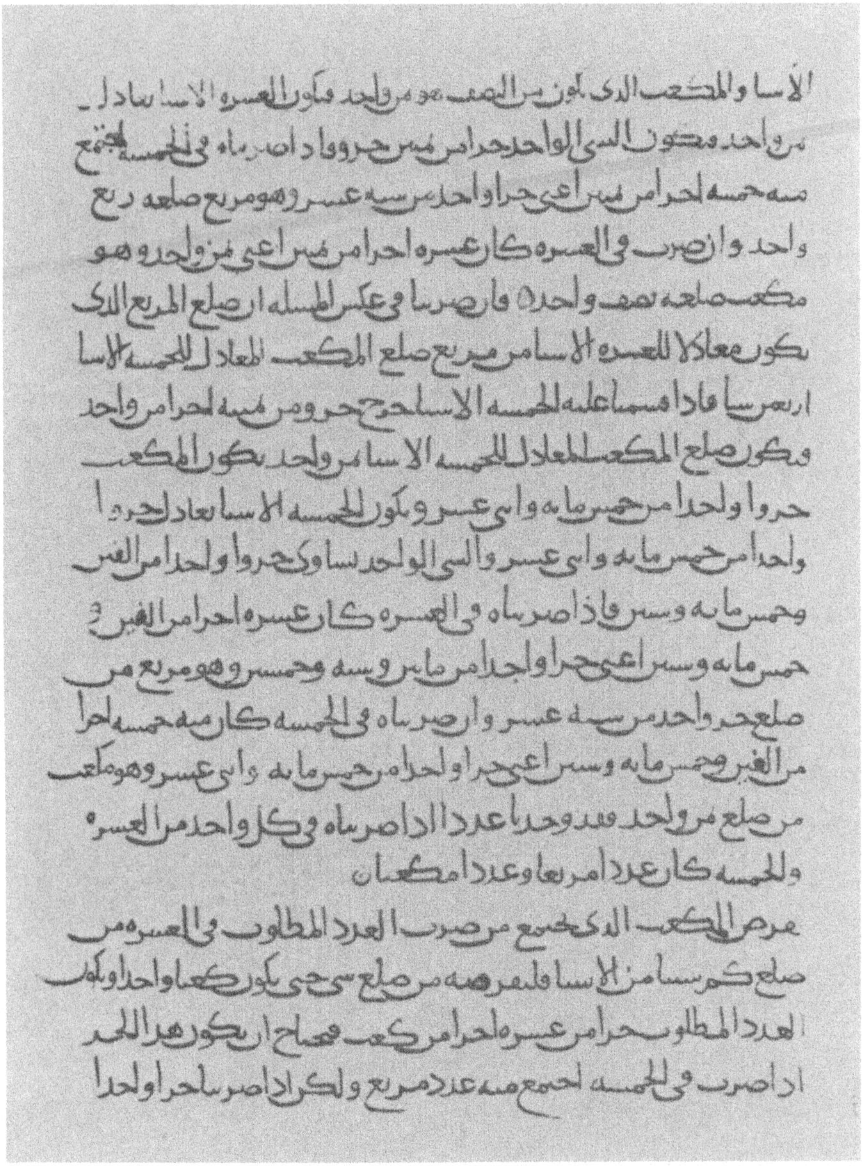

Plate III. Manuscript, page 14.

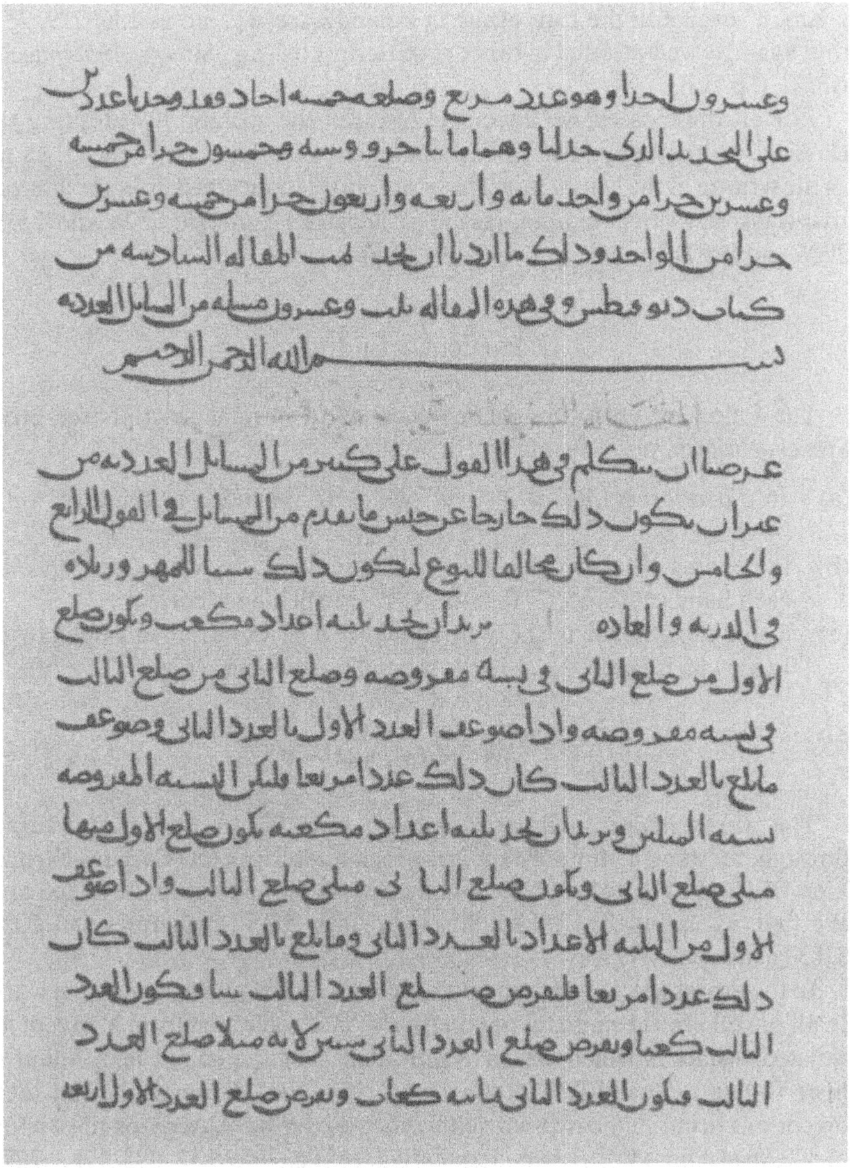

وعشرون جذراً وهو عدد مربع وضلعه خمسة آحاد وقد وجدنا عددين
على الحد الذي حددناه وهما مائتان وستة وخمسون جزءاً من خمسة
وعشرين جزءاً من واحد ومائة وأربعة وأربعون جزءاً من خمسة وعشرين
جزءاً من الواحد وذلك ما أردنا أن نجد تمت المقالة السادسة من
كتاب ديوفنطس وفي هذه المقالة ثلث وعشرون مسئلة من المسائل العددية
بسم الله الرحمن الرحيم

غرضنا أن نتكلم في هذا القول على كثير من المسائل العددية من
غير أن يكون ذلك خارجاً عن جنس ما تقدم من المسائل في القول الرابع
والخامس وإن كان مخالفاً للنوع لتكون ذلك سبباً للمهر والزيادة
في الدربة والعادة نريد أن نجد ثلثة أعداد مكعبة ويكون ضلع
الأول من ضلع الثاني في نسبة مفروضة وضلع الثاني من ضلع الثالث
في نسبة مفروضة وإذا ضوعف العدد الأول بالعدد الثاني وضوعف
ما بلغ بالعدد الثالث كان ذلك عدداً مربعاً فلنكن النسبة المفروضة
نسبة المثلين ونريد أن نجد ثلثة أعداد مكعبة يكون ضلع الأول منها
مثلي ضلع الثاني ويكون ضلع الثاني مثلي ضلع الثالث وإذا ضوعف
الأول من الثلثة الأعداد بالعدد الثاني وما بلغ بالعدد الثالث كان
ذلك عدداً مربعاً فلنفرض ضلع العدد الثالث شيئاً ويكون العدد
الثالث كعباً ونفرض ضلع العدد الثاني شيئين لأنه مثلا ضلع العدد
الثالث فيكون العدد الثاني ثمانية كعاب ونفرض ضلع العدد الأول أربعة

Plate IV. Manuscript, page 130.

notes 330; 124, 770 of the critical apparatus), while the first hand writes *šīnā* instead of *šiʾnā*. Finally, *juzʾ* in the singular (including the indeterminate accusative) frequently has a *waw* as support for the *hamza*$^h$ (with or without a *hamza*$^h$ over it in the case of the first hand): see, e.g., notes 136, 229, 737; this, again, is well-attested in other manuscripts (cf., e.g., Simon, *Anatomie des Galen*, I, p. xxi).

For all these cases, we have standardized the spelling by adopting the classical orthography. The only inconsistency we have reproduced, as far as the writing of the *hamza*$^h$ (and of its support) is concerned, is in the alternative ways of writing *miʾatain* (see, e.g., lines 189 and 216; 2528 and 2529; 2906 and 2907).[13]

## 2. Particular Endings

The following uses, though not peculiar to our manuscript (see Graf, *Sprachgebrauch*, pp. 8–9), are worthy of note:

(a) an *alif otiosum* (*alif al-wiqāya*$^h$) which is appended to the form *yatlū* (note 3);
(b) the ending -*ī* takes the place of the ending -$^{in}$, in two places, once by each hand (notes 15, 771); otherwise the spelling is correct;
(c) again exceptional is the writing of an *alif* where an -*à* ought to be used, which occurs twice in the second handwriting (notes 172, 579).

## 3. Numerals

The words *ṯalāṯa*$^h$, *ṯalāṯūn*, *ṯamāniya*$^h$, *ṯamānūn* are written defectively throughout, as is commonly the case in manuscripts (cf. Caspari–Wright, I, pp. 254 and 257), and we have maintained these spellings in the edited text. We find, exceptionally, full spellings: in lines 2935 (*ṯamāniya*$^h$) and 2026, 3103 (*ṯamānain*).

In the expression of hundreds, 300 and 600 are always written as one word. In all the others, the numeral numbering *miʾa*$^h$ is separated from it, except for isolated instances which we have reproduced in the edited text, found in lines 70 (*400*), 56 and 271 (*500*), 69 and 154 (*700*). Observe that all these exceptions occur in the text written by the first scribe. As regards the writing of 800, *ṯamān miʾat*$^{in}$[14] occurs frequently (see, e.g., lines 817, 923, 1782, 2272, 2776, 3004, 3356); the other form, *ṯamānī miʾat*$^{in}$, is the predominant one in the first half of the manuscript (see, e.g., lines 667, 840, 918, 1045, 1433–34; lastly in 1793).

---

[13] When *miʾatain* occurs in the construct state, *hamza*$^h$ has its own support (e.g., lines 518, 1730; line 943 is an exception).

[14] See Fleischer, *Kl. Schr.*, I, p. 334; also *ibid.*, p. 330 and de Sacy, *Grammaire*, II, p. 324 (line 21).

The plural of *alf* is, as pointed out in note 68, always written defectively (cf. Caspari–Wright, I, p. 259); we have added the supralinear *alif* in the edited text.

## 4. Repeated, Erroneous Spellings

(α) The word *kiltā*, i.e., *kilā(n)* construed with the genitive dual of a feminine substantive, is not written *kiltā* by the first scribe, but *kiltà*—which is admissible (see Caspari–Wright, II, p. 214; Reckendorf, *S.V.*, p. 141)—, or, perhaps, inappropriately, *kiltai*, i.e., with an inflected form (all the occurrences are in the oblique case).

But the spelling *kilà* (or *kilai*), used by the *second scribe* under the same circumstances, is odd. The regularity of its use, however,[15] has led us to keep it, *nolens volens*, in the edition, instead of correcting it each time with a note.

(β) There is another kind of miswriting made by the second copyist, which is found in aggregates in which the initial letters are certain combinations of *alif*(s) and *lām*(s).

The spelling *alif-lām-alif* instead of *alif-lām-alif-alif* (for *illā* + initial *alif*) is found in several passages (with *arbaᶜaʰ*: notes 326, 328, 333, 338, 349, 410, 413, 414, 429, 566; with *iṯnai*: notes 495, 649). The correct spelling is found, e.g., in lines 1771, 2242–46, 2398–2400; 1473, 1695, 2142–46.

Further, *alif-lām-alif* instead of *lām-lām-alif* is found in an unbroken sequence of places (with *arbaᶜaʰ*: notes 344, 348, 353, 355, 439; with *amwāl*: notes 379, 404); otherwise the writing is correct (cf., e.g., lines 571, 1313).

Finally, *lām-lām-alif* instead of *alif-lām-alif* is found in a few cases (with *amwāl*: note 435; with *arbaᶜaʰ*: note 614 (reading *miṯli* as previous word); with *amṯāl*: note 858).

## §5. Additions by Earlier Readers (or Copyists)

Two kinds of minor additions are incorporated in the manuscript's text: those originating from earlier readers' explanations or corrections, which were originally marginal or supralineal, and those resulting from a scribe's mechanical repetition (dittography). Those of the second class have been relegated to the critical apparatus. We shall treat here the more interesting additions of the first sort, which additions we have divided into two groups:

(I) Those which complete or clarify the text in some way, or which, simply, do not render its comprehension difficult; these have been left in the text, for the most part bracketed.

[15] In only three instances does the spelling look like *kiltà* (see note 329).

(II) Those which do not; these have been removed from the text and may be found, like the dittographies, in the critical apparatus.

## I. First Group

**A.** The following passages are *in all probability* interpolated.

1. Lines 21, 24, 25, 26, 27, 29, 32, 34, 49, 51

These remarks were added in order to clarify the meaning of *šai*ʾ. They originated with an Arabic reader, obviously not very familiar with Arabic algebraic terminology—nor, therefore, with Diophantus' earlier Books—, who seems also to have understood *māl*, not in its particular sense of $x^2$ (the only one used in our text), but in its broader meaning of "quantity".[16] He added his explanations rather consistently up to line 51, at which point he either finally grasped the mathematical meanings of *māl* and *šai*ʾ, or simply gave up.

2. Lines 35–37

The bracketed phrase is an explanation of the two Arabic words *jabr* and *muqābala*$^h$, which designate the two operations defined by Diophantus in "Def. XI" (D.G., I, p. 14,11–20), for which no synthetic words exist in Greek. This explanation must have originated with an Arabic reader, and not with the translator: for the latter, in translating the previous Books, must undoubtedly have used these words extensively; or, had he really wished to explain them in the introduction to Book IV, he would have done so at their first occurrence (line 9).

3. Lines 263–264

This addition by an Arabic reader was occasioned by a lacuna in the text. Observe that the whole of the second part of problem IV,13 is confused; this confusion may quite possibly antedate, at least in part, the translation into Arabic.

4. Line 269

Another explanation, again in the second part of IV,13, supplements the text which, in its present form, does not state anywhere that the smaller cube's side was set equal to $x$.

5. Lines 292–293

This explanation certainly goes back to a reader. He is perhaps not responsible for the gross error corrected in note 109 of the app., which could be the doing of a copyist.

6. Lines 607–608

A few words were added by an Arabic reader as a consequence of the unclear formulation of the text (cf. p. 103, n. 49).

[16] Al-Ḫwārizmī uses both senses of *māl* in his *Algebra*, an ambiguity which confused his editor; this led Woepcke to make a rectification in his *Extrait du Fakhrî*, p. 48.

7. Lines 1157–1158

The ineptness of this addition makes it difficult to explain its origin; it is conceivable that it resulted from a copyist's (ours?) attempt to restore an illegible portion of the text. Cf. p. 115, n. 72.

8. Lines 1425–1426

The addition here, intended as a textual elucidation (see app.), was perhaps made by the same Arabic reader who emended, for similar reasons, lines 607–608 (above, no. 6).

9. Lines 2273–2274

The bracketed words, obviously out of place, were originally a marginal note. Concerning the expression used to render the power, see p. 45.

10. Lines 2391–2392

Here, a reader corrected a confusion made in the final statement (see p. 63, no. 6). As in no. 2, the explanation is introduced by *aᶜnī*.

11. Lines 2670–2672

The bracketed words in line 2672 are certainly some reader's addition; this same reader may also have added some other explanations seen in the two previous lines which, notwithstanding their being less suspicious than the one in line 2672, do not seem to be genuine.

12. Line 2972

We find here an excellent example of an *Arabic* interpolation: a distinction is made between *ḍilaᶜ*, "side", and *jiḏr*, "(square) root", a distinction not possible in Greek since the two concepts are rendered by the same word (πλευρά). That the phrase originated with a reader rather than with the translator is hardly questionable: the translator would simply have changed the previous *ḍilaᶜuhū* into *jiḏruhū*, since he also uses *jiḏr* as a translation of πλευρά (see Arabic index).

Observe also that differentiating between the conditions $x^3$ = square and $x$ = square is textually, but not mathematically, relevant.

**B.** Although the following passages do not seem to be as foreign to the text as the previous ones, they are *probably* interpolated.

13. Lines 106–107 (105–109)

What is bracketed in lines 106–7 is truly superfluous, reminding us of no. 5, and must be a reader's addition. In fact, the entire explanation, given in lines 105–9, does not seem to fit in the text (unlike the passage in lines 48–51, for example). But it is also possible that an earlier commentator wished to lay stress on the division of $x^4$ and $x^3$ by $x^3$, an operation which appears in the *Arithmetica* for the first time.

N.B. There are numerous other explanations in the first problems of Book IV which may have been absent in the original translation; see, e.g., lines 80–81, 91–92, and 110–11.

14. Lines 171–172

The bracketed comment looks like an addition; perhaps some reader was baffled by the reasoning, especially if the important lacuna of lines 166–69 was already in his copy.

15. Line 1701

The three values are given abruptly, without any word linking them to the preceding phrase, and may thus be an addition. It is surprising, though, that the original text should merely refer vaguely to "before" without repeating the values. But, after all, there is a somewhat similar instance in line 240, and all of problem V,3 is expressed concisely.

Observe that line 1702 also seems to contain an interpolation (cf. p. 128, n. 9).

16. Lines 2266–2267

It would seem that a reader supplemented the text here; for the statement of the value, although necessary in this place, gives one the impression that it is a later addition.

17. Line 2770

The bracketed words look like a reader's addition caused by the omission of a word shortly before (see line 2768).

C. Our final category lists those phrases which are *possibly* interpolated or mixed with interpolations. Because of this uncertainty, they were not bracketed.

18. Lines 403–405

The statement made in these lines is partly repetitive (cf. lines 399–400) and may be the result of some reader's addition or emendation.

19. Lines 1509–1514

The formulation of problem IV,44 as found in the manuscript (see text and app.) seems odd and may be a mixture of the original version and a reader's additions; but no part can be satisfactorily bracketed.

20. Line 1854.

The second half of the line (see app., note 589) may be an interpolation (cf. the situation in line 1918).

21. Lines 2016–2017

The text seems to be a mixture of the original version and some interpolations. This is also true for other passages in which two consecutive *wa-huwa*'s occur: see app. to lines 2650 and 3339, and also lines 2622–23; cf. note 97.

22. Line 3016

Instead of having the second condition of the text "*wa-yakūn kull wāḥid minhā murabbaʿ*$^{an}$", one would expect to read in line 3015 "let us search for three *square* numbers". This case reminds us of no. 17.

## II. Second Group

The readers' additions listed in this group are those which were senselessly incorporated into the text; they are consequently to be found in the critical apparatus.

**A.** Misused marginal annotations (some marginal annotations were understood by the copyist (of our codex?) to be corrections of words which were in fact—we suppose—correct in his copy).

23. Note 347

The word *al-ūlà* was, probably, written in the margin to mark the position of the *first equation* as the text went on to establish the second one (see lines 1025–26); then, the copyist mistakenly assumed *al-ūlà* to be a correction of *šaiʾ*$^{an}$, missing in the Mashhad manuscript's text.

24. Note 469

The word *wāḥid*, missing in the text (cf. note 470), was added in the margin by a reader, but was taken by the subsequent copyist to replace *kaʿb kaʿb māl*, missing in the Mashhad manuscript.

25. Notes 882 and 883

The errors in these two notes are related. The *wa-arbaʿa*$^{h}$ of note 883, previously written in the margin, was supposed to be inserted where there was the lacuna, that is, after the first *miʾa*$^{h}$ of line 3173. It was, however, inserted after the second *miʾa*$^{h}$ of line 3173 in place of *wa-aḥad wa-arbaʿain.*[17]

Since *wa-arbaʿa*$^{h}$ alone is not sufficient to fill in the gap mentioned in note 882, we must conclude either that the marginal correction was incomplete or that the copyist only partially reproduced it. Other instances make the second possibility plausible (cf. no. 27 and p. 134, n. 28).

**B.** Misplaced marginal corrections.

26. Note 134

The text has a senseless *wa'l-ašyāʾ*, while *al-ašyāʾ* would be perfectly in place just before.

[17] The use of a catchword—i.e., a word from the text repeated in the margin in order to indicate the intended position of an addition—is suggested by the misplacement in the manuscript of marginal additions when the (presumed) catchword appears *twice* within a single passage: in two cases the addition was inserted in the wrong place (see the present example and no. 27 below), and in two others it was inserted in both places (see note 35 (partial repetition), and note 531 and line 1602 (catchword: *sabʿa*$^{h}$)).

27. Notes 400–403

The missing word *(al-)amwāl* (note 400) was presumably written in the margin and said to be an addition to *sabʿa*ʰ; but there are two *sabʿa*ʰ's next to each other (lines 1209 and 1210), and *amwāl* was put by the copyist not after the first, as it should have been, but after the second. Note that if the emendation merely meant to add *(al-)amwāl*, it was not sufficient to correct the passage (cf. the omission in note 401, which is confirmed by the manuscript's reading given in note 402).

C. Simultaneous appearance of error and emendation.

28. Note 124

The manuscript has consecutively two readings of the same clause. The first one (lines 309–310) contains a major error (note 123). The second one (in the app.) no longer has this error but omits two essential words. We may have here a trace of a collation of our manuscript's progenitor with another manuscript (also defective?).

29. Note 229

The erroneous value *wa-ṯumn juzʾ min wāḥid* follows the emendation *wa-ṯamanain juzʾ*ᵃⁿ *min wāḥid*; thus, the correction was simply inserted without the previous version being cancelled.

30. Note 252

Two different versions follow one another, the first being, but for one word (note 251), the better one, and, presumably then, a correction of the second. The problem is that several words are inserted between the two versions. A plausible explanation is the following: the text of note 252 (with or without *iḏ*ᵃⁿ) was a line in the archetype of our manuscript (see §6,1°) and should have been cancelled, the emended version—i.e., the words *tuʿādil* (line 732)... *murabbaʿuhū* (line 734)—being written above. Still, the passage was deficient, and, in order to express the equation, the words *al-muʿādil... illā kaʿb kaʿb* (line 734) were then appended to *murabbaʿuhū*.

It is conceivable that the corrector made his emendation by collating his manuscript with another one; for a transformation of *murabbaʿ wa* into *fa-naqūl* is not the sort of change which would easily have occurred to such readers as those who examined the ancestor(s) of our manuscript seem to have been (see below).

**Remark.** The repetition found in note 746 may be a significant example of dittography (see the other examples in notes 411, 515, 624, 734). But it might also be interpreted as the juxtaposition of two versions: first an erroneous version (namely *aʿnī miʾatain wa-ḫamsa*ʰ *wa-ʿašrain juzʾ*ᵃⁿ *min māl*), and then a corrected version. The first version could easily be the result of a jump made by the copyist of the progenitor from the middle of our line 2511 to the beginning of line 2513, and the second could have been written in the margin by the scribe re-reading his text.

### *Conclusion*

In the first group (nos. 1–22), we have discussed those originally marginal (or supralineal) additions completing in some way the text which are recognizable,[18] and, in the second group (nos. 23–30), we have selected those misused additions which present some interest for the history of the text. Three general remarks can be made from an over-all view of these readers' annotations.

1°. The annotations were not confined to any single part of the text; rather, they were distributed throughout, so that *the whole text of our manuscript's ancestor(s) must have been examined, at one time or another, by one or several persons.*

2°. Despite the fact that this examination was done with some care, so that the text was (or ought to have been) clarified, completed or corrected in many places, a great number of significant omissions and shortcomings—not to mention some serious mistakes considered in §10—were not removed. The readers were unable to grasp completely the procedures in Diophantus' problems, *their mathematical proficiency being limited to elementary algebra.*

3°. We have explicitly attributed only some of these annotations to *Arabic* readers, although it is virtually certain that all were made subsequently to the translation.[19] Furthermore, at least two of them must definitely postdate the separation of Books IV–VII from Books I–III, since they were written by someone who appears to have been unfamiliar with Diophantine resolutions and terminology when reading the first pages of Book IV (cf. nos. 1 and 2). But *all of them must antedate the appearance of the inept scribal errors found throughout the text* (for some examples of these, see §§3 and 6): for, as incompletely as these Arabic readers may have corrected the text, they would surely not have left so many errors of this nature.

All things considered, it seems to me to be probable that the two scribes of the Mashhad manuscript must be responsible for the gross misunderstandings of the text in general and of the marginal additions in particular; consequently, it must have been the immediate predecessor of the Mashhad manuscript which was copiously annotated in its margins. This, incidentally, could account for the manuscript's having been recopied; for it would seem that neither a deteriorated text (illegibility of the immediate predecessor of the Mashhad manuscript was apparently not a source for mistakes), nor any particular purpose of study (our copy was never studied, cf. p. 23), lead to the recopying of the text.

---

[18] One should not forget that there must have been other ones, namely those which happened to be correctly inserted into the text—and are, therefore, not recognizable.

[19] Very few of them could date back to Greek times, and this in theory only (e.g., nos. 14, 17, 18, 22).

# §6. On the Progenitor of Our Manuscript[20]

We have seen in the preceding paragraph that a characteristic of the progenitor of the Mashhad manuscript (presumably its immediate predecessor) was the addition by readers of many marginal or supralineal annotations and that the misinterpretation of part of this extraneous material could account for some errors found in our manuscript. On the other hand, certain gross errors made in the course of the transcription, many of which these readers would surely have corrected had they seen them, are perhaps attributable to characteristics of the progenitor's text.

1°. A few sizeable omissions in our manuscript, which are not, unlike most, explicable by homoeoteleuton, may have arisen from the skipping of a line in the course of the transcription (see lines 396–97, 1554–55, 1911–12, 3546–47, and also above, §5, no. 30).[21]

2°. The script of the progenitor may have caused some of our copyists' misreadings, such as the reading *final ᶜain* instead of *ᶜain* + *hā* (app., notes 357, 709, 970), *min* for *wa* (note 19) or *fā* for *min* (note 788),[22] *nūn* for *rā* (note 751; combined with earlier corruption?); perhaps, also, the reading *final fā* instead of *fā* + *alif* (thus *alf* for *alf*^*an*^ throughout the text, particularly towards the end, and once *alf* for *alfà* (note 550)).

3°. (α) It is evident that the progenitor was not *systematically* provided with diacritical points from such confusions as those made between the radicals *sb*ᶜ and *ts*ᶜ ($\frac{1}{7}$ and $\frac{1}{9}$: note 529; 7 and 9: notes 228, 232, 303, 313, 374, 376, 662, 735, 871, 907; *70* and *90*: notes 230, 249, 305, 570). Other characteristic examples are found in notes 164 and 215, 208 and 433, 212, 498, 521 and 522.

N.B. There is another frequently occurring error which can only have originated from a copyist's misreading in Arabic times (probably our copyist), which is mistaking the radicals *ṯmn* and *ṯlṯ* for one another; thus the confusions between 3 and 8 (notes 619, 644), between *30* and *80* (notes 312, 317, 388, 431, 480), between *33* and *88* (note 528). This error cannot, however, have resulted only from the absence of some diacritical points, but also from the shortness of the medial *lām* (and, in the case of *8* read as *3*, from the copyist's not having seen (the support of) a medial *yā*).

(β) The progenitor must have had, on the other hand, a great many diacritical marks written in. For there are (despite the two scribes' very limited comprehension of the text) many diacritical marks in the section copied by the first hand and, after all, few miswritten words on the whole. Again, the

---

[20] We shall call "progenitor" of the Mashhad manuscript that copy which was copiously annotated by readers and from which, apparently, the Mashhad manuscript's copyists worked.

[21] An interval of comparable length is found between the required and the actual placement of a word (see app., notes 126–127); this word was perhaps written in the margin of the progenitor between two lines and copied by our copyist after the second one instead of after the first.

[22] Such a confusion is understandable with a writing similar to that of our manuscript's first hand; see the third line of the title (plate I) or lines 1, 7, and 18 of the first page (plate II).

presence of some vowel-signs in the first handwriting points to their presence in the progenitor.[23]

4°. Towards the end of the manuscript, the majority of the problems' conclusions states the results obtained without any particle of coordination to connect them (cf. note 703), as if in some earlier copy they had either been separated by red dots or put on different lines. Only in a few cases was the resulting lack of clarity eliminated by the interposition of words (cf. lines 3375–77, 3467–68, 3535–36) or by the writing of supralinear *wa*'s (notes 827, 833), some of which may reproduce (irregularly added) readers' additions in the immediate predecessor of the Mashhad manuscript.

In closing, let us recall that, in all probability, the immediate predecessor of our manuscript did not contain Books I–III either, since the addition of some of the annotations is understandable only coming from readers unacquainted with the content of Books I–III (see §5, Conclusion).

## §7. Grammatical and Lexicological Remarks

### 1. Numbers and Powers

As is usual in Arabic algebraic treatises, numbers are expressed in words. It may be of interest, then, to consider the grammatical rules concerning numbers in a text especially rich in them. Not all irregularities or deviations from the classical norm may be automatically attributed to the translator, who was generally recognized for the quality of his Arabic (cf., e.g., Suter, *Math. u. Astron.*, p. 40).

#### A. Integers

##### *a. Grammatical Peculiarities*

(α) The word *aḥad* in association with tens does not appear anywhere in the manuscript with a final *alif* in the indeterminate accusative[24] (see, e.g., line 296; incidentally, the first hand wrote *aḥad*$^{un}$ here). This is not true for *wāḥid*, which is declined (in most instances), as, e.g., in lines 901, 1036, 1085 *seqq.*, 1266, 1445, 2534, 2603.[25]

(β) The noun denoting the things numbered generally appears after the last numeral. There are some passages, however, in which the noun is repeated

[23] The misunderstanding of a *wāw* of case ending could account for the incongruous presence of three conjunctive *wāw*'s (see notes 174, 593, 732).

[24] We have kept this particular form in the edited text.

[25] Exceptions are in all probability scribal errors, and we have corrected them (app., notes 340, 354, 473, 565, 745, 799).

after some or all numerals (as in lines 2781–82), a case mentioned by grammarians (e.g., Caspari–Wright, II, p. 239). Actually, this second case is encountered only in the later part of the manuscript where it is, nevertheless, exceptional and does not supersede the general usage (cf. lines 2466–67 with 2474–75, 2769–70 with 2775, 3051 with 3049).

(γ) Infrequently the cardinal number does not precede, but follows in apposition, the numbered objects (cf. Caspari–Wright, II, pp. 239–40). Examples of this may be found in lines 2672, 3021, 3112, 3545, but only twice with a number of more than one digit (lines 1366, 1373).

(δ) We find in the manuscript improper agreement in gender of numerals numbering feminine objects; see notes 535, 667, 838, 891, 974. Such errors are common in manuscripts, and there is little doubt that the translator cannot be held responsible for them; the simultaneous appearance in the manuscript of these incorrect forms and of correct ones (e.g., lines 2169 (partly), 2918, 3423, 3428–29) reinforces this impression. Another improper agreement is found in note 551.

(ε) The formulations of certain problems involve the multiplication of an unknown by a given multiplier, the value of which is then stated in the ἔκθεσις (cf. §8,2). There, curiously, the number predicate of the word "multiplier" (*amṯāl*, or *marrāt*) admits of either gender. Thus the masculine form is used with *marrāt* in line 210, but the feminine (absolute) form is found in lines 226, 254. With *amṯāl*, we find the feminine form of the number only in line 1714; otherwise, the masculine form is used (lines 759, 895, 1100–1, 1622). It seems odd that in this last case the understood substantive should be *marrāt*.

### *b. Determination*

Those numerals which take the object numbered in the accusative singular can be separated into two classes, namely (α) the numerals from 11 to 19 and (β) the numerals from 20 to 99.

(α) In classical Arabic (i.e., according to the "best grammarians"), when numerals from 11 to 19 are followed by the name of the object numbered, and the expression is determinate, the article should be prefixed to the units alone (see, e.g., Caspari–Wright, II, p. 245); but other combinations are found (*ibid.* and Reckendorf, *A.S.*, pp. 212–13; Howell, *A.G.*, I, p. 1484).

The following usages occur in our text:

(a) the article is prefixed to the units alone (e.g., in lines 479, 696, 806, 860, 1242);
(b) the article is prefixed to the units and to the tens (e.g., 656, 718, 802–3, 1299, 1312);
(c) the article is prefixed additionally to the numbered object (e.g., 477, 629, 685, 686, 835);

(d) finally, the article is prefixed to the units and to the numbered object (e.g., 695, 783, 863, 865, 866).

($\beta$) The other compounds of tens and units:

(a) either have the article prefixed to both numbers, but not to the numbered object, as in lines 50, 150, 185–86, 404, 450 (this is the classical construction);

(b) or the article is prefixed to both numbers and, in addition, to the numbered object; see, e.g., lines 187, 188, 475, 649, 736.

The same of course holds, *mutatis mutandis*, if only tens are represented (e.g., 440, 1475, and 321, 499–500).

Those numerals which are construed with the genitive of the numbered object ought, in classical Arabic, not to have the article themselves, since they are in the *status constructus*. But this rule is not always observed (cf. Caspari–Wright, II, p. 244).

While the "classical" case is poorly represented in our text (see lines 149, 523, 524–25; cf. note 398), the other two combinations appear frequently. The prefixion of the article to both the numeral and the object (e.g., lines 48, 51, 64, 77, 120) is seen more often than the prefixion of the article to the numeral only (e.g., 46, 297, 385, 604, 1638, 2900).

## B. Fractions

### a. General Fractions

($\alpha$) *Expression*

A general fraction of some magnitude $A$, $(m/n)A$, is expressed in Arabic as "$m$ parts of (*min*) $n$ parts of (*min*) $A$", and our text generally does the same. But it sometimes drops one portion or another of the full expression. Thus:

1°. The first "parts" is missing in lines 402, 2596, 2808. The fact that this occurs rarely suggests that it is accidental.

2°. The second "parts" is missing in lines 1035, 1430, 1452, 1705, 2024, 2084, 2153, 2215, 2464, 2758, 2804, 2807, 3088, 3530. This is not a peculiarity of our text (see, e.g., al-Ḫwārizmī, *Algebra* (ed. Rosen), 83,6; 112,16; 117,15).

3°. "of A" (when A is the unit) may be omitted (and is, more frequently in the beginning of the treatise than elsewhere); see lines 324, 387, 398–99, 400, 617 *passim*. This is an admissible way of formulating numerical fractions (cf. Caspari–Wright, I, p. 264).

4°. Equally common is the reduction of the whole expression to "$m$ parts of $n$", as in lines 325 *seqq*., 401, 404–5.

5°. The omission of the words "of $n$ parts (of $A$)" poses quite another problem:

This omission can occur naturally as when a fraction which has just been mentioned is referred to simply by its numerator, as in lines 2605 and 2609, 3028–29, 3321 and 3325, 3563–64. One may also give the denominator just once when speaking about two fractions having the same denominator (see lines 648–50, 1060–63; 916–17). Similar instances are seen in other treatises: in al-Karajī's *Badī*ᶜ (fol. 102ᵛ, *in initio*) and in al-Ḫwārizmī's *Algebra* (p. 79,7—perhaps accidentally).

When the omission is possibly due to homoeoteleuton,[26] or when the presence of the denominator is required by the previous computations or for the subsequent ones, we have filled in with the missing part (see lines 348–49, 1053, 1440, 1444, 2568–69, 2782–83, 3255, 3289). It is true that the oldest extant manuscript of the Greek *Arithmetica* does not give the denominators systematically (see Tannery's remark in D.G., II, p. xliv, and, e.g., the app. crit. to D.G., I, pp. 90, 254, 256, 266), but, the Arabic version having gone through the hands of a prolix commentator completing the computations (see §8 *seqq.*), it would be surprising to find this kind of omission; and its very rarity speaks rather for inadvertence on the part of the copyist.

**Remarks.** 1°. The denominator is sometimes given, presumably for practical reasons, in the form of the product of the same factor, as in problems IV,12; 37; 39; 42,b.

2°. We find a few examples of fractions in which the numerator contains a fraction. See lines 1453–55, 1459–60, 2880–81, 2882–83, and (with a variant in the expression) lines 1437–41, 1450–53, 1456–59.

(*β*) *Determination*

Whenever a fraction is (grammatically) determined, the article is prefixed to the numerator only (or to part of it, according to the rules of determinations for integers seen above). There are, however, a few instances in which the denominator also takes the article: see lines 925–28, 2485, 2763–64, 2773–74, 2910.

### *b. Aliquot Fractions and Related Cases*

(*α*) *Expression*

The fraction $1/n$, $3 \leq n \leq 10$, is normally expressed by its proper name (form *fu*ᶜ*l*). On some occasions, however, the general, circumlocutory way is used; this occurs for the following fractions in the lines indicated:

$\frac{1}{4}$: <u>2326</u>

$\frac{1}{5}$: 360

$\frac{1}{6}$: 2185–86, 2850

---

[26] Supposing the absence of "of *A*" in the original expression (cf. 3°).

$\frac{1}{8}$: 256, 331–32, 2329–30, <u>2421</u>

$\frac{1}{9}$: 2860

$\frac{1}{10}$: 345–46, 347–48, <u>2249</u>, 2402.[27]

In the underlined references, the short form is also found, either following the circumlocution (2249, 2326), or preceding it (2421).

($\beta$) *Decomposition of Some Fractions*

Grammarians point out two sorts of formulations sometimes used for expressing fractions, one as a product and the other as a sum (cf. Caspari–Wright, I, p. 264; Fleischer, *Kl. Schr.*, I, p. 340).[28]

In our text, only two fractions are expressed as such products.[29] The fraction $\frac{1}{16}$ is represented as $\frac{1}{2} \cdot \frac{1}{8}$ in many problems (e.g., those in the middle of Book VI), though not systematically (see lines 2698, 2728–29, 2801–4, 3088–89 (cf. 3089, end)); the same representation is used in other mathematical works (cf., e.g., al-Karajī's *Badī*ᶜ, fol. 107ʳ). The other fraction expressed likewise is $\frac{2}{3} \cdot \frac{1}{9}$, of which there are only a few occurrences (see below).

Examples of the representations of nonaliquot fractions as sums of fractions are more numerous: $\frac{3}{4} = \frac{1}{2} + \frac{1}{4}$ is common (e.g., in IV,22 and IV,43; mentioned by Caspari–Wright, *loc. cit.*); $\frac{9}{16} = \frac{1}{2} + \frac{1}{2} \cdot \frac{1}{8}$ also occurs (in IV,22, in the middle of Book VI, and in VII,6); $\frac{10}{16} = \frac{1}{2} + \frac{1}{8}$ is used only incidentally (line 402; cf. 2701). Finally, some fractions are decomposed into a sum of fractions having powers of the same number in the denominator: $\frac{24}{25} = \frac{4}{5} + \frac{4}{5} \cdot \frac{1}{5}$ (lines 1318–19), $\frac{21}{25} = \frac{4}{5} + \frac{1}{5} \cdot \frac{1}{5}$ (lines 1042–43), and the several examples of problem IV,40: $\frac{43}{81} = \frac{4}{9} + \frac{7}{9} \cdot \frac{1}{9}$, $\frac{17}{27} = \frac{5}{9} + \frac{2}{3} \cdot \frac{1}{9}$ $(= \frac{5}{9} + \frac{6}{9} \cdot \frac{1}{9})$, $\frac{13}{81} = \frac{1}{9} + \frac{4}{9} \cdot \frac{1}{9}$, $\frac{73}{81} = \frac{8}{9} + \frac{1}{9} \cdot \frac{1}{9}$. All this is seen in other treatises (e.g., the *Badī*ᶜ, cf. fol. 118ᵛ, 124ᵛ, 125ᵛ, 128ᵛ), in some of which it is even quite frequent (as in the *Algebra* of Abū Kāmil).

These decompositions can therefore be considered neither as a peculiarity of our treatise nor, far less, as remnants of the original Greek text even though similar forms are seen in Greek works, most commonly in those of Heron.[30]

($\gamma$) *Grammatical Peculiarities Connected with Aliquot Fractions*

1°. Concerning the expression of $(m/n)A$ as $(1/p + 1/q)A$.

The representation of some fraction $(m/n)A$ as $(1/p + 1/q)A$ occasionally leads to the case mentioned by Nallino (*Opus*, II, p. 320) of a determining noun

---

[27] See also the multiples in lines 3498 ($\frac{1}{4}$), 3293 ($\frac{1}{5}$), 2187 ($\frac{1}{6}$), 3258 ($\frac{1}{7}$), 2361 and 2852 ($\frac{1}{9}$), 2250 and 3523 *seqq.* ($\frac{1}{10}$).

[28] What we call "product", for the sake of simplicity, is actually a subdivision, and we shall represent as $\frac{1}{2} \cdot \frac{1}{8}$ what is in the text expressed as "one half of one eight".

[29] Apart from the fractions having their denominators represented as a product of equal factors (see above).

[30] In fact, such occurrences in the Greek *Arithmetica* are rare: D.G., I, pp. 164,9 and 328,13.

($A$ in the genitive) depending upon two *status constructi* ($1/p$ and $1/q$, $1 \leq p,q \leq 10$). This is what we see in lines 1469–70 and 1472 ($\frac{3}{4} = \frac{1}{2} + \frac{1}{4}$); 1470 ($\frac{5}{4} = 1 + \frac{1}{4}$); 1504 ($\frac{3}{2} = 1 + \frac{1}{2}$).

2°. Concerning the expression of $(m + 1/p)A$.

The expression of $(m + 1/p)A$ ($m, p$ natural, $2 \leq p \leq 10$) can be misleading when, for $A \neq 1$, the text drops the multiplicand in the second term; thus, instead of, say, *al-šaiʾ wa('l)-rubʿ šaiʾin*, we may find *al-šaiʾ wa('l)-rubʿ*, i.e., $mA + 1/p$. For the sake of clarity, we have emended all these passages (see notes 108, 125, 272, 274, 406, 407, 616, 778, 785, 963). But there is no doubt that this omission is not (always) a scribal error; nor does it result from some misreading of the Greek text, for it occurs more or less frequently in other Arabic mathematical works (see, e.g., al-Karajī's *Badīʿ*, fol. 124r, *in fine*; al-Ḫwārizmī's *Algebra* (Rosen), pp. 32,16–18; 33,3 (cf. 33,1); 36,1; 41,8; Abū Kāmil's *Algebra*, fol. 84v,11 and 17).

3°. Concerning the aggregated form of $(1/p)A$.

Nallino has pointed out the possibility of the article being prefixed to the unit fraction $1/p$ ($p$ nat., $2 \leq p \leq 10$), and not to $A$, when the aggregate $(1/p)A$ is determinate (*Opus*, II, p. 320). Occurrences of this phenomenon are seen in lines 288, 290, 294, 548–49 (cf. 547), 794, 1929, 2160, 2449, 3093; 2449 (product of two aliquot fractions); 785, 1668 (duals, i.e., fractions $(2/p)A$).

## C. Grammatical number of a mathematical expression

Generally, the pronoun referring to a mathematical expression can occur either in the singular or in the plural, depending on whether the expression is considered as a whole or as made up of parts. The same holds for verbal agreement.[31]

### *a. Units*

Both singular and plural are found. See, e.g., lines 56, 70, 218, 261, 300, 301, 316, and 82, 97, 112, 133, 217, 246, 247, respectively; compare also lines 671 and 672.

### *b. Multiple of a Power*

The plural is usual (cf. lines 48–50, 51 (interp.), 62, 103, 106, 108 *passim*); but the singular is also found (see lines 425, 695, 1238, 2082 (cf. 2081), 2845, 2941, 3514).[32]

---

[31] On verbal agreement with a mathematical expression, see also below, p. 47.

[32] In all these latter examples, the numbered object is in the accusative singular (but the verb is not necessarily in the singular when the numbered object is: cf. lines 185–86).

### c. *Algebraic Polynomial Expression*

For more than one term the plural is commonly used (see, e.g., lines 241, 1298, 1944, 1994, 2015, 2056, 2070, 2111, 2211, 3234). The singular, however, occurs in many passages (see lines 865–66, 1472, 2301 (dualistic), 2556, 2654, 2675, 2698, 2762, 2794, 2886, 3193, 3272, 3486, 3521).

## D. Powers

### a. *The Greek Power-system*

It is an established fact today that Diophantus did not invent the words used in Greek times to designate the different powers of the unknown up to the sixth. First, the word δυναμοδύναμις appears in Heron's *Metrica*, a work anterior to the *Arithmetica*.[33] Further, according to a text of St. Hippolytus (*ca.* 150), the words δύναμις, κύβος, δυναμοδύναμις, δυναμόκυβος, and κυβόκυβος go back to "Pythagoras", that is to say (taking the usual personalization into account), to the Pythagoreans.[34]

Geometrically, a number $x$ (ἀριθμός) is represented as a segment of a straight line. A geometrical representation is also possible then for the next two powers, the square $x \cdot x$—let us designate it by the symbol $Q$—, a special case of plane number, and the cube $C \equiv Q \cdot x$, a special case of solid number (cf. Euclid, *Elem.*, VII, deff. 16–19). Since any positive integer $N \geq 2$ is representable as $2 \cdot k + 3 \cdot l$ ($k$, $l$ not negative integers), the two fundamental symbols $Q$ and $C$ are sufficient to describe any power $x^N$, then named by using the appropriate repetition of the words δύναμις and κύβος.[35] Thus, $x^4$ is $QQ$ (δυναμοδύναμις), $x^5$ $QC$ (δυναμόκυβος), $x^6$ $CC$ (κυβόκυβος). We do not know with certainty how the higher powers $x^8$, $x^9$ were expressed in Greek (since they occur only in the Arabic version); one would expect them to be δυναμοκυβόκυβος and κυβοκυβόκυβος.

The denominations reported by St. Hippolytus and adopted by Diophantus were not the only ones in use. The Byzantine Psellus mentions two other systems (see D.G., II, pp. 37–38). The first, used by the "Egyptians" (i.e., the Greeks of Egypt), employed the same words for the first four and the sixth powers as Diophantus, but differed in labelling the fifth πρῶτος ἄλογος; it called the seventh δεύτερος ἄλογος, the eighth τετραπλῆ δύναμις, and the ninth κύβος ἐξελικτός. The outstanding rôle played by the squares and the

[33] Heron, however, does not use δυναμοδύναμις in the proper sense of $x^4$, but in its original signification, "fourth power", ἡ ἀπὸ ΒΓ δυναμοδύναμις meaning ΒΓ$^4$ (cf. *Heronis opera*, III, p. 48). This is the meaning found in the beginning of the introduction to Book I of the *Arithmetica* (D.G., I, pp. 2,14–4,7) and in St. Hippolytus' text (see below). Diophantus may have been the first to have used these words absolutely, i.e., as designations of the powers of *the* unknown—whence his subsequent definitions (D.G., I, pp. 4,12–6,8).

[34] The relevant passage of St. Hippolytus is alluded to by Tannery in his *Perte de sept livres*, p. 206 = *Mém. sc.*, II, p. 90, and printed in Diels' *Doxographi graeci*, pp. 556–57.

[35] In this system the values of $k$ are generally limited to 0, 1, 2 only: see below, (*b*).

cubes among the powers is the origin of the designations of the fifth and the seventh powers: ἄλογος is any power which does not belong to a $Q$-class $\{x^{2n}\}$ or to a $C$-class $\{x^{3n}\}$ (in other words any $x^N$ with $N$ not divisible by 2 or 3), and of those $x^5$ is the first and $x^7$ the second ($x^{11}$ being the third and so on). Psellus' text indeed states why $x^5$ is called ἄλογος (πρῶτος): οὔτε γὰρ τετράγωνός ἐστιν οὔτε κύβος.

The other non-Diophantine system consisted simply in naming the powers according to their succession, $x$ being πρῶτος ἀριθμός, $x^2$ δεύτερος ἀριθμός etc. (cf. D.G., II, p. 38,3–12). Had Diophantus employed this system, his usage of abbreviations would certainly have led him to adopt some kind of exponential notation in the manner used in late mediaeval and Renaissance times in Europe (e.g., by N. Chuquet and R. Bombelli).

### *b. The Arabic Power-system*

The power-system commonly used in Islamic countries is obviously taken from the first Greek one, that used by Diophantus, the words *māl* and *kaᶜb* rendering δύναμις and κύβος, respectively. Thus, the combination δυναμόκυβος, for instance, becomes *māl kaᶜb*.[36] The denomination of powers was carried on in the Diophantine way for higher powers, the rules being (in our symbolism) that $x^{3n}$ is represented by $n$ $C$'s, $x^{3n-1}$ $(=x^{3(n-1)+2})$ by one $Q$ followed by $n-1$ $C$'s, and $x^{3n-2}$ $(=x^{3(n-2)+4})$ by two $Q$'s followed by $n-2$ $C$'s. Thus, for instance, $x^8$ is expressed as *māl kaᶜb kaᶜb* and $x^9$ as *kaᶜb kaᶜb kaᶜb*.[37]

N.B. Certain Islamic mathematicians use some different denominations (see below). But quite a different system is that described by Luca Pacioli (*Summa*, fol. 67ᵛ), according to whom the power-names he gives "sonno denominationi de la pratica de algebra secondo li arabi". This system is based on the *multiplication* of the exponents, so that Pacioli's *censo di cubo* is $x^6$ (whereas *mālᵘ kaᶜbⁱⁿ* is $x^5$).[38] Since the basic terms used are *censo* and *cubo*, a special name must be introduced for the powers $x^N$ with $N$ not divisible by 2 or 3: they are the various *relati*, which are numbered in succession as are the ἄλογοι of the "Egyptians".

### *c. The Power-system in Our Text*

Basically, our Arabic translation employs the usual, Diophantine–Arabic, system; there are, however, some singular usages in the denominations

---

[36] Luckey remarked (*Rechenkunst bei al-Kāšī*, p. 55, n. 82) that from the grammatical point of view *mālᵘ kaᶜbⁱⁿ*, used for $x^2 \cdot x^3 = x^5$, ought to mean $(x^3)^2 = x^6$. This is not absolutely correct: neither *māl* nor *kaᶜb* is used in this case as an operator: the words *murabbaᶜ* and *mukaᶜᶜab* play this rôle. It is nevertheless true that the grammatical structure of the denominations of compound powers is odd and suggests foreign origin.

[37] Examples of higher powers are found in the *Badīᶜ* of al-Karajī ($x^{10}$, $x^{11}$, $x^{12}$, $x^{14}$, $x^{16}$ on fol. 78ʳ, 81ʳ); further examples occur in Samawʾal's *Bāhir* (ed., p. 56).

[38] Piero della Francesca, in his *Trattato d'abaco*, adopts a mixed system: see fol. 29ᵛ–30ʳ and 33ᵛ.

of the two powers normally made up of a combination of both $Q$'s and $C$'s, namely $x^5$ and $x^8$.[39]

(α) $x^5$

$x^5$ is used only in Books IV and VI. In Book IV, it is called either *māl kaᶜb*—in accordance with the usual Arabic usage—or *kaᶜb māl* (cf. below, (β)).[40] In Book VI (problems 4–7), $x^5$ is rendered by the circumlocution "$x^3$ multiplied by $x^2$" (*kaᶜb maḍrūb fī māl*) and, when its coefficient is not one, but, say, $m$, it is rendered by the expression "$mx^3$ multiplied by $x^2$".[41] I have seen this decomposition elsewhere only once, in a passage of Abū Kāmil's *Algebra* (fol. 51ʳ,18); the passage is, however, uncertain, since, first, $x^5$ is designated immediately afterwards by *māl māl maḍrūb fī šaiʾ* (which is Abū Kāmil's usual circumlocution; see, e.g., fol. 51ʳ⁻ᵛ), and, second, the Hebrew version does not have the said form.[42] But, whether a reader's addition or not, the denomination did exist—perhaps originally to avoid a misunderstanding of *māl kaᶜb* as $(x^3)^2$.

(β) $x^8$

In Books IV and V, $x^8$ is expressed as *kaᶜb kaᶜb māl*.[43] It is worth observing that this placement of the *māl* after the *kaᶜb*'s, also employed for $x^5$, curiously departs from the usage of practically all known Arabic mathematical texts.[44]

In Books VI and VII, $x^8$ is designated as *QQQQ* (*māl māl māl māl*). The usage of this denomination in Arabic times is confirmed by its repeated appearance in Abū Kāmil's *Algebra* (fol. 57ᵛ, 76ʳ, 77ʳ–78ᵛ).[45] What the Greek *Arithmetica* had in these places we do not know; but one should keep in mind that an expression of $x^8$ by means of $Q$ only is known to have existed in Greek times (the τετραπλῆ δύναμις mentioned above). See also p. 67, N.B.

**Remark.** Powers in the denominator occur in our text in problem VI,23 only. There, the term $m/x^n$ (with $n = 1, 2$) is expressed as "$m$ parts of $x^n$", which is an attested Arabic denomination (see Luckey, *Rechenkunst bei al-Kāšī*, p. 54 *seqq.*). Hence, there are no words corresponding to the Greek ἀριθμοστόν and δυναμοστόν.

[39] Neither $x^7$ nor powers higher than $x^9$ occur in the text.

[40] See the index for the occurrences of the various forms.

[41] The "$x^2$ multiplied by $125x^3$" in lines 2273–74 has been noted as a reader's interpolation (cf. §5, no. 9).

[42] Cf. Levey's edition, p. 173,12. The Latin version, in ms. B.N. lat. 7377 A, has a sizeable gap here.

[43] Note the following incidental occurrences: one "$x^6$ by $x^2$" (line 901) and three "$x^4$ by $x^4$" (in problem IV,42, after the redefinition of the name *kaᶜb kaᶜb māl* in lines 1339–40).

[44] Some of the copies of the *Faḫrī* have the form *kaᶜb kaᶜb māl*; the placement of *māl*, however, varies even within the same manuscript.

[45] Indirectly by Abenbeder (=Ibn Badr, Suter no. 493): he does not use $x^8$, but expresses $x^{10}$ by five *māl*'s, and he gives for a power $x^{6n}$ the denominations $2n$ *kaᶜb*'s and $3n$ *māl*'s as equivalent (*Compendio*, pp. 15–16 of the Arabic text).

### *d. Grammatical Determination of the Powers*

We have seen that the powers $x^n$ $(n \geq 4)$ are generally expressed in Arabic by a sequence of the form $P_1 P_2 P_3 \ldots$, the $P_i$'s being either *māl* or *kaᶜb*. Thus arises the question of the position of the article when such expressions are determined.[46]

In classical Arabic, the article can be prefixed to the first element instead of the last, since the chain of genitives may be considered as a single word (cf. Brockelmann, *Grundriß*, I, p. 475). This (alternative) possibility is found among the combinations used in our text, which follow ($P_i$ underlined indicates prefixion of the article).

(α) *Two Elements*
$\underline{P_1}P_2$: (lines) 629, 685, 734, 804, 869, 1626 *passim.*
$P_1\underline{P_2}$: 23, 25, 31, 123, 128, 149, 187, 453, 685, 686, 735 *passim.*
$\underline{P_1 P_2}$: 452, 455, 694, 695, 739, 760, 761, 783, 1248 *passim.*

(β) *Three Elements*
$P_1 P_2 \underline{P_3}$: 863, 865, 866, 868, 870, 871, 1373, 1379, 1380, 1381, 1483.
$\underline{P_1}P_2 P_3$: 811, 835, 1370, 1395, (1396), 1481, 1490.
$P_1 \underline{P_2 P_3}$: 863–64.
$\underline{P_1 P_2 P_3}$: 1483.[47]

N.B. In three passages, the article is prefixed to the numerical coefficient of the power only; see lines 687, 875, 1366 (cf. 1373).

## 2. Some Grammatical Remarks on Verbs

### *a. Verbal Persons Used*

Except in the introduction to Book IV (where the teacher is addressing the student in the first person singular), the first person plural is employed throughout the text in giving instructions, i.e., when personal style is used. Departures from this usage are rare: once the first person singular is used in a problem (line 123; probably a commentator's addition, cf. p. 179), and the imperative is employed seven times (lines 2280, 2515, 2557–60, 2886).[48]

### *b. Jussives of Weak Verbs*

As is often the case in manuscripts, the jussive forms of defective and hollow verbs may be incorrect. The renown of Qusṭā's Arabic (cf. p. 37) and the

---

[46] The question of the formation of the plural need not be touched on here since the various forms are given in the Arabic index under the corresponding word.
[47] The form $P_1 P_2 P_3 P_4$ $(x^8)$ does not occur with an article in our text.
[48] Note an abrupt use of the imperative in the Greek text also, p. 340,17 *seqq.* in Tannery's edition.

occurrence of correct forms as well as incorrect ones suggest that such mistakes originated with copyists.

These errors are much more common with defective verbs than with hollow ones. Errors occur with *alqà* (notes 74, 282, 583, 594, 598, 615, 652, 661, 665, 729) and with *ġaniya* (note 935); the correct form appears once, with *alqà* (line 875). On the other hand, with the hollow verbs, an incorrect form appears only once, with *zāda* (note 650), all the other instances being correct: lines 835, 2323, 2469, 3562 (*zāda*); 1414 (*qāla*); 2418 (*istaḥāla*); and *kāna* throughout.

**Remark.** The few imperatives of weak verbs are correct; see lines 2559, 2886 (*alqà*); 2558 (*zāda*).

### *c. The Verb* ʿ*adala*

The statement of an equality between two expressions requires, for the sake of clarity, the interposition of a word indicating this equality. If, rather than an adjective like *musāw*$^{in}$, one uses the verb ʿ*adala* (in its first or third form), as our text does consistently, the verb should be preceded by some auxiliary verb at the beginning of the sentence.[49] The auxiliaries used in our text are *kāna* (most frequently), *baqiya*, and (to a lesser extent) *ẖaraja*, *ijtamaʿa*, *ṣāra* (see Arabic index).

(α) *Agreement of* ʿ*adala*

In some Arabic texts, the expression in the "left" side of an equation is considered as a whole, and the verb is thus in the singular (see Luckey, *Richtigkeitsnachweis*, p. 98). We chose in our edition to put the (almost always) unpunctuated verbal form in agreement with the subject *not* taken as a whole, for it would seem that the original text did this (at least sometimes), as is suggested by those two relevant instances in the manuscript where ʿ*adala* is provided with diacritical points (corresponding to our lines 104, 228[50]) and by the use of dual forms (364, 1033, 1948, 1950, 1975–76, 2325, 2888, 3236).

(β) *Agreement of the Auxiliary of* ʿ*adala*

Agreement between ʿ*adala*'s auxiliary and the following subject taken as a plurality is found, e.g., in lines 480, 1042 (*kāna*); 686, 1253 (*baqiya*); 735, 1022 (*ṣāra*); 1374 (*ijtamaʿa*). We see the singular, though, in, e.g., lines 661, 1244 (cf. 1253), 1294, and in those places of the manuscript where the imperfect of the auxiliary is provided with diacritical points (corresponding to our lines 91, 93, 151, 184, 213, 244). Thus, no specific rule seems to have been followed in the manuscript, and, for the edited text, we have chosen to put the originally unpunctuated forms of the imperfect in the singular.

---

[49] The auxiliary is sometimes omitted; see the first problems of Book IV.

[50] This latter agreement is disputable.

# Chapter III

# Tentative Reconstruction of the History of the *Arithmetica*

## §8. Formal Subdivisions of a Problem

### 1. Analysis and Synthesis

In the beginning of the seventh Book of his *Collection*, Pappus mentions two types of analyses and syntheses distinguished by the Greeks.[1] The first, ποριστικόν, type is commonly used by geometers in connection with the *demonstration of a proposition, or of an (already known) solution.* In the corresponding analysis, what is to be proved is supposed to be true (or known), and must be reduced by passing through its successive consequences, either to an identity or to a known proposition. The synthesis then reverses the process. The second kind of analysis, of the ζητητικόν type, is used in the *finding of a solution* to a problem. Supposing the problem solved, the mathematician establishes between the known and the unknown magnitudes some relation, which is then reduced, by elimination, to a final relation containing the smallest number of unknowns possible (one for a determinate problem). This is the analysis. The synthesis simply verifies the exactness of the solution found.

The latter type is used constantly in Diophantus' problems. But, *whereas the Greek Diophantus goes through the analysis and drops the synthesis,*[2] Diophantus contenting himself with some phrase like "καὶ ἡ ἀπόδειξις φανερά", or "(the found numbers) ποιοῦσι τὰ τῆς προτάσεως", *the Arabic text contains, with few exceptions, all the computations pertaining to the synthesis.* This is the most striking difference in form between the Greek and the Arabic texts.

---

[1] *Collectio*, ed. Hultsch, pp. 634–36. The following is a summary of Tannery's account in his *Notions historiques*, pp. 328–31 = *Mém. sc.*, III, pp. 163–66.

[2] Or, more precisely, the *proof* (ἀπόδειξις), since the synthesis includes the computation of the required magnitudes from the value of the chosen unknown (as appears also from our text; see lines 1360–62, 1389–90, 1408–9, 1493 *seqq.*).

## 2. Subdivisions of a Problem

In the full treatment of a geometrical problem there are six constituent parts: πρότασις, ἔκθεσις, διορισμός, κατασκευή, ἀπόδειξις, συμπέρασμα.[3] For a Diophantine, algebraical, problem, the corresponding subdivision into steps is the following:

1°. Πρότασις. Statement of the problem, in terms of required magnitudes and "given" ones (if any).

2°. Διορισμός. Since the solution must be positive and rational, it sometimes happens that one cannot attribute arbitrary values to the given magnitude(s); the diorism then states the limiting condition that it (they) must fulfil. For a condition of *positivity*, the diorism will ultimately result in some restriction in the form of a numerical limit (an inferior and/or a superior one) for one of the numbers; thus, *any* rational value within the continuum defined by the found limits will be suitable for this number. If, however, the given numbers are subject to a condition of *rationality* for the unknown, the set of admissible values for the given numbers, or at least for one of them, will be discrete, and the problem of finding appropriate values will arise. In some cases, this problem is trivial, either because the diorism defines the only given number (problems IV,18; VII,6; "V",9 and 11) or because suitable numbers can be immediately found ("IV", 34–35). The other cases found in the extant *Arithmetica* are simple to solve, since they lead to so-called "constructible problems" (cf. p. 192).[4]

3°. Ἔκθεσις. Numerical setting of the given magnitudes.

4°. Ἀνάλυσις } see above.
5°. Σύνθεσις }

6°. Συμπέρασμα. Final statement, generally recapitulating in abbreviated form what was sought and the found values of the unknowns (appears in the Arabic Books only).

It is of course nothing but a formal requirement for an algebraical problem to include proof and final statement, i.e., the parts absent in the Greek text. But (though hardly the goal of the commentator(s) who added them), their presence has a long-term advantage: by repeating the values and performing in detail the computations, one avoids, to a certain degree, the (progressive) corruption of the numerical results, of which there are numerous examples in the Greek Diophantus—no doubt of quite early origin. An illustration of this may be found in the Arabic Books. There are two problems in which the verification has not been carried out (presumably) because of the size of the

---

[3] See Heron, *Opera*, IV (=*Definitiones*), p. 120,21 *seqq.*, from which Proclus, *In Eucl.*, p. 203,1 *seqq.* (=Heath, *Euclid*, I, pp. 129–31).

[4] Propositions in the *Arithmetica* involving a diorism are marked off by a D (by a $D^p$ if the resulting problem is constructible) in the conspectus of the propositions, pp. 461–483.

resulting numbers;[5] and thus, the second problem has preserved a misreading of a number which must have originated in early Greek times (see p. 246).[6]

The greater prolixity of the Arabic text, the addition of verifications and final statements, and the expression of all numbers and symbols of Diophantus in full words (as is usual in Arabic algebraical treatises) makes an Arabic problem considerably longer than its Diophantine progenitor would, presumably, have been. In the vast majority of the problems, the predominant part of the true increments is formed by the two added steps, i.e., the verification of the solution and the final statement. As to the other additions, their extent depends on the individual problems, and the analysis, which is doubtless diluted and expanded in most problems, may well have been untouched in certain others.[7] We shall come back to this question of the additions later on (§12,2,a).

## §9. Major, Unsystematic Supplements in the *Arithmetica*

We have seen that two sorts of additions were made to the Diophantine text of Books IV–VII, each altering it in quite a different way. The first kind consists of the minor and incidental additions made by readers, the more noticeable of which emended or corrected the Arabic text in points of detail (see §5). The second kind consists of those major and systematically made additions which rendered the Diophantine text more prolix and which enlarged it by the appending of verifications and final statements (see §8). These two kinds of additions were both made long after the composition of

---

[5] Problems IV,44,c and VI,4; the "tolerance of admission" seems to be thirteen digits (in problem IV,42,b).

[6] We deduce from this that the author of the verifications must have had to recalculate some results which, in the course of the verification, he discovered to have been transmitted in an incorrect form.

[7] This is the case, for example, for IV,23. If we leave aside the verification and final statement and rewrite the problem in the concise style of the Greek Diophantus, we obtain a version which corresponds almost literally to the Arabic one (on the Arabic rendering of εὑρεῖν: see index, under *arāda*).

Εὑρεῖν δύο ἀριθμοὺς τετραγώνους ὅπως οἱ ἀπ'αὐτῶν τετράγωνοι συντεθέντες ποιῶσι κύβον.

Τετάχθω ὁ μὲν α$^{ος}$ $\Delta^{Y}\overline{\alpha}$, ὁ δὲ β$^{ος}$ ἀπὸ ϛ$^{ων}$ ὅσων δήποτε. ἔστω δὴ ϛ$\overline{\beta}$. αὐτὸς ἄρα ἔσται ὁ β$^{ος}$ $\Delta^{Y}\overline{\delta}$. ἔσονται ἄρα οἱ □$^{οι}$, ὃς μὲν $\Delta^{Y}\Delta\overline{\alpha}$, ὃς δὲ $\Delta^{Y}\Delta\overline{\iota\varsigma}$, ὁ δὲ συναμφότερος $\Delta^{Y}\Delta\overline{\iota\zeta}$. $\Delta^{Y}\Delta$ ἄρα $\overline{\iota\zeta}$ ἴσαι εἰσὶ κύβῳ.

Πλάσσω τὸν κύβον ἀπὸ ϛ$\overline{\gamma}$ καὶ γίνεται αὐτὸς ὁ κύβος $K^{Y}\overline{\kappa\zeta}$. $\Delta^{Y}\Delta$ ἄρα $\overline{\iota\zeta}$ ἴσαι εἰσὶ $K^{Y}\overline{\kappa\zeta}$ (καὶ πάντα παρὰ $K^{Y}$). ϛ ἄρα $\overline{\iota\zeta}$ ἴσοι $\mathring{M}\overline{\kappa\zeta}$. Καὶ γίνεται ὁ ϛ $\frac{\iota\zeta}{\kappa\zeta}$.

(Καὶ ἔσται ὁ μὲν α$^{ος}$ $\frac{\sigma\pi\theta}{\psi\mu\theta}$, ὁ δὲ β$^{ος}$ $\frac{\sigma\pi\theta}{,\beta϶\iota\varsigma}$, καὶ ποιοῦσι τὰ τῆς προτάσεως).

the text by Diophantus; neither sort added anything important or original to the mathematical content of the problems.

We shall now turn our attention to a type of supplement, found occasionally, which, in contrast with the above-mentioned additions, displays great age and, to a varying extent, originality. Some of these supplements are in fact desirable or necessary additions made no doubt by Diophantus himself, whilst the others originated with (early) Greek scholiasts, as is suggested by the presence of similar interpolations in the extant Greek text.

These supplements may be divided into three categories:

(1) Interpolated problems.
(2) Alternative resolutions.
(3) Other supplements.[8]

The purpose of this paragraph is to provide a general survey of these supplements, of those found in the Arabic as well as in the Greek Books of the *Arithmetica.*

## 1. Interpolated Problems in the *Arithmetica*

We shall briefly indicate here which problems of the *Arithmetica* seem to be interpolated. With respect to the Greek Books, we shall recall Tannery's results, occasionally appending some remarks of our own. A complete discussion of the interpolated problems of the Arabic Books will be found in the mathematical commentary.

*Book I*

None of the problems in this Book was considered by Tannery to have been interpolated. This is to be expected since interpolated problems originating from one Book are found, as the examples in the Greek and Arabic Books show, in some *subsequent* Book.

I shall add only that the genuineness—or, perhaps, the present placement—of I,26 seems open to some suspicion (see *infra*, pp. 195–196).

*Book II*

Tannery considered two groups of problems in Book II to have been interpolated.

1°. He attributed II,1–7 to an older commentator, II,1–5 being "des répétitions absolument inutiles" of problems I,31–34, and II,6–7 being "des variantes sans intérêt d'une question que Diophante avait probablement traitée entre les problèmes I,33 (I,30 in Tannery's own (later) edition) et I,34 (I,31), à savoir: *Trouver deux nombres, connaissant leur différence et la*

[8] The question of the (Diophantine) genuineness concerns of course only the last two categories.

*différence de leurs carrés*, mais qui manque aujourd'hui" (*Perte de sept livres*, p. 198 = *Mém. sc.*, II, p. 81).

Indeed, comparison of the formulations found in II,1–5 with those of their progenitors shows that these interpolated problems simply drop the given condition of proportionality between the two required magnitudes. Thus II,1 corresponds to I,31, II,2 to I,34, II,4 to I,32 and II,5 to I,33; parts (a) and (b) of II,3 treat, once again dropping the condition of proportionality,[9] the two corollaries attached to I,34. Observe that the proportionality between the two required magnitudes, no longer imposed in the formulation, is, in all these problems, chosen to be 2:1.

The origin of II,6–7, or, rather, of II,6 (since II,7 may simply stem from it) is less evident. One may suppose, as Tannery did, that it was a (now lost) problem of Book I. One may also consider II,6 as being a variation on one of the previous types (changing the given multipliers in I,34 to given additive constants), or even as some independent contribution of the scholiast: the Arabic Books offer some examples of interpolated problems without recognizable origin.

2°. Further, Tannery considered II,17–18 as either interpolated (in his edition of Diophantus: I, p. 109, note) or misplaced (*Perte de sept livres*, p. 198 = *Mém. sc.*, II, pp. 80–81). It seems more plausible that they were interpolated, although the idea of there having been some disorder in an earlier Greek manuscript, resulting in a misplacement of problems originally within the group I,21–25 or following I,25, is appealing.[10]

*Book III*

Tannery regards two groups of problems, one at the beginning of the Book and the other at the end, as being later additions.

1°. The first group in question is III,1–4. In his edition of Diophantus, Tannery considered it to be interpolated (cf. D.G., I, p. 139, n. 1). But he had earlier expressed a different opinion, declaring that "les premiers problèmes du Livre III ne sont nullement suspects comme les premiers du Livre II; quoique faciles au fond, ils sont réellement dignes de Diophante, et s'ils ne sont pas de lui, ils appartiennent à un imitateur qui s'était parfaitement rendu maître des procédés du maître. On peut signaler quelques différences de rédaction avec les problèmes analogues du Livre II, mais ces différences sont plutôt en faveur de ceux du Livre III, si l'on considère la concision de l'exposition et l'assurance de la méthode" (*Perte*, p. 199 = *Mém. sc.*, II, p. 82). We are inclined rather to accept Tannery's earlier, more cautious, opinion. See also p. 222.

N.B. There is an allusion in problem VII,7 to the "sixth problem of the third Book", and this reference does in fact apply to Tannery's sixth problem. But we may not take this as a *proof* of the genuineness of the preceding five

[9] See, however, D.G., I, p. 84,16–17.

[10] We have already mentioned that I,26 itself could be out of place.

problems. For, as we shall see (§12), the explicit references to earlier problems seem to date back to the prolix commentary, which was written after the interpolations had been incorporated into (and identified with) the original text.

2°. The case of III,20–21 is clearer: in Tannery's words, these problems represent "des variantes sans intérêt" of II,15–14 (*ibid.*). This opinion is only moderated in the edition, where he says that "elegantius hîc tractata ambo fuisse primo obtutu videntur" (D.G., I, p. 187).

*Book IV*

Book IV contains no interpolated problems, merely extensions and alternative resolutions.

*Book V*

Nor does Book V contain any interpolated problems. It begins with six problems which resemble the ones at the end of Book IV, but there can hardly be any doubt about their genuineness (see pp. 221–222).

*Book VI*

Interpolated problems in our Arabic text begin with Book VI. There are eleven additional problems which can be divided into four groups. VI,1–3 stem from IV,25 and IV,26,a and b, the changes being quite similar to those made for the interpolations at the beginning of Book II: a condition of proportionality is *added* this time, imposing $a = m' \cdot b$ (instead of leaving the choice of $m$ in $b = m \cdot a$ arbitrary) and with the simplest value $m' = 2$. Further, VI,4–7 originate from the corollaries appended to IV,33, just as II,3,a and b arose from the corollaries following I,34. Finally, the group VI,8–10 and the independent problem VI,11 cannot be traced back to any problem of the earlier Books; they seem to be variations on the previous, interpolated propositions. The similarity in the derivation of all these interpolations and those at the beginning of Book II makes it seem possible that both sets were added by the same early commentator.

*Book VII*

The first six problems of Book VII are certainly interpolated. But, among them, only one problem, VII,6, has a traceable origin, namely the last proposition of the previous Book.

*Book "IV"*

The Greek Book "IV" begins with two problems which are simply repetitions (with different values of the given numbers and no diorism) of V,7–8, and are thus interpolated. I have tentatively suggested that "IV",3 might also be an interpolation (see p. 198).

Books "V" and "VI" contain no recognizably interpolated problems.

**Remark.** The greatest distances separating any interpolated problems from their progenitors are those between III,20–21 and their source (II,15–14) and between VI,1–3 and their source (IV,25–26); in both cases, this distance amounts to a separation of about thirty-five problems. Now, counting thirty-five (genuine) propositions from problems V,7–8 on, we come close to the beginning of the Book following Book VII. Since the addition of "IV",1–2 was probably contemporary with the other interpolations and made at a time when the *Arithmetica* was still complete, it may be asked whether the Greek Book "IV" might not be the original Book VIII. This is of course directly connected with the conjectures about the content of the missing Books (see §13).

## 2. Alternative Resolutions (ἄλλως)

### *a. In the Greek Books*

*Book I*

Problems I,18; I,19 and I,21 have two resolutions; Tannery attributes the second one in all three cases to an older scholiast.

*Book II*

Besides the two methods of II,11 (and II,13), both of which probably go back to Diophantus himself, we find a second resolution (virtually identical to the first one) for II,8, and still another for the interpolated (or misplaced) problem II,17.

*Book III*

Of three second resolutions, two (in III,5 and in III,15) are supposed by Tannery to be genuine while one (in III,6) is considered to have been the work of an older scholiast. This last resolution is especially weak, as it merely changes the numerical value of an optional quantity at the very end; it is unquestionably not genuine.

*Book "IV"*

Three problems have two resolutions: "IV",7; "IV",28; "IV",31. All three alternative resolutions look genuine, although Tannery positively asserts this only for the one found in "IV",28.

Books "V" and "VI" do not contain any alternative resolutions.

### *b. In the Arabic Books*

*Book IV*

Problems IV,13; IV,14; IV,15 contain alternative resolutions which are surely later additions. On the other hand, the second resolution found in IV,34 looks genuine, and corresponds to that of the related problem II,11. Finally, IV,42,a outlines three ways of dealing with the problem. Whether

Diophantus himself gave all three or not cannot be ascertained; but if he did so, it was surely in a more concise form.

*Book V*
No alternative resolutions.

*Book VI*
VI,22 is the only problem which truly gives two resolutions. In VI,13, the second part employs another method of resolution; but, since the first treatment did not yield a result fulfilling the conditions, the second is not, strictly speaking, an alternative one.

*Book VII*
Only VII,7 has an alternative resolution.

It is very likely that the alternative interpolated resolutions found in the Arabic Books were added at the same time as those of the Greek Books;[11] they are certainly not inferior to the one in III,6. They are simply written, as is the remainder of the text, more prolixly.

## 3. Other Supplements

### *a. Corollaries*

Corollaries are found in the Greek text at the end of the groups of problems I,31–34 and I,35–38. The banality of the problems formulated in them, especially in the second set,[12] makes their genuineness seem subject to question; but many problems in the elementary Book I are also quite simple, particularly those from which the corollaries stem.[13]

In the Arabic Book IV, the corollaries following group IV,5–9 and the corollary appended to IV,14–15 are also unimpressive. Another set of corollaries is found following the last of the problems of Book IV dealing with a single equation (IV,33); we have mentioned this set in relation to the interpolations in Book VI. It seems genuine, or is evidently early enough to antedate the first interpolations, just as the set appended to I,34 does.

Finally, we find the formulation of VII,15 extended to a larger number of unknowns, together with the statement that the latter case is solved in the same way as the former. This seems to be genuine.

[11] This may in fact apply only to the alternative resolutions in IV,13–15; it is by no means certain that the one in VI,22 and, even less so, the one in VII,7 are interpolated (see mathematical commentary).

[12] It repeats problems I,35–38 with an insignificant change (inverting the rôles of the larger and of the smaller required numbers; cf. p. 464).

[13] A proof of the genuineness (or at least of the great age) of the pair of corollaries following I,34 is that they are the source of an interpolated problem found in Book II.

### *b. Remarks*

Appended remarks are rare in the Greek text, and the genuineness of those found cannot be taken for granted. There is one in "IV",7,2°, concerned with the infinite number of solutions; another one, at the end of problem "IV",19 (restated at the end of the lemma to "IV",34), simply defines the term ἐν τῷ (τῇ) ἀορίστῳ.[14]

On the contrary, in the Arabic text there are several remarks: one in IV,22 (see below, *c*); one in IV,30 (concerning a particularity of the found solution—later used, in IV,42,a); one at the end of IV,42,a (stating that a just used simplified approach can be employed for previous problems); one in V,13 (restricting the application of the resolution); one in VII,11 (stating the insolubility of a problem belonging to the group under consideration). The last two at least must be genuine. To this group of remarks may be added a statement (the purpose of which is unclear) found at the end of IV,36, which looks like a scholiast's addition—certainly made before the systematic addition of the verifications (see p. 210).

### *c. Additional Computations*

There are a few additional computations of some importance in the Greek Books, apparently made in early times.[15] Tannery indicates a minor one in problem I,3 and a major one in III,11. There seem to be other instances, as in II,24 (cf. p. 178, n. 11), perhaps in "IV",28 (ed., pp. 256, 12–258,2) and in "V",8 (330,13–332,13).[16] Other, minor additions of scholiasts are more frequent, and only a few have been put into brackets by Tannery; we noted another example on p. 198. Thus, some supplementary computations found their way into the Greek text.

The situation in the Arabic text is quite different since all computations were performed by the author of the verifications. Hence, nothing can be said about such earlier, additional computations. The only noticeable supplementary computations, of a type not known from the Greek Books, are the deduction of the conditions expressed by the diorisms in IV,21 and 22, and, in this latter case, the resolution of the resulting "constructible problem", to which is appended the remark that the given numbers of the previous

---

[14] A problem is solved ἐν τῷ (τῇ) ἀορίστῳ when the solution is given in terms of (units and) the unknown, i.e., when to any positive and rational value of $x$ corresponds a solution. As concerns the wording, remember that the unknown $x$, prior to its determination, is called *per definitionem* a πλῆθος μονάδων ἀόριστον (D.G., I, p. 6,4). Incidentally, Psellus' reading ἀόριστον, chosen by Tannery, is confirmed by the text of St. Hippolytus (*supra*, p. 43), which, among the many definitions about numbers and powers gleaned by the author, has the phrase ἀριθμὸς δ'ἦν τὸ γένος ἀόριστος (Diels, p. 556,16–17)—as opposed to other powers with defined exponents.

[15] We exclude the late (Byzantine) ones; see, e.g., pp. 106 and 146 (app.) in Tannery's edition.

[16] Shortened resolutions are found in Book "V"; see problems 12–14 and 19 (i.e., Tannery's $XIX_4$)–20.

problems are found similarly. But the core of these supplementary computations may well go back to Diophantus himself.

## Appendix

### A Comparison Between al-Karajī's Version and the Extant *Arithmetica* (complement to the generalities of pp. 10–11)

#### A. *Books I–III*

1°. Diorisms

As a rule, diorisms are placed as remarks at the end of problems in the *Fahrī* since al-Karajī, unlike Diophantus, immediately formulates the problems with the values of the given magnitudes, if any.[17]

Many diorisms for the positivity of the solution, given by Diophantus, do not appear in the corresponding problems of the *Fahrī*: II,46–47 (=D.G. I,8–9);[18] III,30 (=D.G. I,19); III,32 (=D.G. I,21). In one instance there is a single diorism comprising both those of D.G. I,16–17 in a more general form: in *Fahrī* III,25 = D.G. I,17.[19]

2°. Resolutions

Some of the problems taken from the *Arithmetica* were abridged. Thus, only the first case of D.G. I,39 is treated in *Fahrī* III,28; an initial trial made by Diophantus in III,10 (D.G., I, p. 158,5–26) is omitted in *Fahrī* IV,50; the same holds for the next problem (*Fahrī* IV,51 = D.G. III,11), and, in addition, a long passage rejected by Tannery as being a later interpolation is omitted;[20] finally, some intermediate conditions given by Diophantus are missing, as in *Fahrī* III,38 = D.G. II,10 (omission of lines 15–18, p. 94, in Tannery's edition), or in *Fahrī* IV,41 = D.G. II,19 (omission of Tannery's p. 114,1–4).

On the other hand, the *Fahrī* gives some explanations which are missing in the extant Greek text, as in *Fahrī* IV,7 = D.G. II,28 (see Tannery, p. 127, footnote, concerning the unclear step); in *Fahrī* IV,10 = D.G. II,31 (omitted condition, see Tannery, p. 131, note); in *Fahrī* IV,46 = D.G. III,6 (□ in p. 148,5 necessarily $>25$). Further, in the above-mentioned *Fahrī* IV,10 and in *Fahrī* IV,13 = D.G. II,34 the intermediate problem is distinctly stated in the *Fahrī*: see *Extrait*, pp. 107–8).

Finally, one problem from Diophantus is treated with a different (nevertheless Diophantine) method in the *Fahrī*: III,41 (=D.G. II,12).

---

[17] There are a few exceptions, see *Fahrī* V,18–20 (and V,43).

[18] The given numbers first set in *Fahrī* II,46 are changed after the choice is revealed to be inappropriate.

[19] But Samaw al's version of D.G. I,16 (hence the Arabic Diophantus) has the same diorism as does the Greek text (see above, p. 12).

[20] Of all that appears in Tannery's pp. 160,16–164,7 (καὶ), what the *Fahrī* has corresponds to p. 162,8–10 plus the statement of the two values $30\frac{1}{4}$, $12\frac{1}{4}$.

3°. Interpolated Problems

Some of the interpolated problems found in the Greek *Arithmetica* do not appear in the *Faḫrī*: II,1–7 (and 17)—but the progenitors of II,1–7 are missing as well. On the other hand, some problems considered to have been interpolated are reproduced by al-Karajī: *Faḫrī* IV,40 = D.G. II,18;[21] *Faḫrī* IV,59 = D.G. III,20 and 21. Thus, al-Karajī's source for Books I–III contained the interpolations—at least *some* of the interpolations—which Tannery considered to have been early additions. This and the previous considerations on the (early) Greek origin of the interpolated problems in the Greek and Arabic Books lead us to the conclusion that *the extant Greek text as well as the Arabic text both proceed from the same early recension.*

4°. Alternative Resolutions

Al-Karajī does not generally reproduce two resolutions; there are exceptions, namely in *Faḫrī* III,42 = D.G. II,13 and in *Faḫrī* IV,45 = D.G. III,5 (Tannery: *scholiastae vix tribui potest*). The existence of two modes of resolution for II,11 is implied in a remark found in a subsequent problem (see *Extrait*, p. 102).

5°. Additional Problems

A certain number of problems found in sections II–IV of the *Faḫrī*, although Diophantine in type, do not appear in Books I–III of the *Arithmetica*: see Woepcke's *Extrait*, pp. 12–15. The majority of these problems has been taken (directly, presumably) from the *Algebra* of Abū Kāmil, with or without undergoing any change in the choice of constants or (slight) modification in the resolution—as is the case for the problems taken from Diophantus.[22] Of the remaining problems (*Faḫrī* II,30, 33; III,3, 4, 39, 50), one (III,39) is particularly noteworthy in that it falls in the middle of a group of consecutive problems seen in the *Arithmetica*, namely Diophantus' II,8–10 and II,11–16; Woepcke thought that it might be Diophantine in origin.[23]

**Remark.** The absence of some problems of Books I–III in the *Faḫrī* (cf. p. 10) does not at all mean that they were missing in the Arabic Diophantus. Thus, I,26 does not occur in the *Faḫrī*, but does in Samawʾal's *Bāhir* (*supra*, p. 12), while the only problem of Book III missing in the *Faḫrī*, III,4, must have been

---

[21] *Faḫrī* IV,40 gives a resolution actually missing in the Greek text that we possess; but this resolution might well be an Arabic completion: see Woepcke's *Extrait*, pp. 20–21. Although IV,40 is expressed in concrete terms (as the division of a sum of money among three persons), the Diophantine origin is clear.

[22] *Faḫrī* II,22–29 and 31–32 and IV,27–39 correspond to problems nos. 1–3, 5, 6, 10, 11, 13 and 17, 20 and *11*, 23, *13*, 15, *17*, 24, 25, 31, 32, 35–38 of our *Méthodes chez Abū Kāmil*. Two of the problems which are repetitions of previous ones show insignificant changes in the value of one of the given numbers.

[23] In reference to it (and to III,4), he said: "Ils portent entièrement le cachet qui caractérise tant les énoncés que les résolutions de Diophante, et je serais très-porté à croire que ces deux problèmes appartiennent réellement à l'algébriste grec, et font partie des pertes que le texte de Diophante, que nous possédons, a éprouvées dans la suite du temps" (*Extrait*, p. 14).

in the Arabic Diophantus, since III,19 has the same number in Tannery's Greek text[24] and in the Arabic version (*supra*, p. 10).[25]

## B. *Book IV*

The comparison between the problems of Book IV and the corresponding ones in the *Faḫrī* is made in the mathematical commentary. We shall content ourselves here with a brief survey.

1°. Diorisms

The diorisms found in D(iophantus) A(rabicus) IV,17–20 and 22, all of which are concerned with the *rationality* of the solution (and thus essential), are also given by al-Karajī.[26]

2°. Resolutions

A preliminary trial in D.A. IV,6 and 7 is omitted. Conditions missing in D.A. IV,28, 29, 31, 33 are also missing in the *Faḫrī*, and those present, in D.A. IV,10, 37, 39, are repeated by al-Karajī. The resolutions of D.A. IV,14–15 are somewhat modified.

Various additions are also found in the *Faḫrī*: a corollary appended to D.A. IV,15 with general instructions for the resolution is solved in the *Faḫrī* (but in another way); an alternative resolution is added to D.A. IV,16; the diorism is established by al-Karajī for D.A. IV,20; in the counterpart to D.A. IV,22, the problem is fully resolved, unlike in D.A., but at the expense of the establishment of the diorism (which is wrong in al-Karajī's version).

3°. Interpolated Problems

None—or none stemming from other Books—appears in the Arabic Book IV of the *Arithmetica*.

4°. Alternative Resolutions

The one added to D.A. IV,14(–15) by a scholiast was not exactly copied by al-Karajī but it did lead him to add a (confused) alternative resolution of his own. That of D.A. IV,34 is omitted in the *Faḫrī*, but the existence of two modes of resolution is alluded to in the related problems *Faḫrī* V,37–38 (=D.A. IV,36–37).[27]

5°. Additional Problems

A problem not found in the *Arithmetica* occurs after the problem corresponding to D.A. IV,19. Banal subcases of D.A. IV,1–4 occur, quite out of place, in the middle of section V (Woepcke's nos. 23–27).[28]

---

[24] Which, like the Arabic version but unlike Bachet's edition, does not count alternative resolutions as separate problems.

[25] Hence, the Arabic version of Book III had *all* the problems of the Greek text.

[26] The last one contains an error in al-Karajī's version.

[27] The situation is similar to that of the counterpart to D.G. II,11 (see above).

[28] No difference in style is evident enough to allow one to consider them as later additions to the *Faḫrī*.

Finally, let us observe that, as was the case for the first three Books, some Diophantine problems are omitted in the *Fahrī*: D.A. IV,12–13, 21, 25–26(a and b), and 42–44.[29]

**Remarks.** 1°. From this brief survey—as well as from a more extensive comparison between the *Fahrī* and the Diophantine Books—, one is left with the impression, if any general impression can be formed, that al-Karajī clung even more faithfully to the *Arithmetica* near the end than at the beginning.

2°. As for form, the problems in the fifth section of the *Fahrī* are far less verbose than those in Book IV, with few exceptions (most strikingly *Fahrī* V,36 = D.A. IV,35). From the verifications remains at most, and in the first half of the fifth section only, an abbreviated form, while the final statements are never given.

This tendency toward conciseness is even more marked in al-Karajī's counterparts to propositions from Books I to III, his problems often being briefer than their progenitors in the already concise Greek text that we know. It would be unreliable, then, to infer from al-Karajī's text any conclusion about the degree of prolixity of his source for Books I–III of the *Arithmetica*.

3°. Surprisingly, the *Fahrī* (or, at least, manuscripts P, E, K, and L) repeats some problems, merely phrased differently. Thus, D.G. II,22 appears as *Fahrī* II,50 and IV,1, and D.A. IV,20 occurs appropriately as *Fahrī* V,19 but also as the very last problem of the work (V,43). Similar repetitions occur for three problems taken from the *Algebra* of Abū Kāmil (cf. p. 58, n. 22). Since the style of these problems does not give rise *a priori* to suspicion as concerns their authenticity, and since most of these pairs of problems belong to coherent groups of borrowed problems, they may have belonged to the original *Fahrī*.[30]

## §10. Errors in the Problems of the Arabic Books

Since our four Arabic Books are said to be part of Qusṭā's *translation* of the *Arithmetica* (see §2,1), we have every reason to believe that the Greek text from which he made his translation already appeared in the enlarged

[29] All the differences mentioned here are found in the Paris manuscript studied by Woepcke (B.N. arabe 2459—*olim* Suppl. ar. 952), as well as in at least three other copies of the *Fahrī*, namely Esat 3157, Köprülü 950, and Laleli 2714. Variations in the wording in these four manuscripts, incidentally, make it obvious that they were not all copied from the same exemplar.

N.B. We shall henceforth designate these manuscripts as P, E, K, L.

[30] Worth mentioning, though, is a marginal remark written in the Paris manuscript by the second hand (cf. *Extrait*, p. 3, n.), on fol. 98$^r$, concerning the problems of the fourth section: "It is said in some (*baʿḍ*) copies that there are fifty-five problems, whereas I found that there are here sixty". And, indeed, there are five problems in the fourth section which are repetitions of earlier ones found in the second section (IV,1; 15; 27; 29; 31 = II,50; 40; 28; 29; 31).

form discussed in §8.[31] As previously said, this enlarged form ensued from three sorts of additions:

(a) additions within (or rewriting of) the analysis;
(b) addition of the verifications;
(c) addition of the final statements.

At first view, since the addition of the verifications and final statements conforms to the established Greek pattern in the treatment of geometrical problems (see §8), one would be inclined to suppose that the final statements were added at the same time as the verifications and the additions in the analyses. But close study of some of the errors found in the Arabic Books leads one to believe that the completion of the computations—the *major commentary*, as we shall call it—and the addition of the final statements were not made by the same person. These and some other errors also reveal the degree to which the two authors of these supplements understood Diophantus' propositions.

1. IV,8–9

Problems IV,8 and 9 are in reality a single problem. The first (IV,8) reduces the original proposition to a problem already treated (not without some confusion; see the notes in the translation), gives its solution, and ends with the words "this is what we intended to find". Under the heading IV,9 comes, first, the restatement of the original proposition, then its resolution (using the results of the intermediate problem), and, lastly, the synthesis followed by the final statement in its complete (usual) form.

The question who is responsible for this inappropriate separation then arises; one would expect it to have appeared:

($\alpha$) subsequently to the completion of the computations; for the author of the major commentary, having reworked the problem, would presumably have followed the reasoning of its resolution;[32]

($\beta$) subsequently to the addition of the final statements; for if the scholiast who added them had found an already separated problem, he would either have provided IV,8 with a full final statement or have realized that the separation was inappropriate and eliminated it.

Considering then the other possibilities, the separation would have originated with:

1°. an Arabic reader (or copyist);
2°. the translator himself;
3°. a Greek reader (or copyist) reading the already commented text.

[31] In our thesis (and in an article on Diophantus written shortly after its submission for the *DSB*), we held the view that the translator was responsible for the general prolixity. This change in opinion results from closer examination of the Arabic Books.

[32] Note that certain errors considered in this paragraph cast some doubt on his abilities.

The unlikelihood of the first possibility—that is, the likelihood that the separation existed in the original translation—is suggested by two arguments:

(a) From the *wording*, IV,8 and IV,9 really look like two different problems; IV,9 begins as does any other problem and without any reference to the preceding calculations. Since it is hardly credible that ($\alpha$) an Arabic reader altered the original text so as to make two problems out of one, or that ($\beta$) the translator's text was, by chance, and here only, ambiguous, with the result that it misled some later reader or copyist who numbered the problems, we are brought to the conclusion that the translator himself saw (or thought he had before him) two separate problems, and translated accordingly.

(b) The total number of problems given in the colophon of Book IV confirms our manuscript's numbering; unless one supposes that the numbering, and also, therefore, the indication in the colophon, were added after the translation (see above, ($\beta$)), one is again brought to the conclusion that the inappropriate separation appeared in the translation.[33]

Whoever was responsible for the separation, the origin of the mistake is clear: the words "this is what we intended to find", which were merely meant to conclude the intermediate problem before the return to the original proposition, were understood to be the conclusion of a whole, separate proposition.

2. IV,26

Problem IV,26 amounts to solving $|(a^3)^2 - (b^2)^2| = \square$ and is accordingly divided into two parts corresponding to the cases $(a^3)^2 - (b^2)^2 = \square$ and $(b^2)^2 - (a^3)^2 = \square$. Now, the formulation of the first case, that is, practically, the announcement of the problem's subdivision into two parts, *follows* the setting of $a = x$, $b = 2x$, although this choice is valid for the first case only (Diophantus takes $a = x$, $b = 5x$ in the second). The misplacement of the said formulation gives the impression that we are dealing with a scholiast's addition, an impression reinforced by its defective wording (cf. line 712).

**Remark.** The same defective wording is found in the final statement (line 728). Whether we consider that the author of the final statement is also responsible

[33] The statement of the total number of propositions contained in a Book is a not uncommon Arabic practice; see, e.g., Menelaus' *Sphaerica* (ed. Krause, pp. 161 and 192), Ṭūsī's edition of the *Elements* (in headings of Books).

N.B. It would seem that Diophantus himself did not number his problems (thus facilitating the integration of interpolated problems into the text). The oldest Greek manuscript, the Matritensis 4678 (cf. p. 18), does not have any numbering (see D.G., I, p. v), nor does the Vaticanus gr. 191 (except for a few problems at the beginning). In relation to the numbering, note that the subdivision into (numbered) problems in Book IV is inconsistent, for, in similar situations, a single formulation (thus a proposition with a number of its own) may include the cases which are elsewhere presented as distinct problems (cf. IV,26 with IV,30–31; IV,42 with IV,34–35 or IV,40–41; IV,44 with IV,37–39).

for the addition of the formulation of the first case or whether we consider that he simply reproduced in the final statement this formulation with its error, it seems clear that he did not follow the resolution very closely.

3. IV,27(–28)

The solution to proposition IV,27, which is $(a^3)^2 + kb^2 = \square$, with $k = 5$, is found to be $a = 4, b = 32$. The verification is then made by computing the value of the expression $a^3 + 5b^2$. Since *both* $a^3 + 5b^2$ and $(a^3)^2 + 5b^2$ happen to give squares for the found values, the commentator performing the verifications did not realize his mistake. And it seems that in the next problem, similarly, $b^2 + ka^3$ is computed instead of $(b^2)^2 + ka^3$.[34] But, in both cases, the final statements appended to the syntheses restate the two problems with the original, correct formulations.

Errors of this sort support our allegation that the author of the major commentary did not add the final statements. For it is improbable that the same person would have first verified the correctness of the solution while misunderstanding the terms of the problem and then, immediately afterwards, restated the problem in its correct form.

4. IV,40(–41).

In a passage of the analysis of IV,40, $x^2$ (*māl*) is written instead of $x^4$ (*māl māl*) five times, and the same error is repeated in the corresponding places of the next problem, which is its twin proposition (see lines 1256–59 and 1303–5). These mistakes cannot have been made by copyists, nor can they go back to Diophantus' text; thus the author of the major commentary must be responsible for them. The repetition of the error in IV,41 may be due to the commentator's having mechanically followed the sequence of steps used to solve IV,40: these two twin propositions are particularly closely linked (cf. p. 118, n. 81).

5. VI,4

The goal of this (interpolated) problem is to make the expression $(a^3)^2 + a^3b^2$ a square, and the values found are $b = \frac{125}{251}$ and $a = \frac{625}{251}$, which indeed satisfy the condition. The text, however, has a misreading of the value of $a^3$,[35] which is obviously of Greek origin, whether by the author of the major commentary himself or by some earlier copyist (see p. 246). The author of the major commentary did not realize that the value for $a^3$ was incorrect since he did not compute the sum $(a^3)^2 + a^3b^2$—probably because of the large number of digits in the result (cf. pp. 49–50)—; see his final remark.

6. VI,9

We have already pointed out the correction made by a reader to the final statement of the (interpolated) problem VI,9 (see p. 31, no. 10). Instead of

[34] We say "it seems" because the error is revealed only by the wording (lines 790–1): the value found for $b$ being unity, the final result is not affected.

[35] This erroneous value is repeated in the final statement.

giving the values of the required cube and square, the original version gave the values of the cube and the *root* of the square (the latter being the value of the unknown $x$ determined in the analysis). We may infer from this that the individual adding the final statements did not systematically follow the treatment of the problems. Indeed, he may sometimes have been simply glancing through the text to find the values of the required magnitudes.

7. VII,4

We have previously encountered an example of a mistake in powers (in no. 4), the author of the major commentary speaking of $x^2$ instead of $x^4$. Similar, but more serious, is the confusion found in the interpolated problem VII,4: while the required magnitudes had been originally set proportional to $x^4$, they are computed as if they were proportional to $x^2$. It is difficult to attribute this mistake to the author of the problem himself, so that it would seem that the author of the major commentary was responsible for it.

8. VII,14

The final statement mistakenly gives an intermediate result, occurring within the resolution, as one of the three required magnitudes, and this despite the fact that these three required magnitudes are stated *just before*, at the end of the verification. This gives more weight to the opinion expressed earlier (no. 6) that the author of the final statements sometimes only glanced through the resolutions in order to find the numerical values of the required magnitudes; and such an error makes it difficult to believe that the author of the major commentary and the author of the final statements were one and the same person.

## *Conclusion*

(a) We have seen that the author of the major commentary is responsible for two serious mistakes, one of which he did not notice since his verification happened to work (no. 3), and the other because he did not complete the verification (no. 7; cf. no. 5). It may reasonably be supposed, then, that he committed other such mistakes in the course of his verifications, which, however, he discovered when his computations failed to produce the expected result. This would point to a certain carelessness, perhaps resulting from a mechanical performing of the computations.

(b) Some points inclined us to believe that the author of the major commentary did not himself add the final statements: in one case, repetition of the formulation at the end would presumably have drawn his attention to a mistake of his (no. 3), while in three other cases errors in the final statements are hardly compatible with a simultaneous reworking of the resolutions (nos. 2, 6, 8; see also §12,3). In the introductory remarks and in no. 1, we expressed in addition the opinion that both the major commentary and the final statements were added in Greek times.

Thus, the Greek text would have undergone, subsequently to the incorporation of the various interpolations mentioned in §9 and prior to the translation into Arabic, two kinds of additions: first, the additions belonging to the so-called major commentary, often diluting the reasonings and generally completing the computations; then the additions of the final statements—and, perhaps, of some other complements (see no. 2)—by a later scholiast.

## §11. Quality of the Translation

Since the Greek text which reached the Arabs had the prolix form discussed previously, it was hardly necessary to submit it to a critical revision. Accordingly, the translator seems to have done a faithful and (with few exceptions) very careful translation, but without troubling himself unduly about the solving of the problems. This might explain why some gross errors escaped him as they did late Greek readers or scholiasts (cf. §10), and why the translator himself seems to have made some elementary (but unimportant) mistakes in translating. We shall examine, before considering the general character of the translation, these mistakes. Whether they all really originated with the translator, we cannot ascertain. Certain undoubtedly did, and even supposing that many inappropriate formulations already existed in his Greek copy does not modify the general impression of the translator's work which we have formed: that he often paid more attention to the text than to its mathematical content.[36]

### 1. Imperfections in the Translation

(α) There are some errors which, if they go back to the translator, may be easily explained by consideration of the expression probably used in the Greek text. They are the following.

The two operations which in Arabic times were referred to as *jabr* and *muqābala*[h], and which were regularly used before then by Diophantus in setting the final form of an equation, consist respectively in adding the (absolute) value of a negative term of the equation "in common to both sides" and in dropping a common quantity from both sides. The absence of a

[36] When we definitely cannot conjecture whether minor errors go back to the translator or to the Greek manuscript, we do not list them here. Examples of this are references to ratios as numbers, and cases in which the two algebraical operations (*najbur wa-nuqābil* = κοινὴ προσκείσθω ἡ λεῖψις καὶ ἀπὸ ὁμοίων ὅμοια) are said to be applied when in fact only one is necessary in setting the final form of an equation. Such errors are mentioned in footnotes in the translation.

synthetic word in Greek for the designation of each of these operations makes a circumlocution necessary, the usual ones being κοινὸν προσκείσθω τὸ *X*, and κοινὸν ἀφῃρήσθω τὸ *X* respectively (cf., e.g., *Elementa*, II, 11 and 12; D.G., see Tannery's index[37]). The Arabic text (when it does not use the synthetic words) appropriately renders these expressions as *nazīd al-X muštarak*$^{an}$, and *nulqī al-X al-muštarak*. But the proleptic use of κοινόν opens the way to possible confusions between the two cases; thus, our text adds the article to *muštarak* three times in the case of the addition (see notes 381, 415, 436 of the app. crit.).

Another error found a few times in the translation is the use of the verb *baqiya* (= to remain, result from a subtraction) instead of *ijtamaᶜa* (= to result from an addition): see notes 351, 405, 424, 497. The source of the confusion may well lie in the use in Greek of "neutral" verbs like γίνεσθαι, ποιεῖν, εἶναι.

A third indication of the translator's (occasional) inattention is the occurrence a few times of a plural where one would expect a dual in Arabic: notes 425, 453, 632, 921; cf. also note 776.[38]

($\beta$) In the text we find inappropriate formulations, badly constructed phrases, etc., which, again, cannot *all* go back to a deficient Greek text or be the work of an inattentive copyist.

Awkward or unsuitable formulations occur in lines 373–74 (n. 150), 406–7 (n. 163), 415–16 (n. 168), 1059 (n. 363), 2488–91, 2523–27, 3394 (n. 937); we may have other examples in notes 255–56, 502, 920, or in lines 80–81 (see, however, p. 31, no. 13). Finally, a few articles are infelicitously omitted or added: notes 58[39], 147, 541, 837.

When setting or computing the value of a square's or a cube's side, the text sometimes shows some confusion, speaking of the square (or the cube) instead of the side, or of the side instead of the square (cube): see lines 55, 63, 76, 119, 232, 239, 444, 779, 1317, 1336; notes 52, 139, 226. We have kept the manuscript's reading in those passages which are, strictly speaking, incorrect but which are nevertheless clear, namely in lines 60–61, 227–28, 255, 358–59, 769, 838–39, 878, 940, 1138–39, 1183–84, 1466–67, 1537, 2561–62, 2597–98, 2633, 2657–58, 2702–3.

The errors of congruence have been, in part, corrected, as in notes 238, 417 and 455, 420, 623 and 625, 765. Some have been kept, either because they are, *bon gré mal gré*, acceptable, as in lines 691, 1247, 1303–4 (but: 1305–6), 1992–93, or because they are repeated and thus confirmed: see lines 1124 and 1125, 1292 and 1298, and the triplet of lines 126, 148, 185.

---

[37] In the case of the subtraction, the Greek (i.e., uncommented) Diophantus uses a concise ἀπὸ ὁμοίων ὅμοια.

[38] We dismiss the occurrences of *-hā* for *-humā* (notes 11, 151, 181, 365, 903), as they may well have originated with some copyist (as the equally frequent occurrences of *-humā* for *-hā*: notes 43, 245, 712, 923). In notes 171 and 637, we have *hiya* instead of *humā*.

[39] The confusion between *kaᶜb* and *mukaᶜᶜab* is irrelevant here, since it is a common scribal error.

## 2. General Character of the Translation

Notwithstanding the presence in the text of the errors listed above, the Arabic version of Books IV to VII may be considered to be an excellent one; for, the translation contains, in relation to its length, few errors, and poses no difficulties of interpretation whatsoever. This seems to be both because the Greek text was in excellent condition and because the translator was very capable.

($\alpha$) What points to the excellent condition of the Greek text is the fact that we have a *very good translation* which contains *elementary mistakes*—i.e., mistakes caused by inattentiveness on the part of the translator. This suggests that the translator was able to work quickly, without having to reconstitute a damaged or heavily annotated Greek text or a text difficult to read.[40] Since the older codices are said to have been generally in poor condition (cf. p. 15), it is quite possible that the translator's copy was a codex written not long before, perhaps in Leon's time (§2,3,a). The Byzantine copyists of that period, known to have done their work with great care (cf. Impellizeri, *Lett. biz.*, p. 323–24), might well have produced a very readable text. Some minor additions (perhaps the final statements) could also go back to these ninth-century copyists; interpolations which seem to date from that period occur in other Greek mathematical texts.[41]

($\beta$) Even assuming the Greek copy to have been in excellent condition, it would nevertheless be unfair to underestimate the quality of the translator's work. The translation's predominant characteristics are its precision and conciseness of expression which leave no room for uncertainty. In short, a fitting text for a mathematical treatise, written by an individual who obviously knew Greek and Arabic mathematical terminology perfectly.[42]

N.B. We have adopted the policy of referring to "the translator" rather than to "Qusṭā", for we cannot exclude the possibility of such an easily translated text having been left to one of Qusṭā's pupils, as was apparently

[40] Moreover, there are hardly any errors traceable to the script of the Greek text or to misreadings by the translator. Two errors which could, seemingly, be explained by a misreading of the Greek (note 270: διαιρεῖν instead of ἀφαιρεῖν; note 831: Α instead of Δ) must in fact have been made in Arabic times: they are very inept and, from the context, would easily be corrected by any reader of minimal competence. More problematic is the case of *šaiʾ*, employed instead of the usual *ʿadad* in two passages (notes 516, 520); this mistake may have arisen from a confusion between the abbreviated and full forms of ἀριθμός, a confusion found in some Greek manuscripts (cf. Heath, *Diophantus*, p. 34, n. 3).

[41] See the (reconstructed) history of the Eutocius–Apollonius text in Heiberg's Apollonius, II, p. lxviii.

[42] The only inappropriate rendering of a scientific term occurs in line 14, where *jirmī* is used for qualifying a *solid* number. The word *jirmī* is normally used to translate σωματικός, and the correct rendering of στερεός (which is certainly what was in the Greek text) is *mujassam* (cf. Klamroth, *Arab. Euklid*, p. 301; Ṭūsī's Euclid, VII, deff.; Ṯābit's Nicomachus (Kutsch), p. 259).

done on occasion (cf. p. 9). Even the possibility of there having been several translators cannot be entirely dismissed: slight variations in the frequency with which certain words are used (see index), differences in the naming of the powers $x^5$ and $x^8$ (cf. p. 45), and the greater frequency of inappropriate formulations in the last two Books might thus be explained.

## §12. Genealogy of the Mashhad Manuscript

From what we have seen in the previous paragraphs, the history of Diophantus' text emerges as follows.

### 1. Earliest Additions

The earliest additions of importance are interpolated problems and alternative resolutions (cf. §9). The interpolated problems are located, with some disorder,[43] in Books I to "IV" (= VIII?[44]) and stem from problems contained in the first six Books. Of the alternative resolutions, those which do not appear to be genuine[45] must have been added at about the same time.

*The history of the Greek and of the Arabic texts is, up to this point, the same.* With the writing of the major commentary, however, two separate traditions emerged, the earlier of which is preserved in the extant Greek text.

### 2. The Major Commentary

#### *a. Additions Originating with the Major Commentary*

What we have referred to as the "major commentary" consists, in fact, in a *rewriting* of the entire text, *where genuine and interpolated problems, as well as alternative resolutions, undergo the same treatment, thus giving birth*

[43] For example, interpolations originating from problems found at the end of Book I precede interpolations from the middle of Book I, and, similarly, a problem derived from Book VI is found in Book VII, whereas interpolations stemming from Book V appear in Book "IV".

[44] We mentioned on p. 54 the possibility of the Greek Book "IV" having originally been the eighth Book of the *Arithmetica*. But I fail to see any better argument in favour of this hypothesis than what we tentatively inferred from the location of the interpolations. For, comparison of the mathematical content of Books I to VII with that of Book "IV" neither suggests a missing section—no more than did the comparison of Books I to III with "IV" to earlier scholars, none of whom suspected the absence of several Books in between—, nor gives any strong indication of continuity.

[45] The last alternative resolutions which are unquestionably interpolated are found in Book IV. See p. 55, n. 11.

*to a homogeneous whole.* As a consequence of this rewriting, additions are interspersed *throughout* the resolutions of the problems in the Arabic text.

(α) *Additions in the Analysis*

We have already mentioned that the analysis of problems is, in general, more prolix in the extant Arabic Diophantus than in the Greek (cf. p. 50). Thus, statements which are not, or at most incidentally, found in the Greek text occur much more frequently in the Arabic Books. Most noticeable are the following points:

(a) The problems are often fully reformulated after the statement of the given magnitudes.[46]
(b) In those problems in which one of the equations of the given system is identically satisfied by an appropriate choice, the verification of the fulfilment is made explicitly. Some of these additions are easily recognizable: see, e.g., lines 2488–93, 2620–25,[47] and other passages beginning with (*min al-*)*bayyin* (φανερόν?).
(c) Banal identities or theorems used in the course of resolutions (in Book IV particularly) are stated.[48]
(d) References to earlier Books, rarely, if ever, found in the Greek text, appear in several places.[49]

The above are merely specific points; in general, we can say that the commentator explained the treatment in detail, at least when he was capable of doing so (see below).

(β) *Additions in the Synthesis*

Once the value of the unknown has been computed, one proceeds with the synthesis; that is, one returns to the initial hypotheses (i.e., the setting of the required magnitudes in function of $x$) in order to calculate the required magnitudes, and then verifies that the problem is fulfilled by inserting the found values into the given equation(s).

The Arabic text differs from the Greek, first in repeating the initial hypotheses ("since we assumed, etc."), then, and more specifically, in giving the proof of the solution, as has already been mentioned several times.[50]

---

[46] The repeated absence of a word in lines 1513 (formulation) and 1520 (reformulation) must be accidental, and not the commentator's error (the word is found in the final statement).

[47] The first part of this verification (lines 2620–23) is useless, since the first equation is identically satisfied as it stands.

[48] Such as the identities: $a^2/a = a$ (e.g., IV,20), $a^3/a = a^2$ (e.g., IV,21); and the theorems: if $a^2 = b^2$, then $a = b$ (e.g., IV,9), and the same deduction for $a^3 = b^3$ (e.g., IV,18) and for $a^4 = b^4$ (IV,17); if $a/b$ = square, then $(a/b \cdot b^2 =)a \cdot b$ = square (IV,21); the quotient of two squares is a square (e.g., IV,42,a,3° or V,4).

[49] Diophantus satisfies himself with a vague προδέδεικται (see, e.g., pp. 138,14 and 146,11 of Tannery's text). The reference on p. 256,12 may well be a later addition (cf. p. 56), as is that on p. 172,2 (see Tannery's app. crit.). On the Arabic references, see p. 5, n. 10).

[50] These two steps are found only exceptionally in the Greek text, e.g., in I,5 *seqq.*; cf. II,8–9.

N.B. It does not seem that the commentator added particular cases or subcases of the problems treated by Diophantus except, perhaps, in IV,14(–15), to which much has been added (see pp. 190–191).

### b. *Value of This Commentary*

From a purely mathematical point of view, the value of such a rewriting is minimal. The commentary stands very much in the tradition of those, written from the fourth century onwards, which diluted the material of classical treatises for students. Typical are the commentaries of Pappus and Theon on the *Almagest*, or those of Eutocius who reworked the proofs so as "to conform to the scholastic norms of his own time" (Toomer, *Diocles*, p. 18; cf. *ibid.*, p. 177). At that time, the form and content of classical treatises were altered in quite specific ways: as for form, the more prolix a commentary was, the greater was its repute, and, as for content, minor changes, such as the development of particular points and the completion of computations, were viewed with favour. Books IV to VII apparently underwent this kind of revision.

We have seen several errors indicative either of the commentator's mathematical feebleness or of his mechanical performance of computations (see §10, *in fine*), and other minor errors may also go back to him.[51] It is significant that the more difficult steps in some resolutions are not explained (problems IV,44,b or V,1–3), and that none of the results of intermediate problems, directly given by Diophantus and obtainable by methods taught in Book II, is actually computed.[52] All this leads us to believe that the commentator never ventured far from the path traced by Diophantus and that he was doing little more than *diluting an existent reasoning and computing values obtainable by elementary reckoning.*[53]

As limited, mathematically speaking, as he was, the author of the major commentary was, nevertheless, not wholly incompetent. First, he was able to follow, more or less, the reasonings of Diophantus' problems—certainly those in which the explanations are copious. Second, his additions are not always made undiscerningly: the analysis may be more concise when the reasoning has been detailed in preceding problems (see p. 106, n. 55); the computations in the verifications may be abbreviated by the omission of results (either trivial (V,3) or found in neighbouring problems (IV,14,e and 15; V,5 and 9)), by the use of particularities of the equations to verify (IV,43

[51] They are mentioned in the translation; see, e.g., p. 92, n. 21; p. 157, n. 3. See also above, p. 65, n. 36.

[52] Some are obvious, but some others are certainly not easy to obtain (as in problems VII,13–14).

[53] He must also have recomputed some of Diophantus' values when the text was damaged (cf. p. 50, n. 6).

and 44,a and c), or are even in one case dismissed by the quotation of a theorem (IV,7).

As suggested above, the commentary may well have been written for the benefit of students of late antiquity, the too dense and thus difficult original text being replaced by a text which was diluted and which contained verifications of the solutions, in order to put Diophantus' *computations* (not obligatorily his reasonings) at these students' level and to maintain their interest. If this is the case, the commentary is certainly appropriate.

### c. *Possible Authorship of the Major Commentary*

Among the sources still existing today, only one mentions a commentary having been made in antiquity on a Diophantine treatise, namely that of Theon of Alexandria's renowned daughter Hypatia: Suidas' *Lexicon* credits her with a ὑπόμνημα "on Diophantus", that is to say, surely, on his most important work.[54] On the basis of this information, Hypatia's name has been linked in modern times with the extant Greek *Arithmetica*, and this resulted in two hypotheses formulated by Tannery. We shall discuss these preliminarily.

Tannery's first assumption was *that the Greek text which we possess passed through Hypatia's hands.* Since, however, the Greek text bears no trace of any reworking of the problems and since its isolated additions (cf. §9) can hardly be considered to be the result of a systematic commentary, Tannery himself was forced to concede that Hypatia's commentary must have been removed from the text at Planudes' disposal—the source of our Greek text (cf. *Perte de sept livres*, p. 196 = *Mém. sc.*, II, pp. 78–79). But, the idea of Hypatia's having reworked the Greek text was deep-rooted, and the hypothesis was thus retained in Tannery's edition of the text. Tannery's second assumption depended on the first one; by assuming that Hypatia had commented only upon the "first six" (i.e., the Greek) Books of the *Arithmetica*, it was conceivable that *only the commented part had survived*, which could account for the loss of the remaining Books (cf. *ibid.*).

These hypotheses are no longer tenable—if ever they were—and, in the light of the discovery of the Arabic Books, we must endeavour to formulate some other explanation.

We have observed that the Arabic text, unlike the extant Greek text, possesses all the characteristics of a commentary made at about the time of the decline of Greek mathematics. Thus the idea that *our* text might be (part of) Hypatia's commentary arises quite naturally.

One could argue that the relative mediocrity of this commentary is hardly compatible with Hypatia's renown. But, apart from the fact that her

---

[54] See D.G., II, p. 36,20–24; about the necessary emendation, see Tannery, *Art. de Suidas*, p. 199 = *Mém. sc.*, I, pp. 76–77.

fame—enhanced by her outstanding virtue in a time of moral decadence—arose rather from her contributions to philosophy than from her scientific pursuits (see Praechter's article, p. 245), one must keep in mind that being considered a "good mathematician" is a relative thing: she may simply have been a good mathematician in a bad time. Her association with her father in the edition of his commentary on the third Book of the *Almagest* (cf. Rome, *Troisième livre des comm.*, p. 6) shows that she was, for better or for worse, her father's disciple and colleague. And Theon's commentary is considered now to have been "for the most part a trivial exposition of Ptolemy's text, explaining obvious points at excessive length", and to have been "never critical, merely exegetic" (Toomer, art. *Theon*, p. 321; cf. 322–23).

In closing, let us note that Suidas' *Lexicon* gives us no information about the extent of Hypatia's commentary. But, it is unreasonable to suppose that Hypatia would have commented on Books IV to VII without also having commented on the three preceding Books, particularly since Books IV to VII appear to depend so greatly on the fundamental methods taught in Books I to III (see pp. 176, 263). Further, there is a slight, but not negligible, indication that the first three Books reached the Arabs in the commented form as well (see below, 4). Thus, I to VII would all have been covered by the same commentary.

## 3. The Addition of the Final Statements

From §10, it appears that the scholiast who added the final statements (and, perhaps, some other complements) must be someone other than the individual who wrote the major commentary, and that, in fact, he must have made his additions later (possibly in the ninth century, cf. p. 67).

Although the scholiast's task was not demanding, he sometimes satisfied himself with very little, as, for example, when he did not restate the values of the required numbers, contenting himself with repeating the formulation or giving some unprecise conclusion (see problems IV,2, 15 and IV,7, 18, 22, 40–43; V,1, 2, 4, 14–16; VI,5). But he usually repeated the values of the required numbers, and this obliged him to read through the resolutions. From certain examples it appears that he did this with unequal care.

($\alpha$) The scholiast undoubtedly looked at some of the resolutions only superficially or wrote their final statements mechanically: in addition to the examples cited in §10, problem VI,13 illustrates this: after a first attempt at its resolution, it is asserted that we have found two numbers fulfilling the requirements, notwithstanding that the very next phrase states that one of the requirements was *not* satisfied by the found values. Lack of care in going through the text in general may also have occasioned the omission of some of the final statements, sometimes for the subcases found in a problem and sometimes for a problem itself (namely in V,3 and 5).[55] It is unclear whether or not some other errors also go back to this scholiast.[56]

($\beta$) On the other hand, this scholiast, in a few cases, did more than glance through the resolutions in order to find the required magnitudes: in VII,8–10, his final statements depart from the initial formulations of the problems and are adapted to a modification of the requirement of the problem arisen from an initial assumption; in VI,1–3, the scholiast performed some computations since the required numbers are given with a common denominator in the conclusion only.[57]

## 4. The Arabic Diophantus

It is unquestionable that Books I to VII of the *Arithmetica* were translated into Arabic (cf. §2,1). Further, I consider it certain that the Arabic translation of Books I–III contained the early interpolations found in the Greek Books (cf. p. 58). Finally, I have suggested above that the rewriting which probably formed Hypatia's commentary must have covered Books I–III also, and, in this connection, I have alluded to an indication that the first three Books reached the Arabs in the commented form as well. The evidence for this is not found, as one might expect, in the many problems reproduced by al-Karajī in his *Faḫrī*, but in Samawʾal's *Bāhir*.

Of the two problems of Diophantus found in the *Bāhir* (cf. p. 12), the second one follows fairly closely the Greek text of I,16—except that it changes Diophantus' constants and gives two alternative resolutions in the middle (*Bāhir*, p. 231,3 *seqq.*)— ; it is stated after the choice of the constants that they fulfil the diorism, and one may observe that the phrase begins with a *min al-bayyin*, as do many commentator's additions in our text (cf. p. 69). But more revealing is the case of Diophantus' I,26, reproduced by Samawʾal a few lines before. It begins (*Bāhir*, p. 230,9–13): "We wish to find a number such that when we multiply it by two given numbers, the result of the multiplication by the one is a square number, and of the multiplication by the other, the side of that square. Let the two numbers be 200 and 5 (*miʾatain wa-ḫamsa*$^h$, see app.). We wish to find a number such that, when we multiply it by 200 the result is a square, and when we multiply it by 5 the result is the side of that

[55] The final words "this is what we intended to find" are omitted, for no visible reason, in IV,11, 14, 34, 40 and V,2 and 4. The case of IV,37 is different (cf. p. 31, no. 7).

[56] Such as the errors mentioned on pp. 108 (n. 60) and 165 (n. 23).

[57] The end of IV,30 seems to have been rephrased, and this may be the work of the same scholiast.

square". Thus, Samaw'al gives the general formulation of the problem—much in the style of our text[58]—, then the statement of the two given numbers, and, finally, he *restates* the problem with, this time, the values of the given magnitudes. Now, such a complete reformulation, which is a characteristic of the major commentary (cf. p. 69), and which is found in almost all the Arabic problems related to I,26 (cf. D.A. IV,16–18, 20–21, and 22 *in fine*), does not, of course, appear in the Greek text; and the strong resemblance in style which the *Bāhir* text shows to our Arabic version suggests that Samaw'al was repeating essentially what was in the Arabic version of Diophantus at his disposal (remember that he was writing a commentary on Diophantus' treatise—cf. p. 11). Consideration of this passage in the *Bāhir*, then, suggests a prolix, commented Arabic text of the first Books of the *Arithmetica* as well, not just Books IV to VII.

As to the *extant* Arabic text, we have seen (§§5–6) that it derived from some archetype(s) bearing various readers' additions or corrections (a few of which may have resulted from a collation). The immediate predecessor of our manuscript seems also to have contained only Books IV to VII.

## 5. Genealogical Tree of the Mashhad Manuscript

Tannery suggested the following transmission link for the *Greek* Diophantus (cf. D.G., II, p. xxiii):

Text of Diophantus

|

Hypatia's (assumed) recension

|

Lost copy of the eighth or ninth century

| Matritensis 48 (nunc 4678) | (Planudean class) Ambros. Et 157 sup.; Marcianus 308 |
|---|---|
| Further copies | Further copies |

[58] Compare this formulation with those of Diophantus' IV,19–21, particularly with that of IV,20.

In the light of our investigations, the following modifications and completions might be tentatively made:

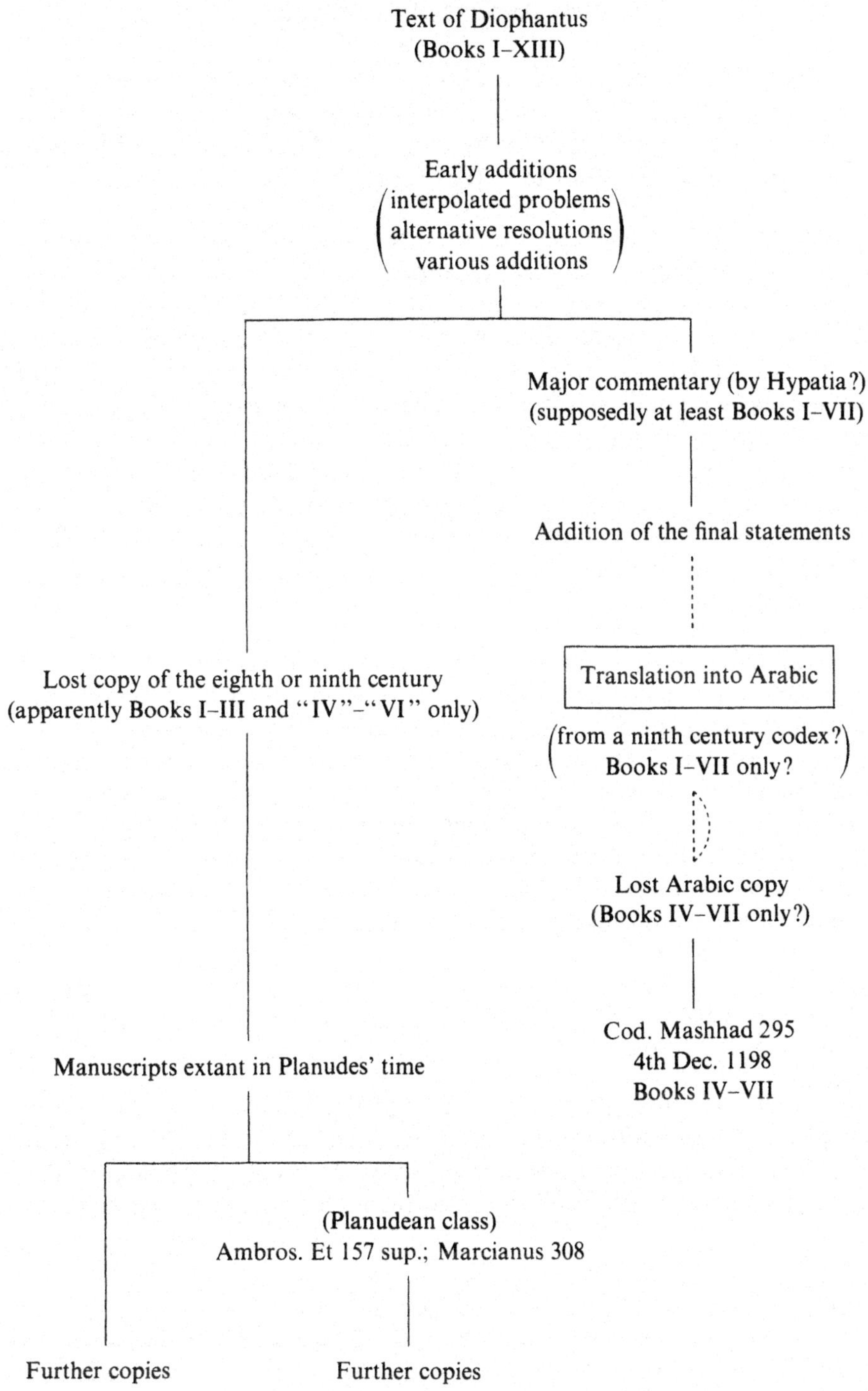

# §13. On the Missing Part of the *Arithmetica*

## 1. New Aspects of the Problem

During the nineteenth and early twentieth centuries, the content of the seven missing Books of the *Arithmetica* provoked much speculation, which led to the formulation of various hypotheses.[59] All of these have been rendered partly or wholly obsolete by the emergence of two new elements.

($\alpha$) Of principal importance, of course, was the discovery in the Mashhad manuscript of four new Books of Diophantus' work. This discovery substantially reduced the *extent* of the missing part, there now being only three Books missing, presumably irretrievably lost. But the question of the *content* of the missing part remains essentially as before, because, contrary to all previous expectations, a significant part of the problems contained in the seven formerly missing Books appears to be aimed at helping the student to acquire "experience and skill" (see pp. 176 and 263); what we encounter in the Arabic Books is a section in which no truly novel methods are presented, and, after the first three Books, Diophantus continues to move in the same circle of artifices. In particular, one always ends up with an equation having just one term on each side.

($\beta$) Diophantus' announcement, in the Greek introduction, that he would later show *how* to solve the case in which two terms are left equal to one term is an important indication, the significance of which has been greatly altered by certain discoveries made in the first part of this century. Earlier scholars had thought the *resolution* of the determinate quadratic equation in the *Arithmetica* to be of central importance. But this cannot have been the case, for the decipherment of mathematical cuneiform texts in the 1930s revealed that as early as in Sumerian times the resolution of the three classical types of quadratic equations having a positive solution, i.e.,

(I) $Ax^2 = Bx + C,$
(II) $Ax^2 + Bx = C,$
(III) $Ax^2 + C = Bx,$

with $A, B, C > 0$ (and $B^2 > 4AC$ in case (III)), was well known. Thus Diophantus must not have considered the resolution itself to be the *arcanum arcanorum* of science but as something with which his readers would probably be acquainted or might easily acquaint themselves.[60] The instances in the

[59] They are summarized in Heath's *Diophantus*, pp. 6–12.

[60] The same impression is left by passages in Heron involving quadratic equations (listed in Heath, *Hist. Gr. Math.*, II, p. 344).

Greek *Arithmetica*, found in Books "IV" to "VI", seem to confirm this.[61] Of course, it is still possible to conjecture that a systematic explanation of the resolution of the second-degree equation was given somewhere in the *Arithmetica*. It would, then, have occurred in a now lost part located between Book VII and Book "IV", whether this part consisted of some Book(s) or, if Book "IV" is the original Book VIII (cf. p. 54 and p. 68, n. 44), of a preface to this Book;[62] the explanations about the resolution found later on would then be given simply as reminders.

## 2. The Announcement in the Greek Introduction

($\alpha$) After defining the various powers and giving the rules for their multiplication, and explaining the nature of the operations of restoration and reduction, Diophantus goes on to say that one has to apply all this skilfully in the propositions' initial hypotheses[63] *so as to be left with, in so far as possible, one term on each side of the equation.*

This final form is the only one found in Books I to VII and is almost always found in Books "IV" to "VI". After setting the required magnitudes indeterminately in function of the unknown and solving, we arrive at an equality of the form

$$\beta x^p = \alpha x^q \qquad (\text{say, } p > q),$$

i.e.,

$$x^{p-q} = \frac{\alpha}{\beta},$$

where $\alpha/\beta = f(k, l, \ldots; m, n, h, \ldots)$ is a rational expression depending on the given quantities $k, l, \ldots$ (if any) and (if the problem is not determinate) on one or more parameters $m, n, h, \ldots$ linking the various required magnitudes.

---

[61] They are the following:

*Type I*: An explanation for the resolution (actually of an inequality) is provided in "IV",39; that it is of little significance is shown by the wording (the explanation begins with a simple ὅταν, as is done for other rules given incidentally; see "IV",33 and 36), and the presence of the rule becomes even less significant when one remembers that knowledge of the resolution of the same type was needed earlier (in "IV",31; I dismiss the trivial case $x^2 = 4x - 4$ in "IV",22). The remaining instances give only the solution: "VI",7, 9, and 11; cf. "V",30, where integral limits to the (irrational) solution are given.

*Type II*: The condition for the rationality of the solution—that the discriminant be a square—is given in "VI",6. Otherwise, the solution alone is indicated ("VI",6, *in fine*, and 10; cf. "VI",8).

*Type III*: An inequality is briefly treated in "V",10. The condition for the discriminant appears in "VI",22, while an approximation to the solution is used in "V",30.

[62] The assumption that there might have been a preface to Book "IV" is certainly not unreasonable, if not probable, considering the presence in the Arabic Books of two intermediate prefaces.

[63] ὑποστάσεις (= *Ansätze*), i.e., the setting of the various required quantities in function of the unknown $x$; cf. D.G., I, pp. 244,21 and 304,18.

Since the solution is supposed to be positive and rational, we may at the outset be given a condition for the given quantities (a "diorism", cf. p. 49), or we may encounter during the resolution some condition restricting the choice of the parameters $m, n, h, \ldots$, most often for positivity, sometimes for obtaining suitable equations (e.g., in III,10; III,15,2°), and, finally, in order to have rational solutions (in IV,8; commonly from "IV",8 onward).

($\beta$) After mentioning this case—in which, the relations between the required magnitudes being suitably set, the mathematical treatment ends with an equality between two single terms—, Diophantus adds: "We shall show you later (ὕστερον) how, in the case also of two terms being left equal to one term, such a problem is solved".

As the rarity of the problems leading to second-degree equations in Books "IV"–"VI" does not seem to justify such an announcement in the main introduction, we are naturally led to suppose that this announcement alludes to the still-missing part of the *Arithmetica*, in the problems of which we would, on the analogy of the previous case, more or less regularly end up with an equation of the form

$$\alpha x^p + \beta x^q = \gamma x^r,$$

with $\alpha$, $\beta$, $\gamma$ positive and rational quantities depending on the given magnitudes and parameters, and $p$, $q$, $r$ exponents all different, but such that the sum of the largest and the smallest is equal to twice the middle one: then we shall end up with one of the three classical types of the complete quadratic equation, that is, with those possessing a positive solution. But, in order that the solution be rational also, the discriminant of this second-degree equation must be a square, and this will lead us to solve first an indeterminate equation linking the given magnitudes and the parameters to choose. No other condition of rationality is involved in the simplest case ($p$, $q$, $r$ consecutive natural numbers). Diophantus probably had this kind of problem in mind, and not the banal resolution of complete quadratic equations, when he wrote his introduction to Book I. We must therefore consider how able Diophantus was to deal with such problems.

## 3. Diophantus and the Equation $Ax^2 + Bx + C = \square$

The cases in which $A$ or $C$ are either nil or square are elementary and regularly found from Book II on (cf. p. 7). In the general case, putting $\square = m^2x^2$, or $\square = m^2$, we shall be obliged to find the solution of

$$4Cm^2 + D = \text{square},$$

or

$$4Am^2 + D = \text{square},$$

where $D = B^2 - 4AC$. In other words, we shall have to solve equations of the type

$$\alpha y^2 + \gamma = \text{square}$$

for given $\alpha$, $\gamma$.

Diophantus undoubtedly knew how to reduce an equation $Ax^2 + Bx + C = \square$ to the form $\alpha y^2 + \gamma = $ square, for he performs this transformation in "IV",31. Let us consider the steps he took. Arriving at $-x^2+3x+18=\square$, he first endeavours to solve by putting $\square = 4x^2$. Since this does not yield a rational result, he indicates that we shall obtain an acceptable solution if (by putting $\square = m^2x^2$), we can find an $m^2$ fulfilling the condition $18(m^2 + 1) + (\frac{3}{2})^2 = $ a square, whence the equation $72m^2 + 81 = $ a square. Thus he establishes quite clearly the condition for the discriminant. We also note: first, that he ends up, as expected, with one of the more easily solved cases, namely the form $\alpha y^2 + \gamma^2 = $ a square; and second that, since he first performs a tentative resolution, he must not have systematically practised before the reduction of the general equation $Ax^2 + Bx + C = \square$ to the form $\alpha y^2 + \gamma = $ a square.

We must now consider what Diophantus knew of equations of the type $Ax^2 + C = \square$. This is of all the more interest in that the general system

$$\begin{cases} A_1x + B_1 = \square, \\ A_2x + B_2 = \square', \end{cases}$$

of which only special cases are solved in the extant Books, is also reducible (putting $\square' = y^2$ and eliminating $x$ from the first equation) to an equation of the above type, namely

$$\frac{A_1}{A_2}y^2 + \frac{A_2B_1 - A_1B_2}{A_2} = \square.$$

Those equations $Ax^2 + C = \square$ actually solved in the extant Books are the simpler cases in which one of the two constants is positive and a square. But Diophantus obviously knew of at least two other cases in which the equation is soluble; this is revealed incidentally in Book "VI".

($\alpha$) In problem "VI",14, Diophantus asserts that it is impossible to solve rationally the equation

$$Ax^2 - C^2 = \square$$

if $A$ is not representable as the sum of two squares.

($\beta$) In the second lemma to "VI",12, Diophantus proves that the equation

$$Ax^2 + C = \square$$

has an infinite number of solutions if $A + C = $ a square. For $x = 1$ satisfies the equation, and taking $x = y + 1$ leads us to a new equation which "onc can solve in an infinite number of ways since the units are (a) square (number)".[64]

($\gamma$) Another lemma, preceding "VI",15, is also concerned with the infinite number of solutions. It states that if $Ax^2 - C =$ square is satisfied for $x = x_0$, one can obtain another solution $y > x_0$ (this is done by putting $x = y + x_0$).

These cases appear to derive from more general theorems, familiarity with which can hardly be denied Diophantus. Thus, the condition given in $\alpha$ is applicable to $C$ in the equation (regarded as different[65]) $-A^2x^2 + C = \square$, while $\beta$ and $\gamma$ both follow from the same proposition, asserting that if $x_0$ satisfies $Ax^2 + Bx + C = \square$, one can find an infinite number of other solutions: putting $x = y + x_0$, we have

$$Ay^2 + y(2Ax_0 + B) + (Ax_0^2 + Bx_0 + C) = \square,$$

and, since the numerical term is a square, the method taught in Book II of taking $\square = (my + \sqrt{Ax_0^2 + Bx_0 + C})^2$ leads to any number of solutions $x(m) = y(m) + x_0$.

Now, since Diophantus uses in Book "VI" some *ad hoc* propositions which are obviously derived from more general theorems relating to indeterminate equations, it is reasonable to suppose that these general theorems were used somewhere in the *Arithmetica*—if so in a now lost part which would follow Book "VI" where these *ad hoc*, particular cases occur.[66]

To suggest that Diophantus treated problems involving equations of the types $Ax^2 + Bx + C = \square$ and $\alpha y^2 + \gamma = \square$ in the lost part of the *Arithmetica* is pure conjecture. But when we consider that Diophantus knew how to reduce the first type to the second one and knew some facts about the solutions of the second type, this seems quite possible—the more so when one considers that some kinds of problems found in early Islamic times and not treated in the *Arithmetica*, but which might well be a remnant of Greek learning, involve the above-mentioned types of equations.[67]

---

[64] In problem III,10 (and III,11), an equation of the type $\alpha x^2 + \gamma = \square$ with $\alpha + \gamma =$ square is *not* solved, Diophantus choosing instead to reformulate the initial hypotheses so as to obtain an εὐχερής equation, that is to say, in this case, one with a square as the coefficient of $x^2$. Diophantus cannot have been unaware of the fact that 1 is solution; simply, he does not acknowledge this for didactic reasons, for we are in the section of the *Arithmetica* dealing only with indeterminate quadratic equations having either the coefficient of $x^2$ or the constant term square, to the exclusion of other cases. Heath's rendering of "(not) εὐχερής" as "impossible" (*Dioph.*, p. 69) is quite misleading.

[65] Diophantus apparently does not consider direct transformations of the form $x = 1/y$ (cf. p. 227, n. 4).

[66] The occurrence of particular cases before the exposition of the general case can be accounted for by the external form of the problems in which they appear; for this form is the unifying characteristic of Book "VI".

[67] That some models of Indian astronomy are dependent on early Greek material has been established recently; but since no such interdependence has been proven for indeterminate algebra we must leave any comparison out of consideration.

## 4. On Some Problems of a Diophantine Nature Found in Islamic Mathematics but Not in the Extant *Arithmetica*

The study of indeterminate equations in Islamic times appears to have depended greatly on Greek material. We have already pointed out the direct influence of the *Arithmetica*, or, more specifically, of its Books I–IV, on al-Karajī's work (see pp. 10–11). We have also conjectured that Abū Kāmil had access to some Greek source other than the *Arithmetica* (see pp. 9–10). We shall now list certain types of indeterminate problems which are treated by these authors but which, as noted above, are not found in the extant *Arithmetica*.

### *a. Problems of Abū Kāmil*[68]

(α) Problems 19, 21, 24, 25 of Abū Kāmil (nos. 24 and 25 appear in the later *Faḫrī*, as IV,32–33) deal with equations of the type

$$-x^2 + 2Bx + C = \square \qquad (B \text{ and/or } C > 0).$$

It is stated that this equation is soluble if $D = B^2 + C$ is representable as the sum of two squares, provided also that $|C| < B^2$ in the particular case $C < 0$. Indeed, putting $\square = m^2$ leads to the intermediate problem of making $B^2 + C - m^2$ a square.

(β) In problems 7–9 (cf. *Badīᶜ*, 52[69]), 22 (=*Badīᶜ*, 49), 23 (=*Faḫrī* IV,28), the resolution of the system

$$\begin{cases} x^2 + kx = \square, \\ x^2 + lx = \square', \end{cases} \qquad k, l \text{ given},$$

passes through the resolution of

$$\begin{cases} u^2 + v = \square_1, \\ u^2 + \dfrac{l}{k} v = \square'_1. \end{cases}$$

First, we put $v = 2mu + m^2$, $m$ arbitrary, and determine $u$ from the second equation $u^2 + (2ml/k)u + m^2 l/k = \square'_1$. This gives a solution $u_0$, $v_0$, and any $u_0 t$, $v_0 t^2$, $t$ rational, will also clearly fulfil the second system. We then obtain a solution to the original system by requiring $(x =) u_0 t = (1/k) v_0 t^2$, whence $t = k u_0 / v_0$ and $x = k u_0^2 / v_0$.

In problems 11 (=*Faḫrī* II,28 or IV,27) and 20 (cf. *Faḫrī* II,32; *Badīᶜ*, 55), the second equation has the form $-x^2 + lx = \square'$, and the intermediate

[68] The numbering of the problems is that employed in our account of Abū Kāmil's methods (and in our forthcoming edition of his *Algebra*, in the third part of which indeterminate problems occur).

[69] The numbering is that employed in our translation.

equation becomes $-u^2 + (2ml/k)u + m^2l/k = \square'_1$. In Abū Kāmil's two examples, the intermediate equation is easy to solve since $l/k$ is a square. Were this not the case, we would find ourselves in the situation of the previous set of problems (α), and the condition would be that $m^2l^2/k^2 + m^2l/k = (m^2l^2/k^2)(1 + (k/l))$, thus $(l + k)/l$, be representable as the sum of two squares.

No systems like nos. 7–9, 11, 20, 22, 23 of Abū Kāmil are treated in this manner in the extant *Arithmetica*. The *method*, however, is employed in three very simple cases (VII,8–10), and this speaks for its being of Greek origin.[70]

(γ) Lastly, Abū Kāmil solves systems of the type

$$\begin{cases} x^2 + lx + k = \square, \\ x^2 + lx + k + h\sqrt{x^2 + lx + k} = \square', \end{cases} \qquad l, k, h \text{ given,}$$

(problems 26–30, 32 (=*Faḫrī* IV,35), 34), and of the general type

$$\begin{cases} x^2 + l_1x + k_1 = \square, \\ x^2 + l_2x + k_2 = \square', \end{cases}$$

(problems 31 (=*Faḫrī* IV,34), 33, 35–38 (=*Faḫrī* IV,36–39)).

The resolution he presents (cf. *Les Méthodes*, pp. 99–103), not seen in any of the extant Books of the *Arithmetica*, is mathematically correct, but may yield an unacceptable solution. Whether or not this might have prevented Diophantus from including such systems in his *Arithmetica* is debatable.

### b. *Problems of al-Karajī*

There is little in al-Karajī which is not taken either from Diophantus or from Abū Kāmil; the only instances worth mentioning here occur in the *Badīʿ*.[71]

(α) Problems 21–22 and 27–33 deal with equations $Ax^2 + C = \square$ and $Ax^2 + Bx + C = \square$, respectively. The condition for rationality in the second case takes al-Karajī back to the first case; hence he is well aware of the transformation. The resolution itself is of little interest, since a solution is found by *istiqrāʾ*, i.e., empirically (see our study on the *Badīʿ*, pp. 303–4).

**Remark.** Problem 21 contains the assertion "And whenever the coefficient of the $x^2$'s forms together with the number of units a square, then $x^2 = 1$ (fulfils the problem)". But this statement (which was perhaps made not by al-Karajī himself but was added later) appears to be an isolated observation,

---

[70] Systems like those of Abū Kāmil would fit best in the first Books of the *Arithmetica*, i.e., in the extant ones, since they are all reducible to an indeterminate equation of the second degree soluble by the methods of Book II. But it appears from the above discussion that this is not always the case.

[71] On some isolated instances of the *Faḫrī*, see pp. 58, 181, and 194.

from which no mathematical inferences—unlike in the *Arithmetica*—are drawn.

($\beta$) Problems 17–20 are considerably more interesting, since they contain an elegant method which enables one to solve equations of the form

$$Ax^2 - C = \square \quad \text{or} \quad C - Ax^2 = \square,$$

when $A \cdot C$ (thus $A/C$) is a square.

Putting $\square = y^2$, we have

$$\frac{C}{A} + \frac{1}{A}y^2 = x^2 \quad \text{and} \quad \frac{C}{A} - \frac{1}{A}y^2 = x^2,$$

and we can solve either case by assuming that

$$x = my - \sqrt{\frac{C}{A}} \quad \left(m^2 > \frac{1}{A}\right),$$

this leading us to

$$x = \sqrt{\frac{C}{A}} \cdot \frac{m^2 + \frac{1}{A}}{m^2 - \frac{1}{A}} \quad \text{and} \quad x = \sqrt{\frac{C}{A}} \cdot \frac{m^2 - \frac{1}{A}}{m^2 + \frac{1}{A}},$$

respectively.[72]

**Remark.** Mention should be made of the problem $x^2 \pm k = \square$, extensively studied, it would seem, in the tenth century (see Woepcke's *Recherches sur* (...) *Léonard de Pise*, 1, III, A and B), and also exhaustively treated by al-Ḫazin (see Anbouba, *Traité d'Abū Jaʿfar*). In al-Ḫazin's tract are found most of the theorems which later appear, some with better proofs, in Leonard of Pisa's *Liber quadratorum*. Both works undoubtedly stem from some common source, which may itself have been based on some Greek work.[73]

## 5. Conclusion

We have seen that Diophantus certainly knew more of general indeterminate equations of the second degree than what actually appears in the extant *Arithmetica*, and that such equations, as well as some other types of problems which strongly recall Diophantine preoccupations, were solved in early Islamic times. We have also conjectured that if any problems leading

[72] The first equation of course goes into the second one by putting $x = 1/z$ and multiplying by $z^2$.
[73] The *form* of the solution of $x^2 \pm k = \square$ is obtainable by elementary Diophantine methods (cf. our study on the *Badīʿ*, p. 336).

to general indeterminate equations of the second degree were treated by Diophantus, as his Greek introduction seems to announce, they would be best placed after Book "VI", thus, perhaps, in the group of three Books now lost which would be Books XI to XIII.

But, as said before, this is pure conjecture, rendered all the more uncertain by the unpredictability of Diophantus' intentions, an unpredictability strikingly demonstrated by the unexpected direction taken by the newly discovered Books. We cannot even guess the number of Books which Diophantus might have devoted to such problems, since the general impression left by the *Arithmetica* is that Diophantus, with but a single method at his disposal, could fill any number of Books by inventing problems *à volonté*—another facet of his mathematical genius.

# Part Two

# Translation

The translation given here is essentially a literal one, and may thus be awkward in some passages; we deliberately chose to translate in this way, however, in order to give readers unacquainted with Arabic an idea of the form and expression of the text.[1] When desirable, the sense of passages has been made clearer by additions which appear in parentheses.

The footnotes consist principally of textual explanations or references to other parts of the *Arithmetica*, or they indicate the presence of those interpolations which we have chosen to leave in the edited text (see pp. 29–33). They are only rarely concerned with mathematical questions or with questions of translation, since all problems are fully discussed in the Mathematical Commentary, while information on the occurrences and meanings of individual words may be found in the Arabic Index. Finally, a few not trivial lacunae of the manuscript are indicated; they are enclosed, as in the edited Arabic text, in angle brackets.

The numerals in the right-hand margin refer to the lines of the printed Arabic text, and those in the left-hand margin indicate the pages of the Mashhad manuscript.

---

[1] We have, however, departed from the text: (α) in rendering some active verbal forms by passive ones or by substantives; (β) in systematically omitting the word "units" which is frequently appended to the expression of constant terms (the same was done by Tannery in his Latin translation of the Greek text); (γ) in systematically using figures instead of words for the numerals.

# Fourth Book of the treatise of Diophantus the Alexandrian on squares and cubes

which (treatise) Qusṭā ibn Lūqā of Baalbek translated from the Greek language into the Arabic language. This is the handwriting of Muḥammad ibn abī Bakr ibn Ḥākir the astronomer, and he wrote in the year 595 of the hegira.

# In the Name of God the Merciful, the Compassionate

# Fourth Book of the Treatise of Diophantus on Squares and Cubes

I have presented in detail, in the preceding part of this treatise on arithmetical problems, many problems in which we ultimately, after the restoration and the reduction[1], arrived at one term equal to one term, (namely) those (problems) involving (either of) the two species of linear and plane number and also those which are composite. I have done that according to categories which beginners can memorize and grasp the nature of.

In order that you[2] miss no thing, in treating which you would acquire ability in that science,[3] I consider it also appropriate to write, once again, for you, in what follows, many problems of this kind, (but now) involving the species of number called solid (alone) as well as in association with (one of) the first two species. In it,[4] I shall follow the same path and advance you along it from one step to another and from one kind to another for the sake of experience and skill. Then, when you are acquainted with what I have presented, you will be able to find the answer to many problems which I have not presented, since I shall have shown to you the procedure for solving a great many problems and shall have explained to you an example of each of their types.

*ef. XII*[5] I say (the following). Every square[6] multiplied by its side gives an $x^3$. When I then divide $x^3$ by $x^2$, the result is the side of $x^3$; if $x^3$ is divided by $x$, namely the root of the said $x^2$,[7] the result is $x^2$.

---

[1] These terms are explained further on, at the end of the introduction.

[2] Diophantus is surely addressing Dionysius, as he does in the Greek introduction.

[3] Namely, the solving of arithmetical problems (cf. also D.G., I, p. 2,3).

[4] That is, "in what follows" (perhaps an inappropriate addition: cf. app. crit., n. 4).

[5] I shall continue with the convenient policy used first in Bachet's edition and then in Tannery's of numbering the presented introductory rules (called "definitions"); there are eleven such rules in the Greek introduction.

[6] The Greek introduction also has "square", not "$x^2$", in similar situations (cf. D.G., I, p. 4,19 and 22). See, though, note 30 of the critical apparatus for a list of confusions between *māl*/δύναμις and *murabbaʿ*/τετράγωνος.

[7] Or "namely the root of the said quantity". This phrase (and those similar to it below) I consider to be Arabic interpolations; see p. 30, no. 1.

When I then multiply $x^3$ by $x$, the result is the same as when $x^2$ is multiplied by itself, and it is called $x^4$. If $x^4$ is divided by $x^3$, the result is $x$, namely the root of $x^2$; if it is divided by $x^2$, the result is $x^2$; if it is divided by $x$, namely the root of $x^2$, the result is $x^3$.

When $x^4$ is then multiplied by $x$, namely the root of $x^2$, the result is the same as when $x^3$ is multiplied by $x^2$, and it is called $x^5$. If $x^5$ is divided by $x$, namely the root of $x^2$, the result is $x^4$; if it is divided by $x^2$, the result is $x^3$;
2 if it is divided by $x^3$, the result is $x^2$; and if it is divided by $x^4$, the result is $x$, namely the root of $x^2$.

When $x^5$ is then multiplied by $x$, the result is the same as when $x^3$ is multiplied by itself and when $x^2$ is multiplied by $x^4$, and it is called $x^6$. If $x^6$ is divided by $x$, namely the root of $x^2$, the result is $x^5$; if it is divided by $x^2$, the result is $x^4$; if it is divided by $x^3$, the result is $x^3$; if it is divided by $x^4$, the result is $x^2$; if it is divided by $x^5$, the result is $x$, namely the root of $x^2$.

*Def. XIII* After the restoration and the reduction—one means by restoration the adding of what is negative to both sides (of the equation) and by reduction the removing of what is equal from both sides—[8] the treatment will result for us in the equality of one of these species—the mutual multiplications and divisions of which we have explained (above)—with another; it will then be necessary to divide the whole by a unit of the side having the lesser degree[9] in order to obtain one species equal to a number.

**1.** We wish to find two cubic numbers the sum of which is a square number.

We put $x$ as the side of the smaller cube, so that its cube is $x^3$, and we put as the side of the greater cube an arbitrary number of $x$'s, say $2x$; then, the greater cube is $8x^3$. Their sum is $9x^3$, which must be equal to a square. We make the side of that square any number of $x$'s we please, say $6x$, so the square is $36x^2$. Therefore, $9x^3$ is equal to $36x^2$. Then, since the side (of the equation) containing the $x^2$'s is lesser in degree than the other, we divide the whole by $x^2$; $9x^3$ divided by $x^2$ gives $9x$, that is 9 roots of $x^2$,[10] and the result from the division of the $36x^2$ by $x^2$ is a number, namely 36. Thus $9x$, that is
3 (nine) roots,[10] equals 36; hence $x$ is equal to 4. Since we assumed the side of the smaller cube to be $x$, the side is 4, and the smaller cube is 64; and since we assumed the side of the greater cube to be $2x$, the side is 8, and the greater cube is 512. The sum of the two cubes is 576, which is a square with 24 as its side.

Therefore, we have found two cubic numbers the sum of which is a square, the lesser being 64 and the larger, 512. This is what we intended to find.[11]

---

[8] The explanation of the two terms seems to be an Arabic addition; see p. 30, no. 2.

[9] That is, by the power of the unknown found in the side of lesser degree.

[10] This again seems to be an Arabic addition (see above note 7).

[11] *bayyana*; from problem IV,7 on it is replaced by *wajada*. We have rendered both by "to find" (see index, *bāna* (II), p. 435).

**2.** We wish to find two cubic numbers the difference of which is a square number.

We put $x$ as the side of the smaller cube, which is then $x^3$, and we put as the side of the larger any number of $x$'s we wish; let us put $2x$ for the side, so that the greater cube is $8x^3$. Their difference is $7x^3$, which is equal to a square number. Let us put for the side of the square $7x$, so that the square is $49x^2$. Thus $7x^3$ is equal to $49x^2$. As the side (of the equation) containing the $x^2$'s is the lesser in degree, we divide the whole by $x^2$, and so obtain $7x$ equal to 49; hence $x$ is equal to 7. Since we assigned to the smaller cube the side $x$, the smaller cube is 343; and, since the greater (cube) has the side $2x$, its side is 14, and the greater cube is 2744. Their difference is 2401, which is a square with 49 as its side.

Therefore, we have found two cubic numbers the difference of which is a square number. This is what we intended to find.

**3.** We wish to find two square numbers the sum of which is a cubic number.

4 We put $x^2$ as the smaller square and $4x^2$ as the greater square. The sum of the two squares is $5x^2$, and this must be equal to a cubic number. Let us make its side any number of $x$'s we please, say $x$ again,[12] so that the cube is $x^3$. Therefore, $5x^2$ is equal to $x^3$. As the side which contains the $x^2$'s is the lesser in degree, we divide the whole by $x^2$; hence $x$ is equal to 5. Then, since we assumed the smaller square to be $x^2$, and since $x^2$ arises from the multiplication of $x$—which we found to be 5—by itself, $x^2$ is 25. And, since we put for the greater square $4x^2$, it is 100. The sum of the two squares is 125, which is a cubic number with 5 as its side.

Therefore, we have found two square numbers the sum of which is a cubic number, namely 125.[13] This is what we intended to find.

**4.** We wish to find two square numbers the difference of which is a cubic number.

We put $x$ as the side of the smaller square and an arbitrary number of $x$'s as the side of the larger, say $5x$; thus, the larger square is $25x^2$ and the lesser, $x^2$. Their difference is $24x^2$, and this is equal to a cube. Let us put for the side of the cube any number of $x$'s we please, say $2x$. Hence $24x^2$ is equal to $8x^3$, for the cube that arises from $2x$ is $8x^3$. We again[14] divide the whole by $x^2$, hence $8x$ is equal to 24; then $x$ is 3. Since we set $x$ as the side of the smaller square and $5x$ as the side of the larger square, the side of the smaller is 3 and

---

[12] As the side of the smaller square.

[13] With a slight change in the text (reading *humā* instead of *huwa*), we have the usual statement of the required magnitudes, i.e., "and these are 100 and 25". The text's reading could well be a scribal mistake (cf. notes 586, 760 of the crit. app.).

[14] As in the preceding problems.

5 that of the larger, 15; (so) the lesser square is 9, the larger square 225, and their difference 216, which is a cubic number having 6 as its side.

Therefore, we have found two square numbers the difference of which is a cubic number, and these are 225 and 9. This is what we intended to find.

**5.** We wish to find two square numbers which comprise[15] a cubic number.

We assume the smaller to be $x^2$ and the side of the larger to be any number of $x$'s we please; let us put $2x$ for the side, so the larger square is $4x^2$. The number they comprise is $4x^4$, which equals a cubic number; we put $2x$ as its side, so that the cube is $8x^3$. Therefore, $4x^4$ is equal to $8x^3$. We divide the whole by $x^3$, hence 8 equals $4x$; for $8x^3$ divided by $x^3$ gives 8—since (the multiplication of) 1 by $x^3$ gives $x^3$, the division of $x^3$ by $x^3$ gives 1—, and the division of $4x^4$ by $x^3$ gives $4x$: therefore $4x$ equals 8.[16] Thus $x$ is equal to 2. Since we took $x^2$ as the lesser square, it is 4, for $x^2$ is yielded by the multiplication of $x$ by itself; and since we took $4x^2$ as the larger square, it is 16. The number comprised by these two squares is 64, which is a cube with 4 as its side.

Therefore, we have found two square numbers which comprise a cubic number, namely 4 and 16. This is what we intended to find.

**6.** We wish to find two numbers, one square and the other cubic, which comprise a square number.

We put as the side of the square an arbitrary number of $x$'s, say $x$, so that the square is $x^2$; likewise, we put as the side of the cube a number of $x$'s of our
6 choice, say $2x$, so that the cube is $8x^3$. The number comprised by them, that is to say by $x^2$ and by $8x^3$, is $8x^5$, and this equals a square. Now, suppose that we put $x$'s as the side of the square; $x^2$'s will result (from the multiplication of the $x$'s by themselves), hence $x^5$'s will equal $x^2$'s, and we shall have to divide both sides by (a unit of) the $x^2$'s; $x^3$'s will then be equal to units, for, as I have mentioned,[17] the division of $x^5$'s by $x^2$'s gives $x^3$'s. Consequently,[18] we put as the side of the square $x^2$'s in any number we please, say $4x^2$, so that the square is $16x^4$. Thus, $8x^5$ is equal to $16x^4$. We divide the whole by $x^4$, since the $x^4$'s are the lower in degree of the two sides; the division of $16x^4$ by $x^4$ yields 16, while the division of $8x^5$ by $x^4$ yields $8x$. Hence $8x$ is

[15] The usual (literal) translation of the Greek περιέχειν. Two numbers "comprise" a third if the product of their multiplication gives the third number. Cf., e.g., Euclid, *Elem.*, VII, def. 19.

[16] These lengthy explanations may be interpolated; see p. 31, no. 13.

[17] See the rules in the introduction to this Book ("def. XII"). Cf. also pp. 178–179 on the genuineness of this reference.

[18] In order to avoid the equating of $x^3$'s to units, which requires a preliminary condition for a rational $x$.

equal to 16, so $x$ is 2. Then, since we set $x$ as the side of the square, the square is 4, and the cube, since we set $2x$ as its side, is 64; the number comprised by them—namely by the square, which is 4, and by the cube, which is 64—is 256, which is a square with 16 as its side.

Therefore, we have found two numbers, one square and the other cubic, which comprise a square number, and these are 4 and 64. This is what we intended to find.

**7.** We now wish to find two numbers, one square and the other cubic, which comprise a cubic number.

We put $x$ as the side of the square, so the square is $x^2$, and we put an arbitrary number of $x$'s as the side of the cube, say $4x$, so that the cube is $64x^3$. The number comprised by them is $64x^5$, which is equal to a cubic number. Now, suppose that we put $x$'s as the side of the cube; then the cube
7 will be made of $x^3$'s, and, after setting them equal to (the) $x^5$'s, we shall have to divide the whole by $x^3$, thus obtaining $x^2$'s equal to units; it will then be necessary for the units which equal $x^2$ to be (a) square (number). If, however, we set $x^2$'s as the side of the cube, the cube will be made of $x^6$'s; after setting that equal to (the) $x^5$'s, we shall have to divide both sides by $x^5$, thus obtaining $x$'s equal to units. Hence we assume the side of the cube to be $2x^2$; the cube is then $8x^6$. Therefore, $8x^6$ is equal to $64x^5$. We divide the whole by $x^5$, for the $x^5$'s are the lower in degree of the two sides; we then obtain, from the division of the $8x^6$ by $x^5$, $8x$, and, from the division of the $64x^5$ by $x^5$, 64. Hence $8x$ is equal to 64, and $x$ is 8. Since we put $x$ as the side of the square, the square is 64; the cube, since we put $4x$ as its side, has the side 32, and the cube itself is 32,768. The result of the multiplication of that by the square, namely (by) 64, is a cubic number, since each one of the two (factors) is a cube.[19]

Therefore, we have found two numbers under the condition we stipulated. This is what we intended to find.

**8.** We wish to find two cubic numbers which comprise a square number.

Suppose that we put, in this problem too,[20] $x$ as the side of the smaller cube, so that the smaller cube is $x^3$, and that we put as the side of the larger whatever number we please of $x$'s, for instance $2x$, so that the larger cube is $8x^3$; the number they comprise is $8x^6$, and that must be equal to a square. Now, it is not correct to put $x$'s as the side of this square; for the square of $x$'s being $x^2$'s, when these have been set equal to (the) $x^6$'s, and (both sides) afterwards divided by the (power of the) side of lesser degree, which consists

[19] Euclid, *Elem.* IX,4.

[20] See what has been done above.

of $x^2$'s, the result is $x^4$('s), equal to units.[21] ⟨But if we set $x^2$'s for the side of the square, the square will be made of $x^4$'s; after having put that equal to $x^6$'s, we shall have to divide the two sides by $x^4$; we shall then obtain $x^2$'s
8 equal to units.⟩[22] Thus it is necessary for the number of units equal to $x^2$ to be a square. Therefore, we are led to seek a square and a cubic number which comprise a square number, because of the convenience of that, which will become clear in the treatment.[23] We then find, as shown above,[24] that one of the two numbers, namely the square, is 4, and the other, namely the cube, is 64; the number these two numbers comprise is 256, which is a square with 16 as its side. This is what we intended to find.[25]

**9.** We wish to find two cubic numbers which comprise a square.

We set $4x$ as the side of the greater cube and $x$ as the side of the smaller cube. Then the greater cube is $64x^3$, the smaller, $x^3$, and the number they comprise is $64x^6$; this must be equal to a square number. We put as its side $x^2$'s, the coefficient of which is equal to the side of the square arising from the multiplication of the 64 by the 4, namely 256, having as its side 16. Therefore, we put as the side of the square $16x^2$, so that the square is $256x^4$. Then $64x^6$ equals $256x^4$. So we divide the whole by $x^4$, since the $x^4$'s are the lower in degree of the two sides; the division of the $64x^6$ by $x^4$ gives $64x^2$, while we obtain 256 from the division of the $256x^4$ by $x^4$. Therefore, $64x^2$ equals 256, hence $x^2$ equals 4; $x^2$ being a square, as well as 4, their sides are thus equal; the side of $x^2$ being $x$, and that of 4 being 2, $x$ is 2. Then, since we set $x$ as the side of the smaller cube, the smaller cube is 8, and since we set $4x$, i.e., 8, as the side of the larger cube, the larger cube is 512. When we
9 multiply it by the smaller cube, the result is the number they comprise, namely 4096, which is a square having 64 as its side.

Therefore, we have found two cubic numbers which comprise a square number, namely 8 and 512. This is what we intended to find.

Suppose now we intend to find a cubic number such that we obtain, after dividing it by a cube, a square number; we shall look for a square number such that, after multiplying it by another cubic number[26]—which

[21] What is not correct is, to begin with, the equation $8x^6$ = square, which is (rationally) impossible, since 8 is not a square; thus, the preliminary condition is that the numerical factor in the side of the larger cube be a square, say $m^2$. No other condition is necessary if we assume the side of the indeterminate square to be proportional to $x^2$ (or $x^4$), whereas one would be if it were taken to be proportional to $x$, say $nx$ (namely that $n \cdot m$ be a square).

There is no doubt that the text as it stands is the result of some reworking, probably that of the major commentary. Its author perhaps misunderstood a statement of impossibility made by Diophantus about the equation $8x^6$ = square.

[22] Note the significant lacuna here.

[23] This last remark about "convenience" seems to be an interpolation (cf. p. 32, no. 14).

[24] Cf. problem 6.

[25] The solution of the proposition continues with what follows. Cf. p. 61.

[26] It is not the one "we intend to find", but the one by which it is divided.

we also seek—a cubic number results from the multiplication. This being found,[27] the result of the multiplication of the one by the other will be the desired cubic number.

Likewise if we intend to find a square number such that the division of it by a square results in a cube: we shall treat it inversely to what precedes.[28]

And similarly for anything we seek involving a division which is of the preceding kind: for these two (cases) are (in reality) one, since division is merely the inverse of multiplication.

**10.** We wish to find a cubic number such that, when we increase it by an arbitrary multiple of the square having the same side, the sum is a square number.

We put $x$ as the side of the cube, so the cube is $x^3$; we put for the multiplicative factor 10, and we add ten times the square of the cube's side, or $x^2$, to $x^3$, thus obtaining $x^3 + 10x^2$, and this is equal to a square. We assume its side to be $x$'s (in) such (quantity) that their square is larger than $10x^2$, thus making the reduction possible.[29] Putting $4x$ as the side of that (square), the square is $16x^2$, hence $x^3 + 10x^2$ equals $16x^2$. Let us remove the common (quantity) $10x^2$, so that $6x^2$ is equal to $x^3$. Dividing that by $x^2$, we obtain $x$ equal to 6. (Thus) $x^3$ is 216. The square of the side is 36; ten times that is 360, and adding this to $x^3$ gives 576, which is a square with 24 as its side.

10 Therefore, we have found a cubic number such that, when we increase it by ten times the square having the same side, it becomes a square number after the addition; the said cube is 216 and its side, 6. This is what we intended to find.

**11.** We wish to find a cubic number such that, when we diminish it by an arbitrary multiple of the square having the same side, the remainder is a square number.

We set $x$ as the side of the cube, so that the cube is $x^3$, and we assume 6 to be the multiplicative factor. We want the remainder of $x^3$ after the subtraction of the $6x^2$ to be a square. We set any number of $x$'s we please for its side, say $2x$, so that the square is $4x^2$. Thus $x^3 - 6x^2$ equals $4x^2$. We restore $x^3$ with the $6x^2$ and add them to the $4x^2$; then $x^3$ equals $10x^2$. Dividing the whole by $x^2$ gives us $x$ equal to 10. Then, since we assumed the side of the cube to be $x$, the cube is 1000. The square of the side is 100, six times which is 600, and the remainder of the 1000 after the subtraction of 600 is 400, which is a square number with 20 as its side.

Therefore, we have found a cubic number such that, when we diminish it by the square of its side taken six times, the remainder is a square number; the said cube is 1000 and its side, 10.

---

[27] See above, problem 7.

[28] We need simply to interchange the words "square" and "cube" in the previous reasoning.

[29] Or: "thus making the equation soluble"; see index, under *muqābala*[h] (p. 450).

**12.** We wish to find a cubic number such that, when we increase it by an arbitrary multiple of the square having the same side, the resulting sum is a cubic number.

We set $x$ as the side of the cube, so that the cube is $x^3$. We increase it by the multiple (of $x^2$) of our choice, which is, (say), the one assumed previously;[30] $x^3$ becomes $x^3 + 10x^2$, which is equal to a cube. We make the side of the cube $2x$, so that $8x^3$ is equal to $x^3 + 10x^2$. Removing $x^3$, which is common, leaves $10x^2$ equal to $7x^3$, and dividing that by $x^2$ results in $7x$ equal to 10;
11 hence $x$ is $\frac{10}{7}$, and the cube is 1000 (units) in the amount of $\dfrac{1}{7 \cdot 7 \cdot 7}$. If we then add to the latter ten times the square, that is, (ten times) $\dfrac{100}{7 \cdot 7}$, or $\dfrac{7000}{7 \cdot 7 \cdot 7}$, the resulting sum is $\dfrac{8000}{7 \cdot 7 \cdot 7}$, which is a cube with $\frac{20}{7}$ as its side.

Therefore, we have found a cube which clearly fulfils the condition imposed upon us, namely $\dfrac{1000}{7 \cdot 7 \cdot 7}$, with side $\frac{10}{7}$. This is what we intended to find.

**13.** We wish to find a cubic number such that, when we diminish it by an arbitrary multiple of the square having the same side, the remainder is a cubic number.

We put $x$ as the side of the cube, so that the cube is $x^3$. We put 7 as the multiplicative factor, so that the remainder (of the subtraction) is $x^3 - 7x^2$; this, then, is equal to a cubic number. We put as its side some fraction of $x$, say $\frac{1}{2}x$, so that the cube is one part of 8 parts of $x^3$; this, then, equals $x^3 - 7x^2$. We restore and reduce; hence $\frac{7}{8}x^3$ is equal to $7x^2$. Dividing then the whole by $x^2$ yields 7 equal to $\frac{7}{8}x$. Thus $x$ is 8, and $x^3$ is 512. Then, if we subtract from the latter seven times the 64,[31] the remainder is 64, which is a cube.

We shall (now) treat this (problem) by another method.[32] We make the side of the first cube any number of $x$'s, say $2x$, so that the cube is $8x^3$. Then, the difference between $x^3$ and $8x^3$, $7x^3$, is equal to seven times the square having the same side as the greater cube. This side being $2x$, its square, $4x^2$, and seven times that being $28x^2$, $28x^2$ is equal to $7x^3$. Dividing the whole by $x^2$ yields 28 equal to $7x$, so $x$ equals 4. Thus, the smaller cube is 64, for its side was $x$, and the greater cube, since $2x$ was set as its side, has the side 8, while the cube is 512.

---

[30] Cf. problem 10.

[31] The presence of the article may indicate that the value of $x^2$ was originally given in the text, together with those of $x$ and $x^3$.

[32] The text of this alternative resolution is somewhat confused and contains two interpolations, as noted in the commentary.

Therefore, it has been found that the other cube, the larger, exceeds the
12 smaller cube by seven times the square of the side of the larger cube; and
this was the condition imposed upon us in this problem. This is what we
intended to find.

**14.** We wish to find a number such that when we multiply it by two given numbers, one of the two (results) is a cube and the other, a square.

We set for the two numbers 5 and 10. We wish to find a number such that when we multiply it by 10 (the result) is a cube and when we multiply it by 5 (the result) is a square. We put $x$ as the number we are seeking; multiplying it by 5 gives $5x$ and afterwards by 10, $10x$. We want to equate the $10x$ with a cubic number and the $5x$ with a square number. We assume the square equal to $5x$ to be any part or any parts we please[33] of the square having the same side as the cube equal to the $10x$, provided that the side of the part is commensurable[34] to the side of the whole, that is to say (provided) that the part is a square. Or, we assume the square of the side of the cube to be any part or any parts of the square equal to the $5x$, provided that this (fraction) is a square. So, let us assume the square of the side of the cube to be one-fourth of the square equal to the $5x$, so that the square of the side of the cube equal to the $10x$ is $1\frac{1}{4}x$. Now, this square—namely $1\frac{1}{4}x$—gives, when multiplied by its side, $10x$; so, if we divide the $10x$ by $1\frac{1}{4}x$, we shall obtain as the result the side of the cube equal to the $10x$. Since the result of the division of $10x$ by $1\frac{1}{4}x$ is 8—for $x$'s when multiplied by units produce $x$'s[35]—, 8 is the side of the cube equal to the $10x$, that is (also) the side of the square equal to $1\frac{1}{4}x$. But,
the cube having side 8 is 512, and this is equal to $10x$; hence $x$ is $51\frac{1}{5}$. Again,
13 the square of 8 is 64, and this is equal to $1\frac{1}{4}x$; so $x$ is four-fifths of 64, or $51\frac{1}{5}$.
If we then multiply the $51\frac{1}{5}$ by 10, we obtain 512, which is a cubic number; and, if we multiply the same number by 5, it becomes 256, which is a square with side 16.

Therefore, we have found a number such that when we multiply it by the two given numbers, namely 10 and 5, the result of its multiplication by 10 is a cubic number and of its multiplication by 5, a square number; and this is (the number) which we intended to find.

But if we want the result of the multiplication of $x$ by 5 to be the cube, and of the multiplication of $x$ by 10 to be the square number, then it is (now) $5x$ which we shall, similarly, equate to a cubic number, and $10x$ to a square number. We assume the square number having the same side as the cube

[33] That is, any aliquot, resp. non-aliquot part; see Euclid, *Elem.* VII, deff. 3,4, or D.G., I, p. 272,18 *seqq.*

[34] Of course linear commensurability, that is, commensurability in the modern sense (cf. *Elem.* X, deff. 1-3).

[35] This explanation looks like an interpolation (cf. p. 30, no. 5); concerning its wording, see note 109 of the app. crit.

equal to the $5x$, again[36] to be one-fourth of the square equal to the $10x$, so that the square of the side of the cube equal to the $5x$ is $2\frac{1}{2}x$. The division by the $2\frac{1}{2}x$ of the $5x$ gives the side of the cube equal to the $5x$. But the result of the division of $5x$ by $2\frac{1}{2}x$ is 2; hence the side of the cube equal to the $5x$ is 2, and thus the cube equal to the $5x$ is 8. So $x$ is $\frac{8}{5}$. The multiplication of $\frac{8}{5}$ by 5 gives $\frac{40}{5}$, or 8, which is a cubic number, and the multiplication of the same by 10 gives $\frac{80}{5}$, or 16, which is a square with 4 as its side.

Let us (now) stipulate in the first problem[37] that the square equal to the $5x$ be to the square having the same side as the cube equal to the $10x$ as one is to four;[38] so the square of the side of the cube equal to the $10x$ is $20x$. The division of the $10x$ by the $20x$ gives $\frac{1}{2}$, and this is the side of the cube
14 equal to the $10x$. As the cube arising from $\frac{1}{2}$ is $\frac{1}{8}$, $10x$ is equal to $\frac{1}{8}$, so $x$ is one part of 80 parts. Then, the multiplication of this last number by 5 results in 5 parts of 80, or one part of 16, which is a square with side $\frac{1}{4}$; the multiplication of the same by 10 gives 10 parts of 80, or $\frac{1}{8}$, which is a cube with side $\frac{1}{2}$.

If we (now) stipulate in the inverse problem that the square equal to the $10x$ be to the square of the side of the cube equal to the $5x$ as one is to four, then the square of the side of the cube equal to the $5x$ is $40x$. Since dividing the $5x$ by the $40x$ results in one part of 8 parts of 1, the side of the cube equal to the $5x$ is $\frac{1}{8}$, and the cube itself is one part of 512. (So) the $5x$ is equal to one part of 512, and $x$ is equal to one part of 2560. Then, if we multiply the latter by 10, it becomes 10 parts of 2560, or one part of 256, which is a square having one part of 16 as its side; the same number, when multiplied by 5, gives 5 parts of 2560, or one part of 512, which is a cube with side $\frac{1}{8}$.

Therefore, we have found a number such that when we multiply it by 10 and by 5 it gives a square number and a cubic number (respectively).

We (now) also use another method. We put an arbitrary number of $x$'s, say $x$, as the side of the cube resulting from the multiplication of the required number by 10; then the cube is $x^3$ and the required number, one part of 10 parts of $x^3$. This fraction must be such that, when multiplied by 5,
15 it results in a square number. But the multiplication of one part of 10 parts of $x^3$ by 5 gives 5 parts of 10 parts of $x^3$, or $\frac{1}{2}x^3$, which is equal to a square number. Let us make the side of that square an arbitrary number of $x$'s, say $2x$, so that the square is $4x^2$. Thus $\frac{1}{2}x^3$ equals $4x^2$. The division of the whole by $x^2$ results in $\frac{1}{2}x$ equal to 4; so $x$ is 8. Since we set $x$ for the side of the cube yielded by the multiplication of the required number by 10, this side is 8 and the cube, 512. The division of 512 by 10 gives the required number, which is $51\frac{1}{5}$.[39]

---

[36] As in the first part of this problem.

[37] That is, the one in which $5x$ is equal to a square and $10x$ to a cube.

[38] Literally: "in the ratio of one-fourth".

[39] No verification of the solution is made, probably because the results are known from the first part of the problem.

We may also set an arbitrary number of $x$'s, say $x$, for the side of the square arising from the multiplication of the required number by 5, so that the square is $x^2$. Then, the required number is one part of 5 parts of $x^2$. This, when multiplied by 10, gives 10 parts of 5 parts of $x^2$, or $2x^2$, which is equal to a cubic number. We assume the side of this cube to be any number of $x$'s, say $x$; so the cube is $x^3$. Then $2x^2$ is equal to $x^3$, and the division of the whole by $x^2$ gives $x$ equal to 2. Since we set $x$ as the side of the square, the side is 2 and the square itself, 4. Then the multiplication of the required number by 5
16 gives 4, so the required number is $\frac{4}{5}$. This last number, when multiplied by 5, gives $\frac{20}{5}$, or 4, which is a square, and the multiplication of the same number by 10 gives $\frac{40}{5}$, or 8, which is a cube.

Therefore, we have found a number such that when we multiply it by 10 and by 5 the results are a square number and a cubic number.[40]

**15.** We wish to find a number such that when we multiply it by two given numbers, the result of its multiplication by the one is a cubic number and by the other, the square having the same side as that cube.

Let one of the two given numbers be 4 and the other, 10. We wish to find a number such that when we multiply it by 10, it gives a cubic number, and when we multiply it by 4, it results in the square having the same side as the cube, or inversely: for the approach in both (cases) is the same.[41]

By the analogy of this with the previous (problem), we (first) take $x$ as the required number.[42] Then the cube is $10x$, and the square having the same side is $4x$. Now, the multiplication of the side of the cube by itself is $4x$, and the whole cube is $10x$; so, since the multiplication of the $4x$ by its side results in $10x$, we divide the $10x$ by the $4x$, thus obtaining $2\frac{1}{2}$ as the side of the cube. The square from it is $6\frac{1}{4}$, so that $4x$ equals $6\frac{1}{4}$. Because of the (occurrence of the) fraction, namely, the fourth, we multiply the whole by 4 and obtain $16x$ equal to 25. So $x$ is 25 parts of 16 parts.[43]

Following the method of the second approach,[44] we assume that the cube which results from the multiplication of the required number by 10 has an arbitrary multiple of $x$ as its side, say $x$, so that the cube is $x^3$. Thus the required number is $\frac{1}{10}x^3$, and the result of the multiplication of it by 4 is
17 $\frac{4}{10}x^3$. Hence $\frac{4}{10}x^3$ is equal to the square of the side of the cube, that is, (to the square of) $x$, or $x^2$; because of the (occurrence of the) fraction, namely the tenths, we multiply the whole of what we have by 10, and obtain $4x^3$ equal to

[40] Properly "the results are a cubic number and a square number", if the correspondence is to be kept (as in the previous final statement).

[41] The two cases were treated separately in IV,14.

[42] See the first part of problem 14.

[43] The verification of the solution is made in the alternative resolution.

[44] Cf. problem 14, penultimate part (we shall put $x$ as the side of the cube, not as the required number).

$10x^2$. The division of the whole by $x^2$ gives $4x$ equal to 10; hence $x$ equals $2\frac{1}{2}$, and the side of the cube is also $2\frac{1}{2}$. So four times the required number, which equals the square of the side of the cube, is $6\frac{1}{4}$. Therefore the required number is 25 parts of 16 parts. It appears that this last number, when multiplied by 4, results in 100 parts of 16 parts, which is a square, and, when multiplied by 10, results in 250 parts of 16, or $15 + \frac{1}{2} + \frac{1}{8}$, which is a cube with side $2\frac{1}{2}$, while the square of this side is $6\frac{1}{4}$. Similarly, the 25 parts of 16, when multiplied by 4, give 100 parts of 16, or $6\frac{1}{4}$, which is a square with $2\frac{1}{2}$ as its side.[45]

Therefore we have found a number such that when we multiply it by two given numbers the results are a cubic number and the square having the same side as that cube.

Suppose that we (now) wish to find two numbers in a given ratio, one being a cubic number and the other, a square, and the ratio taken by us being 3:1. We shall first choose two numbers such that the first is the triple of the second; next, we shall seek by the same method as before a number which when multiplied by each of the two chosen numbers gives a square number and a cubic number. Thus we shall have found two numbers in the ratio 3:1, one
18 being a cube and the other, a square; for the multiplication of any number by two numbers gives products which are in the ratio of the two original numbers.[46]

**16.** We wish to find two numbers such that when we multiply them by a given number, one of the resulting products is a cubic number and the other, the side of that cube.

We put as the (given) number 10. We wish to find two numbers such that when we multiply them by 10, the result of the multiplication of 10 by the one is a cubic number and the result of the multiplication of 10 by the other one is the side of that cube. Let us assume the first number to be an arbitrary number of $x$'s, say $x$. The multiplication of it by 10 gives $10x$, which is the side of the cube; thus, the cube resulting from the multiplication of the second number by 10 is $1000x^3$. (Next,) we assume the second number to be an arbitrary number of $x^2$'s, say $300x^2$; we multiply it by 10 and obtain $3000x^2$. Hence $1000x^3$ equals $3000x^2$. The division of the whole by $x^2$ results in $1000x$ equal to 3000; therefore, $x$ is 3. Since we set $x$ as the first number, it is 3; and, since we set as the second number $300x^2$, where $x^2$ is 9, the second number is 2700. If then we multiply the second number by 10, it becomes 27,000, and if we multiply the first number by 10, the result is 30; and 30 is the side of the cube which is 27,000.

Therefore, we have found two numbers fulfilling the condition imposed upon us, and these are 3 and 2700. This is what we intended to find.

[45] Note that this statement is, in part, repetitive. See p. 32, no. 18. The text has "therefore" instead of "similarly" (cf. app., n. 161).

[46] Euclid, *Elem.* VII,17.

**17.** We wish to find two square numbers having their sides in a given ratio,
19 and such that when each of them is multiplied by a given number, one results in a cube and the other, in the side of that cube.

It is necessary that the number belonging to the given ratio[47] comprise together with the given number a square number. These (problems), from their feasibility, are those called the "constructible" ones.[48]

Let the given ratio be the ratio 20:1 and the given number be 5. We wish to find two square numbers, the side of the one being in the ratio of 20:1 to the side of the other, and such that, when the larger square is multiplied by 5, the result is a cubic number, and when the smaller square is multiplied by 5, the result is the side of that cube. We put $x$ as the side of the smaller square, so that the smaller square is $x^2$; the side of the greater square, then, is $20x$, and the greater square is $400x^2$. The multiplication of the $400x^2$ by 5 gives $2000x^2$, and the multiplication of $x^2$ by 5 gives $5x^2$. Now, the condition in the problem is that the $2000x^2$ is a cube having the $5x^2$ as its side; so we multiply the $5x^2$ by $5x^2$, then by $5x^2$ (again), thus obtaining $125x^6$. Therefore $125x^6$ is equal to $2000x^2$. The division of the whole by a unit of the side of lower degree, namely (by) $x^2$, results in $125x^4$ equal to 2000; hence $x^4$ equals 16. But $x^4$ is a square of square side, and 16 is similarly a square number of square side. The two being equal, the sides of their sides are also equal. As the side of the side of $x^4$ is $x$, and the side of the side of 16 is 2, $x$ equals 2.
20 Since we made the smaller square from the side $x$, the smaller square is 4. And, since we made the larger square from (the side) $20x$, its side is 40 and the larger square itself, 1600. When we multiply 1600 by the given number, or 5, the product is 8000, which is a cube with side 20; and the said 20 results from the multiplication of the smaller square—which was found to be 4—by the given number, or 5.

Therefore, we have found two numbers fulfilling the condition imposed upon us, and these are 4 and 1600. This is what we intended to find.

**18.** We wish to find two cubic numbers having their sides in a given ratio, and such that when each of them is multiplied by a given number, one results in a square and the other, in the side of that square.

It is necessary that the given number be a cube.

[47] That is, the number which the later Greeks called the πηλικότης of the ratio: the word is used by Theon of Alexandria and by Eutocius (Rome, *Comm.* (1–2), p. 533; *Archim. op. cum comment.* (ed. Heiberg), III, p. 120, 18 *seqq.*). On its occurrence in the *Elements*, see Heath, *Euclid*, II, pp. 116–17.

[48] I have changed the text, which one would normally understand as *qadruhū min al-ṯānī allaḏī yudʿā al-muhayyaʾaʰ*, into *fa-hāḏihī* (sc. *al-masāʾil*) *min al-taʾattī allatī tudʿā al-muhayyaʾaʰ*, bringing it thus into accordance with lines 495–96 and my interpretation of the word πλασματικόν (cf. p. 192).

Let the given ratio be 3:1 and the given number be 8. We wish to find two cubic numbers, the side of the one being in the ratio 3:1 to the side of the other, and such that the product of the larger and 8 is a square number and the product of the smaller and 8 is the side of that square. We put $x$ as the side of the smaller cube, which is then $x^3$; (so) the side of the greater cube is $3x$ and the greater cube, $27x^3$. When we multiply $27x^3$ by 8, it becomes $216x^3$, and when we multiply $x^3$ by 8, it becomes $8x^3$. Since $216x^3$ is a square having
21 $8x^3$ as its side, that is, (a square which amounts to) $64x^6$, $216x^3$ is equal to $64x^6$. The division of both by $x^3$, which belongs to the side of lesser degree, gives 216 equal to $64x^3$; hence $x^3$ is $3\frac{3}{8}$. Since $x^3$ is a cube with side $x$, and since $3\frac{3}{8}$ is a cube with side $1\frac{1}{2}$, $x$ is equal to $1\frac{1}{2}$. Therefore the smaller cube is $3\frac{3}{8}$, and the larger cube, with $4\frac{1}{2}$ as its side, is $91\frac{1}{8}$. The result of the multiplication of this larger cube by 8 is 729, which is a square with side 27, this last number being itself the result of the multiplication of the smaller cube—which was found to be $3\frac{3}{8}$—by the given number, or 8.

Therefore, we have found two numbers fulfilling the condition imposed upon us. This is what we intended to find.

**19.** We wish to find a number such that when we multiply it by two given numbers, the result of the multiplication by the one is a cube, and of the multiplication by the other, the side of that cube.

It is necessary that the two given numbers comprise a square number. This (problem) belongs again to the (category of) constructible problems.

Let one of the two given numbers be 5 and the other, 20. We put $x$ as the required number. Multiplying it by 5 gives $5x$ and again multiplying it by 20
22 gives $20x$. Now, $20x$ is a cube with side $5x$, and the multiplication of the side of any cube by the square of that side gives the (said) cube; the cube being $20x$, dividing $20x$ by its side, that is, $5x$, gives the square of the side of $20x$. But $20x$ when divided by $5x$ gives 4; hence 4 is a square with $5x$ as its side. So the side of 4, or 2, is equal to $5x$; thus $x$ is $\frac{2}{5}$. Then, the multiplication of the $\frac{2}{5}$ by 20 gives 8, which is a cube with side 2, while 2 itself results from the multiplication of the required number—which was found to be $\frac{2}{5}$—by the second given number, namely 5.

Therefore, we have found a number such that, when we multiply it by the two given numbers, namely 5 and 20, the result of its multiplication by 20 is a cube and of its multiplication by 5, the side of that cube; and the said number is $\frac{2}{5}$. This is what we intended to find.

**20.** We wish to find a cubic number such that when we multiply it by two given numbers, the result of the multiplication by the one is a square, and of the multiplication by the other, the side of that square.

It is necessary that the square of one of the two given numbers measure the other one by a cubic number.

Let one of the two given numbers be 5 and the other, 200. We wish to find a cubic number such that, when we multiply it by 200 the result is a square and when we multiply it by 5 the result is the side of that square. We put $x$ as the side of the required cube, so that the cube is $x^3$. The multiplication
23 of $x^3$ by 200 and by 5 gives $200x^3$ and $5x^3$, respectively. Now, $200x^3$ is a square with side $5x^3$, and the division of any square by its side gives a result equal to its side; thus, since $200x^3$ when divided by $5x^3$ gives 40, $5x^3$ equals 40. Hence $x^3$ equals 8. As $x^3$ is a cube with side $x$, and 8 is a cube with side 2, so $x$, which we took as the side of the required cube, is 2, and the cube is 8. The multiplication of 8 by 200 results in 1600, and the multiplication of the same by 5 results in 40, which is the side of the square 1600.

Therefore, we have found a cubic number such that when we multiply it by the two given numbers, namely 200 and 5, the result of its multiplication by 200 is a square and of its multiplication by 5, the side of that square; and the said cubic number is 8. This is what we intended to find.

**21.** We wish to find a square number such that when we multiply it by two given numbers, the result of the multiplication by the one is a cube, and of the multiplication by the other, the side of that cube.

It is necessary that the two given numbers comprise a square number having a square side.

Let one of the two given numbers be 2 and the other, $40\frac{1}{2}$; it appears that the plane number comprised by these two numbers, 81, is a square of square side. We wish to find a square number such that, when we multiply it by $40\frac{1}{2}$
24 and by 2, the result of the multiplication by the $40\frac{1}{2}$ is a cube and of the multiplication by the 2, the side of that cube. We assume the square to be $x^2$, and we multiply it by the two given numbers; the two products are then $40\frac{1}{2}x^2$ and $2x^2$. Now, $40\frac{1}{2}x^2$ is a cube having the $2x^2$ as its side, and any cube when divided by its side gives the square of that side; thus, since the quotient of $40\frac{1}{2}x^2$ divided by $2x^2$ is $20\frac{1}{4}$, the square of $2x^2$ is equal to $20\frac{1}{4}$. As the side of $20\frac{1}{4}$ is $4\frac{1}{2}$, $x^2$ is equal to $2\frac{1}{4}$, which is a square with $1\frac{1}{2}$ as its side. When this square, that is, $2\frac{1}{4}$, is multiplied by the first of the two given numbers, that is, (by) $40\frac{1}{2}$, the result is $91\frac{1}{8}$, which is a cube with side $4\frac{1}{2}$; and $4\frac{1}{2}$ arises from the multiplication of the required square number, which was found to be $2\frac{1}{4}$, by the second given number, or 2.

Therefore, we have found a square number fulfilling the condition imposed upon us, and this is $2\frac{1}{4}$. This is what we intended to find.

It was necessary for the two given numbers to fulfil the condition which we indicated; for I say (the following). In setting $x^2$ as the required square, and then multiplying it by each of the two given numbers, the two products are both $x^2$'s, and one of them is a cube with the $x^2$'s forming the other product as its side. Now, if the one which is the cube is divided by the one which is the side, the resulting quotient will be a number, equal to the square of the $x^2$'s

25 which form the side. Consequently, the number resulting from the division has to be a square in order that its side be a number, equal to the $x^2$'s forming the side. Accordingly, it is necessary that the result of the division of one of the two given numbers by the other be a square; now, if this (condition) is met by a pair of numbers, their product is a square also. But the number which is the side of the square number resulting from the division of one of the (given) numbers by the other is equal to the $x^2$'s having their coefficient equal to that one of the two given numbers which is the divisor; thus, we require, in addition, that when the said number is divided by the (coefficient of the) $x^2$'s which are equal to it, the result be a square, in order that $x^2$ be equal to a square number. Therefore it is necessary that, one of the two given numbers being divided by the other and the quotient being a square, the side of this square being divided by the divisor also give a square; in other words, the product of the said side and the given number which is the divisor must be a square. Now, if two numbers are such that when one of them is divided by the other the result is a square, the side of which results, after division by the divisor, in a square, then the product of these numbers will be a square of square side. This is what had to be shown.

**22.** We wish to find a cubic number such that when we multiply it by two given numbers, the results are a cube and the side of that cube.

It is necessary to find first the characteristic of the two given numbers. We then say (the following). Having set $x^3$ for the required cube and multi-
26 plied it by the two given numbers, each of the two products is $x^3$'s, and one of these two products is a cube having the other product as its side. Now, if those $x^3$'s of the two products which form the cube are divided by those which form the side, the resulting quotient is a number, equal to the square of the $x^3$'s forming the side. Consequently, the number resulting from the division must be a square in order that its side (may) be (set) equal to the $x^3$'s forming the side. Thus we shall suppose the two given numbers to be such that the division of the one by the other produces a square. Again, the number which is the side of the square number resulting from the division is equal to the $x^3$'s which are the side and which have their coefficient equal to that one of the two given numbers which is the divisor; so it is necessary that the division of the said number by the (coefficient of the) $x^3$'s equal to it produce a cube, in order that $x^3$ be equal to a cubic number. Hence the characteristic of these two numbers is now in its complete form, which is: the division of the one by the other results in a square and the division of the side of this square by the divisor results in a cube.

We must (now) determine these two numbers. We assume the first to be 2 and we wish to find the second. Since the result of the division of one of these two numbers by the other is a square, the side of which, when divided by the divisor, gives a cube, we have to seek a number which, when divided by 2, gives a cube; such is $6 + \frac{1}{2} + \frac{1}{4}$. Now, $6 + \frac{1}{2} + \frac{1}{4}$ is the side of the square

arising from the division of one of the two (given) numbers by the other; the
27 square generated by the $6 + \frac{1}{2} + \frac{1}{4}$ being $45 + \frac{1}{2} + \frac{1}{2} \cdot \frac{1}{8}$, and the number from which it (i.e., $45 + \frac{1}{2} + \frac{1}{2} \cdot \frac{1}{8}$) arises by division of it (i.e., of the "number" in question) by 2 being $91\frac{1}{8}$,[49] the second number we were looking for is $91\frac{1}{8}$.

By a similar approach one can come to know the characteristics indicated for the given numbers in the preceding problems and find these numbers.[50]

So, one of the two given numbers is 2 and the other, $91\frac{1}{8}$, and we wish to find a cubic number which when multiplied by $91\frac{1}{8}$ gives a cube and which when multiplied by 2 gives the side of that cube. We set $x^3$ as the cube and proceed as we did in the previous problems. Then we shall find that the required cube is $3\frac{3}{8}$. The multiplication of it by $91\frac{1}{8}$ gives a cube, namely 307 and 35 parts of 64 parts; and, the same number when multiplied by 2 gives $6 + \frac{1}{2} + \frac{1}{4}$, which is the side of the cube 307 and 35 parts of 64 parts.

Therefore, we have found a cubic number fulfilling the condition imposed upon us. This is what we intended to find.

**23.** We wish to find two square numbers such that their squares when added give a cube.

We put $x^2$ as the first square and an arbitrary number of $x$'s, say $2x$, as the side of the second, so that the second is $4x^2$. The squares of these two squares are $x^4$ for the smaller and $16x^4$ for the larger; their sum is $17x^4$, which is
28 equal to a cubic number. We form the cube from the side $3x$, so that the cube is $27x^3$. Then $17x^4$ is equal to $27x^3$, and thus $17x$ is equal to 27; hence $x$ is 27 parts of 17 parts of 1. Since we assumed the side of the smaller square to be $x$, the side is 27 parts of 17 parts, and the smaller square is 729 parts of 289 parts of 1; and, since we assumed the side of the larger square to be $2x$, this side is 54 parts of 17 parts, and the larger square is 2916 parts of 289 parts of 1. Accordingly, the square of the smaller square is 531,441 parts of 83,521 parts of 1, and the square of the larger square is 8,503,056 parts of 83,521 parts of 1. The sum of these two squares is 9,034,497 parts of 83,521 parts of 1, or

---

[49] The lack of clarity of the original text gave rise to a reader's remark, later incorporated into the text.

The *original* version can be interpreted either as "and the number which arises by division of it by 2 being $91\frac{1}{8}$" or as "and the number from which it arises by division of it by 2 being $91\frac{1}{8}$". The appropriate translation is the second one.

Now, using this translation and keeping the *manuscript's reading*, one interprets it as: "and the number from which it arises by division of it by 2 being *the number which we have mentioned* (that is) $91\frac{1}{8}$". Although we have already seen $91\frac{1}{8}$ in problems 18 and 21, the words in italics are no doubt an interpolation: some reader, confused by this badly formulated sentence, must have felt the need to specify the subject of "it arises" by referring "it" to the (just) mentioned number, or $45 + \frac{1}{2} + \frac{1}{2} \cdot \frac{1}{8}$. As usual (cf. §5 of the introduction), the gloss was copied undiscerningly, (presumably) by the copyist of the Mashhad manuscript.

[50] See problems 17 and 19–22, and p. 192. Problem 18 has only one given number.

531,441 parts of 4913 parts of 1, which is a cube having 81 parts of 17 parts of 1 as its side.

Therefore, we have found two square numbers fulfilling the condition
29 imposed upon us, and these are 729 parts and 2916 parts of 289 parts. This is what we intended to find.

**24.** We wish to find two square numbers such that the difference of their squares is a cubic number.

We put $x$ as the side of the smaller square and $2x$ as the side of the larger square, so that the smaller (square) is $x^2$ and the larger, $4x^2$; the difference of their squares is $15x^4$, which is (equal to) a cube. Let us take $5x$ as its side. Now, the result of the division of any cube by its side is equal to the square of the said side, and the $15x^4$, which is a cube with side $5x$, gives, when divided by its side, or $5x$, $3x^3$. Thus $3x^3$ is a square with side $5x$. The square arising from $5x$ being $25x^2$, $3x^3$ equals $25x^2$. The division of the two by $x^2$, which belongs to the side of lower degree, results in $3x$ equal to 25. Thus $x$ is equal to $8\frac{1}{3}$. Since we took $x$ as the side of the smaller square and $2x$ as the side of the larger square, the side of the smaller square is $8\frac{1}{3}$ and that of the larger square, $16\frac{2}{3}$; (so) the smaller square is $69\frac{4}{9}$ and the larger square, $277\frac{7}{9}$. The square
30 of the smaller square is 4822 and 43 parts of 81 parts of 1, and the square of the larger square is 77,160 and 40 parts of 81 parts of 1. The difference of these two squares is 72,337 and 78 parts of 81 parts of 1, or (72,337 and) 26 parts of 27 parts of 1, which is a cube having $41\frac{2}{3}$ as its side.

Therefore, we have found two square numbers fulfilling the condition imposed upon us, and these are $69\frac{4}{9}$ and $277\frac{7}{9}$. This is what we intended to find.

**25.** We wish to find two numbers, one square and the other cubic, such that the sum of their squares is a square.

We put $x$ as the side of the cube, so that the cube is $x^3$, and any number of $x$'s, say $2x$, as the side of the square, which is then $4x^2$. The square of the cube is $x^6$ and the square of the square, $16x^4$; their sum is $x^6 + 16x^4$, and this is equal to a square number. It is then necessary to determine the number which is the side of this square. We say then (the following). If we put as the said side $x^2$'s, the square equal to $x^6 + 16x^4$ is $x^4$'s; after the subtraction of the $16x^4$, which is common, from both sides, there remain $x^4$'s equal to $x^6$, and the division of the two by $x^4$, which constitutes the lower in degree of the two
31 sides, gives $x^2$ equal to a number. This number, being equal to $x^2$, must be a square. But the said number is the excess of the (coefficient of the) $x^4$'s in a square number over 16. Thus it is necessary that the coefficient of the $x^4$'s in a square number exceed 16 by a square number. Consequently, we are led to search for two square numbers having 16 as their difference.[51] We then find

[51] *Arithmetica* II,10; the solution here is, however, trivial.

25 for the larger square and 9 for the smaller square. So we put $25x^4$, the side of which is $5x^2$, as the square equal to $x^6 + 16x^4$. Removing the $16x^4$, which is common, from both sides, we obtain $x^6$ equal to $9x^4$. Hence $x^2$ equals 9. As $x^2$ is a square with side $x$ and 9 a square with side 3, $x$ is 3. Since we assumed the side of the cube to be $x$, the side is 3 and the cube, 27. And, since we assumed the side of the square to be $2x$, the side is 6 and the square, 36. The square of 27 is 729 and the square of 36, 1296; the sum of these is 2025, which is a square with 45 as its side.

Therefore, we have found two numbers, one cubic and the other square, such that the sum of their squares is a square; and these are 27 and 36. This is what we intended to find.

**26.** We wish to find two numbers, one cubic and the other square, such that
32 the difference of their squares is a square number.

We set $x^3$ as the cube and $4x^2$ as the square; then the square of the cube is $x^6$ and the square of the square, $16x^4$. We want their difference to be a square number.

Let us first require that the square of the cube exceed the square of the square by a square number.

We say then: $x^6 - 16x^4$ is equal to a square number. We proceed similarly to what has been shown in the problem preceding the present one in the search for the (number of) $x^2$'s which must be set as the side of the said square.[52] We find $3x^2$. The square arising from that is $9x^4$, hence $x^6 - 16x^4$ is equal to $9x^4$. We add $16x^4$ in common to the two sides and so obtain $x^6$ equal to $25x^4$; hence $x^2$ is equal to 25. As $x^2$ is a square with side $x$ and 25 is a square with side 5, $x$ is equal to 5. Since we assumed the side of the cube to be $x$, the side is 5 and the cube, 125; and, since we assumed the side of the square to be $2x$, the side is 10 and the square, 100. The square of 125 is 15,625 and the square of 100, 10,000. Their difference is 5625, which is a square having 75 as its side.

Therefore, we have found two numbers, one cubic and the other square, such that the excess of the square of the cube over the square of the square
33 is a square number; and these are 100 and 125.

Let us next require that the excess of the square of the square over the square of the cube be a square.

We set $x^3$ as the cube and $5x$ as the side of the square, so that we have $625x^4 - x^6$ equal to a square number. Let us seek the side of this square. We say (the following). If we put $x^2$'s for the said side, its square, which is equal to $625x^4 - x^6$, is $x^4$'s. After the addition of $x^6$ in common to both sides, $625x^4$ is equal to $x^6$ plus $x^4$'s; there remains, after the reduction and the division, $x^2$ equal to a number. Thus it is necessary that the said number be

[52] Using, if need be, *Arithmetica* II, 10.

a square. But this number is the excess of the product of the multiplication of 25 by itself over the coefficient of the $x^4$'s forming the square of the side for which we are looking. Hence 625, which is a square number, must be divided into two square numbers in the manner described by us in the second Book.[53] Let the two parts be 400 and 225. We put, as the square equal to $625x^4 - x^6$, $x^4$'s, (the coefficient of which is) equal to one of the two said parts. Let us put it $225x^4$. After the restoration, the reduction, and the division, we obtain $x^2$ equal to 400. Hence $x$, which we set as the side of the cube, is 20, and the cube itself is 8000; the side of the square, set $5x$ by us, is 100, and the square itself is 10,000. The square of the cube, that is, (the square of) 8000, is 64,000,000,
34 and the square of the square, that is, (the square of) 10,000, is 100,000,000;
their difference is 36,000,000, which is a square with 6000 as its side.

Therefore, we have found two numbers, one cubic and the other square, such that the square of the square exceeds the square of the cube by a square number; and these are 10,000 and 8000. This is what we intended to find.

**27.** We wish to find two numbers, one cubic and the other square, such that the square of the cube together with a given multiple of the square number is a square number.

We set $x^3$ as the cube; the multiplication of it by itself gives $x^6$. We set an arbitrary number of $x^2$'s as the side of the square, say $2x^2$, so that the square itself is $4x^4$. Let the given multiplicative factor be 5.[54] The multiplication of $4x^4$ by 5 gives $20x^4$, the addition of which to $x^6$ results in $x^6 + 20x^4$; and this is (equal to) a square. Let us seek two squares having 20 as their difference[55]; such are 36 and 16. So we put the square formed by $x^6 + 20x^4$ equal to $36x^4$. The subtraction of the $20x^4$, which is common, from both sides leaves $16x^4$ equal to $x^6$. Let us divide the whole by $x^4$; then 16 is equal to $x^2$. As 16 is a square with side 4, 4 is equal to the side of $x^2$, that is, $x$. Since we assumed the cube to be $x^3$, its side is 4 and the cube is 64; and, since we assumed the side of the square to be $2x^2$, the side is 32 and the square, 1024. The latter,[56] taken five times, is 5120; the addition of it to the cubic number gives 5184, which is a square number having 72 as its side.

Therefore, we have found two numbers, one square and the other cubic,
35 such that the square of the cube, together with five times the square number,
is a square number; and these are 64 and 1024. This is what we intended to find.

---

[53] Problem II,8.

[54] Observe that the ἔκθεσις (cf. p. 49) is not made, as is normally the case, at the very beginning of the problem; see also problems IV,10–13 and 43.

[55] *Arithmetica* II,10. Note the conciseness of the text: we are not told, as we usually are, to take some multiple of $x^2$ as the side of the square. Such conciseness (a remnant of the original text?) is also seen in some other problems similarly involving an intermediate problem (see IV,36 and 38; V,1–3; VI,1–4 (interpolated), and VI,12–13; cf. p. 70).

[56] What follows does not correspond to the problem as formulated; cf. p. 63, no. 3.

**28.** We wish to find two numbers, one cubic and the other square, such that the square of the square, together with a given multiple of the cubic number, is a square number.

Let the given multiplicative factor be 10. We put $x^3$ as the cube; the multiplication of it by 10 gives $10x^3$. We put $2x$ as the side of the square, so that the square is $4x^2$ and its square, $16x^4$. The addition of the latter to the $10x^3$ gives $16x^4 + 10x^3$, and this equals a square number. We assume the side of this square to be $6x^2$. Now, the result of the division of any square by its side is equal to the said side. So, dividing $16x^4 + 10x^3$ by $6x^2$ gives $2\frac{2}{3}x^2 + 1\frac{2}{3}x$, which is equal to $6x^2$. The subtraction of the $2\frac{2}{3}x^2$, which is common, from both sides leaves $3\frac{1}{3}x^2$ equal to $1\frac{2}{3}x$. Thus $3\frac{1}{3}x$ is equal to $1\frac{2}{3}$, hence $x$ is equal to $\frac{1}{2}$. Then, since we set $x$ as the side of the cube, the side is $\frac{1}{2}$ and the cube, $\frac{1}{8}$; and ten times $\frac{1}{8}$ is $1\frac{1}{4}$. And, since we put $2x$ as the side of the square, the side is 1 and the square, 1 also. If we add that[57] to $1\frac{1}{4}$, that is to say, (to) ten times the cube, we obtain a square number, namely $2\frac{1}{4}$, with side $1\frac{1}{2}$.

Therefore, we have found two numbers, one cubic and the other square, such that the square of the square together with ten times the cubic number, is a square number; and these are 1 and $\frac{1}{8}$. This is what we intended to find.

**29.** We wish to find two numbers, one cubic and the other square, such that the
36 sum of the cube of the cube and of the square of the square is a square number.

We set $x^3$ as the cube, so that its cube is $x^6$ (multiplied) by $x^3$; and this is called $x^9$. We set as the side of the square an arbitrary number of $x^2$'s, say $2x^2$, so that the square is $4x^4$; its square is $16x^4$ (multiplied) by $x^4$, and this is equal to $16x^6$ multiplied by $x^2$, one of which is called $x^8$. Thus $x^9$, together with $16x^8$, is equal to a square number. Let us put for its side a number of $x^4$'s, again arbitrary, say $6x^4$. This, when multiplied by itself, gives $36x^4$ (multiplied) by $x^4$, or $36x^8$. Then $x^9$, together with $16x^8$, equals $36x^8$. Let us remove the $16x^8$, which is common, from both sides; there remains $x^9$ equal to $20x^8$. We divide each of them by a unit of the side of lower degree, namely (by) $x^8$; the $20x^8$ gives, when divided by $x^8$, 20, while $x^9$—which indeed arises from the multiplication of $x^6$ by $x^3$,[58] and is (thus) also the product of the multiplication of $x^8$ by $x$—results, when divided by $x^8$, in $x$. Hence $x$ is equal to 20. Since we put $x$ as the side of the cube, the side of the cube is 20 and the cube, 8000; and, since we put $2x^2$ as the side of the square, and (since) $x^2$ is 400, the side of the square is 800 and the square, 640,000. The cube of the
37 cube is 512,000,000,000, and the square of the square, 409,600,000,000; their sum is 921,600,000,000, which is a square number with 960,000 as its side.

Therefore, we have found two numbers, one cubic and the other square, such that the sum of the cube of the cube and of the square of the square is a

---

[57] Properly: "the square of that".

[58] See the definition of $x^9$ at the beginning of the problem.

square number; and these are 8000 and 640,000. This is what we intended to find.

**30.** We wish to find two numbers, one cubic and the other square, such that the excess of the cube of the cube over the square of the square is a square number.

We put $x^3$ as the cube, so that its cube is $x^6$ (multiplied) by $x^3$, that is, the so-called $x^9$. We put $2x^2$ as the side of the square, so that the square is $4x^4$; its square is $16x^4$ (multiplied) by $x^4$, or $16x^8$. Thus $x^9 - 16x^8$ is equal to a square number. Let us put $2x^4$ as the side of that square, so that the square is $4x^4$ (multiplied) by $x^4$, that is, $4x^8$. Thus $x^9 - 16x^8$ equals $4x^8$. Let us add the $16x^8$ in common to the two sides, then $x^9$ equals $20x^8$; let us divide the whole by $x^8$, which is a unit of the side of lower degree, so we obtain, after the division, $x$ equal to 20. Then, since we put $x$ as the side of the cube, the side is 20 and the cube, 8000; again, since we put $2x^2$ as the side of the square, and (since) $x^2$ is 400, the side of the square is 800, and the square itself is 640,000. So, the cube of the cube, as (previously) found,[59] is
38 512,000,000,000, and the square of the square is 409,600,000,000. Their difference, or, rather, the excess of the cube of the cube over the square of the square, is 102,400,000,000, which is a square number with side 320,000.

And it has already been found in the preceding problem that the sum of these two numbers is a square number as well.[60]

Therefore, we have found two numbers, one cubic and the other square, such that the excess of the cube of the cube over the square of the square is a square number; and these are 8000 and 640,000. This is what we intended to find.

With that, it appears that we have also found two numbers, one cubic and the other square, such that the result of the addition of the square of the square to the cube of the cube is a square number and the result of the subtraction of the square of the square from the cube of the cube is a square number; and these are, again, the two said numbers.

**31.** We wish to find two numbers, one square and the other cubic, such that the excess of the square of the square over the cube of the cube is a square number.

We set $x^3$ as the cube, so that its cube is $x^6$ (multiplied) by $x^3$, which is the so-called $x^9$. We put the side of the square $2x^2$, so that the square is $4x^4$ and its square, $16x^4$ (multiplied) by $x^4$, that is to say, (16 times) the so-called $x^8$. Thus the $16x^8$, which is the square of the square number, (must) exceed $x^9$ by a square number. Let us put $2x^4$ as the side of that square; as the result

[59] See problem 29.

[60] The placement of the coming final statement is inappropriate.

of the division of any square by its side equals the said side, the result of the
39 division of $16x^8 - x^9$ by $2x^4$ equals $2x^4$. But as $16x^8$ results from the multiplication of $16x^4$ by $x^4$, the division of it by $2x^4$ gives $8x^4$; and, as $x^9$ results from the multiplication of $x^6$ by $x^3$, while $x^6$ is the product of $x^4$ and $x^2$, $x^9$ is the product of $x^4$ and $x^5$, and, thus, the result of the division of $x^9$ by $2x^4$ is $\frac{1}{2}x^5$.[61] Hence we obtain, from the division (of $16x^8 - x^9$ by $2x^4$), $8x^4 - \frac{1}{2}x^5$, and this is equal to $2x^4$. We make $\frac{1}{2}x^5$ common by adding it to both sides, so that we have $8x^4$ equal to $2x^4 + \frac{1}{2}x^5$. Let us remove the $2x^4$, which is common, from both sides, so $\frac{1}{2}x^5$ equals $6x^4$; after the division, we obtain $\frac{1}{2}x$ equal to 6, hence $x$ is equal to 12. Since we put $x$ as the side of the cube, the side is 12 and the cube, 1728; and, since we put $2x^2$ as the side of the square, and (since) $x^2$ is 144—for $x$ is 12—, the side of the square is 288 and the square, 82,944. The cube of the cube is 5,159,780,352 and the square of the square, 6,879,707,136;
40 the excess of the latter number over the cube of the cube is 1,719,926,784, which is a square number with 41,472 as its side.

Therefore, we have found two numbers fulfilling the condition required by us, and these are 1728 and 82,944. This is what we intended to find.

**32.** We wish to find two numbers, one cubic and the other square, such that the cube of the cube together with a given multiple of the product of the multiplication of the square by the cube is a square number.

Let the given multiplier be 5. We put $x^3$ as the cube, so that its cube is $x^9$; we put $2x^3$ as the side of the square, so that the square is $4x^6$. Multiplying $4x^6$ by the cubic number—which we put $x^3$—gives $4x^9$; five times that is $20x^9$, which when added to the cube of the cube yields $21x^9$; and this is equal to a square number. Let us put $7x^4$ as its side, so that the square is $49x^8$;[62] and this equals $21x^9$. Let us divide each of them by $x^8$, so $21x$ is equal to 49; hence $x$ is equal to $2\frac{1}{3}$. Since we set $x$ as the side of the cube, the said side is $2\frac{1}{3}$, and the cube, since its side is $\frac{7}{3}$, is 343 parts of 27 parts; and, since we assumed the side of the square to be $2x^3$, the said side is 686 parts of 27 parts of 1, and the
41 square is 470,596 parts of 729 parts of 1. Then, the cube of the cube is 40,353,607 parts of 19,683 parts of 1. The product of the multiplication of the square number by the cubic number is 161,414,428 parts of 19,683 parts of 1, which, taken five times, yields 807,072,140 parts (of 19,683 parts). The addition of that to the cube of the cube results in 847,425,747 parts of 19,683 parts of 1, or 282,475,249 parts of 6561 parts of 1, which is a square number with 16,807 parts of 81 parts of 1 as its side.

Therefore, we have found two numbers fulfilling the condition which we stipulated, and these are 343 parts of 27 parts of 1 and 470,596 parts of 729 parts of 1. This is what we intended to find.

---

[61] Observe that this passage (though certainly not going back to Diophantus) is, in so far as the decomposition of higher powers is concerned, quite in the spirit of the given "definitions"; cf. pp. 176–177.

[62] "$49x^6$ by $x^2$" in the text.

**33.** We wish to find two numbers, one cubic and the other square, such that the cube of the cube exceeds a given multiple of the product of the multiplication of the square number by the cubic number, by a square number.

42 Let the given multiplier be 3. We put $x^3$ as the cube, so that its cube is $x^9$. We put as the side of the square $\frac{1}{2}x^3$, so that the square is $\frac{1}{4}x^6$. The multiplication of the latter by the cubic number—put by us $x^3$—gives $\frac{1}{4}x^9$, which, taken three times, is $\frac{3}{4}x^9$; the subtraction of $\frac{3}{4}x^9$ from the cube of the cube results in $\frac{1}{4}x^9$, (which is) equal to a square number. Let us put as the side of this square an arbitrary number of $x^4$'s, say $x^4$; then $x^8$ is equal to $\frac{1}{4}x^9$ and, after the division, $\frac{1}{4}x$ equals 1; hence the whole $x$ equals 4. Then, since we assumed the side of the cube to be $x$, the side is 4 and the cube, 64; and, since we assumed $\frac{1}{2}x^3$ to be the side of the square, the side of the square is 32, and thus the square is 1024. The cube of the cube is 262,144, and the product of the multiplication of the square number by the cubic number is 65,536; the latter, taken three times, gives 196,608, the subtraction of which from the cube of the cube results in 65,536, which is a square with side 256.

Therefore, we have found two numbers fulfilling the condition which we stipulated, and these are 64 and 1024. This is what we intended to find.

In the manner described above, we solve the remaining types of this kind of problem, for instance finding two numbers, one cubic and the other square, such that the square of the square together with a given multiple of the product of the multiplication of the square number by the cubic number is a square
43 number; (or) also, such that the cube of the square together with a given multiple of the product of the multiplication of the square number by the cubic number is a square number; and the corresponding inverse and similar (problems).

**34.** We wish to find two numbers, one cubic and the other square, such that the cube when increased by the square gives a square number and when decreased by the square also gives a square number.

We put $x^3$ as the cube and $4x^2$ as the square; then, $x^3 + 4x^2$ is equal to a square number and $x^3 - 4x^2$ is likewise equal to a square number.

We treat that (firstly) by the method of the double-equation.[63] We take the difference between the two said squares, namely $8x^2$, and seek two numbers (of $x$'s) such that the multiplication of the one by the other give $8x^2$; such are $2x$ and $4x$. Their difference is $2x$, half of which is $x$. The square of $x$ is $x^2$, and this equals $x^3 - 4x^2$. Adding then the $4x^2$ in common to both sides, we obtain $x^3$ equal to $5x^2$. Again, if we add the $2x$ to the $4x$, we obtain $6x$; half of it is $3x$, the square of which is $9x^2$, and this equals $x^3 + 4x^2$. Removing then the $4x^2$, which is common, from both sides, we obtain $x^3$ equal to $5x^2$. Thus the

[63] Cf. *Arithmetica* II,11.

(resulting) equation for the two equations (of the proposed system)[64] turned out to be the same, ending in each one with $x^3$ equal to $5x^2$. Let us divide all of this by $x^2$; we obtain $x$ equal to 5. Thus the side of the cube is 5 and the cube, 125, and the side of the square is 10 and the square, 100. The addition of the 100 to the cubic number results in 225, which is a square number with side 15; and, the subtraction of the same from the cubic number gives 25,
44 which is a square with side 5.

We (now) also treat this (problem) by the procedure avoiding the double-equation.[65]

We say: $x^3 + 4x^2$ is equal to a square number. If we put $x$'s for its side, the square is $x^2$'s, (which are) equal to $x^3 + 4x^2$. The subtraction of the $4x^2$, which is common, from both sides leaves $x^3$ equal to $x^2$'s, and the division of both by $x^2$ results in $x$ for $x^3$ and a number for the $x^2$'s. Consequently, the number assumed to be $x$ in the problem equals the coefficient of the $x^2$'s left over. Again, (we say): $x^3 - 4x^2$ is equal to a square number. If we also put $x$'s for its side, the square is $x^2$'s. The addition of the $4x^2$ in common to both sides results in $x^3$ equal to $x^2$'s, and, consequently, the number assumed to be $x$ in the problem equals the coefficient of the $x^2$'s added up. Therefore, it is necessary that the coefficient of the $x^2$'s left over in the first equation be equal to the coefficient of the $x^2$'s added up in the second equation. But the (coefficient of the) $x^2$'s left over in the first equation is the remainder of a square number after subtracting 4, while the (coefficient of the) $x^2$'s added up in the second equation is a number formed by the addition of a square number and 4. Thus we shall seek two square numbers such that the larger diminished by 4 and the smaller increased by 4 be equal. So we must look for two square numbers having 8 as their difference.[66] Such are $12\frac{1}{4}$ and $20\frac{1}{4}$. We put for the greater square, which is equal to $x^3 + 4x^2$, $20\frac{1}{4}x^2$, and for the smaller square, which is equal to $x^3 - 4x^2$, $12\frac{1}{4}x^2$. Thus, in both equations,
45 we shall end up with $x^3$ equal to $16\frac{1}{4}x^2$; hence $x$ is equal to $16\frac{1}{4}$. Since we set $x$ as the side of the cube, the side of the cube is $16\frac{1}{4}$ and the cube, 4291 and one part of 64 parts of 1; and, since we set $2x$ as the side of the square, the said side is $32\frac{1}{2}$ and the square, $1056\frac{1}{4}$. The addition of the latter to the cubic number results in 5347 and 17 parts of 64, which is a square number with side $73\frac{1}{8}$, and the subtraction of the same from the cubic number gives 3234 and 49 parts of 64 parts of 1, which is a square with side $56\frac{7}{8}$.

Therefore, we have found two numbers, one cubic and the other square, such that the cubic number when increased by the square number gives a

[64] The text has two distinct words which can be rendered by "equation": one (*ṭaraf*) designates a given expression in $x$ equal to "a square" (hence *ṭarafān* is the system to be solved), the other (*muʿādala*^h^) is the resulting equation in $x$.

N.B. This explanation of the word *ṭaraf* supplants my former explanation, *Eq. ind. dans le Badīʿ*, p. 377.

[65] *Arithmetica* II,11, second part, follows the same principle.

[66] *Arithmetica* II,10.

square number, and when decreased by the square number also gives a square number.[67]

**35.** We wish to find two numbers, one cubic and the other square, such that the square number when increased by the cubic number gives a square number and when decreased by the cubic number leaves a square number.

We put $x^3$ as the cube and $4x^2$ as the square; thus $4x^2 + x^3$ equals a square number and $4x^2 - x^3$ equals a square number. If we then assume the side of the square equal to $4x^2 + x^3$ to be $x$'s, the square is $x^2$'s (which) equal $4x^2 + x^3$; the subtraction of the $4x^2$, which is common, from both sides results in $x^3$ equal to $x^2$'s, and the number taken as $x$ in the problem equals the
46 coefficient of the $x^2$'s left over. Again, if we assume the side of the square equal to $4x^2 - x^3$ to be $x$'s, the square is $x^2$'s (which) equal $4x^2 - x^3$; the addition of $x^3$ in common to both sides gives $x^2$'s plus $x^3$ equal to $4x^2$, and the subtraction of the $x^2$'s which are common from both sides results in $x^3$ equal to $x^2$'s; (so) the number taken as $x$ in the problem is equal to the coefficient of the $x^2$'s left over once again. Hence the (coefficient of the) $x^2$'s left over in the first equation must equal the coefficient of the $x^2$'s left over in the second equation. But the (coefficient of the) $x^2$'s left over in the first equation is a square number minus 4, while the (coefficient of the) $x^2$'s left over in the second equation is 4 minus a square number. So we say: a square minus 4 equals 4 minus another square. The addition of the 4 subtracted from the first square in common to both sides gives a square equal to 8 minus a square, and the addition of the second square in common to both sides results in two squares, equal to 8. But 8 is composed of two equal square numbers, so we have to divide 8 into two other square numbers, in the manner expounded in the second Book.[68] Let the two said parts be 4 parts of 25 (parts) of 1, and 7 and 21 parts of 25 parts of 1. We put, as the square equal to $4x^2 + x^3$, $7x^2$ plus 21 parts of 25 parts of $x^2$, and, as the square equal to $4x^2 - x^3$, 4 parts of 25 parts of $x^2$. Then, in each of the
47 two equations, we shall arrive after the restoration and the reduction[69] at $3x^2$ plus 21 parts of 25 parts of $x^2$ equal to $x^3$; after dividing both by $x^2$, we obtain $3 + \frac{4}{5} + \frac{1}{5} \cdot \frac{1}{5}$ equal to $x$. Since we put as the side of the cube $x$, the said side is 96 parts of 25 parts, and the cube is 884,736 parts of 15,625 parts of 1; and, since we put $2x$ as the side of the square, the said side is 192 parts of 25 parts of 1 and the square, 36,864 parts of 625 parts of 1, which is also 921,600 parts of 15,625 parts of 1. The addition of this to the cubic number gives 1,806,336 parts of 15,625 parts, which is a square number with side 1344 parts of 125 parts of 1; this same number when diminished by

[67] The found values are not restated here, presumably because there is just one final statement for the two sets of solutions.

[68] *Arithmetica* II,9 (any number which is the sum of two squares can be resolved into two squares in any number of ways).

[69] The restoration, of course, only in the second equation.

the cubic number leaves 36,864 parts of 15,625 parts of 1, which is a square number with side 192 parts of 125 parts of 1.

Therefore, we have found two numbers, one cubic and the other square, such that the square when increased by the cube results in a square number and when decreased by the cube leaves a square number; and these are
48 884,736 parts, and 921,600 parts, (both) of 15,625 parts of 1. This is what we intended to find.

**36.** We wish to find a cubic number such that when we increase it by a given multiple of the square having the same side the result is a square number, and when we decrease it by another given multiple of the said square the remainder is a square number.

Let the positive multiplier be 4 and the negative multiplier be 5. We wish to find a cubic number such that when we increase it by four times the square having the same side the result is a square number, and when we decrease it by five times the said square the remainder is a square number. We put $x^3$ as the cube, so that the square having the same side is $x^2$. We seek two square numbers such that the larger diminished by 4 and the smaller increased by 5 be equal; in other words, we look for two square numbers having 9 as their difference.[70] We then find 16 for the one square and 25 for the other. We add to the cube four times the square arising from the multiplication of the side of the cube by itself, so the cube becomes $x^3 + 4x^2$, which is (equal to) a square number. Let us make the square equal to $x^3 + 4x^2$ $x^2$'s, (the coefficient of) which equals the larger of the two squares having 9 as their difference; that is to say, (we make it) $25x^2$. We remove the $4x^2$, which is common, from both sides; so $x^3$ is equal to $21x^2$. Again, we subtract from the cube five times the square having the same side, or $5x^2$; we obtain $x^3 - 5x^2$, and this equals a square number. Let us make the square equal to $x^3 - 5x^2$ $x^2$'s, (the coefficient of) which equals the lesser of the two squares having 9 as their difference; that is, (we make it) $16x^2$. We add the $5x^2$ subtracted
49 from $x^3$ in common to both sides, hence $x^3$ is equal to $21x^2$. Thus we have arrived, in the two equations, at $x^3$ equal to $21x^2$. Let us divide both by $x^2$; hence $x$ is equal to 21. Since we set $x$ as the side of the cube, the said side is 21 and the cube, 9261; the square arising from the multiplication of the side of the cube by itself is 441, and four times that is 1764. The addition of the latter to the cubic number gives 11,025, which is a square with side 105. And, five times the square of the side of the cube is 2205; the subtraction of this from the cubic number leaves 7056, which is a square number with side 84.

Therefore, we have found a cubic number such that when we increase it by four times the square having the same side the result is a square number, and when we decrease it by five times the said square the remainder is a square number; and this is 9261. This is what we intended to find.

---

[70] *Arithmetica* II,10. Note, here again (cf. p. 106, n. 55), the conciseness of the text.

We notice also that, if we had required that the positive multiplier be 5 and the negative multiplier be 4, the side of the cube would be 20 and the cube, 8000. The addition to 8000 of five times the square having the same side, that is, (the addition of) 2000, gives 10,000, which is a square number with side 100, while the subtraction from the same of four times the said square, that is, (the subtraction of) 1600, gives 6400, which is a square number with side 80.

**37.** We wish to find a cubic number such that, when we multiply the square
50 having the same side by two given numbers and add each of the two products to the cubic number, the result is (in both cases) a square number.

Let the two (given) numbers be 5 and 10. We wish to find a cubic number such that, when we multiply the square having the same side by 5 and by 10 and add the two products to the cubic number, the result is (in both cases) a square number. We put $x^3$ as the cube, and multiply the square of its side, or $x^2$, by 5 and by 10; we obtain $5x^2$ and $10x^2$, each of which we add to $x^3$. So $x^3 + 5x^2$ is equal to a square number, and $x^3 + 10x^2$ is equal to a square number. If we (now) make the side of the square consisting of $x^3 + 5x^2$ $x$'s, the square is $x^2$'s; next, the subtraction of the $5x^2$, which is common, from both sides gives $x^3$ equal to $x^2$'s, and it appears that the number assumed to be $x$ in this problem is equal to the coefficient of the $x^2$'s left over. Again, if we make the side of the square consisting of $x^3 + 10x^2$ $x$'s, the square is $x^2$'s; the subtraction of the $10x^2$, which is common, from both sides gives $x^3$ equal to $x^2$'s, and thus the number assumed to be $x$ in this analysis is equal to the coefficient of the $x^2$'s left over. Consequently, the (coefficient of the) $x^2$'s left over in the first equation has to be equal to the (coefficient of the) $x^2$'s left over in the second equation. But the (coefficient of the) $x^2$'s left over in the first equation is a square number minus 5, while the (coefficient of the) $x^2$'s left over in the second equation is a square minus 10; thus we have to find two square numbers such that the larger diminished by 10 and the
51 smaller diminished by 5 are equal. We then say: a square minus 5 equals another square minus 10. We add the 10 in common to both sides and obtain a square plus 5 equal to a square. Hence we must seek two squares having 5 as their difference and with the lesser being greater than 5.[71] Let the smaller square be $53\frac{7}{9}$, with side $7\frac{1}{3}$, and the larger be $58\frac{7}{9}$, with side $7\frac{2}{3}$. We set, for the square equal to $x^3 + 5x^2$, $53\frac{7}{9}x^2$, and, for the square equal to $x^3 + 10x^2$, $58\frac{7}{9}x^2$. Then, in each of the two equations, we shall arrive at $x^3$ equal to $48\frac{7}{9}x^2$. Dividing both by $x^2$, we obtain $x$ equal to $48\frac{7}{9}$. Since we put $x$ as the side of the cube, the side is $\frac{439}{9}$, and the cube, $\frac{84{,}604{,}519}{9 \cdot 9 \cdot 9}$, or $\frac{761{,}440{,}671}{9 \cdot 9 \cdot 9 \cdot 9}$; the square of the side of the cube is $\frac{192{,}721}{9 \cdot 9}$, or $\frac{15{,}610{,}401}{9 \cdot 9 \cdot 9 \cdot 9}$.

---

[71] *Arithmetica* II,10. The condition amounts to a restriction on the choice of the parameter.

52 This when multiplied by 5 gives $\frac{78{,}052{,}005}{9\cdot 9\cdot 9\cdot 9}$, which, when added to the cubic number, results in $\frac{839{,}492{,}676}{9\cdot 9\cdot 9\cdot 9}$, which is a square with side $\frac{28{,}974}{9\cdot 9}$. Again, if we multiply the square of the side of the cube by 10, the result is $\frac{156{,}104{,}010}{9\cdot 9\cdot 9\cdot 9}$, the addition of which to the cubic number gives $\frac{917{,}544{,}681}{9\cdot 9\cdot 9\cdot 9}$, which is a square with side $\frac{30{,}291}{9\cdot 9}$.

Therefore, we have found a cubic number fulfilling the condition stipulated by us; and these are the two numbers mentioned by us.[72]

**38.** We now wish to find a cubic number such that, when we multiply the square having the same side by two given numbers and subtract each of the two (products) from the cubic number, the result is (in both cases) a square number.

Let the two (given) numbers be 5 and 10. We wish to find a cubic number such that, when we multiply the square of its side by 5 and by 10 and subtract each of the two products from the cubic number, the result is (in both cases) a square number. Once again,[73] we put $x^3$ as the cube, and we multiply the square of its side, or $x^2$, by 5 and by 10; we obtain $5x^2$ and $10x^2$. Subtracting these two (results) from the cubic number, we have $x^3 - 5x^2$ and $x^3 - 10x^2$, and each is equal to a square number. If (now) one adds $5x^2$ to the square of
53 the $x$'s forming the side of the square equal to $x^3 - 5x^2$, the result is $x^2$'s having as their coefficient the number assumed to be $x$ in the problem.[74] And, if one adds $10x^2$ to the square of the $x$'s forming the side of the square equal to $x^3 - 10x^2$, the result is $x^2$'s having as their coefficient the number assumed to be $x$ in the problem. Therefore, we shall have to take two square numbers such that the larger when increased by 5 and the lesser when increased by 10 are equal. We then say: a larger square plus 5 equals a small(er)[75] square plus 10. We remove the 5, which is common, from both sides, and obtain a small(er) square plus 5 equal to a large(r) square. Thus the difference of the two squares is 5. Let us then seek two square numbers having 5 as their difference, no matter what the two (square) numbers are.[76] Let the smaller be 4 and the larger be 9. We put $9x^2$ as the square equal to $x^3 - 5x^2$, and $4x^2$

[72] *Sic*. Not only the sense but also the unusual wording of this final statement suggest that we have here an interpolation made by some reader or copyist (cf. p. 31, no. 7); did he have in mind the two forms given for $x^3$?

[73] See the previous problems.

[74] Note here again the conciseness of the text.

[75] The positive form is used when no strict comparison is involved. The "larger" (*aʿẓam*) written before may well be a scribal error.

[76] There is no restricting condition, unlike in the preceding problem.

as the square equal to $x^3 - 10x^2$. Then we shall arrive in each of the two equations at $x^3$ equal to $14x^2$; (so) $x$ is equal to 14. Since we assumed the side of the cube to be $x$, the said side is 14 and the cube, 2744. The square arising from the side of the cube is 196; and the multiplication of 196 by 5 gives 980, the subtraction of which from the cubic number results in 1764, which is a square with 42 as its side. Again, the multiplication of the same square by 10
54 gives 1960, the subtraction of which from the cubic number results in 784, which is a square with 28 as its side.

Therefore, we have found a cubic number fulfilling the condition stipulated by us, and this is 2744. This is what we intended to find.

**39.** We wish to find a cubic number such that, when we multiply the square of its side by two given numbers and subtract the cube from each of the two (products), the remainder is in both cases a square number.

Let the two given numbers be 3 and 7. We wish to find a cubic number such that, when we multiply the square of its side by 3 and by 7 and subtract the cube from each of the two products, the remainder is in both cases a square number. Let us put $x^3$ as the cube. We multiply the square of its side, or $x^2$, by 3 and by 7; the subtraction of $x^3$ from each of the two (products) gives $3x^2 - x^3$, (which is) equal to a square, and $7x^2 - x^3$, (which is) equal to a square. We assume the side of the square equal to $3x^2 - x^3$ to be $x$'s. Multiplying these $x$'s by themselves, they become $x^2$'s, (which) equal $3x^2 - x^3$. We add $x^3$ in common to both sides, whence $x^2$'s plus $x^3$ equal to $3x^2$. Then, the subtraction of the $x^2$'s, which are common, from the $3x^2$ leaves $x^3$ equal to $x^2$'s; (so) $x$ is equal to the coefficient of the remaining $x^2$'s. Again, if we assume the side of the square equal to $7x^2 - x^3$ to be $x$'s, and (if) we multiply these $x$'s by themselves, (equate the result to $7x^2 - x^3$), restore and reduce, we then likewise obtain $x^3$ equal to the remainder of the $7x^2$, while $x$ is likewise equal to the remainder of 7. Thus the (coefficient of the) $x^2$'s remaining from the $3x^2$ must equal the (coefficient of the) $x^2$'s remaining from the $7x^2$. But
55 the (coefficient of the $x^2$'s) remaining from the $3x^2$ is 3 minus a square number, while the (coefficient of the $x^2$'s) remaining from the $7x^2$ is 7 minus a square number. Then 3 minus a square number is equal to 7 minus a square number. We add each one of the two squares in common to both sides; so 7 plus a small(er) square is equal to 3 plus a large(r) square. The removing of 3, which is common, gives a large(r) square equal to a small(er) square plus 4. Consequently we must seek two square numbers having 4 as their difference; and let the smaller be less than 3.[77] Such are $2\frac{1}{4}$ and $6\frac{1}{4}$. We set $2\frac{1}{4}x^2$ as the square equal to $3x^2 - x^3$ and $6\frac{1}{4}x^2$ as the square equal to $7x^2 - x^3$. Then, in each of the two equations, we shall arrive at $x^3$ equal to $\frac{3}{4}x^2$. Hence $x$ is $\frac{3}{4}$. (So) $x^3$ (hence the required cube) is $\dfrac{27}{8 \cdot 8}$, and the square of the side of the cube is

[77] *Arithmetica* II,10 (the condition amounts to a restriction on the choice of the parameter).

$\frac{36}{8\cdot 8}$. The multiplication of the latter by 3 gives $\frac{108}{8\cdot 8}$, which, when diminished by the cubic number, results in $\frac{81}{8\cdot 8}$, which is a square with $\frac{9}{8}$ as its side. Again, the multiplication of the square of the side of the cube, or $\frac{36}{8\cdot 8}$, by 7 gives $\frac{252}{8\cdot 8}$, which, when diminished by the cubic number, results in $\frac{225}{8\cdot 8}$, which is a square number with $\frac{15}{8}$ as its side.

Therefore, we have found a cubic number fulfilling the condition stipulated by us, and this is $\frac{27}{8\cdot 8}$. This is what we intended to find.

**40.** We wish to find two numbers, one square and the other cubic, such that the square of the square, when increased by the cube, results in a square
56 number, and when decreased by the cube, results in a square number.

Let us put $2x$ as the side of the square, so that the square is $4x^2$, and its square, $16x^4$. Let us put as the side of the cube any number of $x$'s we please, say $4x$, so that the cube is $64x^3$. Adding this cube to $16x^4$ and subtracting it from $16x^4$, we obtain $16x^4+64x^3$, (which is) equal to a square number, and $16x^4-64x^3$, (which is) equal to a square number. Next, we begin to search for what will make one and the same the equation (resulting from) the two (proposed) equalities, in the way we did before.[78] So we say (the following). If we put $x^2$'s as the side of the square equal to $16x^4+64x^3$, the square is $x^4$'s, (which) equal $16x^4+64x^3$; the subtraction of the $16x^4$, which is common, from both sides results in $64x^3$ equal to $x^4$'s, and the division of all that by $x^3$ gives $x$'s equal to 64; thus the number taken as $x$ in the problem equals the result from the division of the coefficient of the $x^3$'s, of which there are 64, by the coefficient of the remaining $x^4$'s. Again, if we put $x^2$'s as the side of the square equal to $16x^4-64x^3$, the square is $x^4$'s, (which) equal $16x^4-64x^3$; the addition of the $x^3$'s subtracted (from $16x^4$) in common to both sides results in $x^4$'s plus $64x^3$ equal to $16x^4$; if we then subtract the $x^4$'s which are common, there remains
57 $64x^3$ equal to $x^4$'s, and the division of all that by $x^3$ gives 64 equal to $x$'s; thus $x$ is the number resulting from the division of 64 by the coefficient of the remaining $x^4$'s.[79] Consequently, the coefficient of the $x^4$'s left over in the first equation is equal to the coefficient of the $x^4$'s left over in the second equation. But the coefficient of the $x^4$'s left over in the first equation is a square number minus 16, and the (coefficient of the) $x^4$'s left over in the second equation is 16 minus a square number. Thus a large(r) square number minus 16 is equal to 16 minus a small(er) square number. The addition of the small(er) square, and also of the

[78] See problems IV,34,2° and 35–39.

[79] Instead of having "$x^4$" here and in the following four places, the text has "$x^2$". Cf. p. 63, no. 4.

16 subtracted from the large(r) square, in common to both sides, results in a large(r) square plus a small(er) square equal to 32. But 32 is composed of two equal squares; it is therefore possible to divide it into two different square numbers.[80] Let it be divided, and let the two squares be $\frac{16}{5\cdot 5}$ and $31\frac{9}{5\cdot 5}$. We put $31\frac{9}{5\cdot 5}x^4$ as the square equal to $16x^4 + 64x^3$, and $\frac{16}{5\cdot 5}x^4$ as the square equal to $16x^4 - 64x^3$. Then in each of the two equations we shall arrive at $64x^3$ equal to $15\frac{9}{5\cdot 5}x^4$. The division of all that by $x^3$ gives $15\frac{9}{5\cdot 5}x$ equal to
58 64; hence $x$ is the result of the division of 1600 by 384, which is $4\frac{1}{6}$. Then, since we assumed the side of the square to be $2x$, this side is $8\frac{1}{3}$ and the square, $69\frac{4}{9}$, and the square of the square is $4822 + \frac{4}{9} + \frac{7}{9\cdot 9}$. And, since we assumed the side of the cube to be $4x$, this side is $16\frac{2}{3}$ and the cube, $4629 + \frac{5}{9} + \frac{2}{3}\cdot\frac{1}{9}$. When the latter is added to the number resulting from the multiplication of the square number by itself, the result is $9452 + \frac{1}{9} + \frac{4}{9\cdot 9}$, which is a square number with $97\frac{2}{9}$ as its side; when the same number is subtracted from the square of the square number, the remainder is $192 + \frac{8}{9} + \frac{1}{9\cdot 9}$, which is a square number with $13\frac{8}{9}$ as its side.

Therefore, we have found two numbers fulfilling the condition imposed by us, and these are the two numbers which we have determined.

**41.** We wish to find two other[81] numbers, one cubic and the other square, such that the cubic number, when increased by the square of the square, results in a square number, and when decreased by the square of the square, results in a square number.

In the manner described (above), we say (firstly): $64x^3 + 16x^4$ is equal to a square number. If we then put as the side of $64x^3 + 16x^4$ $x^2$'s, the square is $x^4$'s, (which) equal $64x^3 + 16x^4$; the subtraction of the $16x^4$, which is common, from both sides gives $64x^3$ equal to $x^4$'s, and the division of both by $x^3$ results in 64 equal to $x$'s. Hence $x$ is the result of the division of 64 by the
59 coefficient of the remaining $x^4$'s. Again, (we say): $64x^3 - 16x^4$ is equal to a square number. If we then make the side of $64x^3 - 16x^4$ $x^2$'s, the square is $x^4$'s, (which) equal $64x^3 - 16x^4$; the addition of $16x^4$ in common to both sides gives $x^4$'s equal to $64x^3$, and the division of the two by $x^3$ results in $x$'s equal to 64. So $x$ here is likewise the result of the division of 64 by the co-

[80] *Arithmetica* II,9.

[81] The presence of this "other" and the abrupt beginning of the problem may indicate that nos. 40 and 41 were once considered to be a single problem with two subdivisions.

efficient of the $x^4$'s added up.[82] Then, the number of $x^4$'s left over in the first equation, of which there is a square number minus 16, must equal the coefficient of the $x^4$'s added up in the second equation, which is a square number plus 16. Thus a square number minus 16 is equal to another square plus 16. We add the 16 subtracted in common to both sides, and obtain a square plus 32 equal to a large(r) square. Thus we shall seek two square numbers with 32 as their difference;[83] hence the larger square will be (*ipso facto*) larger than 16. Let the smaller square be 4 and the larger, 36. We put $36x^4$ as the square equal to $64x^3 + 16x^4$, and $4x^4$ as the square equal to $64x^3 - 16x^4$. Then in each of the two equations we shall arrive at $64x^3$ equal to $20x^4$. Let us divide each by $x^3$, so we obtain $20x$ equal to 64; hence $x$
60 is $3\frac{1}{5}$. Since we assumed the side of the square to be $2x$, the side is $6\frac{2}{5}$ and the square, $40 + \frac{4}{5} + \frac{4}{5 \cdot 5}$, and the square of the square is 1677 and 451 parts of 625 parts of 1. And, since we assumed the side of the cube to be $4x$, this side is $12\frac{4}{5}$ and the cube, 2097 and 95 parts of 625 parts of 1. The latter, when increased by the square of the square, gives 3774 and 546 parts of 625 parts of 1, which is a square with 61 and 11 parts of 25 parts of 1 as its side. And, when the square of the square number is subtracted from the said cube, the remainder is 419 and 269 parts of 625 parts of 1, which is a square number with 20 and 12 parts of 25 parts of 1 as its side.

Therefore, we have found two numbers fulfilling the condition imposed by us, and these are the two numbers which we have determined. This is what we intended to find.

**42.** We wish to find two numbers, one cubic and the other square, such that both the sum and the difference of the cube of the cube and of the square of the square are square numbers.

We put an arbitrary number of $x$'s as the side of the cube, say $2x$, so that the cube is $8x^3$, and the cube of that cube, $512x^9$. We put as the side of the square a number, again arbitrary, of $x^2$'s, say $4x^2$, so that the square is $16x^4$
61 and the square of the square, $256x^4$ (multiplied) by $x^4$, that is to say (256 times) the so-called $x^8$.[84]

Let us first require that the cube of the cube, when increased by the square of the square, result in a square, and when decreased by the square of the square, leave a square.

We have found previously two numbers having that characteristic, by chance, without aiming to find that characteristic.[85] (Thus) we shall (merely)

[82] Here and in the next two places, the text once again has "$x^2$'s" instead of "$x^4$'s".

[83] *Arithmetica* II,10.

[84] Defined in problem 29.

[85] See problems 29 and 30.

mention here the manner in which their discovery occurs. So we say: $512x^9 + 256x^8$ is equal to a square number, and also: $512x^9 - 256x^8$ is equal to a square.

We may treat this by the method of the double-equation, which is (as follows). We take the difference between the said squares, which is $512x^8$,[86] and seek two numbers of $x^4$'s such that the product of their multiplication give $512x^8$.[87] Then, we take half of the sum of the two numbers, multiply the result by itself, and equate with that the larger square, that is, $512x^9 + 256x^8$. Next, we take half of the difference between the two numbers, multiply the result by itself, and equate with that the lesser square, that is, $512x^9 - 256x^8$.[88] Then we shall arrive in each of the two equations at $512x^9$ equal to the same number of $x^8$'s. The division of each (side) by a unit of the one of lower degree, that is, (by) $x^8$, will result in $512x$ equal to a number, whence we shall know $x$. $x$ being known, we shall return to the initial hypotheses adopted by
62 us,[89] performing then subsequently to the knowledge of $x$ the synthesis of all the elements of the problem.

We may (also) use the method of seeking identicalness of the equation (resulting from) the two (proposed) equalities, as expounded in the preceding problems.[90] It consists in saying (the following). If we set as the side of the larger square $x^4$'s, their square is $x^8$'s, (which are) equal to the larger square. Then, the subtraction of the $256x^8$, which is common, from both sides gives $512x^9$ equal to $x^8$'s, and the division of both by $x^8$ results in $512x$ equal to a number. Thus, the number equal to the coefficient of the remaining $x^8$'s, when divided by 512, gives the number assumed to be $x$ in the problem. Again, if we put as the side of the smaller square $x^4$'s, their square is $x^8$'s, (which are) equal to the smaller square. Then, the addition of the $256x^8$ in common to both sides gives $x^8$'s equal to $512x^9$, and the division of both by a unit of the side of lower degree, namely (by) $x^8$, results in $512x$ equal to a number. Thus, the division of this number by 512 gives as a result the number assumed to be $x$ in the problem. Consequently, it is necessary that the coefficient of the $x^8$'s left over in the first equation be equal to the coefficient of the $x^8$'s added up in the second equation. But the coefficient of the $x^8$'s left over in the first equation is a square number minus 256, while the coefficient of the $x^8$'s added up in the second equation is a square number plus 256. Therefore, we have to seek two (square) numbers having the double
63 of 256, or 512, as their difference.[91] Having found them, we make the larger (number) $x^8$'s and equate to that the larger square, and make the lesser (number) $x^8$'s and equate to that the lesser square. After that, we shall

[86] Here as in two other places below, $x^8$ is represented as "$x^4$ by $x^4$" in the text.

[87] "$512x^4$ by $x^4$" in the text.

[88] "$512x^9 - 256x^4$ by $x^4$" in the text.

[89] The Greek must have been something like ἀνατρέχομεν ἐπὶ τὰς ὑποστάσεις.

[90] See problems 34,2° and 35–41.

[91] *Arithmetica* II,10.

arrive in each of the two equations (after division by $x^8$) at $512x$ equal to one and the same number, whence we shall know $x$, the amount of which we aimed to find. Next, we shall go back and undertake the synthesis of the problem.

We may (also) say: $512x^9 + 256x^8$ is equal to a square, and $512x^9 - 256x^8$ is equal to a square. Now, any square which is divided by a square gives a square as a result; thus, if we divide the $512x^9 + 256x^8$ by a square— say $x^8$, or $4x^8$, or $9x^8$, or $16x^8$, or by any arbitrary square numbers provided that we make each of them $x^8$'s—(the) $x^8$'s result in a number and (the) $x^9$'s, in $x$'s. Suppose we divide the two (terms) by $16x^8$, then the result of the division is $32x + 16$. In the same way as we have divided this square, let us divide the other one, namely $512x^9 - 256x^8$, which becomes then $32x - 16$. Hence $32x + 16$ and $32x - 16$ are (equal to) squares. So let us seek a number which,
64 when increased by a given number, namely 16, gives a square, and when diminished by a given number, namely 16, leaves a square.[92] Having found that number, we divide it by 32; the result of the division will be $x$. Once we know $x$, we shall come back, and then make the synthesis of the problem according to the way adopted by us in its analysis.

In the same manner which we have (just) described, one can treat most of the previously presented problems involving (a system of) two equations.

Let us (now) require that the square of the square, when increased by the cube of the cube, result in a square, and when decreased by the cube of the cube, leave a square number.

Similarly, let us say, as we did previously: $256x^8 + 512x^9$ is equal to a square, and $256x^8 - 512x^9$ is equal to a square. We treat this by seeking identicalness of the (resulting) equation in the (proposed system of) two equations, as already expounded in the preceding problems of this kind.[93] Then we shall end up with the division of the double of 256, that is to say, (of the double of) a square number, namely 512, into two unequal square numbers.[94] Let the smaller of these two square numbers be 10 and 6 parts of 25 parts of 1, with side $3\frac{1}{5}$, and the larger one be 501 and 19 parts of 25 parts of 1, with side $22\frac{2}{5}$. Putting the lesser of these two squares equal to the lesser of the two first (mentioned) squares, and the larger equal to the larger,[95] we shall
65 end up in each of the two equations with $512x^9$ equal to $245x^8$ and 19 parts of 25 parts of $x^8$. Let us divide both by $x^8$, then $512x$ is equal to 245 and 19 parts of 25 (parts) of 1; thus $x$ is 12 parts of 25. Since we assumed the side of the cube to be $2x$, the said side is 24 parts of 25 parts of 1 and the cube, 13,824 parts of the cube of 25; the cube of that cube is 2,641,807,540,224 parts of the cube of the cube of 25, or 105,672,301,608 parts and 24 parts of 25 parts of

[92] This may be done with either of the two previous methods.

[93] See IV,34,2°, 35–41 and the second method in the first part of the present problem.

[94] *Arithmetica* II,9, applicable to any number which is the sum of two squares.

[95] It should have been said that the two found squares were made $x^8$'s.

one part of the square of the square of 625. And, since we put as the side of the square $4x^2$, and (since) $x^2$ is 144 parts of 625—for $x$ is 12 parts of 25—, the side of the square is 576 parts of 625 parts and the square, 331,776 parts of the square of 625; the square of the said square is 110,075,314,176 parts of the
66 square of the square of 625. This last number, when increased by the cube of the cubic number, gives 215,747,615,784 parts and 24 parts of 25 (parts) of one part of the square of the square of 625, which is a square with $464{,}486\frac{2}{5}$ parts of the square of 625 as its side. And, when the cube of the cubic number is subtracted from the square of the said square, the remainder is 4,403,012,567 parts and one part of 25 parts of one part of the square of the square of 625, which is a square with side $66{,}355\frac{1}{5}$ parts of the square of 625.

Therefore, we have found two numbers in accordance with our requirement, and these are the two numbers which we have determined. This is what we intended to find.

**43.** We wish to find two numbers, one cubic and the other square, such that when we add a given multiple of the square of the square to the cube of the cube the result is a square number, and when we subtract a given multiple of the square of the square from the same the remainder is a square number.

Let us put $x^3$ as the cube, so that its cube is $x^9$, and an arbitrary number of $x^2$'s as the side of the square, say $2x^2$, so that the square is $4x^4$ and the square of the square, $16x^8$. Let the given multiplier for the addition be $1\frac{1}{4}$ and the one for the subtraction be $\frac{1}{2} + \frac{1}{4}$. We add to the cube of the cube $1\frac{1}{4}$ times the
67 square of the square, namely $20x^8$, whence $x^9 + 20x^8$, which is equal to a square number; let us (now) subtract from the cube of the cube $\frac{1}{2} + \frac{1}{4}$ of the square of the square, namely $12x^8$, whence $x^9 - 12x^8$, (which is) equal to a square number. Then, if we put $x^4$'s as the side of the square equal to $x^9 + 20x^8$, their square is $x^4$'s (multiplied) by $x^4$'s, that is, those of which one is called $x^8$; if we equate to them $x^9 + 20x^8$, then subtract the $20x^8$ which is common, we shall have $x^9$ equal to $x^8$'s having their coefficient equal to a square minus 20: and this (coefficient) is the number taken as $x$ in the present treatment. Again, if we put $x^4$'s as the side of the square equal to $x^9 - 12x^8$, their square is $x^8$'s; if we add to them the $12x^8$ subtracted from $x^9$, making them a common increment to both sides, we shall have $x^9$ equal to $x^8$'s having their coefficient equal to a square plus 12: and this (coefficient) is the number taken as $x$ in this problem. Thus a square minus 20 equals a small(er) square plus 12. We add the 20 in common to both sides; so a small(er) square plus 32 equals a large(r) square. Hence[96] the small(er) square is 4; when increased by 32, it results in 36, which is the large(r) square. Then, we put $36x^8$ as the square equal to $x^9 + 20x^8$ and $4x^8$ as the square equal to the second square (or $x^9 - 12x^8$). So, in each of the two equations
68 we shall arrive, after the restoration, the reduction,[97] and the division, at $x$

[96] See problem 41.

[97] Only reduction in the first equation, and only restoration in the second one.

equal to 16. We (now) make the synthesis of the problem according to the way adopted by us in its analysis. We assumed the side of the cube to be $x$, so the side is 16 and the cube, 4096; we took as the side of the square $2x^2$, so, $x^2$ being 256, the side of the square is 512 and the square, 262,144. Hence the cube of the cube is 68,719,476,736; and the square of the square is also equal to this number. Thus the cube of the cube is a square, equal to the result of the multiplication of the square number by itself. Therefore, when the cube of the cube is increased by $1\frac{1}{4}$ times the square of the square, the result is $2\frac{1}{4}$ times the square of the square, and this is a square number with $1\frac{1}{2}$ times the square number as its side; again, if the cube of the cube is diminished by $\frac{3}{4}$ of the square of the square, the remainder is $\frac{1}{4}$ of the square of the square number, and this is a square with half of the square number as its side.

Therefore, we have found two numbers having the indicated characteristic, and these are the two numbers which we have determined. This is what we intended to find.

**44.** We wish to find two numbers, one cubic and the other square, such that when we multiply the square of the square number by two given numbers: (either) adding the cube of the cube to each of the two (products) gives in both cases a square number; or, subtracting each of the two (products) from the cube of the cube gives (in both cases) a square number; or, subtracting the cube of the cube from each of the two (products) gives in both cases a square number.

Let one of the two given numbers be 3 and the other be 8. We wish to
69 find two numbers, one cubic and the other square, such that, when we multiply the square of the square by 3 and by 8: (either) adding each of the two products to the cube of the cube gives in both cases a square number; or, subtracting each of the two products from the cube of the cube leaves in both cases a square number; or, subtracting the cube of the cube from each of the two products leaves in both cases a square number.

Let us examine the first of the three (cases). We put $x$ as the side of the cube, so that the cube is $x^3$ and the cube of the cube, $x^9$; we put $2x^2$ as the side of the square, so that the square is $4x^4$ and the square of the square, $16x^8$. The multiplication of the $16x^8$ by 3 and by 8 gives $48x^8$ and $128x^8$, the addition of which to the cube of the cube gives $x^9 + 48x^8$ and $x^9 + 128x^8$, and each is (equal to) a square. As the division of any square by a square is a square, let us divide each of them by a square, and let the said square be $x^8$; then the first quotient is $x + 48$, which (again) equals a square number, since it resulted from the division of a square by a square; the second quotient is $x + 128$, which equals a square number, since it resulted from the division of a square by a square. Thus $x$ is such that, when increased by 48, it results in
70 a square, and when increased by 128, it also results in a square. So let us seek a number which, when added to the two said numbers, gives in both cases a

square.[98] Such is 16. Hence $x$ is 16. Since we assumed the side of the cube to be $x$, the side is 16, and the cube is the cubic number found by us in the preceding problem;[99] again, its cube is the number which was the cube of the one in the previous problem. Accordingly, again, the square of the square is equal to the cube of the cube. The multiplication of the square of the square by 3, and the addition to the (result) of the cube of the cube gives four times the square of the square, which is a square with twice the square number as its side; again, the multiplication of the same by 8 and the addition of the result to the cube of the cube gives nine times the square of the square, which is a square with three times the square number as its side.

Therefore, we have found two numbers, one cubic and the other square, such that when we multiply the square of the square by 3 and by 8, then add each of the two (products) to the cube of the cube, the result is in both cases a square number; and this cube and this square are 4096 and 262,144, respectively.

Let us also examine the second of the three (cases). If, similarly, we put $x^3$ as the cube and $4x^4$ as the square, we obtain as (the) two squares $x^9 - 48x^8$ and $x^9 - 128x^8$. Now, any square which is divided by a square gives again a square. Let the square by which we divide $x^9 - 48x^8$ and $x^9 - 128x^8$ be $x^8$, which is the result of the multiplication of $x^4$ by itself. Then, the first quotient is $x - 48$ and the second, $x - 128$, and each is (equal to) a square.
71 Let us then seek a number which, when diminished by 48 and by 128, leaves in both cases a square number;[100] and this number will be the one taken as $x$ in the treatment of the problem. Such is 192.

Now, since the side of the cube found in the preceding problem[101] is 16 and the side of the (present) cube is 192, the side of the present cube is to the side of the previous one in the ratio 12:1, and the present cube is to the previous one in the ratio $12^3:1$. And, since we put as the side of the square $2x^2$, and (since) the present $x$ is to the previous one—(that is, to the one) occurring in the preceding problem—in the ratio 12:1, the present $x^2$ is to the previous $x^2$ in the ratio $12^2:1$, and so it will be for (the ratio of) the side of the (present) square to the side of the previous square; hence, the ratio of the two squares is the ratio $144^2:1$. Then, the cube of the present cube is to the cube of the previous cube as $(12^3)^3$ is to 1, and the square of the (present) square is to the square of the previous one as $(144^2)^2$ is to 1; and the square of the previous square was equal to the cube of the previous cube, so that the cube of the previous cube was a square. Now, 1 is also a square cube.[102]

[98] *Arithmetica* II,11.

[99] One would expect to see an analogous statement here for the required square.

[100] *Arithmetica* II,13.

[101] Allusion to the first part of this problem (but IV,43 also has this solution).

[102] The reasoning is abstruse here, and the text, as it stands at present, may have passed, as noted in the commentary, through a commentator's hands.

Consequently, the side of the present cube is 12,[103] the cube, 1728, the side of the square, 144, and the square, 20,736. The cube of the present cube is
72 5,159,780,352, and the square of the present square, 429,981,696. Three times the square of the square is 1,289,945,088, and the subtraction of that from the cube of the cube leaves 3,869,835,264, which is a square with 62,208 as its side; and, eight times the square of the square is 3,439,853,568, the subtraction of which from the cube of the cube leaves 1,719,926,784, which is a square having 41,472 as its side.

Therefore, we have found two numbers, one cubic and the other square, such that, when we multiply the square of the square by 3 and by 8, and subtract each (product) from the cube of the cubic number, the remainder is (in both cases) a square number; and these are the two numbers which we have found.

Let us now examine the remaining aspect, of the three (aspects of the problem), defined by us. We say: $48x^8 - x^9$ is equal to a square, and $128x^8 - x^9$ is equal to a square. Let us divide both by $x^8$; the two quotients are then $48 - x$ and $128 - x$, and each is (equal to) a square. So let us seek a number which, when subtracted from 48 and from 128, leaves in both cases a
73 square.[104] Let it be 47; and this is the number assumed to be $x$ in the treatment of the present problem. Since we put $x$ as the side of the cube, the side is 47, so that the cube is 103,823; and, since we put $2x^2$ as the side of the square, and (since) $x^2$ is 2209, the side of the square is 4418, and the square is 19,518,724. The cube of the cube, when subtracted from three times the square of the said square, leaves a square, the side of which is 4,879,681, and, when subtracted from eight times the square of the square, leaves a square having 43,917,129 as its side.

Therefore, we have found two numbers, one cubic and the other square, such that the multiplication of the square of the square by 3 and by 8 and the subtraction of the cube of the cube from each of the two (products) gives in both cases a square number; and these are the two numbers which we have determined. This is what we intended to find.

End of the fourth Book of the treatise of Diophantus on squares and cubes, and it contains forty-four problems.

[103] The denomination "side of the present cube" is odd (since it has already been used for the value 192) and should be understood to mean "side of a second cube satisfying the present problem". Similarly for the square.

[104] *Arithmetica* II,12.

# In the Name of God the Merciful, the Compassionate

# Fifth Book of the Treatise of Diophantus the Alexandrian on Arithmetical Problems

**1.** We wish to find two numbers, one square and the other cubic, such that when we add to the square of the square a given multiple of the cubic number, the result is a square number, and when we subtract from the same another given multiple[1] of the cubic number, the remainder is a square number.

Let the positive multiplier be 4 and the negative one, 3. We wish to find two numbers as indicated by us. We put $x$ as the side of the square, so that the square is $x^2$ and the square of the square, $x^4$; the latter, together with four
74 times a certain cube, is equal to a square, and minus three times the same cube, is again equal to a square. Hence the cube is equal to a certain quantity, having to $x^4$ a given ratio, and such that four times it when added to $x^4$ gives a square and three times it when subtracted from $x^4$ leaves a square. So we shall seek three square numbers such that the excess of the largest over the middle be to the excess of the middle over the smallest as four is to three.[2] Let these (three) numbers be 81, 49 and 25. $x^4$ being put 49 parts,[3] the quantity given in ratio to $x^4$ such that four times it—i.e., 32 parts of 49 parts (of $x^4$)—when added to $x^4$ gives a square and three times it—i.e., 24 parts of 49 parts (of $x^4$)—when subtracted from $x^4$ leaves a square, is 8 parts of 49 parts of $x^4$. So the required cube is equal to 8 parts of 49 parts of $x^4$. Let us put as the side of the cube an arbitrary number of $x$'s, say $2x$; so the cube is $8x^3$. Hence $8x^3$ is equal to 8 parts of 49 parts of $x^4$. Let us divide both by $x^3$, so 8 parts of 49 parts of $x$ equals 8; hence $x$ is equal to 49. Thus the side of the square is 49 and the square is 2401. Since we put $2x$ as the side of the cube, the said side is 98 and the cube, 941,192. So the square of the square is 5,764,801. When
75 increased by four times the cubic number, that is, (by) 3,764,768, it results in 9,529,569, which is a square with 3087 as its side; and, when the same is

[1] If the multiple is the same, the treatment is similar to that in IV,40.

[2] *Arithmetica* II,19.

[3] That is, we take one "part" as $\frac{1}{49}x^4$.

decreased by three times the cubic number, that is, (by) 2,823,576, it results in 2,941,225, which is a square number with 1715 as its side.

Therefore, we have found two numbers fulfilling the condition required by us. This is what we intended to find.

**2.** We wish to find two numbers, one square and the other cubic, such that, when we multiply the cubic number by two given numbers and add each of the two (products) to the square of the square, the result is in both cases a square.

We take 12 and 5 as the given numbers. We put $x$ as the side of the square, so that the square is $x^2$ and its square, $x^4$. The latter, together with twelve (times a certain) cube, is equal to a square, and together with five times the said cube, is again equal to a square. Therefore, let us look for the quantity given in ratio to $x^4$ such that twelve times it when added to $x^4$ gives a square, and also five times it when added to $x^4$ gives a square. Thus we are led to the search for three square numbers such that the excess of the largest over the middle be to the excess of the middle over the smallest as the excess of 12 over 5 is to 5, i.e., (as) $1\frac{2}{5}$ is to 1.[4] Let these (three) numbers be 16, 9, and 4. $x^4$ being put 4 parts,[5] it appears that the quantity given by its ratio to $x^4$ such that five times it—i.e., 5 parts—when added to $x^4$ gives
76 a square and twelve times it—i.e., 12 parts—when added to $x^4$ gives a square, is $\frac{1}{4}x^4$. Hence $\frac{1}{4}x^4$ is equal to a cubic number. Let us put $2x$ as its side, so that the cube is $8x^3$; this is equal to $\frac{1}{4}x^4$. Let us divide both by $x^3$, so $\frac{1}{4}x$ equals 8; thus $x$ is equal to 32. So the side of the square is 32, the square is 1024, and the square of the square, 1,048,576. Since we assumed the side of the cube to be $2x$, the side of the cube is 64 and the cube, 262,144. The multiplication of the latter by 12 gives 3,145,728, the addition of which to the square of the square results in 4,194,304, which is a square having 2048 as its side; again, the multiplication of the cubic number by 5 gives 1,310,720, the addition of which to the square of the square results in 2,359,296, which is a square having 1536 as its side.

Therefore, we have found two numbers fulfilling the condition required by us; and these are the two numbers which we have determined.

**3.** We wish to find two other[6] numbers, one cubic and the other square, such that, when we multiply the cube by two given numbers and subtract each of the two (products) from the square of the square, the remainder is (in both cases) a square.

---

[4] *Arithmetica* II,19.

[5] One "part" being taken as $\frac{1}{4}x^4$.

[6] This "other" seems to be meaningless; were the formulation not a standard one, and therefore confusion in translation unlikely, one might see the origin of the "other" in a misunderstood ἕτερος in "Δυσὶ δοθεῖσιν ἀριθμοῖς προσευρεῖν δύο ἑτέρους ἀριθμούς κ.τ.λ." (cf., e.g., D.G., I, p. 76,26).

Let the two given numbers be 12 and 7. We again put $x^2$ as the square,[7]
77 so that the square of the square is again $x^4$. Hence $x^4$ minus 12 (times the) cube equals a square, and ($x^4$) minus 7 (times the) cube also equals a square. (So) let us seek the quantity given in ratio to $x^4$ such that twelve times it when subtracted from $x^4$ leave a square, and seven times it when subtracted from the same also leave a square. This amounts to the search for three square numbers such that the excess of the largest over the middle be to the excess of the middle over the smallest as 7 is to the subtraction of 7 from 12. Such are the numbers which we have mentioned previously, (namely) 16, 9, and 4.[8] Therefore, the quantity given in ratio to $x^4$ which we have defined[9] is one part of 16 parts of $x^4$. Hence the cube equals one part of 16 parts of $x^4$. We assume the side of the cube to be $\frac{1}{2}x$, so that the cube is $\frac{1}{8}x^3$. So $\frac{1}{8}x^3$ is equal to one part of 16 (parts) of $x^4$. Hence $\frac{1}{2} \cdot \frac{1}{8}x$ equals $\frac{1}{8}$, so $x$ equals 2. Thus the square is 4 and the square of the square, 16; and, since we set $\frac{1}{2}x$ as the side of the cube, the side of the cube is 1 and the cube, 1, again. The multiplication of it by 12 and by 7, and the subtraction of each one of the two (products) from the square of the square leaves (in both cases) a square.

**4.** We wish to find two numbers, one square and the other cubic, such that, when we increase the square of the square by a given multiple of the cube of the cube, the result is a square number, and when we decrease the same by another given multiple[10] of the cube of the cube, the remainder is again a square number.

Let the positive multiplier be 5 and the negative multiplier be 3. Let us set $x$ as the side of the cube, which is then $x^3$, and its cube is $x^9$. We set $2x^2$ as the
78 side of the square, so that the square is $4x^4$ and its square, $16x^8$. Then $16x^8 + 5x^9$ is equal to a square, and $16x^8 - 3x^9$ is equal to a square. Now, the division of any square by a square results in a square. So let us divide each of the above two squares by the square $x^8$; the two quotients are then $16+5x$ and $16-3x$, and each is (equal to) a square. (But,) any square number to which is added five times its quarter, and from which is subtracted three time its quarter, gives in both cases a square. Hence $x$ is the quarter of 16, or 4. Since we set $x$ as the side of the cube, the said side is 4 and the cube, 64; and, since we set $2x^2$ as the side of the square, and (since) $x^2$ is 16, the side of the square is 32 and the square, 1024. The square of the square is 1,048,576, and the cube of the cube is 262,144. The addition of five times the latter to the

---

[7] Cf. problems V,1 and 2.

[8] See the preceding problem; the statement of the three values seems to be an interpolation (cf. p. 32, no. 15).

[9] The words "which we have defined" may refer to the conditions for $x^4$ given previously. But they might also be (part of) a marginal gloss, now incorporated into the text, added in order to emend the reasoning. For the text does not state, as it should have done (cf. V,1 and 2), that $x^4$ is put 16 "parts", with one part taken as $\frac{1}{16}x^4$. On partial reproductions of glosses, see p. 33, no. 25.

[10] If the given multiple were the same, the treatment would be similar to that in IV,42,b.

square of the square gives 2,359,296, which is a square with side 1536, and the subtraction of the same taken three times from the square of the square leaves 262,144, which is a square with 512 as its side.

Therefore, we have found two numbers fulfilling the condition stipulated by us; and these are the two numbers which we have determined.

79 **5.** We wish to find two numbers, one cubic and the other square, such that, when we multiply the cube of the cube by two given numbers and add each of the two (products) to the square of the square, the result is (in both cases) a square number.

We take for the two given numbers 12 and 5. We wish to find two numbers in accordance with what we have indicated. We put $x$ as the side of the cube, so that the cube is $x^3$ and its cube, $x^9$. We assume the side of the square to be $2x^2$, so that the square is $4x^4$ and the square of the square, $16x^8$. Thus $16x^8 + 12x^9$ is equal to a square and $16x^8 + 5x^9$ is equal to a square. As any square divided by a square results in a square, let us divide both by the square $x^8$. Therefore, $16 + 12x$ and $16 + 5x$ are both (equal to) a square. But any square which is increased by five times its quarter, and also by 12 times its quarter, gives in both cases a square. Hence $x$ is the quarter of 16, or 4. Thus the cube is 64 and the square, 1024. It appears that, adding to the square of the said square twelve times the cube of the said cube,[11] that is, 3,145,728, gives 4,194,304, which is a square with side 2048. And it has been found in the preceding problem that adding to the same five times the cube of the cube also gives a square.

**6.** We wish to find two numbers, one cubic and the other square, such that,
80 when we multiply the cube of the cube by two given numbers and subtract each one of the two (products) from the square of the square, the remainder is (in both cases) a square.

Let the two given numbers be 7 and 4. We make the side of the cube $x$, so that the cube is $x^3$ and its cube, $x^9$. We make the side of the square $3x^2$, so that the square is $9x^4$ and the square of the square, $81x^8$. Thus $81x^8 - 7x^9$ equals a square, and $(81x^8) - 4x^9$ also equals a square. Let us divide both by the square $x^8$, so $81 - 7x$ equals a square and $81 - 4x$ also equals a square. Let us seek the given (fractional) quantity of any square such that seven times it when subtracted from the square and also four times it when subtracted from the same square leave in both cases a square. One seeks this in the previous manner.[12] Let the said quantity be $\frac{8}{9 \cdot 9}$; then, after subtracting from 81 seven times its $\frac{8}{9 \cdot 9}$th, or 56, the remainder is a square,

---

[11] The values of the square of the square and of the cube of the cube are known from the preceding problem.

[12] Using the method of II,19, as in the first problems of this Book.

namely 25, and, after subtracting from 81 four times its $\frac{8}{9 \cdot 9}$th, or 32, the remainder is a square, namely 49. Hence $x$ is the $\frac{8}{9 \cdot 9}$th of 81, i.e., 8. Since we set $x$ as the side of the cube, the cube is 512; and, since we set $3x^2$ as the side of the square, and (since) $x^2$ is 64, the said side is 192 and the square is 36,864.
81 The square of the square is 1,358,954,496, and the cube of the cube is 134,217,728. When seven times the latter is subtracted from the square of the square, the remainder is 419,430,400, which is a square with side 20,480; and, when four times the same is subtracted from the square of the square, the remainder is 822,083,584, which is a square with side 28,672.

Therefore, we have found two numbers fulfilling the condition stipulated by us, and these are 512 for the cube and 36,864 for the square. This is what we intended to find.

**7.** We wish to find two numbers such that their sum and the sum of their cubes are equal to two given numbers.

It is necessary that four times that one of the two (numbers) which is given for the sum of the cubes of the two (required) numbers exceed the cube of the number given for their sum by a number which, when divided by three times the number given for the sum of the two numbers, gives a square, and which, when multiplied by three quarters of the number given for the sum of the two numbers, gives a square. This (problem) belongs to the (category of) constructible problems.

Let the number given for the sum of the two numbers be 20 and the number given for the sum of their cubes be 2240. We wish to find two numbers such that their sum is 20 and the sum of their cubes, 2240. We put $2x$ as the difference of the two numbers, so that one is $10 + x$ and the other, $10 - x$.
82 We form from each of them a cube.[13] Now, whenever we wish to form a cube from (some) side made up (of the sum) of (say) two different terms—so that a multitude of terms does not make us commit a mistake—, we have to take the cubes of the two different terms, and add to them three times the results of the multiplication of the square of each term by the other; then, the result is composed of four terms, and this is the cube arising from the sum of the two different terms. (But) when the two terms are such that one is subtracted from the other, we take the cube of the larger,[14] add to it three times the result of the multiplication of the square of the smaller term by the larger term, and subtract from them the cube of the smaller term and three times the result of the multiplication of the square of the larger term by the smaller; the result is then the cube arising from the difference between the two different terms. Hence the cube arising from the side $10 + x$ is the sum

---

[13] That is, we raise each of them to the third power (Gr. πλάσσειν).

[14] Since the expression must be positive, the positive term is the larger.

of the cube of 10, or 1000, and of the cube of $x$, or $x^3$, plus three times the result of the multiplication of 10 by the square of $x$, or $30x^2$, plus, again, three times the result of the multiplication of $x$ by the square of 10, or $300x$; thus, the cube arising from $10 + x$ is $1000 + x^3 + 300x + 30x^2$. Again, the cube arising from the side $10 - x$ is also equal to the cube of 10, or 1000, and to three times the result of the multiplication of 10 by the square of $x$, $x^2$, i.e., (to) $30x^2$, minus the cube of $x$, or $x^3$, and minus three times the result of the multiplication of $x$ by the square of 10, or $300x$; thus, the cube arising from $10 - x$ is $1000 + 30x^2 - x^3 - 300x$. The sum of these two cubes is 2000 +
83 $60x^2$, because the subtracted $x^3 + 300x$ in the one cube is cancelled by the added $x^3 + 300x$ in the other. Then, $2000 + 60x^2$ is equal to 2240. Let us subtract the 2000 which is in one side from the number which is in the other side, whence $60x^2$ equals 240; thus $x^2$ is 4. And, each of these being a square, their sides are also equal; but the side of $x^2$ is $x$, and the side of 4 is 2, so that $x$ is 2. Since we put as the larger of the two required numbers $10 + x$, the said number is 12; and, since we put as the smaller number $10 - x$, it is 8. The cube of the larger number is 1728 and the cube of the smaller number, 512; and their sum is 2240.

Therefore, we have found two numbers such that their sum is 20 and the sum of their cubes, 2240; and these are 12 and 8. This is what we intended to find.

**8.** We wish to find two numbers such that their difference and the difference of their cubes are equal to two given numbers.

It is necessary that four times the number given for the difference of the two cubes exceed the cube of the number given for the difference of the two (required) numbers by a number which, when divided by three times the number given for the difference of the two numbers, gives a square, and which, when multiplied by three quarters of the number belonging to the said difference, gives a square.

Let the number given for the difference of the two numbers be 10 and the
84 number given for the difference of the two cubes be 2170. We wish to find two numbers such that their difference is 10 and the difference of their cubes, 2170. We put $2x$ as the sum of the two numbers, so that one is $x + 5$ and the other, $x - 5$: this, in order that their difference amount to 10.[15] We form from each of them a cube. So the cube with side $x + 5$ is, as explained (before),[16] equal to the cube of $x$, or $x^3$, plus the cube of 5, or 125, plus three times the product of the multiplication of the square of $x$ by 5, or $15x^2$, plus three times the product of the multiplication of the square of 5 by $x$, or $75x$; hence the cube arising from the side $x + 5$ is $x^3 + 125 + 15x^2 + 75x$. The cube having $x - 5$ as its side equals the cube of $x$, or $x^3$, plus three times the

[15] This last phrase could come from a marginal remark; see p. 32 (no. 20).

[16] In the preceding problem.

result of the multiplication of the square of the 5 subtracted from $x$ by $x$, or $75x$, minus the cube of 5, or 125, and minus three times the product of the multiplication of the square of $x$ by 5, or $15x^2$; hence the cube arising from the side $x - 5$ is $x^3 + 75x - 15x^2 - 125$. Let us subtract this cube from the first one, so we obtain $250 + 30x^2$, for the $15x^2 + 125$ subtracted in the latter cube will, because of the subtraction, (become) positive and be added to the positive $15x^2 + 125$ in the other cube, while the $x^3 + 75x$ will be eliminated
85 from both. Hence $250 + 30x^2$ is equal to 2170. Let us remove the 250, which is common, from both sides, so there remains 1920 equal to $30x^2$; thus $x^2$ is 64. And, each of these being a square, their sides are equal; that of $x^2$ being $x$, and that of 64 being 8, $x$ is 8. Since we put $x + 5$ as the larger number, it is 13; and, as we had put $x - 5$ as the smaller number, the smaller is 3. The cube of the larger is 2197 and the cube of the smaller, 27; and their difference is 2170.

Therefore, we have found two numbers such that their difference is 10 and the difference of their cubes, 2170; and these are 13 and 3. This is what we intended to find.

**9.** We wish to divide a given number into two parts such that the sum of their cubes is a given multiple of the square of their difference.

It is necessary that the given multiplier be greater than three quarters of the given number by a number comprising, together with the cube of the given number, a square number.

Let the given number be 20 and the (given) multiplier be 140. We wish to divide 20 into two parts such that the sum of their cubes is 140 times the square of their difference. Let us assume the difference of the two parts to be again[17] $2x$, so that one of the two parts is $10 + x$ and the other, $10 - x$. The sum of their cubes is, according to what has been explained above,[18] $2000 + 60x^2$. But the square of the difference of the two numbers is $4x^2$.
86 Hence $2000 + 60x^2$ equals 140 times $4x^2$, that is, $560x^2$. Removing the $60x^2$, which is common, from both sides gives 2000 equal to $500x^2$; so $x^2$ is equal to 4. As the side of $x^2$ is $x$ and the side of 4 is 2, $x$ is equal to 2. Since we assumed the first of the two parts to be $10 + x$, it is 12; and, since we assumed the second part to be $10 - x$, it is 8. The cube of 12 (or 1728), when increased by the cube of 8 (or 512)[19], results in 2240; the difference of the two parts is 4, the square of which is 16, and 2240 is 140 times 16, or (140 times) the square of the difference of the two parts found by us.

Therefore, we have divided 20 into two parts in the desired manner, the larger part being 12 and the smaller, 8. This is what we intended to do.

---

[17] Cf. problem 7 (and, *mutatis mutandis*, 8).

[18] Cf. problem 7.

[19] These two values are known from V,7.

**10.** We wish to find two numbers such that their difference is a given number and the difference of their cubes is to the square of their sum in a given ratio.

It is necessary that the number belonging to the given ratio[20] be greater than three quarters of the number given for the difference of the two numbers ⟨by a number comprising, together with the cube of the number given for the difference of the two (required) numbers, a square number⟩.[21]

Let the number given for the difference of the two required numbers be 10 and the number corresponding to the given ratio be $8\frac{1}{8}$.[22] We wish to find two numbers such that their difference is 10 and the ratio of the difference of their cubes to the square of their sum is the ratio $8\frac{1}{8}:1$. We put $2x$ as their sum,
87 and we set as one of the two numbers $x + 5$ and as the other $x - 5$ in order that their difference be 10. We take the difference between their cubes, namely[23] $250 + 30x^2$. The square of the sum of the two numbers being $4x^2$, $250 + 30x^2$ equals $8\frac{1}{8}$ times $4x^2$, i.e., $32\frac{1}{2}x^2$. Let us remove the $30x^2$, which is common, from both sides, so 250 is equal to $2\frac{1}{2}x^2$; thus $x^2$ equals 100, and therefore $x$ is 10. Since we set as the first number $x + 5$, it is 15; and, since we set as the second number $x - 5$, it is 5. The cube of 15 is 3375 and the cube of 5, 125, the difference of which is 3250; the square of the sum of the two numbers is 400, and the ratio of 3250 to 400 is the ratio $8\frac{1}{8}:1$.

Therefore, we have found two numbers such that their difference is 10 and the difference of their cubes is $8\frac{1}{8}$ times the square of their sum; and these are 15 and 5. This is what we intended to find.

**11.** We wish to find two numbers such that their difference is a given number and the sum of their cubes is to their sum in a given ratio.

It is necessary that the number belonging to the given ratio exceed three quarters of the square of the number given for the difference of the two numbers by a square number.

Let the difference of the two numbers be 4 and the number belonging to the given ratio be 28.[24] We wish to find two numbers such that their difference is 4 and the sum of their cubes is to their sum in the ratio 28:1. We set $2x$
88 as the sum of the two numbers, so the first is $x + 2$ and the second, $x - 2$. The cube of the larger is $x^3 + 8 + 6x^2 + 12x$, and the cube of the smaller is $x^3 + 12x - 6x^2 - 8$. Their sum is $2x^3 + 24x$, for the negative $6x^2 + 8$ in the cube of the smaller number is eliminated by the positive $8 + 6x^2$ in the cube of the larger number. Hence $2x^3 + 24x$ is equal to 28 times the sum of the two numbers, $2x$, which is $56x$. We remove the $24x$, which is common, from both sides, and obtain $2x^3$ equal to $32x$; the division of both by $x$ gives $2x^2$ equal

[20] That is, the πηλικότης of that ratio (see above, p. 99, n. 47).
[21] This omission no doubt originated with a copyist.
[22] Literally: "$8\frac{1}{8}$ times".
[23] Cf. problem 8.
[24] Literally: "28 times"; see previous problem.

to 32, hence $x^2$ is equal to 16. As $x^2$ is a square with side $x$ and 16 is a square with side 4, $x$ equals 4. Since we set as the larger number $x + 2$, it is 6; and, since we set as the smaller number $x - 2$, it is 2. The cube of the larger is 216 and the cube of the smaller, 8; the sum of these two cubes is 224, which is 28 times the sum of the two numbers, or 8.

Therefore, we have found two numbers such that their difference is 4 and the sum of their cubes is 28 times their sum, and these are 6 and 2. This is what we intended to find.

**12.** We wish to divide a given number into two parts such that the difference of their cubes is a given multiple of their difference.[25]

89 It is necessary that the number belonging to the given ratio exceed three quarters of the square of the given number, here too,[26] by a square number.

Let the given number be 8 and the multiplier corresponding to the given ratio be 52.[27] We wish to divide 8 into two numbers such that the difference of their cubes is 52 times their difference. We put $2x$ as the difference of the two numbers, so the larger part is $4 + x$ and the smaller, $4 - x$. The cube of the larger part is $64 + x^3 + 48x + 12x^2$ and the cube of the smaller part, $64 + 12x^2 - x^3 - 48x$. Their difference is $2x^3 + 96x$; so $2x^3 + 96x$ equals 52 times the difference between the two numbers, $2x$, which is $104x$. We remove the $96x$, which is common, from both sides, so $2x^3$ is equal to $8x$; the division of each by $x$ gives $2x^2$ equal to 8, hence $x^2$ equals 4 and $x$, 2. Since we put $4 + x$ as the larger part, it is 6; and, since we put $4 - x$ as the smaller part, it is 2. The cube of the larger part is 216, and the cube of the smaller part, 8; their difference is 208, which is 52 times the difference of the two parts, or 4.

Therefore, we have divided 8 into two parts[28] such that the difference of their cubes is 52 times their difference; and these are 6 and 2. This is what we intended to do.

**13.** We wish to find a cubic number such that, when we add to a given multiple
90 of the square of its side a given number, the result is equal to the sum of two numbers, each of which gives, when added to the cube, a cube.

---

[25] Observe the various ways in which such problems are formulated: compare this enunciation with that of V,7 and of the Greek "IV",1.

[26] See previous problem.

[27] "be the ratio of the 52 times" in the text.

[28] The manuscript has "into two *different* parts". We put "different" in the apparatus, for it is not possible that the scholiast who added the final statements ever considered that the same value for the two parts, i.e., 4, fulfils the conditions of the problem. Perhaps some reader pointed out in the margin that the solution of this problem is the same as that of the previous one but with two *different* given ratios, and only the word "different" was inserted by a (our?) copyist in the text (cf. p. 33, no. 25, *in fine*).

Let the given number be 30 and the given multiplier be 9. We wish to find a cubic number such that, when we add nine times the square having the same side to 30, the result is equal to (the sum of) two numbers, each of which gives, when added to the cubic number, a cube. Let us put $x$ as the side of the cube, so that the cube is $x^3$; let us take nine times the square arising from its side, that is, $9x^2$, which we add to 30, so we obtain $9x^2 + 30$. Now, this $9x^2 + 30$ is equal to (the sum of) two numbers, each of which, when added to the cube, or $x^3$, yields a cube. So we shall form two cubes from two sides consisting each of $x$ and a certain (number of) units, (then) take the excess of each of these cubes over the (required) cube (that is, $x^3$), replace the two numbers by the(se) excesses, add them and equate their (sum) to $9x^2 + 30$; at that point, we shall have reached our goal. But these excesses are made up of $x^2$'s, $x$'s and a number; so it is necessary that those $x^2$'s which are contained in the sum of the two excesses amount to $9x^2$ and that the number which is with them be less than 30 in order that we arrive at a number equal to $x$. So,[29] we have to form the two cubes from two sides consisting each of $x$ plus a number in such a way that the sum of the $x^2$'s of the two cubes amount to $9x^2$ and the (total number of) units be less than 30, which is the given number. But the (number of) positive $x^2$'s found in each of the two cubes is three times the number added to $x$ in the (corresponding) side, and the total number of units found in the two cubes is the sum of the cubes of the said numbers.
91 Thus it is necessary that the sum of the two numbers added to $x$ be 3 in order that three times the said sum give the number of $x^2$'s, that is, 9. (So) we have to divide 3 into two parts such that the sum of their cubes be less than 30.[30] Such are 2 and 1. We form one of the cubes from the side $x + 2$, so that it is $x^3 + 6x^2 + 12x + 8$, and the other from the side $x + 1$, so that it is $x^3 + 3x^2 + 3x + 1$. Then, the $6x^2 + 12x + 8$, when added to $x^3$, gives a cube, and so does the $3x^2 + 3x + 1$; thus we shall make their sum, or $9x^2 + 15x + 9$, as mentioned, equal to $9x^2 + 30$.[31] Removing the $9x^2$, which is common, from both sides, we have $15x + 9$ equal to 30; we then remove the 9, which is common, from both sides, thus obtaining $15x$ equal to 21. Hence $x$ is $1\frac{2}{5}$. Since we put $x$ as the side of the required cube, the said side is $\frac{7}{5}$ and the cube, 2 and 93 parts of 125 parts of 1. So the square of the side of the cube is 1 and 24 parts of 25 (parts) of 1, and nine times that is 17 and 16 parts of 25 parts, or (17 and) 80 parts of 125 parts; the addition of that to 30 gives 47 and 80 parts of 125 parts. We had assumed one of the two parts of this last resulting number to be $6x^2 + 12x + 8$; the $6x^2$ being 11 and 95 parts of 125 parts of 1, and
92 the $12x$ being 16 and 100 parts of 125 parts of 1, the whole first number is 36 and 70 parts of 125 parts of 1. The second number is the remainder of the 47

---

[29] The text has here a causal clause, beginning with *li-annā* (line 2002), the apodosis of which is found far below, at line 2008. Similar situations also occur in the Greek text (see D.G., I, p. 262,2 *seqq.*).

[30] Cf. problem V,7.

[31] Some words in this passage seem to be interpolated; see p. 32, no. 21.

and 80 parts of 125 (after the subtraction of the first number), that is, 11 and 10 parts of 125. If the first of the said numbers is added to the cubic number, that is, (to) 2 and 93 parts of 125, the sum of that is 39 and 38 parts of 125 parts, which is a cubic number with side $3\frac{2}{5}$. And, if the second number is added to the cubic number, the sum of that is 13 and 103 parts of 125 parts of 1, which is a cubic number with $2\frac{2}{5}$ as its side.

Therefore, we have found a cubic number such that when we add nine times the square of its side to 30 the result is equal to (the sum of) two numbers, each of which gives, when added to the cubic number, a cube; and this is the cube determined by us. This is what we intended to find.

It is necessary to know that this problem is soluble by this treatment whenever the cube of the third of the multiplier is less than four times the given number.

**14.** We wish to find a cubic number such that, when we subtract from a given multiple of the square of its side a given number, the result is equal to (the sum of) two numbers, each of which leaves, when subtracted from the cube, a cube.

Let the given number be 26 and the given multiplier, 9. We wish to find a cubic number as indicated by us. We put $x$ as the side of the cube, which is then $x^3$. We take nine times the square of the side of the cube, or $x^2$, which is
93 $9x^2$; the subtraction from it of the given number results in $9x^2 - 26$, which is equal to (the sum of) two numbers such that each of them leaves, when subtracted from the cube, a cube. Following the way described in the previous problem, let us form two cubes, each having as its side $x$ minus a number (chosen) in such a way that the sum of the $x^2$'s subtracted in the cubes amount to $9x^2$. It is not necessary in the present problem that the sum of the two numbers contained in the cubes be less than the given units; they are (in this respect) arbitrary. Let us form the first cube from the side $x - 2$, so the said cube is $x^3 + 12x - 6x^2 - 8$; and, let us form the second one, from the side $x - 1$, so that the cube is $x^3 + 3x - 3x^2 - 1$. Then, the subtraction of the $6x^2 + 8 - 12x$ from the (required) cube (that is, from $x^3$) gives a cube, and so does the subtraction of the $3x^2 + 1 - 3x$ from the said cube also. Thus let us put the sum of these two numbers equal to $9x^2 - 26$. But their sum is $9x^2 + 9 - 15x$, so this is equal to $9x^2 - 26$. Let us add 26, and similarly $15x$, to both sides, and we remove the $9x^2$, which is common, from both sides. There remains, after the restoration and the reduction, $15x$ equal to 35; hence $x$ is $2\frac{1}{3}$. Since we put $x$ as the side of the cube, the said side is $2\frac{1}{3}$ and the cube 12 and 19 parts of 27 parts of 1. The square of the side of the cube is 5 and 12 parts of 27 parts; and nine times that is 49. Let us
94 subtract from it the given 26, so the remainder is 23. We had assumed the first of the two parts of the said 23 to be $6x^2 + 8 - 12x$; now, $6x^2$ is 32 and 18 parts of 27, and $12x$ is 28, so the larger of the two numbers is 12 and 18 parts of 27 (parts) of 1. Thus, the smaller number is 10 and 9 parts of 27 parts of 1.

Now, when the larger of the two said numbers is subtracted from the cube, found by us to be 12 and 19 parts of 27 parts of 1, the remainder is one part of 27 parts of 1, which is a cube with side $\frac{1}{3}$; and, when we subtract the smaller number from the cubic number, the remainder is 2 and 10 parts of 27 parts of 1, which is a cube with side $1\frac{1}{3}$.

Therefore, we have found a cubic number fulfilling the condition stipulated by us. This is what we intended to find.

**15.** We wish to find a cubic number such that, when we subtract from a given multiple of the square of its side a given number, the result is equal to (the sum of) two numbers, one of which gives, when added to the cube, a cube, and the other leaves, when subtracted from the cube, a cube also.

Let the given multiplier be 9 and the given number be 18. We wish to find a cubic number such that when 18 is subtracted from nine times the square of its side, the remainder is (the sum of) two numbers, one of which gives, when added to the cube, a cube, and the other leaves, when subtracted from the cube, a cube. We put $x$ as the side of the cube, so that the cube is $x^3$. We take nine times the square of its side, or $9x^2$, and then subtract from it 18. Next, we form two cubes with sides $x$ plus a number and $x$ minus a number, in such a way that the positive $x^2$'s in the one cube give, together
95 with the $x^2$'s subtracted in the other cube, $9x^2$. (So) let the side of the first[32] cube be $x - 2$, so that the first cube is $x^3 + 12x - 8 - 6x^2$, and let the side of the second cube be $x + 1$, so that the second cube is $x^3 + 3x^2 + 3x + 1$. So, the subtraction of the $6x^2 + 8 - 12x$ from the required cube, that is, (from) $x^3$, results in a cube, and the addition of the $3x^2 + 1 + 3x$ to the required cube results in a cube. Then, let us make the sum of the two said numbers equal to $9x^2 - 18$; their sum being $9x^2 + 9 - 9x$, this is equal to $9x^2 - 18$. Let us restore and reduce that, so there remains, after the restoration and the reduction, 27 equal to $9x$; hence $x$ is equal to 3. Since we put $x$ as the side of the cube, the side is 3 and the cube, 27. The square of the side of the cube is 9, and nine times that is 81; let us subtract from it the given number, or 18, so the remainder is 63. We had assumed one of the two numbers to be $6x^2 + 8 - 12x$; hence it is 26, and the second number is the remainder of 63 (after the subtraction of 26), or 37. The subtraction of 26 from the cube, that is, (from) 27, gives 1, which is a cube, and the addition of 37 to the cube, that is, (to) 27, gives 64, which is a cube with side 4.

Therefore, we have found a cube fulfilling the condition stipulated by us. This is what we intended to find.

**16.** We wish to find a cubic number such that, when we subtract from a
96 given multiple of the square of its side a given number, the result is equal to (the sum of) two numbers such that the subtraction of the one from the cube

[32] The one mentioned secondly in the previous sentence.

results in a cube, and the subtraction of the cube from the other number results in a cube.

Let the given multiplier again[33] be 9 and the given number be 16. We wish to find a cubic number such that when we subtract 16 from nine times the square of its side the result is equal to (the sum of) two numbers such that the subtraction of the one from the cube results in a cube, and the subtraction of the cube from the other number results in a cube. We again put[34] $x^3$ as the cube, and we subtract 16 from nine times the square of its side. We form two cubes having as their sides $x$ minus a number and a number minus $x$, and let the (sum of the) $x^2$'s occurring in them amount to $9x^2$. Thus we form the first cube from the side $x - 1$, so that it is $x^3 + 3x - 3x^2 - 1$, and the second cube from the side $2 - x$, so that it is $8 + 6x^2 - x^3 - 12x$. Then, when the $3x^2 + 1 - 3x$ is subtracted from the (required) cube (or $x^3$), the result is a cube, which is, as already said, $x^3 + 3x - 3x^2 - 1$; and, when the (required) cube, or $x^3$, is subtracted from $6x^2 + 8 - 12x$, the result is a cube, which is, likewise as already said, $8 + 6x^2 - 12x - x^3$. So let their sum be (put) equal to $9x^2 - 16$. But their sum is $9x^2 + 9 - 15x$, hence this is equal to $9x^2 - 16$. Let us restore and reduce that. We arrive, after the restoration and the reduction, at $15x$ equal to 25; hence $x$ is $1\frac{2}{3}$, and this is the side of the cube,
97 so that the cube is 4 and 17 parts of 27 (parts) of 1. The square of the side of the cube is 2 and 21 parts of 27 parts, and nine times that is 25. Let us subtract the 16 from it; the remainder is 9. And we had assumed that that one of the two numbers having 9 as their sum which is subtracted from the cube is $3x^2 + 1 - 3x$; as $3x^2$ is $8\frac{1}{3}$ and $3x$ is 5, the aforesaid number is $4\frac{1}{3}$, and the second number is the remainder of the 9 (after the subtraction of $4\frac{1}{3}$), namely $4\frac{2}{3}$. If $4\frac{1}{3}$ is subtracted from the cube, that is, (from) 4 and 17 parts of 27 parts, the remainder is 8 parts of 27 parts of 1, which is a cube with side $\frac{2}{3}$; and if the second number, or $4\frac{2}{3}$, is diminished by the cube, the remainder is one part of 27 parts of 1, which is a cube with side $\frac{1}{3}$.

Therefore, we have found a cubic number fulfilling the condition stipulated by us. This is what we intended to find.

End of the fifth Book of the treatise of Diophantus on arithmetical problems, and it contains sixteen problems.

[33] As in the preceding problems of this group.

[34] As in all the problems of this group.

# In the Name of God the Merciful, the Compassionate

# Sixth Book of the Treatise of Diophantus

**1.** We wish to find two numbers, one cubic and the other square, having their sides in a given ratio, such that when their squares are added, the result is a square number.

Let the given ratio be the ratio 2:1. We wish to find two numbers, one cubic and the other square, the side of the cube being twice the side of the square, such that when their squares are added, the result is a square number. Let us take $x$ as the side of the square, so that the square is $x^2$, and (therefore)
98 $2x$ as the side of the cube, so that the cube is $8x^3$. The sum of the square of the cube and of the square of the square is $64x^6 + x^4$, and it must be a square. Let us then seek a square number which, when diminished by 64, leaves a square. Finding that is easy on the basis of what has been shown previously in our treatise[1]. Such is 100. So let us make it $x^6$'s, so that it becomes $100x^6$; we equate $64x^6 + x^4$ with the $100x^6$ and remove the common (term), thus obtaining $36x^6$ equal to $x^4$. The division of the two sides by the one of lower degree, namely $x^4$, gives 1 equal to $36x^2$; hence $x^2$ is one part of 36 parts of 1 and $x$ is one part of 6 parts of 1, and it is the side of the square number. The side of the cubic number is twice that, which is two parts of 6 parts of the unit, and the cube is 8 parts of 216 parts of the unit. When its square, that is, 64 parts of 46,656 parts of the unit, is added to the square of the square number, which is 36 parts of 46,656 parts, the result is 100 parts of 46,656, which is a square number with 10 parts of 216 parts of the unit as its side.

Therefore, we have found two numbers fulfilling the condition imposed upon us, and these are 8 parts of 216 parts of the unit, and 6 parts of 216 parts of the unit. This is what we intended to find.

**2.** We wish to find two numbers, one cubic and the other square, having their sides in a given ratio, such that when the square of the square is subtracted from the square of the cube, the remainder is a square.

[1] *Arithmetica* II,10.

Let the given ratio be the ratio 2:1. We wish to find two numbers, one cubic and the other square, the side of the cube being twice the side of the square, such that when the square of the square is subtracted from the
99 square of the cube, the remainder is a square. Let us put $x$ as the side of the square number and (therefore) $2x$ as the side of the cubic number; hence the square number is $x^2$ and its square, $x^4$, and the cubic number is $8x^3$ and its square, $64x^6$. If we subtract $x^4$ from $64x^6$, we obtain $64x^6 - x^4$, which must be a square number. Let us then look for a square number which, when subtracted from 64, leaves a square number. Finding that is easy on the basis of a previous exposition;[2] the (required number) is 40 and 24 parts of 25 parts of the unit. Let us make it $x^6$'s, so it becomes $40x^6$ and 24 parts of 25 parts of $x^6$, which is equal to $64x^6 - x^4$. We restore that[3] and drop the common (term), thus obtaining $23x^6$ and one part of 25 parts of $x^6$ equal to $x^4$. The division of the two sides by $x^4$ gives 1 equal to $23x^2$ and one part of 25 (parts) of $x^2$. Hence $x^2$ is 25 parts of 576 parts of the unit, and $x$ is 5 parts of 24 parts of 1. We had set as the side of the cubic number $2x$, which is 5 parts of 12 parts of the unit; so the cube is 125 parts of 1728 parts of the unit, and its square, 15,625 parts of 2,985,984 parts. If we subtract from the latter the square of the square number, namely 5625 parts of 2,985,984, we obtain
100 10,000 parts of 2,985,984 parts, which is a square number with side 100 parts of 1728 parts.

Therefore, we have found two numbers fulfilling the condition imposed upon us, and these are 125 parts of 1728 parts of the unit and 75 parts of 1728 parts of the unit. This is what we intended to find.

**3.** We wish to find two numbers, one cubic and the other square, having their sides in a given ratio, such that when we subtract the square of the cube from the square of the square number, the remainder is a square.

Let the given ratio be the ratio 2:1. We wish to find two numbers, one cubic and the other square, the side of the cube being to the side of the square in the ratio 2:1, such that when the square of the cubic number is subtracted from the square of the square number, the remainder is a square. We put $x$ as the side of the square number; hence the side of the cube is $2x$ and the cube, $8x^3$, and the square of the latter is $64x^6$. Since we had taken $x$ as the side of the square number, the square is $x^2$ and its square, $x^4$. We subtract from it the square of the cube, or $64x^6$, and obtain $x^4 - 64x^6$, which must be a square. (So) we shall seek a square number which, when increased by 64, results in a square;[4] such is 36, with side 6. Thus we put, as the side of $x^4 - 64x^6$, $6x^3$, and multiply that by itself; we obtain $36x^6$, which equals $x^4 - 64x^6$. We restore[5] and obtain $x^4$ equal to $100x^6$; the division of all that

[2] *Arithmetica* II,8.
[3] *fa-nuqābil bi-hā* in the text.
[4] *Arithmetica* II,10.
[5] *fa-najbur wa-nuqābil* in the text.

by $x^4$ results in 1 equal to $100x^2$. Hence $x^2$ is one part of 100, or $\frac{1}{10} \cdot \frac{1}{10}$, and $x$ is
101 one part of 10, or $\frac{1}{10}$. We had put $2x$ as the side of the cube, so the said side is 2 parts of 10 and the cube, 8 parts of 1000; its square is 64 parts of 1,000,000. When we subtract it from the square of the square number, that is, (from) 100 parts of 1,000,000, the remainder is 36 parts of 1,000,000, which is a square number with side 6 parts of 1000 parts of the unit.

Therefore, we have found two numbers fulfilling the condition imposed upon us, and these are 8 parts of 1000 parts and 10 parts of 1000 parts. This is what we intended to find.

**4.** We wish to find two numbers, one cubic and the other square, the side of the cube being to the side of the square in a given ratio, such that, when the number comprised by them is increased by the square of the cube, the result is a square.

Let the (given) ratio be the ratio 5:1. We wish to find two numbers, one cubic and the other square, the side of the cube being five times the side of the square, such that, when the number comprised by them is increased by the square of the cubic number, the result is a square. Let us put $x$ as the side of the square, so that the square is $x^2$; the side of the cube is (then) $5x$ and the cube, $125x^3$. Hence the number which they comprise is $125x^5$.[6] We increase the $125x^5$ by the square of the cube—and the square of the cube is $15{,}625x^6$—[7] and obtain $15{,}625x^6 + 125x^5$, which must be a square. Let us then seek a square number which, when diminished by 15,625, gives a small number; and we do not need the remainder to be a square number. Such a number is
102 15,876, the side of which is 126. Let us make it—that (is, the side of) $15{,}625x^6 + 125x^5$—[8] ($x^3$'s, so it becomes) $126x^3$; we multiply $126x^3$ by itself, and obtain $15{,}876x^6$, which is equal to $15{,}625x^6 + 125x^5$. We remove the common $15{,}625x^6$ from the two sides, whence $251x^6$ equals $125x^5$. Divide[9] the two sides by $x^5$, hence 125 is equal to $251x$; then $x$ is 125 parts of 251, and this is the side of the square, and the square is 15,625 parts of the square of 251, that is, (of) 63,001. The side of the cubic number was five times the side of the square number, which is 625 parts of 251; (thus) the cube is 244,140,625 parts of 2,563,001.[10] And we shall content ourselves with the correctness of the treatment of the present problem on the basis of the related problems.

Therefore, we have found two numbers fulfilling the condition imposed upon us, and these are 15,625 parts of 63,001 and 244,140,625 parts of 2,563,001. This is what we intended to find.

---

[6] "$125x^3$ multiplied by $x^2$" in the text; see p. 45.

[7] This seems to be a later addition (cf. p. 32, no. 16).

[8] There is, clearly, an interpolation here (see p. 31, no. 9).

[9] The imperative, found here for the first time, occurs only seven times in our manuscript (see p. 46).

[10] Here and further below, 2,563,001 is given instead of 15,813,251. See commentary.

**5.** We wish to find two numbers, one cubic and the other square, the side of the cube being equal to the side of the square, such that when the number which they comprise is increased by the square of the square number, the result is a square.

103 Let us put $x$ as the side of the square, so that the square is $x^2$; again, the side of the cube is $x$, and the cube is $x^3$. The number which they comprise is $x^5$. We increase it by the square of the square number, or $x^4$; it becomes $x^5 + x^4$, which equals a square number. Let us put $2x^2$ as its side, hence $4x^4$ is equal to $x^4 + x^5$. We remove the common $x^4$, so $x^5$ is equal to $3x^4$. The division of the whole by $x^4$ results in $x$ equal to 3, and this is the side of the square, and the square is 9. Again, the side of the cube was equal to the side of the square, so it is 3, and the cubic number is 27. The number which they comprise is the result of the multiplication of 9 by 27, namely 243. When 243 is increased by the square of the square number, or 81, the result is 324, which is a square number with 18 as its side.

Therefore, we have found two numbers fulfilling the condition imposed upon us. This is what we intended to find.

**6.** We wish to find two numbers, one square and the other cubic, the side of the cube being equal to the side of the square, such that when we subtract from the number which they comprise the square of the cubic number, the remainder is a square number.

Let us put $x$ as the side of the square number, so the square number is $x^2$; again, the side of the cubic number is $x$, and the cubic number is $x^3$. We have to subtract the square of $x^3$ from the number comprised by $x^3$ and $x^2$. But the number which they comprise is $x^5$. So, when we subtract from it the square of the cubic number, that is, $x^6$, we have $x^5 - x^6$, and this must be a square. Let us put $x^3$ as its side; we multiply $x^3$ by itself and obtain $x^6$, which equals $x^5 - x^6$. Let us add $x^6$ (in common) to the two sides, and we divide
104 the two (resulting) sides by the one of lower degree, which is $x^5$; then $2x$ is equal to 1, so $x$ is equal to $\frac{1}{2}$. We had put $x$ as the side of the square, so the square is one part of 4, or $\frac{1}{4}$; again, the side of the cube is $\frac{1}{2}$, and the cube is $\frac{1}{8}$. Since the number which they comprise is one part of 32, subtracting from it the square of the cube, or one part of 64 parts, leaves one part of 64, which is a square with one part of 8 as its side.

Therefore, we have found two numbers fulfilling the condition imposed upon us, and these are $\frac{1}{4}$ and $\frac{1}{8}$. This is what we intended to find.

**7.** We wish to find two numbers, one cubic and the other square, having their sides equal, such that when the number which they comprise is diminished by the square of the square number, the remainder is a square.

Let us put $x$ as the side of the square number, which is then $x^2$; since the side of the cube is equal to the side of the square number, the cubic number must be $x^3$. So the number which they comprise is $x^5$. Now, if we subtract from $x^5$ the square of the square number, or $x^4$, the remainder is $x^5 - x^4$, and this must be a square. Let us set as its side $x^2$; the multiplication of $x^2$ by itself gives $x^4$, (which is) equal to $x^5$ diminished by $x^4$. We restore and solve,[11] hence $x$ is 2. We had assumed the side of the square to be $x$, so the said side is 2 and the square, 4; again, the side of the cube is 2 and the cube, 8. The square being 4 and the cube being 8, the number which they comprise is 32. If we subtract from 32 the square of the square number (namely 16), we obtain 16, which is a square number with side 4.

Therefore, we have found two numbers fulfilling the condition imposed upon us, and these are 8 (and)[12] 4. This is what we intended to find.

**8.** We wish to find two numbers, one cubic and the other square, such that
105 when the number which they comprise is increased by its side, the result is a square.

We put 64 as the cubic number and $x^2$ as the square number, so that the number which they comprise is $64x^2$. Now, if we add to $64x^2$ its side, namely $8x$, the result is $64x^2 + 8x$, which must be a square. Let us put as its side any multiple of $x$ we please provided that it is greater than $8x$, say $10x$; we multiply that by itself, thus obtaining $100x^2$, and this is equal to $64x^2 + 8x$. We remove (the common) $64x^2$ from the two sides, so $36x^2$ equals $8x$. Dividing $36x^2$ by $x$ gives $36x$ and dividing $8x$ by $x$ gives 8. Hence 8 is equal to $36x$, and $x$ is two parts of 9. We had put $x$ as the side of the square, so the square is 4 parts of 81 parts of the unit; that is the square number, and the cubic number is 64. (So) the number which they comprise is 256 parts of 81 parts of the unit. If we increase it by its side, namely (by) 16 parts of 9, or 144 parts of 81, the result is 400 parts of 81, which is a square number with 20 parts of 9 as its side.

Therefore, we have found two numbers fulfilling the condition imposed upon us, and these are 64, (and) 4 parts of 81 parts of 1. This is what we intended to find.

**9.** We wish to find two numbers, one cubic and the other square, which comprise a number such that when it is diminished by its side, the remainder is a square.

We put 64 as the cubic number and $x^2$ as the square number, so that the number which they comprise is $64x^2$. Diminishing it by its side results in
106 $64x^2 - 8x$, which must be a square. Let us put for its side any number of $x$'s we wish, provided that it is less than $8x$, say $7x$; the multiplication of $7x$ by itself gives $49x^2$, which equals $64x^2 - 8x$. We restore and reduce, whence $15x^2$ equals $8x$. The division of that by $x$ gives $15x$ equal to 8; so $x$ is 8 parts

[11] *fa-najbur wa-nuqābil* in the text.

[12] The final results are, from here on, generally stated without the conjunction. See p. 37.

of 15 parts of the unit. We had assumed the side of the square number to be $x$, so the square number is 64 parts of 225 parts of the unit. Since the cubic number is 64, the number which they comprise is 4096 parts of 225 parts; diminishing it by its side, that is, (by) 64 parts of 15, or 960 parts of 225, results in 3136 parts of 225, which is a square number with 56 parts of 15 as its side.

Therefore, we have found two numbers fulfilling the condition imposed upon us, and these are 64, and 8 parts of 15 parts of the unit, that is to say 64 parts of 225.[13] This is what we intended to find.

**10.** We wish to find two numbers, one cubic and the other square, such that, when the number which they comprise is subtracted from its side, the remainder is a square number.

Let us put 64 as the cubic number and $x^2$ as the square number; so the number which they comprise is $64x^2$. Subtracting $64x^2$ from the side of $64x^2$, i.e., $8x$, we obtain $8x - 64x^2$, which must be a square. We take as its side any number of $x$'s we wish, say $4x$. Thus $16x^2$ equals $8x - 64x^2$. Restoring,[14]
107 we have $8x$ equal to $80x^2$. The division of the two sides by $x$ gives 8 equal to $80x$, hence $x$ is one part of 10. We had assumed the side of the square to be $x$; so the square is one part of 100 parts of the unit. And, as the cubic number is 64, the number which they comprise is 64 parts of 100 parts. If we subtract that from its side, namely 8 parts of 10, or 80 parts of 100, the remainder is 16 parts of 100, which is a square number with 4 parts of 10 as its side.

Therefore, we have found two numbers fulfilling the condition imposed upon us, and these are 64, and one part of 100 parts of the unit. This is what we intended to find.

**11.** We wish to find a cubic number such that if we add it to its square, the result is a square number.

We put $x$ as the side of the cubic number, so that the cubic number is $x^3$. Adding $x^3$ to its square, that is, (to) $x^6$, we obtain $x^6 + x^3$, which must be a square. Let us put for its side a number ⟨of $x^3$'s such that, when we subtract from their square $x^6$, the remainder is a cube; such is⟩[15] $3x^3$: when we subtract $x^6$ from the square of $3x^3$, we obtain $8x^6$, which is a cubic number. Hence, if we equate $8x^6$ with a cubic number,[16] the problem will be soluble and the treatment will not be impossible. Let us multiply the $3x^3$ by themselves, so we obtain $9x^6$, which then equals $x^6 + x^3$. We remove the $x^6$ which is common, so $8x^6$ equals $x^3$. The division of the two sides by $x^3$ gives

[13] This rectification certainly arose from a reader's gloss intended to correct the final statement (cf. p. 31, no. 10).

[14] *fa-najbur wa-nuqābil* in the text.

[15] We assume that there is a gap in the text here. See commentary.

[16] *Sic*, instead of "$x^3$".

$8x^3$ equal to 1; hence $x^3$ is $\frac{1}{8}$, or one part of 8. If we increase this by its square, that is, (by) one part of 64 parts of the unit, the result is 9 parts of 64 parts of the unit, which is a square number with 3 parts of 8 as its side.

Therefore, we have found a number fulfilling the condition imposed upon us, and this is one part of 8 parts of the unit. This is what we intended to find.

108 **12.** We wish to find two square numbers such that the quotient of the larger divided by the lesser, when added to the larger, gives a square, and also when added to the lesser, gives a square.

Let us put $x^2$ as the smaller number; we take $\frac{1}{2}x^2 + \frac{1}{2}\cdot\frac{1}{8}x^2$ as the quotient of the larger divided by the lesser. Thus, the addition of this quotient to $x^2$ gives a square. (So) the larger number is $\frac{1}{2}x^4 + \frac{1}{2}\cdot\frac{1}{8}x^4$. Then, when we increase it by $\frac{1}{2}x^2 + \frac{1}{2}\cdot\frac{1}{8}x^2$, we obtain $\frac{1}{2}x^4 + \frac{1}{2}\cdot\frac{1}{8}x^4 + \frac{1}{2}x^2 + \frac{1}{2}\cdot\frac{1}{8}x^2$, which has to be a square number. Hence, let us seek a square number which, when diminished by $\frac{1}{2} + \frac{1}{2}\cdot\frac{1}{8}$, gives a square number, and let us keep in mind that the remaining square be less than 81 parts of 256 parts of 1.[17] Finding that is easy from what has been explained in the second Book.[18] The said number is 169 parts of 256 parts of the unit, with side 13 parts of 16 parts of the unit. It appears that, when we subtract $\frac{1}{2} + \frac{1}{2}\cdot\frac{1}{8}$, or 144 parts of 256 parts, from 169 parts of 256 parts of the unit, the remainder is 25 parts of 256 parts of the unit, which is a square number with side 5 parts of 16 parts. So let us put, as the side of $\frac{1}{2}x^4 + \frac{1}{2}\cdot\frac{1}{8}x^4 + \frac{1}{2}x^2 + \frac{1}{2}\cdot\frac{1}{8}x^2$, 13 parts of 16 parts of $x^2$; we multiply it by itself, whence 169 parts of 256 parts of $x^4$, which then equal $\frac{1}{2}x^4 + \frac{1}{2}\cdot\frac{1}{8}x^4 + \frac{1}{2}x^2 + \frac{1}{2}\cdot\frac{1}{8}x^2$. Let us remove the $\frac{1}{2}x^4 + \frac{1}{2}\cdot\frac{1}{8}x^4$ which is
109 common, so 25 parts of 256 parts of $x^4$ equal $\frac{1}{2}x^2 + \frac{1}{2}\cdot\frac{1}{8}x^2$, and let us multiply the whole by 10 and 6 parts of 25; we obtain $x^4$ equal to $5x^2$ and 19 parts of 25 parts of $x^2$. We divide the two sides by $x^2$, hence $x^2$ is equal to 5 and 19 parts of 25 parts of the unit. We had put $x^2$ as the smaller number, so it is 5 and 19 parts of 25 parts of the unit; let us multiply that by 25, it then becomes 144, (which is) parts of 25 parts. And, since we set for the larger number $\frac{1}{2}x^4 + \frac{1}{2}\cdot\frac{1}{8}x^4$, it is 11,664 parts of 625 parts of the unit. Let us make the 144 parts of 25 parts, which form the smaller square, parts of 625, in other words (let us) multiply them by 25; then the smaller square is 3600 parts of 625. The quotient of the larger square divided by the smaller square is 3 and 6 parts of 25 (parts) of the unit. Let us make that parts of 625, so it becomes 2025 parts of 625. The addition of this to the larger square, that is, (to) 11,664 parts of 625, gives 13,689 parts of 625 (parts) of the unit, which is a
110 square number with side 117 parts of 25 parts. Again, let us add the 2025 parts of 625 to the smaller square, that is, (to) 3600 parts of 625, so the result is 5625 parts of 625, which is a square number with side 75 parts of 25.

[17] The text has only: "that the remaining square be less than 1". See commentary.

[18] *Arithmetica* II,10.

Therefore, we have found two numbers fulfilling the condition imposed upon us, and these are 11,664 parts of 625 parts of the unit (and) 3600 parts of 625 parts of the unit. This is what we intended to find.

**13.** We wish to find two square numbers such that the quotient of the division of the larger by the smaller, when subtracted from each of them, leaves (in both cases) a square.

Let us put $x$ as the side of the smaller square, so the smaller square is $x^2$; we set for the quotient of the division of the greater square by the smaller square—which is $x^2$—something which, when subtracted from $x^2$, leaves a square; it is further necessary that the (term) subtracted from $x^2$ be a square. So let us divide $x^2$ into two square parts;[19] such are 16 parts of 25 parts of $x^2$ and 9 parts of 25 parts of $x^2$. Then, let us set the 9 parts of 25 parts of $x^2$ as the quotient of the division of the larger square by $x^2$. The multiplication of the 9 parts of 25 parts of $x^2$ by $x^2$ gives 9 parts of 25 parts of $x^4$, which is the larger number. It appears that, subtracting the quotient of the division of the larger number—or 9 parts of 25 parts of $x^4$—by the smaller number—or $x^2$—, i.e., subtracting 9 parts of 25 parts of $x^2$ from the smaller number,
111 that is, (from) $x^2$, the remainder is 16 parts of 25 parts of $x^2$, which is a square with side $\frac{4}{5}x$. We now have to subtract 9 parts of 25 parts of $x^2$ from the larger number, or 9 parts of 25 parts of $x^4$, so that a square number remain. But, when we subtract 9 parts of 25 parts of $x^2$ from 9 parts of 25 parts of $x^4$, we obtain 9 parts of 25 parts of $x^4$ minus 9 parts of 25 parts of $x^2$, which is equal to a square number. So let us seek a square number which, when subtracted from 9 parts of 25, leaves a square number.[20] Such is 81 parts of 625 parts of the unit; the subtraction of it from 9 parts of 25, or 225 parts of 625 parts of the unit, results in 144 parts of 625, which is a square number with side 12 parts of 25 parts of the unit.

Now that we have reached our goal, let us put, for the root of 9 parts of 25 parts of $x^4$ minus 9 parts of 25 parts of $x^2$, 9 parts of 25 parts of $x^2$; we multiply this by itself, whence 81 parts of 625 parts of $x^4$, which is equal to 9 parts of 25 parts of $x^4$ minus 9 parts of 25 parts of $x^2$, or 225 parts of 625 parts of $x^4$ minus 225 parts of 625 parts of $x^2$. We restore[21] and remove the common (term); then 144 parts of 625 parts of $x^4$ is equal to 225 parts of
112 625 parts of $x^2$. Divide the two sides by $x^2$, this gives 144 parts of 625 parts of $x^2$ equal to 225 parts of 625 parts of 1; (so) $x^2$ is equal to 1 and 81 parts of 144 parts of the unit, or 1 and 9 parts of 16 parts. We had assumed the smaller square to be $x^2$, so it is 25 parts of 16 parts of the unit; the larger (square) number is 9 parts of 25 parts of the square of the smaller (square) number, that is to say, 225 parts of 256 parts of the unit. The division of the larger (square) number, that is, 225 parts of 256 parts of 1, by the smaller (square)

[19] *Arithmetica* II,8 (but the result is trivial).

[20] *Arithmetica* II,8.

[21] *fa-najbur wa-nuqābil* in the text.

number, that is, 25 parts of 16, or 400 parts of 256 parts of 1, gives the result $\frac{1}{2}+\frac{1}{2}\cdot\frac{1}{8}$, or 144 parts of 256 parts. The subtraction of this last number from the first of the two squares, that is, from 400 parts of 256 parts of the unit, gives 256 parts of 256 parts, or 1, which is a square with side 1. Again, the subtraction of the quotient of the division, that is, (of) 144 parts of 256 parts of the unit, from the square which is 225 parts of 256, results in 81 parts of 256, which is a square with side 9 parts of 16.

Therefore, we have found two numbers fulfilling the condition imposed
113 upon us, and these are 400 parts of 256 parts of 1 (and) 225 parts of 256 parts of the unit. This is what we intended to find.

Our purpose in this problem (however) was that the dividend be the larger number; but the treatment left us with the larger number being the divisor. Since our treatment was correct—there can be no doubt about it—, we have recorded it.

We shall (now) solve this problem by a second treatment leading to our requirement regarding the quotient of the larger square divided by the smaller square. And let it be a treatment easier than the preceding one.

We put $1\frac{2}{3}$ as the side of the smaller square, so that the square is $2\frac{7}{9}$. We put $x$ as the side of the larger square, so that the larger square is $x^2$. Dividing the larger square, or $x^2$, by the smaller square, or $2\frac{7}{9}$, gives the quotient 9 parts of 25 parts of $x^2$. Then, if we subtract it from the larger square, that is, (from) $x^2$, the remainder is 16 parts of 25 parts of $x^2$, which is a square number with side $\frac{4}{5}x$. Again, subtracting the result of the division, that is, 9 parts of 25 parts of $x^2$, from the smaller square, or $2\frac{7}{9}$, leaves $2\frac{7}{9}$ minus 9 parts of 25 parts of $x^2$, which has to be a square. We put as its side $1\frac{2}{3}-1\frac{1}{5}x$, and we multiply that by itself; we obtain $2\frac{7}{9}$, plus $x^2$, plus 11 parts of 25 parts of $x^2$, minus $4x$, which is then equal to $2\frac{7}{9}$ minus 9 parts of 25 parts
114 of $x^2$. Restore each of the two sides with its subtractive (term), add its amount to the other side and remove the similar common (term); there remains then $1\frac{4}{5}x^2$ equal to $4x$. Divide both sides by $x$, hence $1\frac{4}{5}x$ is equal to 4; so $x$ is $2\frac{2}{9}$. As the side of the larger square was $x$, the said side is $2\frac{2}{9}$ and the larger square, 400 parts of 81 parts of 1. The division of the latter by the smaller square, that is, (by) $2\frac{7}{9}$, or 225 parts of 81 parts of the unit, gives the quotient $1\frac{7}{9}$, or 144 parts of 81 parts; if we subtract that from the larger square, that is, (from) 400 parts of 81 parts, the remainder is 256 parts of 81 parts, which is a square number with side 16 parts of 9. And, if we subtract the same from the smaller square, or 225 parts of 81 parts, the remainder is 81 parts of 81, or 1, which is a square with side 1.

Therefore, we have found two numbers fulfilling the condition imposed upon us, and these are 400 parts of 81 parts of the unit (and) 225 parts of 81 parts of the unit. This is what we intended to find.

**14.** We wish to find two square numbers such that, when the larger is divided by the smaller, the division results in something which, when diminished by the larger square, leaves a square, and also when diminished by the smaller square, leaves a square.

Let us put $x$ as the side of the larger square, so the larger square is $x^2$;
115 again, we put, as the side of the smaller square, $\frac{4}{5}$, so the smaller square is 16 parts of 25 parts of 1. It appears that, if we divide the larger square, or $x^2$, by the smaller square, or 16 parts of 25 parts of 1, the result of the division is $x^2$ plus 9 parts of 16 parts of $x^2$, and (thus) if we diminish it by the larger square, or $x^2$, the remainder is 9 parts of 16 parts of $x^2$, which is a square with side $\frac{3}{4}x$. We now subtract from the quotient the smaller square, that is, 16 parts of 25 parts of the unit; the remainder is then 25 parts of 16 parts of $x^2$ minus 16 parts of 25 parts of 1, which has to be a square. Let us take as its side $1\frac{1}{4}x - 2$; we multiply it by itself, so we obtain 25 parts of 16 parts of $x^2$, plus 4, minus $5x$, and this equals 25 parts of 16 parts of $x^2$ minus 16 parts of 25 parts of 1. We restore each side with its subtractive (term), add its amount to the other side, and remove the common (term). There remains $5x$ equal to 4 and 16 parts of 25 parts of 1; hence $x$ is a fifth of 4 plus 16 (parts) of 25 parts of 1, which is 116 parts of 125 parts of the unit. We had put $x$ as the side of the larger square, so the side is 116 parts of 125 parts of the unit, and thus the square is
116 13,456 parts of 15,625. The division of that by the smaller square, that is, (by) 16 parts of 25, or 10,000 parts of 15,625, results in 1 and 3456 parts of 10,000, or 21,025 parts of 15,625. If we diminish that by the larger square, that is, (by) 13,456 parts, the remainder is 7569 parts of 15,625, which is a square number with 87 parts of 125 parts as its side. Again, if we diminish the same by the smaller (square) number, that is, (by) 10,000 parts, the remainder is 11,025 parts of 15,625, which is a square number with side 105 parts of 125 parts of the unit.

Therefore, we have found two numbers fulfilling the condition imposed upon us, and these are 13,456 parts of 15,625 parts of the unit (and) 10,000 parts of 15,625 parts of the unit. This is what we intended to find.

**15.** We wish to find two square numbers such that when the excess of the larger over the smaller is added to each of them, the result of that is (in both cases) a square.

Let us put $x$ as the side of the larger square, so the larger square is $x^2$. We take as the excess of it over the smaller square $2x + 1$; hence the smaller (square) number is $x^2 - 2x - 1$. It appears that, if we add the excess of the larger of the two numbers over the smaller, that is, $2x + 1$, to the smaller,
117 that is, (to) $x^2 - 2x - 1$, the result is $x^2$, which is the larger square number and (therefore) is a square.[22] If we now add the excess of the larger square over the smaller square, or $2x + 1$, to the larger square, or $x^2$, $x^2$ becomes $x^2 + 2x + 1$, which is a square number with side $x + 1$. It is then necessary that the

[22] The words "is the larger square number and" may be interpolated (cf. p. 32, no. 21).

smaller number, or $x^2 - 2x - 1$, be a square. Let us put $x - 2$ as its side; the multiplication of this by itself results in $x^2 + 4 - 4x$, which equals $x^2 - 2x - 1$. We add $4x$ to $x^2 + 4 - 4x$, which becomes $x^2 + 4$, and we also add $4x$ to $x^2 - 2x - 1$, which becomes $x^2 + 2x - 1$. Then, we add 1 to both sides and remove the common (term) so as to have a single term equal to a single term. Hence 5 is equal to $2x$, and $x$ is $2\frac{1}{2}$. We had set $x$ as the side of the larger square; so the side is $2\frac{1}{2}$ and the larger square is $6\frac{1}{4}$. If we diminish it by $2x + 1$, that is, (by) 6, the remainder is $\frac{1}{4}$, and this is the smaller square. It appears that the excess of the larger square, or $6\frac{1}{4}$, over the smaller square, or $\frac{1}{4}$, is 6, the addition of which to the larger square gives $12\frac{1}{4}$, which is a square number with $3\frac{1}{2}$ as its side, and, also, the addition of 6 to the smaller square gives $6\frac{1}{4}$, which is a square number, having $2\frac{1}{2}$ as its side.

Therefore, we have found two numbers fulfilling the condition imposed upon us, and these are $6\frac{1}{4}$ (and) $\frac{1}{4}$. This is what we intended to find.

**16.** We wish to find two square numbers such that the excess of the larger
118 over the smaller, when subtracted from the larger, leaves a square number, and also, when subtracted from the smaller, leaves a square number.

Let us put $x$ as the side of the larger square, so that the larger square is $x^2$. We take, as the excess of it over the smaller square, $2x - 1$; hence the smaller square is $x^2 + 1 - 2x$. It appears that, if we subtract the excess of the larger square over the smaller square, namely $2x - 1$, from the larger square, namely $x^2$, the remainder is $x^2 + 1 - 2x$, which is the smaller square and (therefore) is a square.[23] If we now subtract the excess of the larger square over the smaller square, or $2x - 1$, from the smaller square, or $x^2 + 1 - 2x$, the remainder is $x^2 + 2 - 4x$, and this has to be a square. Let us put $x - 4$ as its side; we multiply $x - 4$ by itself and obtain $x^2 + 16 - 8x$, and this is equal to $x^2 + 2 - 4x$. We add $8x$ to both sides and remove the $x^2 + 2$ which is common; hence $4x$ is equal to 14, and $x$ is $3\frac{1}{2}$. We had set $x$ as the side of the larger square, so the said side is $3\frac{1}{2}$ and its square, $12\frac{1}{4}$. Diminishing $12\frac{1}{4}$ by twice its root minus one, that is, (by) 6, results in $6\frac{1}{4}$; and this number is the smaller square. (So) the excess of the larger square over the smaller square is 6. Then, subtracting 6 from the larger square results in $6\frac{1}{4}$, which is a square number with side $2\frac{1}{2}$; again, subtracting 6 from the smaller square gives $\frac{1}{4}$, which is a square number with side $\frac{1}{2}$.

Therefore, we have found two numbers fulfilling the condition imposed
119 upon us, and these are $12\frac{1}{4}$ (and) $6\frac{1}{4}$. This is what we intended to find.

**17.** We wish to find three square numbers which, when added, give a square, and such that the first of these (three square) numbers equals the side of the second, and the second equals the side of the third.

---

[23] The words "is the smaller square and" may be an interpolation (cf. p. 32, no. 21).

Let us put $x^2$ as the first, so that the second is $x^4$—for $x^4$ is the square of $x^2$, and $x^2$ is equal to the side of the second—, and the third is $x^8$, —which equals the square of the second, and the second is its side.[24] The three numbers, when added, give $x^8 + x^4 + x^2$, and this has to be a square number. Let us put as its side $x^4 + \frac{1}{2}$; this when multiplied by itself gives $x^8 + x^4 + \frac{1}{4}$, which is equal to $x^8 + x^4 + x^2$. We remove the identical common (terms); so $x^2$ is equal to $\frac{1}{4}$. We had put $x^2$ as the first of the three numbers, so it is $\frac{1}{4}$. This $\frac{1}{4}$ is equal to the side of the second, (so) the second is $\frac{1}{2} \cdot \frac{1}{8}$. Again, the second equals the side of the third, (so) the third is one part of 256 parts of 1. These three numbers, when added, give 81 parts of 256 parts of the unit, which is a square number with side 9 parts of 16.

Therefore, we have found three numbers fulfilling the condition imposed upon us, and these are $\frac{1}{4}$, $\frac{1}{2} \cdot \frac{1}{8}$, (and) one part of 256 parts of 1. This is what we intended to find.

**18.** We wish to find three square numbers such that when we multiply the first number by the second number, and then the product by the third number, and add to the result of that the number formed by the sum of the three numbers, the result is a square number.

120 Let us put 1 as the first number, $\frac{1}{2} + \frac{1}{2} \cdot \frac{1}{8}$ as the second and $x^2$ as the third. Then, we multiply the first, or 1, by the second, or 9 parts of 16; we obtain 9 parts of 16 parts of the unit, which we multiply by the third number, or $x^2$, so that we obtain 9 parts of 16 parts of $x^2$. We increase that by the number formed by the sum of the three numbers, that is, (by) $x^2$ plus 25 parts of 16 parts of 1; the result is 25 parts of 16 parts of $x^2$, plus 25 parts of 16 parts of the unit, and this has to be a square number. Let us put as its side $1\frac{1}{4}x + \frac{1}{4}$, which we multiply by itself; hence we obtain 25 parts of 16 parts of $x^2$ plus 10 parts of 16 parts of $x$, plus one part of 16 parts of 1, and this is equal to 25 parts of 16 parts of $x^2$, plus 25 parts of 16 parts of 1. We remove the identical common (terms); so 24 parts of 16 parts of 1 is equal to 10 parts of 16 parts of $x$, so that the whole $x$ is equal to $2\frac{2}{5}$. We had put $x$ as the side of the third number, so the said side is $2\frac{2}{5}$ and the third number, 144 parts of 25 parts of the unit; the first number is, as we had assumed, 1, and the second number is, (also) as we had assumed, 9 parts of 16 parts of 1. The multiplication of the first number by the second number and then of the product by the third number gives 81 parts of 25 parts of the unit, or 1296 parts of 400 parts of the unit. We then increase that by the number formed by the (sum of the)
121 three numbers, namely (by) 144 parts of 25 parts, plus 1, plus 9 parts of 16 parts of the unit, or 2929 parts of 400; we obtain 4225 parts of 400 parts of the unit, which is a square number with side 65 parts of 20 parts of the unit.

[24] The text seems to contain some interpolations here (cf. p. 31, no. 11).

Therefore, we have found three numbers fulfilling the condition imposed upon us, and these are 144 parts of 25 parts of the unit, 1, and 9 parts of 16 parts of the unit. This is what we intended to find.

**19.** We wish to find three square numbers such that when the first is multiplied by the second and the product by the third, and the number formed by the sum of the three numbers is subtracted from the result, the remainder is a square.

Let us put 1 as the first number, 1 and 9 parts of 16 parts as the second, and $x^2$ as the third. We multiply the first by the second and the result by the third, so we obtain $x^2$ plus 9 parts of 16 parts of $x^2$. We diminish that by the number formed by the sum of the three numbers, namely (by) $x^2$ plus 2 and 9 parts of 16 parts of 1; the remainder is 9 parts of 16 parts of $x^2$, minus 2 and 9 parts of 16 parts of 1, and this has to be a square. We assume its side to be $\frac{3}{4}x - \frac{1}{4}$, and multiply that by itself; it becomes 9 parts of 16 parts of $x^2$, plus one part of 16 parts of 1, minus 6 parts of 16 parts of $x$. This, then, equals 9 parts of 16 parts of $x^2$, minus 2 and 9 parts of 16 parts of 1. We add to both sides 2 and 9
122 parts of 16 parts of 1, plus 6 parts of 16 parts of $x$; so, after the addition, 9 parts of 16 parts of $x^2$, plus 6 parts of 16 parts of $x$, are equal to 9 parts of 16 parts of $x^2$, plus 2 and 10 parts of 16 parts of 1. We remove the 9 parts of 16 parts of $x^2$, which are common, from the two sides; then 42 parts of 16 parts of 1 are equal to 6 parts of 16 parts of $x$. So $x$ is equal to 7. We had put $x$ as the side of the third square, so the said side is 7 and the third square, 49; the first square is, as we had assumed, 1, and the second square, (also) as we had assumed, is 1 and 9 parts of 16 parts of 1. Multiplying the first square by the second square, then the result by the third square, gives 76 and 9 parts of 16 parts of 1. When this last number is diminished by the number formed by the sum of the three numbers, or 51 and 9 parts of 16 parts of 1, the remainder is 25, which is a square number with side 5.

Therefore, we have found three numbers fulfilling the condition imposed upon us, and these are 49, 1, (and) 1 and 9 parts of 16 parts of 1. This is what we intended to find.

**20.** We wish to find three square numbers such that when the first is multiplied by the second and the product by the third, and the result is subtracted from the number formed by the sum of the three numbers, the remainder is a square.

Let us put 4 as the first square, 4 parts of 25 parts of 1 as the second, and $x^2$ as the third. Next, we multiply the first square by the second square and then the result by the third square; this gives 16 parts of 25 parts of $x^2$. Let
123 us subtract it from the number formed by the sum of the three numbers, that is, (from) $x^2$, plus 4, plus 4 parts of 25 parts of 1; the remainder is 9 parts of 25 (parts) of $x^2$, plus 4, plus 4 parts of 25 parts of 1, and this has to be a square. Let us put as its side $\frac{3}{5}x + 1$; multiplying that by itself, it becomes 9 parts of

25 parts of $x^2$, plus $1\frac{1}{5}x$, plus 1, which equals 9 parts of 25 parts of $x^2$, plus 104 parts of 25 parts of 1. We remove the 9 parts of 25 parts of $x^2$ plus 1 which are common, so as to have a single term equal to a single term; hence 30 parts of 25 parts of $x$ equal 79 parts of 25 parts of 1, so $x$ is equal to 79 parts of 30 parts of the unit. We had assumed the third square number to be $x^2$; so its side is 79 parts of 30 parts of the unit, and the square is 6241 parts of 900 parts of the unit. It is, then, the third number.[25] The first number is, as we had assumed, 4, and the second, (also) as we had assumed, 4 parts of 25 parts of the unit. Then, when we multiply the first number, or 4, by the second number, or 4 parts of 25 parts of the unit, then the product by the third number, or 6241 parts of 900 parts of 1, we obtain 99,856 parts of 22,500. If we subtract that from the number formed by (the sum of) the three
124 numbers, namely (from) 4, plus 4 parts of 25 parts of 1, plus 6241 parts of 900 parts of 1, or 249,625 parts of 22,500, the remainder is 149,769 parts of 22,500, which is a square number with side 387 parts of 150 parts.

Therefore, we have found three numbers fulfilling the condition imposed upon us, and these are 4, 4 parts of 25 parts of the unit, (and) 6241 parts of 900 parts of the unit. This is what we intended to find.

**21.** We wish to find two square numbers such that when the number formed by their sum is added to the square of each, the result of that is (in both cases) a square.

Any square number which is increased by its side plus $\frac{1}{4}$ gives a square. Hence we shall set as one of the two numbers $x^2$; so its square is $x^4$ and, when one increases it by its side plus $\frac{1}{4}$, the result is $x^4 + x^2 + \frac{1}{4}$, that is, a square number with side $x^2 + \frac{1}{2}$. It appears then that the number formed by the sum of the two numbers is $x^2 + \frac{1}{4}$. And, since we had put $x^2$ for the first number, the second is $\frac{1}{4}$. Now, if we add to the square of $\frac{1}{4}$, or $\frac{1}{2} \cdot \frac{1}{8}$, the number formed by the sum of the two numbers, that is, $x^2 + \frac{1}{4}$, the result is $x^2$ plus 5 parts of 16 parts of 1, and this must be a square. We assume its side to be $x + \frac{1}{2}$; multiplying that by itself, it becomes $x^2 + x + \frac{1}{4}$, which then equals $x^2$ plus 5 parts of 16 parts of 1. We remove $x^2 + \frac{1}{4}$ from both sides, so one part of 16 parts of 1 equals $x$; hence $x$ is one part of 16 parts of 1. We had put for one of
125 the two squares $x^2$, so its side is one part of 16 parts of 1 and the square, one part of 256 (parts) of 1. The other number is, as assumed, $\frac{1}{4}$. (So) the number formed by their sum is 65 parts of 256 parts of 1. If this is added to the square of one of the two numbers, namely[26] (to) 16 parts of 256 (parts) of 1, the result is 81 (parts) of 256 parts of 1, which is a square number with side 9 parts of 16 parts of 1; again, if we add the same to the square of the other number, that is, (to) one part of 65,536, the result is 16,641 parts of 65,536, which is a square number with side 129 parts of 256.

[25] Perhaps an interpolation; see p. 32, no. 17.

[26] The two numbers are now taken in the reverse order.

Therefore, we have found two numbers fulfilling the condition imposed upon us, and these are $\frac{1}{4}$ (and) one part of 256 parts of 1. This is what we intended to find.

**22.** We wish to find two square numbers such that when they are added (the result) is a square number and when one is multiplied by the other this gives a cubic number.

Any cubic number results from multiplying a number by itself and the product again by the same number. Hence we put for the first square number $x^2$; multiplying it by itself gives $x^4$, so let us put $x^4$ as the second number. It appears that if we multiply the first number, or $x^2$, by the second number, or $x^4$, the result is $x^6$, which is a cubic number, since it is produced by the multiplication of a number by itself and of the product by the same number. Now, the addition of the two square numbers gives $x^4 + x^2$, which must be a
126 square. Let us put as its side $1\frac{1}{4}x^2$. The multiplication of this by itself gives $x^4$ and 9 parts of 16 parts of $x^4$, which equals $x^4 + x^2$. We remove the $x^4$, which is common, from the two sides; so 9 parts of 16 parts of $x^4$ equal $x^2$. Let us divide the two sides by $x^2$, hence 9 parts of 16 parts of $x^2$ are equal to 1; thus the whole $x^2$ is equal to 16 parts of 9 parts of 1. We had put $x^2$ as the first number, so it is 16 parts of 9 parts of 1, and the second number (being the square of the preceding) is 256 parts of 81 parts of the unit. The multiplication of 16 parts of 9 parts of 1 by 256 parts of 81 parts of 1 results in 4096 parts of 729 parts of 1, which is a cubic number having 16 parts of 9 parts of 1 as its side. Again, the addition of the two square numbers gives 400 parts of 81, which is a square number with side 20 parts of 9.

Therefore, we have found two numbers fulfilling the condition imposed upon us, and these are 16 parts of 9 (and) 256 parts of 81. This is what we intended to find.

We (now) want to treat this problem by another method, which is easier than the first one. We (first) seek two square numbers such that their sum be a square. Such are $16x^2$ and $9x^2$. Then, we multiply them, thus obtaining $144x^4$, and this equals a cubic number. Let the cubic number be $8x^3$, so $144x^4$
127 equals $8x^3$. The division of both sides by $x^3$ results in $144x$ equal to 8, so $x$ is one part of 18 parts of 1. We had put $9x^2$ for one of the two square numbers; so its side is $3x$, which is one part of 6 parts of 1; the multiplication of that by itself gives one part of 36 parts of 1, and this is the first of the two numbers. The other number was put $16x^2$; its side is $4x$, which is 2 parts of 9 parts of 1, and the multiplication of that by itself gives 4 parts of 81 parts of 1, which is the second number. It appears that the addition of the two square numbers results in 25 parts of 324, which is a square number having 5 parts of 18 as its side, and (that) the multiplication of the first number, or one part of 36 parts of 1, by the second number, or 4 parts of 81, results in 4 parts of 2916, or one part of 729, which is a cubic number having one part of 9 parts of 1 as its side.

Therefore, we have found two numbers fulfilling the condition imposed upon us, and these are one part of 36 parts of 1 (and) 4 parts of 81 parts of 1. This is what we intended to find.

**23.** We wish to find two square numbers such that, a given square number being divided by each of them and the results of the two divisions being added, the result is a square number, and such that when the three numbers—that is to say, the two required numbers and the given number—are added, the result is a square.

Let the given square number be 9. We wish to find two square numbers such that, 9 being divided by each of them and the results of the division(s) being added, this gives a square number, and such that when the three numbers—that is to say, the two required numbers and the given 9—are
128 added, the result is a square number. Now, whenever we divide a square number into two square parts and then divide a square number by each of the two parts, the sum of the results of the division(s) is a square number. So let us take a square number, and (let us) divide it into two square parts. The number we take is $x^2$, and we divide it into two square parts, which are (say) 9 parts of 25 parts of $x^2$ and 16 parts of 25 parts of $x^2$; let these two parts be the two required numbers. We divide 9 by 9 parts of 25 parts of $x^2$; it becomes 25 parts of $x^2$. We also divide 9 by 16 parts of 25 parts of $x^2$, thus obtaining as a quotient 14 parts and $\frac{1}{2} \cdot \frac{1}{8}$ of a part of $x^2$. The addition of the results of the two divisions gives 39 parts and $\frac{1}{2} \cdot \frac{1}{8}$ of a part of $x^2$, which is a square number with side 6 parts and $\frac{1}{4}$ of a part of $x$. Now, if we add the three numbers, namely the two required numbers and the given 9, the result is $x^2 + 9$, which has to be a square. Let us put $x + 1$ as its side; we multiply it by itself and obtain $x^2 + 2x + 1$, and this equals $x^2 + 9$. Remove $x^2 + 1$ from the two sides so as to have a single term equal to a single term; so $2x$ is equal to 8, hence $x$ is 4. One of the two required numbers was 16 parts of 25 parts of $x^2$, and its side is $\frac{4}{5}x$, so its side is $\frac{4}{5}$ of 4, or $\frac{16}{5}$. This, when multi-
129 plied by itself, results in 256 parts of 25, which is one of the two required numbers. Again, the other number was 9 parts of 25 parts of $x^2$, and its side is $\frac{3}{5}x$; $x$ being 4, the side is $\frac{12}{5}$. This, when multiplied by itself, results in 144 parts of 25 parts of 1, which is the second required number. If we divide the given number, that is, 9, or 225 parts of 25 parts, by the first number, that is, (by) 256 parts of 25, the result of the division is 225 parts of 256 parts; again, dividing the 9, that is, the 225 parts of 25, by the other number, that is, (by) the 144 parts of 25 parts, gives as a quotient 225 parts of 144 parts, or 400 parts of 256 parts. The addition of that to the result of the division of 9 by the other (first-mentioned) number, that is, (to) 225 parts of 256, gives 625 parts of 256, which is a square number, with side 25 parts of 16 parts of 1. Then, the addition of the three numbers, namely 256 parts of 25 parts of the unit, 144
130 parts of 25, and 9, or 225 parts of 25, gives 625 parts of 25, or 25, which is a square number with side 5.

Therefore, we have found two numbers fulfilling the condition imposed upon us, and these are 256 parts of 25 parts of 1 (and) 144 parts of 25 parts of the unit. This is what we intended to find.

End of the sixth Book of the treatise of Diophantus, and this Book contains twenty-three arithmetical problems.

# In the Name of God the Merciful, the Compassionate

# Seventh Book of the Treatise of Diophantus

Our intention is to expound in the present Book many arithmetical problems without their departing from the type of problems seen previously in the fourth and fifth Books—even if they are different in species[1]—in order that it be an opportunity for (acquiring) proficiency and an increase in experience and skill.

**1.** We wish to find three cubic numbers such that the side of the first is to the side of the second in a given ratio and the side of the second is to the side of the third in a given ratio,[2] and such that when the first number is multiplied by the second number and the product by the third number, the result is a square number.

Let the given ratio be 2:1. We wish to find three cubic numbers, such that the side of the first is twice the side of the second and the side of the second is twice the side of the third, and such that when the first of the three numbers is multiplied by the second number and the product by the third number, the result is a square number. Let us put $x$ as the side of the third number, so that the third number is $x^3$; we put $2x$ as the side of the second number—for it is twice the side of the third number—so that the second number is $8x^3$;
131 we put as the side of the first number $4x$—for it is twice the side of the second number—so that the first number is $64x^3$. Now, the multiplication of the first number, or $64x^3$, by the second number, or $8x^3$, and of the result by the third number, or $x^3$, gives $512x^9$, which must be a square. Let us put as its side $32x^4$; this when multiplied by itself yields $1024x^8$, which is equal to $512x^9$. We divide the $512x^9$ by $x^8$, and obtain $512x$, and we divide (the) $1024x^8$ by $x^8$, and obtain 1024. Then, 1024 is equal to $512x$, so that $x$ is 2.

[1] Or: appearance.

[2] The formulation is misleading since the given ratio is the same in both cases.

We had put $x$ as the side of the third number, and $x$ is 2, so the side of the third number is 2 and the third number, 8. And, we had put as the side of the second number $2x$—for it is twice the side of the third number—, and $2x$ is 4, so the second number is 64. (Finally,) we had put as the side of the first number $4x$—for it is twice the side of the second number—, and $x$ is 2, so the side of the first number is four times 2, or 8, and the first number is 512. If we multiply the first number, 512, by the second number, 64, the result is 32,768; multiplying that by the third number, 8, gives 262,144, which is a square number with 512 as its side.

Therefore, we have found three numbers fulfilling the condition imposed upon us, and these are 512, 64, and 8. This is what we intended to find.

132 **2.** We wish to find three cubic numbers which are also square, such that when the first of these numbers is multiplied by the second number and, again, the product is multiplied by the third number, the result is a square of square side.

Let us put as the first number one part of 64 parts, which is a cubic number with side $\frac{1}{4}$, and it is also a square number with side $\frac{1}{2}$; we put as the second number 64, which is a cubic number with side 4, and it is also a square number with side 2; we put as the third number $x^6$, which is a cubic number with side $x^2$, and it is also a square number with side $x^3$.[3] The multiplication of the first number, or one part of 64 parts of the unit, by the second number, or 64, gives 1, and the multiplication of 1 by the third number, or $x^6$, gives $x^6$; its side must be a square—and by "its side" is meant in this place "its root".[4] Now, the side of $x^6$ is $x^3$; so we equate to $x^3$ a square number, say $4x^2$. Dividing the two sides by $x^2$ gives $x$ equal to 4; such is $x$, and it is the side of $x^3$, and $x^3$ is (therefore) 64. We had put as the third number $x^6$, which arises from the multiplication of $x^3$ by itself; $x^3$ being 64, we multiply 64 by itself, and obtain 4096, which is the third number. The multiplication of the first number, or one part of 64 parts of 1, by the second number, or 64, gives 1; then, the multiplication by 1 of the third number, or 4096, gives 4096, which is a square number with side 64, and it is also a square number, with side 8.

Therefore, we have found three numbers fulfilling the condition imposed
133 upon us, and these are one part of 64 parts of 1, 64, and 4096. This is what we intended to find.

[3] Note the discrepancy between this passage and the two previous ones: whereas the first two give the square roots of the *sides* of the cubes, the last one gives the square root of the unknown cube itself. This last formulation, concerning the unknown cube, probably belongs to the original text, while the other two must have originated with the author of the major commentary.

[4] This is certainly an Arabic addition (cf. p. 31, no. 12).

**3.** We wish to find a square number of square side such that when we divide it into three parts, each of these parts is a cube.[5]

Let us put $x^2$ as the side of this number, so the (said) number is $x^4$. We wish to divide $x^4$ into three parts such that each of them is a cube. Let us put $x^3$ as the first part, $8x^3$ as the second part, and $64x^3$ as the third part. It appears that each of these parts is a cube. Now, the sum of the three parts is $73x^3$, so this is equal to the number to be divided, namely $x^4$. The division of $x^4$ by $x^3$ results in $x$, and the division of $73x^3$ by $x^3$ results in 73; so this equals $x$, and $x$ is 73. We had put $x^2$ as the side of the number to be divided, so the side is the square of 73, or 5329, and the number to be divided is the square of 5329, that is, 28,398,241. And, we had put as the first part $x^3$, and $x^3$ arises from the multiplication of 73 by 73, and then of the result by 73; this gives 389,017, which is a cubic number, namely the first of the three parts. The second part is eight times that, for we assumed it to be $8x^3$; it is (thus) 3,112,136. The last part is $64x^3$, so it equals the first part taken sixty-four times, which is 24,897,088. It appears that adding these three parts, each
134 of which is a cube, gives as the number formed by their sum 28,398,241; this is the number to be divided, and it is a square number of square side.

Therefore, we have found a number fulfilling the condition imposed upon us, and this is 28,398,241. This is what we intended to find.

**4.** We wish to divide a cubic number of square side into three parts such that each of them is a square.

Let us put $x^2$ as the side of the cube, so that the cube is $x^6$. We wish to divide $x^6$ into three parts such that each of them is a square. Let us then seek three numbers such that, when added, the result is a square and such that each is a square.[6] Finding that is easy from what precedes;[7] one of the numbers is 1, the second, 4, and the third, $\frac{4}{9}$. Let us put each of these numbers $x^4$'s, so the first number is $x^4$, the second, $4x^4$, and the third, $\frac{4}{9}x^4$; and, since we want to divide a cubic number into three square parts, let us set for each of the three parts one of these three numbers, their sum being (then) the said cubic number. The number formed by their sum is 49 parts of 9 parts of $x^4$, so this equals the cubic number, that is, $x^6$. The division of all that by $x^4$ gives $x^2$ equal to 49 parts of 9 parts of 1. Since we put $x^2$ as the side of the cubic number, and (since) $x^2$ is 49 parts of 9 parts of the unit, this is the side of the cube; the cubic number results from the multiplication of (the) 49 parts by themselves, the product being multiplied (again) by 49 parts; and this is

---

[5] The formulation, as it stands, seems to imply that *any* division into three parts will give three cubes. The problem should be stated thus: "We wish to divide a square number of square side into three parts such that each of them is a cube". Observe that the (shortened) formulation found just below is correct.

[6] The phrase "and such that each is a square" might be an interpolation (cf. p. 33, no. 22).

[7] The problem is incidentally solved in III,5, ἄλλως, and is altogether trivial (see commentary).

117,649 parts of 729 parts of the unit. And, since we put $x^2$ as the first of the parts,[8] the said part is 49 parts of 9 parts of the unit; since we put as the second
135 part $4x^2$, it is 196 parts of 9; the third part, being $\frac{4}{9}x^2$, is 196 parts of 729[9] parts of the unit. The sum of these three parts is equal to the cubic number.

Therefore, we have found a number fulfilling the condition imposed upon us, and this is 117,649 parts of 729. This is what we intended to find.[10]

**5.** We wish to find a cubic number of cubic side such that, when it is multiplied by two numbers, one cubic and the other square, and the products are added, the result is a square number.[11]

Let us put, as the side of the cubic number, a cubic number, say 8, so that the cubic number is 512. We wish to find two numbers, one cubic and the other square, such that, when each one is multiplied by 512 and the products are added, the result is a square. Let us assume the cubic number to be $x^3$ and the square to be $x^2$. We multiply $x^3$ and $x^2$ by 512; the sum of this is $512x^3 + 512x^2$, which must be a square number. We put as its side $64x$; the multiplication of the $64x$ by itself gives $4096x^2$, which equals $512x^3 + 512x^2$. We remove $512x^2$ from the two sides, hence $512x^3$ equals $3584x^2$; the division of the two sides by $x^2$ results in 3584 equal to $512x$, so $x$ is 7. Since we put, as the square number, $x^2$—with side $x$, which is 7—, and (since) $x^2$ is 49, the square number is 49. Again, since we put as the cubic number $x^3$, and (since)
136 $x^3$ is produced by the multiplication of $x^2$ by $x$, the cubic number is 343. Then, the multiplication of the cubic number for which we have put a cube as side, namely 512, by the cubic number which is 343, gives 175,616; again, the multiplication of 512 by the square number, that is, (by) 49, gives 25,088. This, then, when increased by the 175,616, results in 200,704, which is a square number with side 448.

Therefore, we have found a number fulfilling the condition imposed upon us, and this is 512. This is what we intended to find.

**6.** We wish to find two square numbers such that the number formed by their sum is a square and such that when the one is multiplied by the other, the result is to the number formed by their sum in a given ratio.

Now, the given ratio[12] can only be a square number: because for any pair of square numbers, the ratio of the larger to the smaller can only be a square

---

[8] The three parts having been set $x^4$'s, the coming figures are wrong.

[9] *Sic*! Perhaps the number was unreadable and some reader or copyist attempted to restore it.

[10] This final statement does not really correspond to the formulation of the problem.

[11] The following formulation would have been better: "We wish to find a cubic number of cubic side and two numbers, one cubic and the other square, such that when the cubic number (of cubic side) is multiplied by each of the two numbers and the products are added, the result is a square number".

[12] Properly, here and below: "the number belonging to the given ratio". See p. 99, n. 47.

number, and, likewise, the quotient of the smaller divided by the larger can only be a square.

So let the given ratio be the ratio 9:1. Let us put as the number formed by
the sum of the two numbers $x^2$; we divide $x^2$ into two square parts.[13] Let the
first be 16 parts of 25 parts of $x^2$ and the second, 9 parts of 25 parts of $x^2$.
Now, the multiplication of the two parts gives 144 parts of 625 parts of $x^4$;
hence this must be equal to nine times the number formed by (the sum of)
137 the two square numbers, i.e., $9x^2$. The division of 144 parts of 625 parts of
$x^4$ by $x^2$ results in 144 parts of 625 parts of $x^2$, and the division of $9x^2$ by $x^2$
results in 9; so 9 is equal to 144 parts of 625 parts of $x^2$, hence the whole $x^2$
is equal to 39 and one part of 16 parts of 1. One of the two numbers was 16
parts of 25 parts of $x^2$, and this is 25; the other number was 9 parts of 25
parts of $x^2$, and this is 14 and one part of 16 (parts) of 1. The sum of the two
numbers is $39 + \frac{1}{2}\cdot\frac{1}{8}$, which is a square number with side $6\frac{1}{4}$, and multi-
plying one of the two numbers by the other, that is, (multiplying) 25 by
$14 + \frac{1}{2}\cdot\frac{1}{8}$, gives $351 + \frac{1}{2} + \frac{1}{2}\cdot\frac{1}{8}$, which is nine times the sum of the two
numbers, i.e., $39 + \frac{1}{2}\cdot\frac{1}{8}$.

Therefore, we have found two numbers fulfilling the condition imposed upon us, and these are 25 (and) $14 + \frac{1}{2}\cdot\frac{1}{8}$. This is what we intended to find.

**7.** We wish to divide a square number of cubic side into three parts such that the sum of any two is a square.

Let us put $x^3$ as the side of the square number, so that the square number is
$x^6$. We wish to divide $x^6$ into three parts such that the sum of any two is a
square. Let us then seek three numbers such that any two when added give
a square, and such that the number formed by (the sum of) the three numbers
be a square. Finding that is easy on the basis of what we have expounded in
the sixth problem of the third Book. So the first number is 80, the second, 320,
and the third, 41; the sum of the three numbers is 441. Let us take $x^4$'s instead
140 of the units. Then the sum of the three numbers is $441x^4$, which is equal to $x^6$.
We divide the two sides by $x^4$; the division of $x^6$ by $x^4$ results in $x^2$ and the
division of $441x^4$ by $x^4$ results in 441. So 441 is equal to $x^2$, thus $x^2$ is 441.
Hence $x^4$ is the product of the multiplication of 441 by itself, that is, 194,481.
Since we put for the first of the three parts $80x^4$, this (part) is 15,558,480;
again, since we put for the second part $320x^4$, it is 62,233,920; again, since
we put for the third part $41x^4$, it is 7,973,721. As the number which had
to be divided into these three parts is the number formed by their sum,
it is 85,766,121, which is a square number with side 9261, and the said side
is a cubic number with side 21. Since the first of the three parts is 15,558,480
and the second part is 62,233,920, the number formed by their sum is
141 77,792,400, which is a square number with side 8820; again, since the second

[13] *Arithmetica* II,8 (but the result is trivial).

part is 62,233,920 and the third part is 7,973,721, the number formed by their sum is 70,207,641, which is a square number with side 8379; again, since the third part is 7,973,721 and the first part is 15,558,480, the number formed by their sum is 23,532,201, which is a square number with side 4851.

Therefore, we have found a number fulfilling the condition imposed upon us, and this is 85,766,121. This is what we intended to find.[14]

Having (thus) completed the treatment of this problem, we want (now) to solve it by a second treatment which is easier than the first one. Let us begin, prior to the treatment, with the formulation (of the problem). We wish to find a square number of cubic side such that, when it is divided into three parts, the sum of any two parts is a square number.[15]

Let us put, for the square number which we wish to divide, 64, which is a square number having a cube as its side. We wish to divide 64 into three parts such that the sum of any two is a square. So let us seek three numbers such
138 that their sum be a square and the sum of any two be a square. We have already expounded that in the sixth problem of the third Book; this exposition allows us to dispense with the repetition. The first of these three required numbers is 320, the second one, 80, and the third one, 41. The number formed by the sum of these numbers is 441, which is a square number.

If the last had been the number which we wish to divide, we would have reached our goal. But the number which we wish to divide is 64. So let us take, from each of the three numbers having 441 as their sum, a number such that the quotient of it divided by the number from which it is taken equals the quotient of 64 divided by 441.[16] In other words, we multiply each of the three numbers by 64, (each) result being then parts of 441. The first number being 320, it becomes, when multiplied by 64, 20,480, which is then parts of 441; the second part being 80, it becomes, when multiplied by 64, 5120, which is then parts of 441 parts of the unit; again, the third part being 41, it becomes, when multiplied by 64, 2624, which is (then) parts of 441. Thus, we have divided the 64 into three parts such that: (firstly,) the addition of the first and the second gives 25,600 parts of 441, which is a square number with side 160 parts of 21; (secondly), the addition of the second and the third gives
139 7744 parts of 441, which is a square number with side 88 parts of 21; (thirdly,) the addition of the third and the first gives 23,104 parts of 441, which is a square number with side 152 parts of 21.

Therefore, we have found a number fulfilling the condition imposed upon us, namely 64, and we have divided it into three parts, which are 20,480 parts of 441, 5120 parts of 441, (and) 2624 parts of 441. This is what we intended to find.

---

[14] This final statement does not correspond to the formulation of the problem (as in the case of VII,4).

[15] Here again (cf. VII,3), the formulation is inappropriate.

[16] This means that $u_i$ being the three known numbers, we seek $a_i$ such that $a_i : u_i = 64:441$.

**8.** We wish to find a square number of cubic side such that the addition to it of a certain number gives a square, and also the addition to it of twice this number gives a square.

Let us put as the square number 64, which is a square number of cubic side. We wish to find a number such that the addition of it to 64 gives a square and the addition of twice it to 64 gives again a square. Let us look for this (property) for a square number other than 64: we seek a square number such that when we increase it by a certain number (the result) is a square and when we increase it by twice this number the result is a square. Now, any square number to which we add twice its root, plus 1, results in a square. Hence we assume the square number to be $x^2$.[17] We add to it twice its root plus 1; $x^2$ results in $x^2 + 2x + 1$, which is a square number with side $x + 1$. Now, if we add to $x^2$ twice $2x + 1$, that is, $4x + 2$, the result is $x^2 + 4x + 2$;
142 we want this to be a square. Let us assume its side to be $x - 2$; the multiplication of $x - 2$ by itself gives $x^2 + 4 - 4x$, which thus equals $x^2 + 4x + 2$. We add $4x$ to the whole, and cancel out the one $x^2$ against the other, so $8x + 2$ is equal to 4; removing 2 from the two sides leaves $8x$ equal to 2. Hence $x$ is $\frac{1}{4}$, and $x^2$, $\frac{1}{2} \cdot \frac{1}{8}$. The number added to $x^2$ was $2x + 1$; this is $1\frac{1}{2}$, and the other number added to $x^2$ is the double of $1\frac{1}{2}$, or 3. Let us then multiply all that by 16 in order that $x^2$ be an integral number; so $x^2$ is (now) 1, the (first) number added to $x^2$ is 24, and the other number added to $x^2$ is the double of 24, or 48. It appears that adding 24 to 1 results in 25, which is a square number; again, adding to 1 the double of 24, or 48, results in 49, which is a square number.

If the assumed square number had been 1, we would have reached our goal. But it is 64. Since 64 numbers 1 sixty-four times, we have to multiply each of the two added numbers, that is, 24 and 48, by 64. The multiplication of 24 by 64 gives 1536, which is the (first) number added to 64, and the multiplication of 48 by 64 gives 3072, which is the double of the first (added) number.
143 Adding 1536 to 64 gives 1600, which is a square number with 40 as its side, and adding the double of 1536, or 3072, to 64 gives 3136, which is a square number with 56 as its side.

Therefore, we have found two numbers, one of which is the double of the other, such that the addition of each to a square number of cubic side gives a square; and these are 1536 (and) 3072. This is what we intended to find.[18]

**9.** We wish to find a square number of cubic side such that when we subtract from it a certain number, the remainder is a square, and also when we subtract from it twice this number, the remainder is a square.

Let us put as the square number 64, which is a square number of cubic side. Now, we wish to find a number such that, when we subtract it from 64

---

[17] And the other number to be $2x + 1$.

[18] In this problem and the next two, the final statement is adapted to the changed goal of the problem (we immediately assumed the main required number).

the remainder is a square, and when we subtract twice it from 64 the remainder is a square. Hence we shall look for this (property) for a square number other than 64: we seek a square number such that when we diminish it by a certain number the remainder is a square and when we diminish it by twice this number it becomes a square. Now, any square number from which we subtract twice its root minus 1 leaves a square. So let us put $x^2$ for the square;[19] the subtraction from it of twice its root minus 1 gives a square. Now, if we subtract from it the double of twice its root minus 1, that is, four of its roots minus 2, the remainder is $x^2 + 2 - 4x$; and this must be a square. Let us assume its side to be $x - 3$; we multiply it by itself, and obtain $x^2 + 9 - 6x$,
144 which is equal to $x^2 + 2 - 4x$. Removing $x^2 + 2 - 4x$ from both sides, there remains $2x$ equal to 7;[20] hence $x$ is $3\frac{1}{2}$, $x^2$, $12\frac{1}{4}$, and the two numbers subtracted from $x^2$, 6 and 12. We then multiply the whole by 4 in order that $x^2$ be an integral number; so $x^2$ is (now) 49 and the two subtracted numbers, 24 and 48.

If the square number had been 49, we would have reached our goal; but it is 64. And 64 numbers 49 one time and 15 parts of 49 parts of one time. Consequently, we have to add to the two subtracted numbers, that is, (to) 24 and 48, 15 parts of 49 parts of each. So let us multiply 24 by 64; the result is 1536, which is parts of 49 parts of the unit; this is the (first) number subtracted from the 64. Again, we multiply 48 by 64, and obtain 3072, which is parts of 49; and this is the other number subtracted from the 64, and it is the double of the first number. The first number being 1536 parts of 49, the subtraction of it from 64, or 3136 parts of 49, results in 1600 parts of 49, which is a square number with side $\frac{40}{7}$; again, the second number being the double of the first
145 number, that is, 3072 parts of 49, it leaves, when subtracted from 64, or 3136 parts of 49, 64 parts of 49, which is a square number with side 8 parts of 7.

Therefore, we have found two numbers, one of which is the double of the other, such that subtracting each from a square number of cubic side leaves a square; and these are 3072 parts of 49 (and) 1536 parts of 49. This is what we intended to find.

**10.** We wish to find a square number of cubic side and a number such that, when we add it to the square number, the result is a square, and, when we subtract it from the square number, the remainder is a square.

Let the square number be 64. We look for a square number other than 64 which, when increased by a certain number, gives a square, and when diminished by the said number, leaves a square. As any square number from which we subtract twice its root minus 1 results in a square, let us put $x^2$

[19] And $2x - 1$ for the subtracted number.

[20] The performed operation is simply wrong in terms of ancient mathematics, and certainly goes back to a commentator; the restoration with $6x$ should have preceded the removal of $x^2 + 2$, in order to avoid arriving at an expression equal to zero.

as the square number and $2x - 1$ as the number subtracted from it. Now, if we add $2x - 1$ to $x^2$, we obtain $x^2 + 2x - 1$; this, then, must be a square. Let us assume its side to be $x - 3$; the multiplication of $x - 3$ by itself gives $x^2+9-6x$, which is then equal to $x^2+2x-1$. We remove $x^2-6x-1$ from the two sides and obtain $8x$ equal to 10;[21] hence $x$ is equal to $1\frac{1}{4}$, and $x^2$ is 25 parts of 16 parts of the unit. The number subtracted from $x^2$ was $2x - 1$; this is 24 parts of 16 parts of the unit, and such is also the (value of the)
146 number added to $x^2$. Let us then multiply all that by 16 in order that $x^2$ be an integral number. So $x^2$ is (now) 25, the number added to it, 24, and the subtracted number, 24 (also).

If the assumed (square) number had been 25, we would have reached our goal. But the assumed number is 64. And, 64 numbers 25 twice and 14 parts of 25 parts of one time. So we have to multiply the added number—which is also the subtracted one—, namely 24, by 64; this gives 1536, which is parts of 25; and this is the number which we add to 64 and subtract from 64. It appears that, if we add 1536 parts of 25 to 64, that is, (to) 1600 parts of 25, we obtain 3136 parts of 25, which is a square number with side 56 parts of 5; and, if we subtract 1536 parts of 25 parts from 64, that is, (from) 1600 parts of 25, we obtain 64 parts of 25, which is a square number with side 8 parts of 5.

Therefore, we have found a number such that subtracting it from a square number of cubic side gives a square and adding it to the same gives a square; and this is 1536 parts of 25 parts of the unit. This is what we intended to find.

**11.** We wish to divide a given square number into two parts such that the addition to the said square of one of them gives a square and the subtraction from the same of the other one gives a square.

147 Let the given number be 25. We wish to divide 25 into two parts such that adding the one part to 25 gives a square number and subtracting the other part from 25 gives a square. Let us aim to find a certain square which we shall divide into two parts such that adding the one to it and subtracting the other from it give, after the addition and the subtraction, a square. But if we add to $x^2$ twice its root plus 1, the result is $x^2 + 2x + 1$, which is a square number, and if we subtract from $x^2$ twice its root minus 1, the remainder is $x^2$, plus 1, minus two roots (of $x^2$), which is a square number.[22] Now, we want the sum of the added and of the subtracted numbers to be $x^2$; their sum being $4x$, $4x$ equals $x^2$. The division of the whole by $x$ gives $x$ equal to 4; and, since $x$ is the side of $x^2$, $x^2$ is 16. The number added to $x^2$ was $2x + 1$, which is 9, and the number subtracted from $x^2$ was $2x - 1$, which is 7; and the addition of 9 and 7 results in 16. Hence we have attained the object of our investigation.

---

[21] This operation, although not wrong (as is the one in the preceding problem), is nonetheless expressed rather unconventionally.

[22] Hence, we set $x^2$ as the required square, and $2x + 1$ and $2x - 1$ as the parts added and subtracted.

But the given (square) number was 25; so let us multiply 9 by 25, which gives 225, and then divide that by 16; we obtain 225 parts of 16 parts, which is one of the two parts of 25, namely the added part. Again, let us multiply 7 by 25, so we obtain 175, and we divide that by 16; this gives 175 parts of 16, which is the second part, namely the one subtracted from 25. It appears that, adding the 225 parts to 25, that is, (to) 400 parts of 16, gives 625 parts of 16,
148 which is a square number with side 25 parts of 4; again, subtracting the other part, namely 175 parts of 16, from the 400 parts gives 225 parts of 16, which is a square number with side 15 parts of 4; and the sum of the two parts is 25.

Therefore, we have divided the 25 into two parts fulfilling the condition imposed upon us, and these are 225 parts of 16 (and) 175 parts of 16. This is what we intended to do.[23]

And since it is not possible to find a square number such that, dividing it into two parts and increasing it by each of the parts, we obtain (in both cases) a square, we shall (now) present something which is possible.

**12.** So we say: We wish to divide a given square number into two parts such that when we subtract each from the said square the remainder is (in both cases) a square.

Let the given number be 25. We wish to divide 25 into two parts such that when we subtract each from 25 the remainder is a square. So let us seek this condition in some square. Now, for any square which is divided into two square parts, the subtraction of each of the two parts from the square gives a square, which is the other part;[24] and the way of performing that[25] has been seen earlier in this treatise of ours. One of the two parts is 16 and the other, 9.

Therefore, we have divided the 25 into two parts such that when we subtract each from 25 the remainder is a square, and these are 9 (and) 16. This is what we intended to do.

**13.** We wish to divide a given square number into three parts such that the addition of each to the said square gives a square.

149 Let the given number be 25. We wish to divide 25 into three parts such that the addition of each to 25 gives a square. Now, the division of a square number into three parts and the addition of each one to the divided number produce three numbers such that the number formed by their sum equals four times the divided number;[26] therefore, if we divide 25 into three parts

[23] "This is what we intended to find" in the text, an error repeated in three other places (cf. p. 448, *ʿamila*, 3°). The same confusion occurs in other Arabic translations: see Ḥajjāj's *Euclid* (*Cod. Leid.*, footnote to prop. VI,10), as well as Klamroth, *Arab. Euklid*, p. 286.

[24] The words "which is the other part" may be an interpolation (cf. p. 32, no. 21).

[25] Namely, the division of a square into two square parts (*Arithmetica* II,8).

[26] If $u^2 = u_1 + u_2 + u_3$, $(u^2 + u_1) + (u^2 + u_2) + (u^2 + u_3) = 4u^2$.

and add each part to 25, the sum of the three (resulting) numbers is 100. Hence, let us divide 100 into three square parts and let each part be larger than 25. It has been seen earlier in this treatise of ours how to divide any square number into square parts,[27] and we shall dispense with the repetition of the treatment. So, the first part is 36, the second, 30 and 370 parts of 841 parts of the unit, and the third, 33 and 471 parts of 841 parts of the unit. Since each of these three parts is composed of 25 and of one of the parts of 25, if we subtract 25 from each of these three parts, the remainder of each part is one of the parts of 25. Now, subtracting 25 from 36 gives 11, which is the first of the parts of 25. Again, let us subtract 25 from the second[28] (found)
150 part, that is, (from) 33 and 471 parts of 841; the remainder is 8 and 471 parts of 841, which is the second of the parts of 25. Again, subtracting 25 from the third (found) part, that is, (from) 30 and 370 parts of 841, gives 5 and 370 parts of 841, which is the third of the parts of 25. Indeed, adding these three parts together gives 25, while increasing 25 by each of them results in a square number.

Therefore, we have divided the 25 into three parts such that the addition of each to the 25 gives a square number; and these parts are the following: the first is 11; the second is 8 and 471 parts of 841 parts of the unit; the third is 5 and 370 parts of 841. This is what we intended to do.

**14.** We wish to divide a given square number into three parts such that the subtraction of each from the said square gives a square.

Let us put 25 as the square number. We wish to divide 25 into three parts such that the subtraction of each part from the 25 gives a square. Now, if we divide 25 into three parts and subtract each part from 25, we shall thereby have found three numbers such that the number formed by their sum is 50.[29] So let us divide 50 into three square parts, and let each one be less than 25. It has been seen earlier in this treatise of ours how to divide a number into square parts;[30] so, given what precedes, let us dispense with the repetition.
151 Thus, the first part is 16, the second, 22 and 3 parts of 169 parts of the unit, and the third, 11 and 166 parts of 169 parts of the unit. Since each of these parts is equal to 25 diminished by each of its parts, we have to subtract each of these three parts (which we have found) from 25: thus (each) remainder of 25 is (one of) the (required) parts of 25. Now, the subtraction of 16 from 25 gives 9, which is the first of the parts of 25. Again, the subtraction of 22 and 3 parts of 169 parts from 25 gives 2 and 166 parts of 169, which is the second part. Again, the subtraction of 11 and 166 parts of 169 from 25 gives 13 and 3 parts of 169 parts of the unit, which is the third part. The number formed by

---

[27] *Arithmetica* II,8 (iterated).

[28] The parts given above are now taken in order of magnitude.

[29] If $u^2 = u_1 + u_2 + u_3$, $(u^2 - u_1) + (u^2 - u_2) + (u^2 - u_3) = 2u^2$.

[30] Provided that the number is a square (II,8) or that we already know a representation of this number as a sum of two squares (II,9); the second case is applicable here.

(the sum of) these three parts, that is to say (of) 9, 2 and 166 parts of 169, and 13 and 3 parts of 169, is 25; and the subtraction of each of these three parts from 25 results in a square number.

Therefore, we have divided the 25 into three parts such that the subtraction of each from 25 gives a square number; and these are 13 and 3 parts of 169
152 parts of 1, 9, (and) 11 and 166 parts of 169.[31] This is what we intended to do.

**15.** We wish to divide a given square number into four parts such that two of the four parts each leave, when subtracted from the given square number, a square, and the other two also each give, when added to the given square number, a square number.

Let the given square number be 25. We wish to divide 25 into four parts such that two of the four parts each leave, when subtracted from 25, a square, and (the other) two each give, when added to 25, a square. Let us seek this condition in some square number. Since, if we add to a square number, say (to) $x^2$, its side taken twice plus 1, the result is a square, we take as the first (additive) part $2x + 1$; again, since if we add to $x^2$ its side taken four times plus 4, the result is a square, let us set as the second additive part $4x + 4$. The number formed by (the sum of) the two additive parts is $6x + 5$. Likewise, since if we subtract from $x^2$ its side taken twice minus 1, that is, $2x - 1$, the remainder is a square, we set as one of the two subtractive parts $2x - 1$; again, since if we subtract from the square number, that is, (from) $x^2$, its side taken four times minus 4, the remainder is a square number, we take as the second subtractive part $4x - 4$. The number formed by (the sum of) the two subtractive parts is $6x - 5$. Now, the number formed by (the sum of) the two additive parts was $6x + 5$. Hence the number formed by (the sum of)
153 the four parts is $12x$, and this is then equal to $x^2$, since our goal was to divide $x^2$ into four parts. As the division of $x^2$ by $x$ gives $x$, and the division of $12x$ by $x$ gives 12, $x$ is equal to 12; $x$ being the side of $x^2$, $x^2$ is 144. The first of the two parts added to $x^2$ was $2x + 1$, so it is 25; the second additive part was $4x + 4$, so it is 52. Again, the first subtractive part was $2x - 1$, so it is 23; the second (subtractive) part, being $4x - 4$, is 44. Therefore, we have completed the requisite search for the said square number; but we have not reached the desired end of the problem.

For, if the given (square) number had been 144, we would have reached our goal; but it is 25. Consequently, we have to multiply each of the parts of 144 by 25 and divide the results by 144. Now, if we multiply the first of the (four) parts, namely 25, by 25, the result is 625, which, when divided by 144, becomes 625 parts of 144; and this is the first of the two parts added to 25. Again, since the other additive part is 52, we multiply 52 by 25; the result is 1300, which, when divided by 144, becomes 1300 parts of 144; and this is the other part added (to 25). Again, since the first subtractive part is 23, we

[31] *Sic*, instead of 2 and 166 parts of 169. See p. 64, no. 8.

154 multiply 25 by 23; it becomes 575, which, when divided by 144, gives 575 parts of 144; and this is the (first) number subtracted from 25. Again since the other subtractive part is 44, we multiply 44 by 25; it becomes 1100, which, when divided by 144, gives 1100 parts of 144; and this is the second part subtracted from 25. It appears that the addition of the four said parts results in 25, and that the addition to 25 of each of the two additive parts gives a square while the subtraction from 25 of each of the two subtractive parts leaves a square.

Therefore, we have divided 25 into four parts under the condition imposed upon us; and they are (as follows): the two additive (parts), 625 parts of 144 (and) 1300 parts of 144; (and) the two subtractive (parts), 575 parts of 144 (and) 1100 parts of 144. This is what we intended to do.

By an analogous treatment we (would) solve the problem with the (following) formulation: We wish to divide a given square number into eight parts such that four (of the eight) parts each give, when added to the given square, a square, and the other four each leave, when subtracted from the given (square) number, a square number.

**16.** We wish to find three square numbers which are also in proportion[32] such that the subtraction of the first from the second gives a square and the subtraction of the second from the third gives a square.

155 It is in the nature of (any) three square numbers which are also in proportion and are such that the subtraction of the first from the second gives a square, that the subtraction of the second from the third (also) gives a square. Let us then put as the first number 1 and as the third number $x^4$; thus the second number is $x^2$.[33] Now, the subtraction of the first number, or 1, from the second number, or $x^2$, gives $x^2 - 1$, which must be a square number. Let us then put as its side $x - 2$, which we multiply by itself; hence we obtain $x^2 + 4 - 4x$. This, then, equals $x^2 - 1$. We add to the two sides $4x + 1$, so $x^2 + 4x$ equals $x^2 + 5$; removing the $x^2$, which is common, gives 5 equal to $4x$, hence $x$ is $1\frac{1}{4}$. Since we assumed the second number to be $x^2$,[34] and (since) the side of $x^2$ is $x$, which is $1\frac{1}{4}$, or 5 parts of 4, $x^2$ (hence the second number) is 25 parts of 16 parts of the unit. And, the third number was assumed to be $x^4$, which is the product of the multiplication of $x^2$ by itself, or 625 parts of 256; thus the third number is 625 parts of 256 parts of the unit. The first number is as set by us, i.e., 1. Now, the subtraction of the first number, namely 1, from the second number, namely 25 parts of 16, gives 9 parts of

---

[32] That is, in continuous proportion.

[33] The proposition "if $a{:}b = b{:}c$, so $ac = b^2$" is not quoted here, whereas its extension to four terms is given in the next problem and it is itself found later on in the *Arithmetica* (see D.G., I, pp. 236,5–7; 310,8–9—perhaps an interpolation).

On the presence of this (spurious) proposition in the *Elements*, see Heiberg's ed., II, pp. 229–31; Heath's transl., II, p. 320.

[34] Actually, it was deduced from the assumptions made for the two others.

16 parts of the unit, which is a square number with 3 parts of 4 as its side. Again, the subtraction of the second number, namely 25 parts of 16 parts, or 400 parts of 256 parts, from the third number, namely 625 parts of 256, gives
156 225 parts of 256 parts of the unit, which is a square number with side 15 parts of 16.

Therefore, we have found three numbers fulfilling the condition imposed upon us, and these are 1, 400 parts of 256, (and) 625 parts of 256. This is what we intended to find.

**17.** We wish to find four square numbers which are also in proportion such that the number formed by their sum is a square.

If four numbers are in proportion, the (result of the) multiplication of the first by the fourth equals (the result of the multiplication of) the second by the third.[35] We put 1 as the first square number, $16x^2$ as the fourth, and (we put) as the second number $x^2$'s (in a quantity) such that, when we add them to $16x^2$, the result is $x^2$'s having a square as their coefficient; such is $9x^2$, since adding $9x^2$ to $16x^2$ gives $25x^2$, which is a square number with side $5x$. Now, the (result of the) multiplication of the second number by the third equals the (result of the) multiplication of the first number by the fourth, and the (result of the) multiplication of the first number by the fourth is $16x^2$; hence we divide $16x^2$ by $9x^2$, which gives $1\frac{7}{9}$, and this is the third number. Consequently, the number formed by (the sum of) the four numbers is $25x^2 + 2\frac{7}{9}$; this, then, has to be a square. Let us assume its side to be $5x + \frac{1}{3}$; the multiplication of $5x + \frac{1}{3}$ by itself gives $25x^2 + 3\frac{1}{3}x + \frac{1}{9}$, which equals $25x^2 + 2\frac{7}{9}$. We remove the common (terms) from the two sides, so $3\frac{1}{3}x$ is equal to $2\frac{2}{3}$; hence $x$ is 8 parts of 10 parts of the unit. Since the side of the
157 second number is $3x$, and the second number, $9x^2$, the side is 24 parts of 10, and the second number is 576 parts of 100. Again, since the fourth was put $16x^2$, with side $4x$, and (since) $x$ is 8 parts of 10, $4x$ is 32 parts of 10, and it is the side of the fourth number, and the fourth number is 1024 parts of 100 (parts) of the unit. Since we assumed the first number to be 1, it is 1, as assumed. We has assumed the third number to be $1\frac{7}{9}$, so it is $1\frac{7}{9}$, as assumed.[36] Each of these four numbers is a square, and the number formed by their sum is 16,900 parts of 900, which is a square number with side 130 parts of 30 parts of the unit.

Therefore, we have found four numbers fulfilling the condition imposed upon us, and these are successively: 1, 576 parts of 100, $1\frac{7}{9}$, 1024 parts of 100 parts of 1. This is what we intended to find.

**18.** We wish to find four square numbers which are also in proportion such that the subtraction of the first from the second gives a square, the subtraction

[35] Euclid, *Elem.* VII,19. This theorem is used a few lines below.

[36] As *deduced* from the other initial hypotheses; see also the previous problem.

of the second from the third gives a square, and the subtraction of the third from the fourth gives a square.

We have already found that it is in the nature of numbers that, for any four numbers in proportion which are also squares and are such that the subtraction of the first number from the second number gives a square, the subtraction of the third number from the fourth number also gives a square.[37] Therefore, we shall seek four square numbers in proportion ⟨such that the subtraction of the first from the second gives a square and the subtraction of the second from the third gives a square⟩.[38] Let us put for the first one any number, but square, of units, say 9. Since subtracting the first from the second
158 results in a square, let us put for the second any square number which, when diminished by 9, leaves a square, say 25. Let us put for the fourth number any number, but square, of $x^2$'s, say $x^2$. (So,) since the multiplication of the first (number), or 9, by the fourth number, or $x^2$, gives $9x^2$, the (product of the) multiplication of the second number, or 25, by the third number, must also give $9x^2$; hence the third number is 9 parts of 25 parts of $x^2$. Now, if we subtract the second number, or 25, from the third number, or 9 parts of 25 parts of $x^2$, we obtain 9 parts of 25 parts of $x^2$, minus 25, which must be a square. Let us put as its side $\frac{3}{5}x - 1$; the multiplication of it by itself gives 9 parts of 25 parts of $x^2$, plus 1, minus $\frac{6}{5}x$. So this equals 9 parts of 25 parts of $x^2$, minus 25. Let us add to both sides $\frac{6}{5}x + 25$, and we remove the 9 parts which are common, so $\frac{6}{5}x$ equals 26; hence $x$ is 130 parts of 6. Since the fourth number was assumed to be $x^2$—the side of which is $x$, which is 130 parts of 6—, the fourth number is 16,900 parts of 36 parts of the unit. Again, since the third number is 9 parts of 25 parts of $x^2$, it is 6084 parts of 36 parts of the unit.[39] So, the subtraction of the first number, or 9, from the second number, or
159 25, gives 16, which is a square number with side 4. The subtraction of the second number, or 25, that is, 900 parts of 36 parts of the unit, from the third number, or 6084 parts of 36 parts of the unit, gives 5184 parts of 36 parts of the unit, which is a square number with side 72 parts of 6 parts of the unit. Again, the subtraction of the third number, or 6084 parts of 36, from the fourth number, or 16,900 parts of 36, gives 10,816 parts of 36 parts of the unit, which is a square number with side 104 parts of 6.

Therefore, we have found four numbers fulfilling the condition imposed upon us, and these are 9, 25, 6084 parts of 36, and 16,900 parts of 36 parts of the unit. This is what we intended to find.

---

[37] The text seems to assimilate the present proposition with the one seen in VII,16, *in initio*. In fact, the one of VII,16 is merely a particular case of the present one.

[38] On a possible explanation of this important lacuna, see p. 36.

[39] Unlike in the preceding problems, the assumed values of the remaining numbers are not restated here.

End of the seventh Book of the treatise of Diophantus on the restoration and the reduction, and it contains eighteen problems.

End of the treatise. Praise be to God, Lord of the Universe. The completion of the copy took place on the date of Friday, the third of Ṣafar, in the year 595. Praised be God, the Most High, and blessed be His prophet Muhammad and all his family.

# Part Three

# Mathematical Commentary

# Book IV

## The Introduction

The introduction to Book IV can be divided into three distinct parts.

### *a. Generalities*

As was done occasionally in Greek scientifical treatises,[1] Diophantus begins by recapitulating what the reader has already encountered. First he mentions that the previous problems were all reducible, after the restoration and the reduction,[2] to an equality between two terms. Further, he states that these problems involved linear numbers (γραμμικοὶ ἀριθμοί, thus unknowns in the first degree: $a$, $b$), or plane numbers (ἐπίπεδοι ἀριθμοί: $a^2$, $a \cdot b$, $b^2$), or, lastly, the two kinds combined.[3] Finally, he observes that the problems were arranged by categories in order that the beginner might better remember what he was learning.

These three observations do indeed apply to the problems found in the Greek Books I–III. Note in particular that the first and third points had been previously formulated in the Greek introduction ("Def. XI", and D.G., I, pp. 14,27–16,6, respectively).

---

[1] See, e.g., Heron, *Mechanics* (= *Opera*, II), introduction to Book III; Ptolemaeus, *Synt. math.*, particularly II,1 and III.introd.; see also the beginning of the Books in Pappus' and Theon's commentaries on the Almagest.

[2] Remember that these are the synthetic Arabic denominations used for the two basic operations defined by Diophantus in the (Greek) "Definition XI"; see also, below, "Definition XIII".

[3] Of course additively, so as not to deal with numbers other than linear or plane ones.

There is, though, a pair of propositions in Book II in which the formulation involves a *product* of two unknown squares (II,28–29); but, by assuming a numerical value for one of the unknowns, Diophantus immediately reduces the problem to one of the second degree. See also p. 178, n. 11.

After this brief survey, Diophantus goes on to explain[4] that in the coming part of the *Arithmetica* many problems of a similar kind[5] will be presented, as before, in order, which will allow the reader to acquire "*experience and skill*". These two words, which appear again in the preface to Book VII, evoke what may be the dominant characteristic of the Arabic Books: knowledge previously acquired is employed and consolidated, and nearly half of the problems will lead to the resolution of intermediate problems or types of equations studied in the previous Books (see pp. 5–7).

We do, however, encounter an important novelty which consists in the introduction of the cubic power of the unknown. This has two consequences. The first becomes tangible in the subsequent "Definition XII": since the Diophantine power-system is generated by two powers, the *square* and the *cube* (see p. 43), the introduction of the cubic power of the unknown allows us to construct higher powers also, which many of the coming problems will involve. The second consequence is revealed in "Definition XIII": in order to arrive at an equality between a certain power of $x$ and some number (as in the first three Greek Books), we shall be obliged to make regular use of the division of powers. Thus in Book IV the reader will learn how to deal with higher powers and, in particular, how to choose, in the initial assumptions, the powers best suited for the required magnitudes.

### b. "*Definition XII*"

(α) *Content*

In addition to giving the definitions of $x^3$, $x^4$, $x^5$, and $x^6$, "Definition XII" expounds the rules of divisions of these powers by lower ones. For the sake of convenience, we have used both here and in the remainder of the translation modern symbolism, denoting the $n$th power of the unknown by $x^n$. But by doing so, the system underlying the denominations as well as the rules given in the introduction, though obvious in Diophantine notation, is no longer evident. Let us, then, consider the explanations of the text in the light of the two-symbol system $Q$, $C$ (see p. 43). In the first Books, we became familiar with the power $Q \equiv x \cdot x$.

**Definition of $x^3$:** $Q \cdot x \equiv C$ ($\mathrm{K}^{\Upsilon}$, *ka*ʿ*b*).

Rules of division

$$\frac{C}{Q}\left(\equiv\frac{Q\cdot x}{Q}\right)=x;\qquad \frac{C}{x}\left(\equiv\frac{Q\cdot x}{x}\right)=Q.$$

[4] Observe the correspondence between the points of the preceding survey and what follows it in Diophantus' text.

[5] Leading likewise to an equality between two terms.

**Definition of $x^4$:** $C \cdot x = Q \cdot Q \equiv QQ$ ($\Delta^{Y}\Delta$, *māl māl*).

Rules

$$\frac{QQ}{C}\left(\equiv \frac{C \cdot x}{C}\right) = x; \quad \frac{QQ}{Q} = Q; \quad \frac{QQ}{x}\left(\equiv \frac{C \cdot x}{x}\right) = C.$$

**Definition of $x^5$:** $QQ \cdot x = Q \cdot C$[6] $\equiv QC$ ($\Delta K^{Y}$, *māl kaʿb*[7]).

Rules

$$\frac{QC}{x} = QQ;^{8} \quad \frac{QC}{Q} = C; \quad \frac{QC}{C} = Q; \quad \frac{QC}{QQ} = x.$$

**Definition of $x^6$:** $QC \cdot x = Q \cdot QQ = C \cdot C \equiv CC$($K^{Y}K$, *kaʿb kaʿb*).

Rules

$$\frac{CC}{x} = QC; \quad \frac{CC}{Q} = QQ; \quad \frac{CC}{C} = C; \quad \frac{CC}{QQ} = Q; \quad \frac{CC}{QC} = x.$$

The eighth and ninth powers, found only in the Arabic Books, are defined when they first appear, in IV,29:

**Definition of $x^8$:** $QQ \cdot QQ = CC \cdot Q \equiv CCQ$ (*kaʿb kaʿb māl*[9]; $\Delta K^{Y}K$?).

The following rule of division is explained when needed in the text (in IV,31):

$$\frac{CCQ}{QQ} = \frac{QQ \cdot QQ}{QQ} = QQ.$$

**Definition of $x^9$:** $CC \cdot C \equiv CCC$ (*kaʿb kaʿb kaʿb*; $K^{Y}K^{Y}K$?).

The following two rules are given in an incidental way:

$$\frac{CCC}{CCQ} = \frac{CC \cdot C}{CCQ} = \frac{CCQ \cdot x}{CCQ} = x \quad \text{(IV,29)}$$

$$\frac{CCC}{QQ} = \frac{CC \cdot C}{QQ} = \frac{QQ \cdot Q \cdot C}{QQ} = \frac{QQ \cdot QC}{QQ} = QC \quad \text{(IV,31).}$$

---

[6] $C \cdot Q$ in the text.

[7] *kaʿb māl, kaʿb maḍrūb fī māl* elsewhere in the text; see p. 45.

[8] For some reason, the dividing powers are now taken in the reverse order.

[9] *māl māl māl māl* elsewhere in the text; see p. 45.

(*β*) *On the Genuineness of "Definition XII"*

The presentation in "Definition XII" of the powers $x^3$ to $x^6$ is surprising, for they are already known to the reader from the Greek introduction.[10] Further, the divisions of powers are not really new to the reader, since he became acquainted with them indirectly when he learned how to multiply the various powers (up to the sixth) by their inverses ("Def. VIII"); and Diophantus himself explicitly states in the Greek introduction that the divisions of powers are obvious (φανεροί) from the rules of multiplication just seen (D.G., I, p. 14,1–2). Since it is unlikely that Diophantus would repeat himself without at least pointing out this reiteration, one must seriously consider the possibility of interpolation, either of "Def. XII", or, alternatively, of some of the Greek introduction.

It would seem possible, at first sight, that the definitions and rules concerning $x^3$, $x^4$, $x^5$, and $x^6$ (none of which is needed for Books I–III), might have been added on to the beginning of Book I later. But such an hypothesis is untenable, for it would oblige us to consider as extraneous entire passages of the Greek introduction, leaving, at best, a very disconnected text.

Our suspicions, then, must fall upon "Definition XII", and this leads us to ask who might have been responsible for its addition. It may well have been the author of the major commentary, although the possibility of its having been added earlier cannot be excluded: after all, early interpolators performed the resolutions of corollaries which were considered by Diophantus to be straightforward given previous explanations—as Diophantus stated explicitly in connection with the rules for division. Whatever their origin, the location of these definitions and rules in the introduction to Book IV is understandable since the use of higher powers was, as said, not required in the earlier Books, in which the knowledge of $x$ and $x^2$ only was needed.[11]

**Remark.** These definitions and rules cannot have been merely taken from the introduction to Book I and added on to the introduction to Book IV by an Arabic copyist when Books IV–VII were copied as a separate entity, for a formal definition of $x$ and $x^2$ would surely have been added as well. At least one reader was a victim of this omission (cf. p. 30, no. 1).

We shall then consider it highly probable that Diophantus did not write these definitions and rules twice, and that "Definition XII" is therefore a later

---

[10] Observe that the rules of multiplication stated in the definitions of $x^3$ to $x^6$ are exactly those found in "Def. IV".

[11] I dismiss the $\Delta^Y\Delta$ appearing in II,24 (D.G., I, p. 120,2 and 4) because I consider lines 2–4 and half of line 5 in Tannery's edition to be interpolated. One can easily restore the text and bring it into conformity with lines 21–23 of the same page: this agreement with problem II,25 (and with similar cases, as in III,2) is no doubt desirable. Observe that the raising of $x$ to the fourth power does not appear in *Faḫrī* IV,3 which reproduces (though not *verbatim*) Diophantus' II,24.

addition, as must be, of course, the passages in Book IV alluding to the rules (lines 37–38 and 123–24).[12]

### *c.* "*Definition XIII*"

In the Greek "Definition XI", the reader was told to eliminate first the negative terms of an equation (the Arabic *al-jabr*) and then the magnitudes common to both sides (the Arabic *al-muqābala*[h]). One was always left with a single term equal to some number, either immediately or after a division by $x$ (see problems I,26 and 31 *seqq.*). Because we shall be dealing with higher powers in the coming part of the *Arithmetica* this kind of final division must be used more systematically and is thus worthy of mention in one of these "definitions". Thus "Definition XIII", after repeating the essence of "Definition XI", completes it in requiring that the division by the power of lesser degree follow the performance of the two basic algebraical operations. This will leave us, as before, with the equality of some power of $x$ and some number, the implication being that we shall not (yet) be left with any complete quadratic equations (cf. Part One, §13).

---

## The Problems in Book IV

Preliminary remark on the mathematical commentary:

In some of the propositions which do not lead to intermediate problems involving the basic numerical methods taught in Book II, we have given the resolutions in algebraic notation, the values of the constants being inserted after the establishment of the final formula for the unknown. Otherwise, the computations of the text have been reproduced, often elaborated upon. In some cases, computations have been supplemented—particularly the resolutions of the intermediate problems (not performed in the text) when some explanation is required. Finally, the general methods of resolution, if any, have been commented on or summarized at the end of each group of problems. This has been done in the hope of making Diophantus accessible to a broader circle of readers.

---

**Problem IV,1.** $b^3 + a^3 = \square.$

We put $a = x$, $b = mx$; hence

$$(m^3 + 1)x^3 = \square.$$

[12] As to the definitions of $x^8$ and $x^9$, found in the middle of Book IV (problems 29 and 31), clearly they were kept here, the commentator respecting an order going back to Diophantus himself.

Taking $\square = (nx)^2$, we obtain $(m^3 + 1)x^3 = n^2x^2$; so

$$x = \frac{n^2}{m^3 + 1}.$$

For $m = 2, n = 6$: $\quad x = 4.$

Hence $\quad a^3 = x^3 = 64, \quad b^3 = 8^3 = 512, \quad \square = 576 = 24^2.$

Al-Karajī, in the corresponding problem of the *Faḫrī*, takes $m = 2, n = 3$, and thus has $a^3 = 1, b^3 = 8, \square = 9$ (see *Extrait du Fakhrî*, V,1).

**Problem IV,2.** $\quad b^3 - a^3 = \square.$

We put $a = x, b = mx$ $(m > 1)$; hence

$$(m^3 - 1)x^3 = \square.$$

Taking $\square = (nx)^2$, we obtain $(m^3 - 1)x^3 = n^2x^2$; so

$$x = \frac{n^2}{m^3 - 1}.$$

For $m = 2, n = 7$: $\quad x = 7.$

Hence $\quad a^3 = x^3 = 343, \quad b^3 = 14^3 = 2744, \quad \square = 2401 = 49^2.$

Al-Karajī (*Extrait*, V,2) takes the same $m, n$, hence he has the same results.

**Problem IV,3.** $\quad b^2 + a^2 = \text{❒}.$

We put $a = x, b = mx$; hence

$$(m^2 + 1)x^2 = \text{❒}.$$

Taking $\text{❒} = (nx)^3$, we obtain

$$x = \frac{m^2 + 1}{n^3}.$$

For $m = 2, n = 1$ (thus $\text{❒} = a^3$): $\ x = 5.$

Hence $\quad a^2 = x^2 = 25, \quad b^2 = 4x^2 = 100, \quad \text{❒} = 125 = 5^3.$

Al-Karajī (*Extrait*, V,3) also reduces the problem to $b^2 + a^2 = a^3$, and takes the same value for $m$.

**Problem IV,4.** $\quad b^2 - a^2 = \text{❒}.$

We put $a = x, b = mx$ $(m > 1)$, hence

$$(m^2 - 1)x^2 = \text{❒}.$$

Taking $\boxdot = (nx)^3$, we obtain

$$x = \frac{m^2 - 1}{n^3}.$$

For $m = 5, n = 2$: $x = 3.$

Hence $a^2 = 9, \quad b^2 = 15^2 = 225, \quad \boxdot = 216 = 6^3.$

Al-Karajī (*Extrait*, V,4) takes (as above) $m = 2, n = 1$, thus obtaining also $a = 3$, but the smaller values $b^2 = 36$, $\boxdot = 27$.

---

The method used in the group of problems IV,1–4 is clear and does not require any further explanation. One may, however, add the following remarks about the remaining possible combinations of squares and cubes:

1°. The problems involving one square and one cube on the left side can all be reduced to one form or another of IV,1–4 by moving a term from one side to the other. These cases, although banal, are treated in the middle of the fifth section of the *Faḫrī* (see *Extrait*, V,23–27[13]). Thus, they are obviously out of place, as already remarked (p. 59).

2°. The two pairs of problems which involve only squares or only cubes on both sides are:

(a) $b^2 \pm a^2 = \square$, which is soluble for any numerical value of $\square$ (cf. II,8 and 10).

(b) $b^3 \pm a^3 = \boxdot$, which is impossible; this appears to have been well known in the tenth century (see Woepcke, *Recherches sur (...) Léonard de Pise*, p. 301), so that it is difficult to imagine the Greeks not having been aware of it.

The next group of problems involves, this time, products of squares and cubes.

---

**Problem IV,5.** $b^2 \cdot a^2 = \boxdot.$

We put $a = x$, $b = mx$; hence

$$m^2x^4 = \boxdot.$$

Taking $\boxdot = (nx)^3$, we obtain

$$x = \frac{n^3}{m^2}.$$

For $m = 2, n = 2$: $x = 2.$

Hence $a^2 = 4, \quad b^2 = 4x^2 = 16, \quad \boxdot = 64 = 4^3.$

---

[13] These present the cases: $a^2 + b^3 = \square$ (23); $b^3 - a^2 = \square$ (24); $a^2 - b^3 = \square$ (25); $b^3 + a^2 = \boxdot$ (26); $b^3 - a^2 = \boxdot$ (27). Missing is $a^2 - b^3 = \boxdot$.

Diophantus has reduced the problem to $b^2 \cdot a^2 = b^3$; al-Karajī, taking $m = 2, n = 1$, reduces it to $b^2 \cdot a^2 = a^3$ and obtains the results $x = \frac{1}{4}, a^2 = \frac{1}{16}$, $b^2 = \frac{1}{4}$, and ⧉ $= \frac{1}{64}$ (cf. *Extrait*, V,5).

**Problem IV,6.** $b^3 \cdot a^2 = \square$.

We put, say, $a = x, b = mx$; hence

$$m^3x^5 = \square.$$

Putting $\square = (nx)^2$, we would have $m^3x^5 = n^2x^2$. The subsequently necessary condition for the rationality of $x$ can be avoided if we take $\square = (nx^2)^2$. Doing so, we obtain

$$x = \frac{n^2}{m^3}.$$

For $m = 2, n = 4$: $x = 2$.

Hence $a^2 = 4, \quad b^3 = (2x)^3 = 64, \quad \square = 256 = 16^2$.

Al-Karajī (*Extrait*, V,6) has the same values. Neither here nor in the next problem does he make a preliminary choice, as does our text. Arriving in this problem at $8x^5 = \square$, he simply states that it is necessary (*yajib*) to put $\square = (4x^2)^2$ "in order that the equation (*muʿādala*[h]) be possible and lead to (something) known (i.e., rational: *maʿlūm*)".

**Problem IV,7.** $b^3 \cdot a^2 =$ ⧉.

We put $a = x, b = mx$; hence

$$m^3x^5 = \text{⧉}.$$

Taking ⧉ $= (nx)^3$ would give $m^3x^2 = n^3$ (and the condition of rationality $n^3/m^3 =$ square is precisely the problem to be solved). Thus, we shall put ⧉ $= (nx^2)^3$, whence

$$x = \frac{m^3}{n^3}.$$

For $m = 4, n = 2$: $x = 8$.

Hence

$a^2 = 64, \quad b^3 = 32^3 = 32{,}768, \quad$ ⧉ $= 64 \cdot 32{,}768 =$ cube (namely $(4 \cdot 32)^3$).

Al-Karajī, taking $m = 1, n = 1$, obtains $x = 1$, thus $a^2 = 1, b^3 = 1$ and ⧉ $= 1$(!) (cf. *Extrait*, V,7).

**Problem IV,8.** $b^3 \cdot a^3 = \square$.[14]

We put $a = x, b = mx$ ($b = 2x$ in the text); hence

$$m^3x^6 = \square.$$

Putting $\square = (nx)^2$ leads to $x^4 = n^2/m^3$, which is impossible for the assumed value $m = 2$, and which is inconvenient in general.[15] Thus we shall take $\square = (nx^2)^2$, whence

$$x^2 = \frac{n^2}{m^3}.$$

We are led to the intermediate problem of finding $x, n, m$ fulfilling $m^3x^2 = n^2$, a solution of which has been found in IV,6, namely $x^2 = 4$, $m^3 = 64$, $n^2 = 256$.

**Problem IV,9.** This is not a new problem, but merely the return to the original problem (by the insertion of the values of the coefficients just obtained), the "determination" of $x$, and the verification (cf. pp. 61–62).

$x$ being 2, we have

$$a^3 = 8, \quad b^3 = 8^3 = 512, \quad \square = 4096 = 64^2.$$

Al-Karajī (*Extrait*, V,8) has a single problem. Having arrived at $8x^6 = \square$, he simply states: "It is not correct[16] to take for that (*ḏālika*) anything but the result of the multiplication of a square number by a cubic number which comprise a square number. We have shown the method for (finding) that. After seeking these two numbers, you will find for the one 64 and for the other 4, and the number comprised by these two numbers is 256". There then follows the reconstruction of the problem. The reasoning, defective in our text, is hardly any better in al-Karajī's version.

• **Corollary.**[17] One can reduce to the above group of problems the following set.

(a) $$\frac{b^3}{a^3} = \square.$$

The problem is tantamount to solving $a^3 \cdot c^2 = ⧆$, which has been treated in IV,7.

---

[14] This would have been the auxiliary problem to solve in IV,6 had we continued with the original assumption $\square = (nx)^2$.

[15] The explanations in the text are not altogether clear; cf. p. 92, n. 21.

[16] *lā yastaqīm*; cf. line 164 of our Arabic text.

[17] Not in the *Faḫrī*.

(b) $$\frac{b^2}{a^2} = \boxdot.$$

The equivalent problem $c^3 \cdot a^2 = \square$ has been solved in IV,6.

(c) The text then states that the same holds for the remaining problems of this kind. What must be understood is these two sets of problems:

$\frac{b^3}{a^2} = \square$, or $\frac{b^2}{a^3} = \square$, reducible to IV,5 and IV,6, respectively;

$\frac{b^3}{a^2} = \boxdot$, or $\frac{b^2}{a^3} = \boxdot$, reducible to IV,7 and IV,8–9, respectively.

---

Problems IV,5 to IV,8–9, namely

$$b^2 \cdot a^2 = \boxdot, \quad b^3 \cdot a^2 = \square, \quad b^3 \cdot a^2 = \boxdot, \quad b^3 \cdot a^3 = \square$$

(and the problems with the corresponding divisions on the left sides), present no difficulties. One will notice that the first problem does not involve any condition of rationality, while the next two lead to a condition which can be avoided,[18] and, finally, in the last one a condition must be fulfilled.

---

**Problem IV,10.** $a^3 + k \cdot a^2 = \square, \quad k = 10.$

Putting $a = x$, we have $x^3 + 10x^2 = \square.$

We take $\square = (nx)^2$, with $n^2 > k = 10$; thus

$$x = n^2 - 10.$$

For $n = 4$, we have $x = 6;$

hence $a^3 = x^3 = 216, \quad \square = 576 = 24^2.$

Al-Karajī (*Extrait*, V,9) has the same values and the condition for $n^2$.

**Problem IV,11.** $a^3 - k \cdot a^2 = \square, \quad k = 6.$

Putting $a = x$, we have $x^3 - 6x^2 = \square.$

We take $\square = (nx)^2$, so

$$x = n^2 + 6.$$

For $n = 2$, we have $x = 10;$

hence $a^3 = x^3 = 1000, \quad \square = 400 = 20^2.$

---

[18] They are, incidentally, reducible to the banal forms $b^3$ = square and $a^2$ = cube (i.e., $b^3$ and $a^2$ are sixth powers), and, further, solutions of IV,6 and IV,7 are known from the two other problems of the group. The reason for their presence obviously lies in the method of their resolution.

Al-Karajī (*Extrait*, V,10) has $k = 10$ (as in the previous problem); he takes $n = 1$, thus obtaining $x = 11$, $a^3 = 1331$, $\square = 121$.

**Problem IV,12.** $\qquad a^3 + k \cdot a^2 = ⧉, \qquad k = 10.$

Putting $a = x$, we have $\quad x^3 + 10x^2 = ⧉.$

We set $⧉ = (nx)^3$ $(n > 1)$; hence

$$(n^3 - 1)x = 10, \qquad x = \frac{10}{n^3 - 1}.$$

For $n = 2$, we have $\qquad x = \frac{10}{7};$

hence $\qquad a^3 = x^3 = \frac{1000}{7 \cdot 7 \cdot 7}, \quad ⧉ = \frac{8000}{7 \cdot 7 \cdot 7} = \left(\frac{20}{7}\right)^3.$

Neither this problem nor the following one is found in al-Karajī's *Faḫrī*.

**Problem IV,13.** $\qquad a^3 - k \cdot a^2 = ⧉, \qquad k = 7.$

1°. First method.
We put $a = x$, so $\qquad x^3 - 7x^2 = ⧉.$

Taking $⧉ = (nx)^3$, with $n < 1$, we have $(1 - n^3)x = 7$, hence:

$$x = \frac{7}{1 - n^3}.$$

For $n = \frac{1}{2}$, we have: $\qquad x = 8.$

So $\qquad a^3 = x^3 = 512, \quad ⧉ = 64.$

2°. Second method.
The text takes $⧉ = x^3$ (implicitly[19]) and puts $a = 2x$. Thus we have

$$a^3 - ⧉ = 7x^3 = k \cdot a^2 = 28x^2,$$

hence $\qquad x = 4, \quad a^3 = 8^3 = 512, \quad ⧉ = 64.$

The second method, doubtless interpolated, does not differ substantially from the previous one: it simply takes $⧉$, instead of $a^3$, as $x^3$ (and the choice of the factor of proportionality of $a$ to $x$ gives the same results as previously). The treatment itself is rather carelessly done,[20] and the only distinctive feature of the method—choosing the side of the indeterminate cube as the unknown $x$—is not even made evident. The final statement is formulated in an unusual

[19] This omission, however, can be accounted for by supposing a lacuna by homoeoteleuton (see note 91 of the app. crit.).

[20] Hence a couple of readers' glosses (now incorporated into the manuscript; cf. p. 30, nos. 3 and 4).

way and restates the problem in the form $a^3 - \boxdot = k \cdot a^2$; the scholiast might have been confused by the disorganized presentation of the problem when he added this final statement.

---

IV,10 to IV,13 form another group of problems which are easily solved.[21] At most, we have to fulfil a condition for positivity of the unknown when choosing the parameter.

---

**Problem IV,14.** $\begin{cases} k \cdot a = \boxdot, \\ l \cdot a = \square', \end{cases}$ $k, l$ given numbers.

(a) $k = 10, l = 5$.

We put $a = x$, so $\begin{cases} 10x = \boxdot \equiv u^3, \\ 5x = \square' \equiv v^2. \end{cases}$

We assume, the text says in effect, that $u^2 = (n^2/m^2) \cdot v^2$, where $n$, $m$ are any natural numbers.[22]

Thus $$u^2 = \frac{n^2}{m^2} \cdot v^2 = \frac{n^2}{m^2} \cdot 5x$$

and $$u \equiv \frac{u^3}{u^2} = \frac{10x}{(n^2/m^2) \cdot 5x} = \frac{2m^2}{n^2}.$$

Choosing $n^2/m^2 = \frac{1}{4}$, we have $u = 8$, hence $u^3 = 512$. Thus

$$a = x = \frac{u^3}{10} = 51\tfrac{1}{5}, \quad \boxdot = 512, \quad \square' = 256.$$

(b) $k = 5, l = 10$.[23]

With $a = x$: $\begin{cases} 5x = \boxdot \equiv u^3, \\ 10x = \square' \equiv v^2. \end{cases}$

With $u^2 = (n^2/m^2) \cdot v^2$, we have

$$u^2 = \frac{n^2}{m^2} \cdot v^2 = \frac{n^2}{m^2} \cdot 10x, \quad \text{hence} \quad u \equiv \frac{u^3}{u^2} = \frac{5x}{(n^2/m^2) \cdot 10x} = \frac{m^2}{2n^2}.$$

Taking again $n^2/m^2 = \frac{1}{4}$, we obtain $u = 2$. So

$$a = x = \frac{u^3}{5} = \frac{8}{5}, \quad \boxdot = 8, \quad \square' = 16.$$

---

[21] Observe that IV,10–11 are reducible to the trivial forms $a + k = \square'$ and $a - k = \square'$.

[22] Which amounts simply to setting a proportionality (with a positive rational factor) between the sides of $u^3$ and $v^2$.

[23] This problem is later called the "inverse" (ʿ*aks*) of the preceding one (text, line 328).

(c) $\begin{cases} 10x = \boxdot \equiv u^3, \\ 5x = \square' \equiv v^2. \end{cases}$

Same case as (a), but we take $n^2/m^2 = 4$; then

$$u = \frac{2m^2}{n^2} = \frac{1}{2}.$$

Therefore $$a = x = \frac{u^3}{10} = \frac{1}{80}, \quad \boxdot = \frac{1}{8}, \quad \square' = \frac{1}{16}.$$

(d) $\begin{cases} 5x = \boxdot \equiv u^3, \\ 10x = \square' \equiv v^2. \end{cases}$

Same case as (b), but we take $n^2/m^2 = 4$; then

$$u = \frac{m^2}{2n^2} = \frac{1}{8}, \qquad u^3 = \frac{1}{512}.$$

Therefore $$a = x = \frac{u^3}{5} = \frac{1}{2560}, \quad \boxdot = \frac{1}{512}, \quad \square' = \frac{1}{256}.$$

(e) Another method, applied to the problem

$$\begin{cases} 10a = \boxdot \equiv u^3, \\ 5a = \square' \equiv v^2, \end{cases}$$

is to put $u^3 = (nx)^3$, say $u^3 = x^3$; then $a = \frac{1}{10}x^3$. Inserting this into the second equation gives $\frac{1}{2}x^3 = v^2$.

Putting $v^2 = (mx)^2$, we obtain $x = 2m^2$.

For $m = 2$: $$x = 8,$$

so $$x^3 = \boxdot = 512, \quad a = 51\tfrac{1}{5}, \quad (\square' = 256).$$

(f) In the same problem, we may of course put (firstly)

$$v^2 = (mx)^2, \quad \text{say} \quad v^2 = x^2; \quad \text{then} \quad a = \tfrac{1}{5}x^2.$$

Inserting this into the first equation gives $2x^2 = u^3$.

Putting $u^3 = (nx)^3$, we have $x = 2/n^3$.

For $n = 1$: $$x = 2,$$

so $$x^2 = \square' = 4, \quad a = \tfrac{4}{5}, \quad \boxdot = 8.$$

Al-Karajī also treats the two forms of the problem, but not without some confusion. The summary given by Woepcke in his *Extrait* (problems V,11–12) being somewhat inappropriate for our purposes, we shall now give a more suitable presentation of al-Karajī's cases.

(1) $\begin{cases} 10 \cdot a = \boxdot, \\ 5 \cdot a = \square. \end{cases}$

So $\square/5 = \text{❒}/10$. With $\square = x^2$, $\text{❒} = x^3$, we have $x^3 = 2x^2$. Hence $x = 2$, $\square = 4$, $\text{❒} = 8$ and $a = \frac{4}{5}$.

$$(2)\quad \begin{cases} 10 \cdot a = \text{❒}, \\ \ \ 5 \cdot a = \square'. \end{cases}$$

With $\square' = (2x)^2 = 4x^2$ and $\text{❒} = x^3$, we have $x^3 = 8x^2$, hence $(x = 8,)$ $\text{❒} = 512$, $\square' = 256$ and $a = 51\frac{1}{5}$.

$$(3)\quad \begin{cases} \ \ 5 \cdot a = \text{❒}, \\ 10 \cdot a = \square'. \end{cases}$$

So $\square' = 2\text{❒}$. With $\square' = (2x)^2 = 4x^2$ and $\text{❒} = x^3$, we have $x^3 = 2x^2$, hence $x = 2$, $\text{❒} = 8$, $\square' = 16$ and $a = 1\frac{3}{5}$.

Under the denomination "other treatment" is found the following approach, solving (2) and (3) again, with little change:

$$(4)\quad \begin{cases} 10 \cdot a = \text{❒}, \\ \ \ 5 \cdot a = \square'. \end{cases}$$

With $\text{❒} = x^3$ and $\square' = (nx)^2$, $n$ arbitrary,[24] say $\square' = (2x)^2 = 4x^2$, we have $\square'/5 = \frac{4}{5}x^2$; hence $10 \cdot \frac{4}{5}x^2 = 8x^2 = \text{❒} = x^3$, and $x = 8$. Thus $\square' = 256$, $\text{❒} = 512$ (and $a = 51\frac{1}{5}$).

$$(5)\quad \begin{cases} \ \ 5 \cdot a = \text{❒}, \\ 10 \cdot a = \square'. \end{cases}$$

The same choice as made above gives $\square'/10 = \frac{2}{5}x^2$, hence $5 \cdot \frac{2}{5}x^2 = 2x^2 = \text{❒} = x^3$ and $x = 2$. Thus $\square' = 16$, $\text{❒} = 8$ (and $a = \frac{8}{5}$).

The two following approaches are briefly explained, without numerical computation.

$$(6)\quad \begin{cases} 10 \cdot a = \text{❒}, \\ \ \ 5 \cdot a = \square', \end{cases} \quad \text{and} \quad \begin{cases} \ \ 5 \cdot a = \text{❒}, \\ 10 \cdot a = \square'. \end{cases}$$

We put this time $\text{❒} = (mx)^3$ and $\square' = (nx)^2$. Hence

$$5 \cdot \frac{m^3x^3}{10} = \square' \equiv n^2x^2 \quad \text{and} \quad 10 \cdot \frac{m^3x^3}{5} = \square' \equiv n^2x^2,$$

respectively.

(7) In the more restricted case, in which the square and the cube have the same side, we put $k \cdot (\square/l) = \text{❒}$. This is nothing but a repetition of problem (1).

---

[24] "*Ijʿal ḍilaʿ al-murabbaʿ al-muʿādil li-ḍarb al-maṭlūb fī ẖamsa*[h] *baʿd dilaʿ al-murabbaʿ al-kāʾin min ḍilaʿ al-mukaʿʿab au aḍʿāfahū*".

N.B. Manuscripts E, L (cf. p. 60, n. 29) have, instead of *li-ḍarb al-maṭlūb fī ẖamsa*[h], the reading *li-ẖamsa*[h] *ašyāʾ*; K has the latter version, with the Paris manuscript's version as a correction in the margin. The use of *šaiʾ* is of course inappropriate here, since *a* is not taken as the unknown.

It is clear that this set of problems takes its inspiration from the one in the *Arithmetica*. Al-Karajī's version, however, shows some disorder in the presentation, and begins and ends with two similar cases, namely "Diophantus"'s IV,14,f/IV,15. Observe also that the required number, $a$, is never chosen as the unknown $x$ in the *Faḫrī*'s version. All in all, al-Karajī's set surpasses that of Book IV in terms of banality, a banality for which, as we shall see, Diophantus himself must not be held responsible.

**Problem IV,15.**
$$\begin{cases} k \cdot a = \text{❒}, \\ l \cdot a = \square. \end{cases}$$

with $k = 10$ and $l = 4$ (or inversely, the text says, for the method is the same; cf. IV,14).

1°. First method (as in IV,14,a but now with $n^2/m^2 = 1$).

We put $a = x$, so
$$\begin{cases} 10x = \text{❒} = u^3, \\ 4x = \square = u^2. \end{cases}$$

Hence

$$u = \frac{10x}{4x} = 2\tfrac{1}{2}, \quad \square = u^2 = 6\tfrac{1}{4}, \quad (\text{❒} = 15\tfrac{5}{8}) \quad \text{and} \quad x = \tfrac{25}{16}.$$[25]

2°. Second method (as in IV,14,e).

We put $\text{❒} = (nx)^3$, say $= x^3$, so

$$a = \tfrac{1}{10}x^3.$$

The second equation gives $\frac{4}{10}x^3 = \square = x^2$, so $x = 2\frac{1}{2}$; then

$$\square = 6\tfrac{1}{4}, \quad \text{❒} = 15\tfrac{5}{8} \quad \text{and} \quad a = \tfrac{25}{16}.$$

• **Corollary.**[26]
$$\frac{b^3}{a^2} = \frac{k}{l}, \quad \text{given ratio.}$$

The text gives the ratio (3:1) but explains only the method for solving: taking any two numbers in the given ratio, say 3h and h, we are led to the problem of finding an $x$ fulfilling the system

$$\begin{cases} 3h \cdot x = \text{❒}, \\ h \cdot x = \square', \end{cases}$$

which we know how to solve from IV,14. Then $b^3 = \text{❒}$ and $a^2 = \square'$.

---

[25] Observe that the problem is determinate ($a = k^2/l^3$).

[26] Actually, this corollary depends on the set of resolutions now numbered IV,14.

**Remark.** There is here much ado about nothing: when such a problem is proposed, one can put $a = x$, $b = mx$ ($m$ arbitrary), thus obtaining

$$x = \frac{k}{lm^3}.$$

It is in this way that the auxiliary problem $b^3 = 2a^2$ is solved in the remainder of the *Arithmetica*.[27]

Al-Karajī's problems V,13–14 are simply numerical computations based on the initial statement of the corollary above. He considers:

(1) $\frac{b^3}{a^2} = \frac{3}{1}$ and puts $b = a = x$ (V,13)

(2) $\frac{b^3}{a^2} = \frac{1}{3}$ and puts also $b = a = x$ (V,14,a)

(3) $\frac{b^3}{a^2} = \frac{3}{1}$ with $b = x$, $a = 2x$, thus $a \neq b$ (V,14,b)

(4) $\frac{b^3}{a^2} = \frac{1}{3}$ with $b = x$, $a = 2x$ (V,14,c).

We see that, while taking the same ratio as does our text,[28] al-Karajī applies the simpler resolution and does not assume a number $h$, as is directed in our version.

---

The problems given as IV,14 and 15 in the Arabic translation present far too many banal cases to be wholly attributable to Diophantus himself.

1°. The last two parts of IV,14, (e) and (f), were doubtless added by a scholiast. That their author is the same as the one of the second resolution of IV,13 is suggested by the similarity of the treatments: as in IV,13,2°, one of the indeterminate magnitudes of the right sides, instead of $a$, is taken as the unknown $x$. There is in fact no substantial difference between the resolution of (a)–(d) and that of (e) and (f), since (e) and (f) ultimately amount to setting a proportion between $u^2$ and $v^2$, as was done before. As to part (f), the scholiast unfortunately chose $n = m = 1$, thus anticipating the following problem (IV,15).

2°. The second method in IV,15, similar to the ones seen in IV,13,2° and IV,14,e and f, must also be a later addition, which most probably goes back to the same scholiast.

---

[27] See, just above, IV,14,e and f, as well as "VI", 1 and 24 (in problems 2 and 19 one is directly given the numerical solution).

[28] And also its inverse, in the manner seen in the previous problems.

3°. Even without these alternative resolutions, the text presents too many banal aspects of a single problem to be wholly genuine: ($\alpha$) parts (c) and (d) of IV,14—in which the ratio assumed for $u^2 : v^2$ in parts (a) and (b) is simply inverted—are certainly interpolations; ($\beta$) part (b) of IV,14—in which the values of the (given) constants of part (a) are merely interchanged—can hardly be genuine either;[29] ($\gamma$) as to IV,15 (which appears as a separate problem in the Arabic text despite the fact that it is a particular case of IV,14) and the subsequent corollary with its intricate method of resolution, they are highly suspect, although IV,15 itself could be the result of a remark made in the original text.

Thus, I am inclined to consider the genuine portion of group IV,14–15 to be essentially part (a); that is, a single, basic problem—which indeed deserves no more attention than is given it in part IV,14,a.

---

**Problem IV,16.** $\begin{cases} k \cdot b = \square, \\ k \cdot a = \sqrt[3]{\square}, \end{cases} \qquad k = 10.$

We put $a = nx$, say $a = x$, so that the cube is $k^3x^3$, and $b = mx^2$, so that the cube is also $kmx^2$. Hence

$$x = \frac{m}{k^2}.$$

For $k = 10$, $m = 300$, we have $x = 3$; so

$$a = 3, \quad b = 2700, \quad \square = 27{,}000 = 30^3.$$

One would have expected a preliminary remark indicating that the choice $a = x$, $b = mx$ necessitates $m$ being a square.

Al-Karajī (*Extrait*, V,15) takes $k = 10$, $m = 200$, whence $x = 2$. He also suggests the possibility of taking any cube, say $h^3$, and then forming

$$b = \frac{h^3}{10}, \quad a = \frac{h}{10}.$$

**Problem IV,17.** $\begin{cases} k \cdot b^2 = \square, & k = 5, \\ k \cdot a^2 = \sqrt[3]{\square}, & \\ b = ma, & m = 20, \text{ given.} \end{cases}$

*Condition*: $m \cdot k =$ square. This condition is said to represent a constructible problem (see below).

---

29 This "inverse case" is alluded to at the beginning of IV,15.

We put $a = x$, so $$b = mx.$$

We have: $$❒ = kb^2 = km^2x^2 = (ka^2)^3 = k^3x^6,$$

so $$x^4 = \frac{km^2}{k^3} = \frac{m^2}{k^2} \quad \text{and} \quad x = \sqrt{\frac{m}{k}}\left(= \frac{1}{k}\sqrt{mk}\right).$$

With $m = 20$, $k = 5$, we have $x = 2$. So

$$a^2 = 4, \quad b^2 = 1600, \quad \text{and} \quad ❒ = 8000 = 20^3.$$

Al-Karajī (*Extrait*, V,16) has $k = 4$, $m = 9$; hence $x = a = \frac{3}{2}$, $a^2 = 2\frac{1}{4}$, $b^2 = 182\frac{1}{4}$ and $❒ = 729$. He gives the condition, but does not allude to its "constructibility" (neither does the word *muhayya*ʾ appear anywhere in the *Faḫrī*).

The rationality of the solution is subject to a condition which prevents an arbitrary choice of the given numbers $k$ and $m$. Thus, the question of how to find acceptable values for these two numbers arises. In the present case, the difficulty is easily overcome since the condition represents a "constructible" (*muhayya*ʾ = πλασματικόν) problem.[30]

A constructible problem is one of the form

$$f(k, l) = r^n \quad (n \text{ natural, known}),$$

where $r$ must be a rational number and $f$ a rational function linear in one of the two variables, say $l$. The solution appears immediately: we choose any rational value for $k$, take the $n$th power of any rational number and then determine $l$ from the resulting linear equation.

An example of such a construction is given in IV,22, but an even better illustration is found in the first (Greek) Book (problems 27, 28, 30):

**I,27**, computed *in indeterminato*, leads to the condition

$$k^2 - 4l = \Box.$$

Taking for $\Box$, the smaller square, 16, and for $k^2$, the larger square, 400, we have $l = 96$.

**I,28** gives the condition

$$2l - k^2 = \Box.$$

With the same choice as above, we obtain $l = 208$.

**I,30** depends on the condition

$$k^2 + 4l = \Box.$$

[30] The word "constructible", though perhaps not the best translation of πλασματικόν, has been chosen by us since it characterizes the resolution of such questions.

Taking again 16 as the smaller square (now $k^2$) and 400 as the larger, we have $l = 96$.

These examples show clearly the way to treat such problems.

The constructible problems occurring in the *Arithmetica* are thus the following:

(1) $k^2 - 4l =$ square (in I,27);

(2) $2l - k^2 =$ square (in I,28);

(3) $k^2 + 4l =$ square (in I,30);

(4) $\frac{k}{l}$ (or $k \cdot l$) $=$ square (in IV,17 and 19);

(5) $\frac{k}{l^2}$ (or $k \cdot l$) $=$ cube (in IV,20);

(6) $\frac{k}{l^3}$ (or $k \cdot l$) $=$ fourth power (in IV,21);

(7) $\frac{k}{l^3}$ (or $k \cdot l^3$) $=$ sixth power (in IV,22);

(8) $\frac{4l - k^3}{3k} =$ square (in V,7 and 8 = "IV",1 and 2);

(9) $\frac{k^3}{l - \frac{3}{4}k} =$ square (in V,9 and 10);

(10) $l - \frac{3}{4}k^2 =$ square (in V,11 and 12).

Since the constructible problems arise from a condition for rationality of the unknown as function of the given numbers, they are associated with problems which are determinate.

**Problem IV,18.**
$$\begin{cases} k \cdot b^3 = \square, & k = 8, \\ k \cdot a^3 = \sqrt{\square}, & \\ b = ma, & m = 3. \end{cases}$$

*Condition*: $k =$ cube.

We put $a = x$, so $b = mx$.

Then:
$$\square = kb^3 = km^3x^3 = (ka^3)^2 = k^2x^6,$$

so
$$x^3 = \frac{km^3}{k^2} = \frac{m^3}{k} \quad \text{and} \quad x = \frac{m}{\sqrt[3]{k}}.$$

With $m = 3, k = 8$: $x = \frac{3}{2}$.

So $a^3 = 3\frac{3}{8},\quad b^3 = 91\frac{1}{8},\quad \square = 729 = 27^2.$

Al-Karajī (*Extrait*, V,17) has the same values and the condition.

**Problem IV,19.**
$$\begin{cases} k \cdot a = \boxdot, & k = 20, \\ l \cdot a = \sqrt[3]{\boxdot}, & l = 5. \end{cases}$$

*Condition*: $k \cdot l =$ square, which is a constructible problem (the same as the one of problem 17).

Putting $a = x$, we have:

$$\boxdot = ka = kx = (la)^3 = l^3x^3.$$ [31]

Hence:
$$x = \sqrt{\frac{k}{l^3}} = \frac{1}{l}\sqrt{\frac{k}{l}}.$$

With $k = 20, l = 5$: $\quad x = \frac{2}{5}$, so $\boxdot = 8$.

Woepcke, in his *Extrait*, does not give a problem corresponding to IV,19; but there is, in point of fact, one in the *Faḫrī*, which has the diorism and the same values for $k$ and $l$ (thus the same results).[32]

Then follows (*Extrait*, V,18) a problem not found in our text of the *Arithmetica*, namely

$$\begin{cases} k \cdot a^2 = \square, \\ l \cdot a^2 = \sqrt{\square}. \end{cases}$$

*Condition*: $k/l^2 =$ square.

Putting $a = x$, we have: $\square = kx^2 = (lx^2)^2$,

thus
$$x^2 = \frac{k}{l^2}.$$

With $k = 64, l = 2$:

$$x = 4, \quad a^2 = 16, \quad \square = 1024 = 32^2$$

(al-Karajī gives the verification).

We do not know whether al-Karajī himself added this problem or whether he found it as an addition in his version of Diophantus. The problem fits perfectly where it is, and is not a trivial case. One might suspect *our* manuscript to have a lacuna here and the problem to be genuinely Diophantine. But, for one thing, the numbering of the problems (which does not seem to be

[31] What the text actually does is to divide $kx$ by $lx$ without cubing the latter; the square root of the result gives $\sqrt[3]{\boxdot}$. A similar procedure is used in problems IV,20, 21, 24, 28, 31, and (partly) taken over by al-Karajī.

[32] See, e.g., B.N., fonds arabe 2459, fol. 102ʳ,1–8.

posterior to the translation, cf. p. 62) shows that there is no missing proposition. Furthermore, the diorism cannot possibly have been formulated in the above way by Diophantus.[33]

**Problem IV,20.** $\begin{cases} k\cdot a^3 = \square, & k = 200, \\ l\cdot a^3 = \sqrt{\square}, & l = 5. \end{cases}$

*Condition*: $k/l^2$ = cube (or also $l/k^2$ = cube; both are equivalent to $k\cdot l$ = cube). This condition is, again, a constructible problem.

Putting $a = x$, we have:

$$\square = ka^3 = kx^3 = (la^3)^2 = (lx^3)^2.$$

Hence:
$$x = \sqrt[3]{\frac{k}{l^2}}.$$

With $k = 200$, $l = 5$:

$$x = 2, \quad a^3 = 8, \quad \square = 1600 = 40^2.$$

Al-Karaǰī (*Extrait*, V,19) has $k = 32$, $l = 2$; hence $x = 2$, $a^3 = 8$ (but $\square = 256 = 16^2$). He establishes the condition in the same manner as our text does in the next two problems.

The last proposition of the *Faḫrī* (*Extrait*, V,43) is the same problem, merely phrased differently (cf. p. 60).

Diophantus' problem IV,20 has been solved, as a matter of fact, in *Arithmetica* I,26, where it appeared in the form

$$\begin{cases} k\cdot a = \square \\ l\cdot a = \sqrt{\square}, \end{cases}$$

with the same $k = 200$, $l = 5$, hence the same solution.

Let us consider this problem. Firstly, observe that it is the only one in Book I in which an expression has to be made a square. Secondly, I see no reason why Diophantus should have taken such particular values as $k = 200$, $l = 5$, when any pair of simple values would have been sufficient to obtain a solution. Thirdly, the occurrence of the verb τετραγωνίζειν is odd. We find this word in only two other places (D.G., I, p. 162,13–14 and 17) which Tannery rightly considers to be interpolated;[34] also, the Arabic word corresponding to

---

[33] We can be certain that the condition was indeed stated in the above form, and not as $k$ = square, since the problem ends with the assertion: "If the *two* given numbers of the present problem do not possess the indicated property, the problem is not soluble".

[34] Despite the *etc.* following the indication of these three passages in Tannery's *index graecitatis*, the word does not occur elsewhere in the Greek text.

τετραγωνίζειν, namely *rabba*ʿ*a*, never appears in the (extant) Arabic Diophantus, which always uses, as does the Greek text, the verb "multiply" with a reflexive expression. Hence, I strongly suspect that Diophantus himself never used τετραγωνίζειν.

In view of these facts, I am inclined to question the genuineness of I,26, though I find it difficult to explain how and why it was placed where it now is.[35]

**Problem IV,21.** $\begin{cases} k \cdot a^2 = \text{❒}, & k = 40\tfrac{1}{2}, \\ l \cdot a^2 = \sqrt[3]{\text{❒}}, & l = 2. \end{cases}$

*Condition*: $k \cdot l$ = fourth power (constructible).

Putting $a = x$, we have:

$$\text{❒} = ka^2 = kx^2 = (la^2)^3 = (lx^2)^3.$$

Hence
$$x = \sqrt[4]{\frac{k}{l^3}}\left(= \frac{1}{l}\sqrt[4]{kl}\right).$$

With $k = 40\tfrac{1}{2}$, $l = 2$:

$$x = \tfrac{3}{2}, \quad a^2 = 2\tfrac{1}{4}, \quad \text{❒} = 91\tfrac{1}{8} = (4\tfrac{1}{2})^3.$$

After the problem comes the deduction of the condition: since $\text{❒}/\sqrt[3]{\text{❒}} \equiv \square = k/l$, we must have $k/l$ = square, that is, $kl$ = square; but $la^2 = \sqrt[3]{\text{❒}} = \sqrt{\square} = \sqrt{k/l}$, hence $\sqrt{k/l}/l$, and therefore $\sqrt{k/l} \cdot l = \sqrt{kl}$ must be a square. Thus, the complete condition is

$$kl = \text{fourth power}.$$

The problem is not found in the *Faḫrī*.

**Problem IV,22.** $\begin{cases} k \cdot a^3 = \text{❒}, & k = 91\tfrac{1}{8}, \\ l \cdot a^3 = \sqrt[3]{\text{❒}}, & l = 2. \end{cases}$

The problem begins with the establishment of the condition:

Since $\text{❒}/\sqrt[3]{\text{❒}} \equiv \square = k/l$, $k/l$ (or $k \cdot l$) must be a square; but $\sqrt{\square} = l \cdot a^3$, so $\sqrt{\square}/l = \sqrt{k/l}/l$ = cube. Thus the two conditions given by the text:

$$\frac{k}{l} = \text{square} \quad \text{and} \quad \frac{\sqrt{k/l}}{l} = \text{cube.}^{36}$$

This is a constructible problem. So, we take firstly $l = 2$; to be found is the side of a square, which when divided by 2 gives a cube. Putting for the said

---

[35] What is disturbing here is not that the problem is found in the middle of the Book but that it is found in a Book preceding the presumed source; we know of another example of the former case, but none of the latter (cf. pp. 51–53).

[36] Which is to say, $k/l^3$ must be a sixth power.

cube $(\frac{3}{2})^3$, we have $\sqrt{k/l} = 2(\frac{3}{2})^3 = \frac{27}{4} = 6\frac{3}{4}$. So $k/l = 45\frac{9}{16}$, hence $k = 91\frac{1}{8}$.

One finds similarly, the text states, the characteristics and the values of the "given numbers" in the previous problems.

The resolution of the problem itself is not carried out in our text; we shall find, it is said, $a^3 = x^3 = (\frac{3}{2})^3$, the multiplication of which by $k$ gives $\text{⧉} = 307\frac{35}{64} = (6\frac{3}{4})^3$.

Al-Karajī has the full resolution of this problem (*Extrait*, V,20), but with the values $k = 64$, $l = 1$, so that $x = 2$ and $a^3 = 8$. His only condition (at least in four manuscripts—cf. p. 60, n. 29) is not $\sqrt{k/l^3} =$ cube, as it should be, but $\sqrt{k/l^2} =$ cube. Since he takes $l = 1$, one cannot say whether his condition (which he does not establish) was originally stated correctly or not.

---

The method of solving the problems of the group IV,14–22 presents no difficulty. In all the problems—with the exception of the first two, in which the right sides of the given equations are formed by a cube and a square—one of the two given expressions is a square or a cube and the other, its side. We have either a pair of required magnitudes in a known ratio and one given multiplier,[37] or one required magnitude and two given multipliers.

Let us consider all the problems of these two kinds.

I. (a) $\left\{\begin{array}{l} k\cdot b = \square, \\ k\cdot a = \sqrt{\square}, \\ b = ma. \end{array}\right.$ (b) $\left\{\begin{array}{l} k\cdot b^2 = \square, \\ k\cdot a^2 = \sqrt{\square}, \\ b = ma. \end{array}\right.$ (c) $\left\{\begin{array}{l} k\cdot b^3 = \square, \\ k\cdot a^3 = \sqrt{\square}, \\ b = ma. \end{array}\right.$

(d) $\left\{\begin{array}{l} k\cdot b = \text{⧉}, \\ k\cdot a = \sqrt[3]{\text{⧉}}, \\ b = ma. \end{array}\right.$ (e) $\left\{\begin{array}{l} k\cdot b^2 = \text{⧉}, \\ k\cdot a^2 = \sqrt[3]{\text{⧉}}, \\ b = ma. \end{array}\right.$ (f) $\left\{\begin{array}{l} k\cdot b^3 = \text{⧉}, \\ k\cdot a^3 = \sqrt[3]{\text{⧉}}, \\ b = ma. \end{array}\right.$

II. (g) $\left\{\begin{array}{l} k\cdot a = \square, \\ l\cdot a = \sqrt{\square}. \end{array}\right.$ (h) $\left\{\begin{array}{l} k\cdot a^2 = \square, \\ l\cdot a^2 = \sqrt{\square}. \end{array}\right.$ (i) $\left\{\begin{array}{l} k\cdot a^3 = \square, \\ l\cdot a^3 = \sqrt{\square}. \end{array}\right.$

(j) $\left\{\begin{array}{l} k\cdot a = \text{⧉}, \\ l\cdot a = \sqrt[3]{\text{⧉}}. \end{array}\right.$ (k) $\left\{\begin{array}{l} k\cdot a^2 = \text{⧉}, \\ l\cdot a^2 = \sqrt[3]{\text{⧉}}. \end{array}\right.$ (l) $\left\{\begin{array}{l} k\cdot a^3 = \text{⧉}, \\ l\cdot a^3 = \sqrt[3]{\text{⧉}}. \end{array}\right.$

I.

(a) is elementary, since we have at once $a = m/k$.

(b) $k$ has to be a square, as one readily sees in the first equation.

(c) $k$ has to be a cube. This is problem IV,18.

N.B. Generally, if in a system of the previous type the unknowns $a$ and $b$ occur with a $n$th power, $k$ must be a $n$th power (for $a^n = m^n/k$).

(d) $m$, the factor of proportionality, ought to be a square. Since Diophantus does not impose $b = ma$, he avoids the condition by simply taking, initially, $b = ma^2$ (IV,16).

---

[37] The ratio is not given in IV,16, which is therefore the only one of IV,16–22 not determinate.

(e) The condition is $m \cdot k =$ square. This is problem IV,17. The condition could have been avoided by dropping, as above, the imposed ratio and putting $b = ma^2$.

(f) Besides requiring that $k$ be a cube, $m$ has to be a square.

N.B. Generally, if in a system of the type (d)–(f) the unknowns $a$ and $b$ occur with a $n$th power, $m^n/k^2$ must be a $2n$th power (for $a^{2n} = m^n/k^2$). Only the condition $k^2 = n$th power will remain if one is allowed to put, as in IV,16, $b = ma^2$.

II.

(g) This is the simple case treated in I,26 (cf. *supra*, p. 195).

(h) $k$ has to be a square. This is the problem found in the *Faḫrī*, but not in the *Arithmetica* (cf. p. 194).

(i) is problem IV,20.

N.B. Generally, if in a system of the type (g)–(i) $a$ occurs with a $n$th power, $k/l^2$ $(=a^n)$ has to be a $n$th power. Similarly, in the next group, $k/l^3$ $(=a^{2n})$ has to be a $2n$th power.

(j) is problem IV,19.

(k) is problem IV,21.

(l) is problem IV,22.

A problem showing some resemblance to those above is the Greek "IV",3, which is

$$\begin{cases} b \cdot a = \square, \\ b \cdot a^2 = \sqrt[3]{\square}. \end{cases}$$

It does certainly differ from IV,14–22 in that the multiplier is not a given magnitude and the main unknown is raised to a different power in the two equations. But, since the two previous problems "IV",1 and 2 are interpolated (see p. 233) and "IV",3 does not really fit in with the subsequent group "IV",4–9—in which the relation is additive instead of multiplicative[38]—, we could admit the possibility of its having been added later. In that case, it could have been suggested to a commentator by consideration of both groups, IV,14–22 and "IV",4–9, taking from the first the multiplicative relation and from the second the dissimilar powers.

**Remark.** This problem is among those of the Greek *Arithmetica* which give evidence of scholiasts' additions (see D.G., I, pp. 192,22–23 and 194,2–3).

---

**Problem IV,23.** $(b^2)^2 + (a^2)^2 = \square$.

We put $a = x$, $b = mx$, hence

$$(m^4 + 1)x^4 = \square.$$

[38] The additive relation requires an entirely different method of solving.

Taking $\square = (nx)^3$, we have

$$x = \frac{n^3}{m^4 + 1}.$$

With $m = 2, n = 3$: $\qquad x = \frac{27}{17}.$

So

$$a^4 = \left(\frac{27}{17}\right)^4 = \left(\frac{729}{289}\right)^2 = \frac{531{,}441}{83{,}521}, \quad b^4 = \left(\frac{54}{17}\right)^4 = \left(\frac{2916}{289}\right)^2 = \frac{8{,}503{,}056}{83{,}521},$$

$$\square = \frac{531{,}441}{4913} = \left(\frac{81}{17}\right)^3.$$

Al-Karajī (*Extrait*, V,21) takes the same $m$, $n$ and thus obtains the same results.

**Problem IV,24.** $\qquad (b^2)^2 - (a^2)^2 = \square.$

We put $a = x$, $b = mx$, hence

$$(m^4 - 1)x^4 = \square.$$

Taking $\square = (nx)^3$, we have

$$x = \frac{n^3}{m^4 - 1}.$$

With $m = 2, n = 5$: $\qquad x = \frac{125}{15} = 8\frac{1}{3}.$

So

$$a^4 = (8\tfrac{1}{3})^4 = (69\tfrac{4}{9})^2 = 4822\tfrac{43}{81}, \quad b^4 = (16\tfrac{2}{3})^4 = (277\tfrac{7}{9})^2 = 77{,}160\tfrac{40}{81},$$

$$\square = 72{,}337\tfrac{26}{27} = (41\tfrac{2}{3})^3.$$

Al-Karajī (*Extrait*, V,22) takes $m = 2, n = 3$, as above, whence $x = 1\frac{4}{5}$. The *Faḫrī* has then (*Extrait*, V,23–27) the five problems spoken of before (p. 181).

---

These two problems of Diophantus form a group by themselves, in which either the sum or the difference of two fourth powers is equal to a cube. They are also the last two problems of Book IV in which a proposed expression has to be made a cube.

Did Diophantus realize that the sum and the difference of fourth powers (not nil) can never be equal to squares, as was proved by Euler (*Algebra*, II, 2, §202 *seqq.*)? Perhaps. He must, in any event, have considered these two problems.

**Problem IV,25.** $(a^3)^2 + (b^2)^2 = \square.$

We put $a = x$, $b = mx$, say, $b = 2x$; hence

$$x^6 + 16x^4 = \square.$$

Putting $\square = (nx^2)^2$, we shall arrive at $x^2 = n^2 - 16$, or $n^2 - x^2 = 16$. How to solve this problem, namely, finding two square numbers having a given difference, has been shown in II,10. In our case, an obvious solution is $n^2 = 25$, $x^2 = 9$. The problem is then "reconstructed" with the value found for $n^2$. So

$$(a^3)^2 = (3^3)^2 = 27^2 = 729, \qquad (b^2)^2 = (6^2)^2 = 36^2 = 1296,$$

$$\square = 2025 = 45^2.$$

Neither this nor the following problem has a counterpart in the *Faḫrī*.

**Problem IV,26.** $|(a^3)^2 - (b^2)^2| = \square.$

(a) $(a^3)^2 - (b^2)^2 = \square.$

We put $a = x$, $b = 2x$; hence

$$x^6 - 16x^4 = \square.$$

Taking $\square = (nx^2)^2$, we arrive at $x^2 = n^2 + 16$, or $x^2 - n^2 = 16$, with a solution $x^2 = 25$, $n^2 = 9$.

The problem is then reconstructed. So

$$(a^3)^2 = (5^3)^2 = 125^2 = 15{,}625, \quad (b^2)^2 = (10^2)^2 = 100^2 = 10{,}000,$$

$$\square = 5625 = 75^2.$$

**Remark.** With a solution to IV,34, which is the system

$$\begin{cases} a^3 + b^2 = \square, \\ a^3 - b^2 = \square', \end{cases}$$

we have at once a solution to the present problem. The same correspondence holds between IV,35 and the next part of the present problem.

(b) $(b^2)^2 - (a^3)^2 = \square.$

We put $a = x$, $b = 5x$;[39] we arrive at

$$625x^4 - x^6 = \square.$$

With $\square = (nx^2)^2$, we shall have $625 - x^2 = n^2$, or $x^2 + n^2 = 625$, which amounts to dividing a square number into two square numbers. One readily

[39] Diophantus surely departs from the simple choice $b = 2x$ to obtain an integral solution.

With $k = 3, m = \frac{1}{2}, n = 1$: $\qquad x = 4.$

So

$$(a^3)^3 = (4^3)^3 = 64^3 = 262{,}144, \quad b^2 = 32^2 = 1024, \quad \square = 65{,}536 = 256^2.$$

Al-Karajī (*Extrait*, V,34) chooses the same values; he does not have any stated condition for the magnitude of the parameter $m$ either.

---

The last representative of the group,

$$k \cdot a^3 \cdot b^2 - (a^3)^3 = \square$$

would be solved in the very same way.

---

Before leaving these types of problems and proceeding to systems of two equations, the text states the following

• **Corollary.**[44] We would solve in the same manner the problem

$$(1^a) \qquad (b^2)^2 + k \cdot a^3 \cdot b^2 = \square$$

(and the other members of the group, namely:

$$(1^b) \qquad (b^2)^2 - k \cdot a^3 \cdot b^2 = \square$$

$$(1^c) \qquad k \cdot a^3 \cdot b^2 - (b^2)^2 = \square);$$

as also the problem

$(2^a)$

$$(b^2)^3 + k \cdot a^3 \cdot b^2 = \square \quad \text{—and its “inverse”} \ (a^3)^2 + k \cdot a^3 \cdot b^2 = \square\text{—}$$

(and the other members of the group, namely:

$$(2^b) \qquad (b^2)^3 - k \cdot a^3 \cdot b^2 = \square, \quad \text{and} \quad (a^3)^2 - k \cdot a^3 \cdot b^2 = \square$$

$$(2^c) \qquad k \cdot a^3 \cdot b^2 - (b^2)^3 = \square, \quad \text{and} \quad k \cdot a^3 \cdot b^2 - (a^3)^2 = \square).$$

Indeed, it is sufficient to take $b = mx$, $a = x$, as was done in previous problems,[45] and further $\square = n^2x^4$ and $\square = n^2x^6$, respectively, in order to arrive at a linear equation.[46]

Just as the corollary appended to I,34 had been the source for interpolated problems (cf. p. 52), so the above corollary has also inspired a scholiast (perhaps the same one) and his resolutions were afterwards incorporated into

---

[44] Not in the *Faḫrī*.

[45] See, e.g., IV,28; in the next problems one is obliged to take either $b = mx^2$ (29–31) or $b = mx^3$ (32–33).

[46] Note that the first problem of the pair under $2^a$ differs from IV,28 only by the factor $b^2$, which does not affect the solution. We could similarly drop a factor $b^2$ in all the other cases except in those of the inverse forms.

the main text: ($2^a$) and ($2^c$) (in the inverse form) gave rise to VI,4 and VI,6, respectively, while ($1^a$) and ($1^c$) are the source of VI,5 and VI,7, respectively. Not surprisingly, the scholiast simplified the original problems so as to make the resolutions of the derived ones even easier.

---

The text now leaves single equations and goes on to indeterminate systems of two equations of degree three or more, a category which will extend to the beginning of Book V. Observe that (not unexpectedly) almost all the types of equations involved from here on have already been solved, but singly. Thus, we can associate

IV,34–35 with IV,3–4 and *similia* (see p. 181)
IV,36–39 with IV,10–11 and *sim.*
IV,42 with IV,29–31
V,1–3 with IV,28 and *sim.* (see p. 202).

---

**Problem IV,34.** $$\begin{cases} a^3 + b^2 = \square, \\ a^3 - b^2 = \square'. \end{cases}$$

We put $a = x$, $b = 2x$, so we have

$$\begin{cases} x^3 + 4x^2 = \square, \\ x^3 - 4x^2 = \square'. \end{cases}$$

1°. Using the method of the double-equation taught in II,11,1°:

$\square - \square' = 8x^2 = d_1 \cdot d_2$ (where $d_1, d_2$ must be taken proportional to $x$ in order to obtain a linear equation[47]); we have then

$$\square = \left\{\frac{1}{2}\left(\frac{\square - \square'}{d_i} + d_i\right)\right\}^2 = \{\tfrac{1}{2}(d_1 + d_2)\}^2 \qquad (i = 1, 2)$$

and

$$\square' = \left\{\frac{1}{2}\left(\frac{\square - \square'}{d_i} - d_i\right)\right\}^2 = \{\tfrac{1}{2}(d_1 - d_2)\}^2.$$

Thus, with $8x^2 = 4x \cdot 2x$:

$$\square' = x^3 - 4x^2 = \left\{\frac{1}{2}\left(\frac{8x^2}{2x} - 2x\right)\right\}^2 = x^2$$

or else, $$\square = x^3 + 4x^2 = \left\{\frac{1}{2}\left(\frac{8x^2}{2x} + 2x\right)\right\}^2 = 9x^2.$$

In both cases, we obtain $x^3 = 5x^2$, so $x = 5$. Hence

$$a^3 = 5^3 = 125, \quad b^2 = 10^2 = 100, \quad \square = 225 = 15^2, \quad \square' = 25 = 5^2.$$

---

[47] This is not specified in the text, but a similar condition is stated in problem IV,42,a (which is the only other problem in the Arabic Books using the method of the double-equation).

2°. Avoiding the double-equation (cf. II,11,2°):

We put $\square = (mx)^2$, so

$$x^3 + 4x^2 = m^2x^2 \quad \text{and} \quad x = m^2 - 4,$$

and we put $\square' = (nx)^2$, so

$$x^3 - 4x^2 = n^2x^2 \quad \text{and} \quad x = n^2 + 4.$$

Hence $m^2 - 4 = n^2 + 4$, or $m^2 - n^2 = 8$. Using II,10,[48] we set $m = n + h$; so $n^2 + 2nh + h^2 - n^2 = 8$, and

$$n = \frac{8 - h^2}{2h} \quad \left(\text{hence } m = \frac{8 + h^2}{2h}\right) \quad (h^2 < 8 \text{ assumed}).$$

With $h = 1$,[49] we have

$$n = \tfrac{7}{2}, \qquad n^2 = \tfrac{49}{4} = 12\tfrac{1}{4}, \quad \text{and} \quad m = \tfrac{9}{2}, \qquad m^2 = \tfrac{81}{4} = 20\tfrac{1}{4}.$$

Inserting the two values $m^2, n^2$ leads then to the same equation $x^3 = 16\frac{1}{4}x^2$; hence $x = 16\frac{1}{4}$. So

$$a^3 = (16\tfrac{1}{4})^3 = 4291\tfrac{1}{64}, \quad b^2 = (32\tfrac{1}{2})^2 = 1056\tfrac{1}{4},$$

$$\square = 5347\tfrac{17}{64} = (73\tfrac{1}{8})^2, \quad \square' = 3234\tfrac{49}{64} = (56\tfrac{7}{8})^2.$$

Al-Karajī (*Extrait*, V,35) uses only the first method and obtains Diophantus' results.

**Remark.** Between the method of the double-equation and this alternative one leading to II,10, there is the following (external) difference: in the first method, we form immediately the final equation, whereas in the second method, we must first solve the intermediate problem, the aim of which is to make the (final) equation resulting from each of the two proposed equations the same (this is done by determining the appropriate values of the coefficients $m^2$ and $n^2$, the difference of which we know).

In such problems, these two methods ultimately amount to the same thing, since the parameter that we choose to begin with in the first method (by setting $\square - \square' = 8x^2 = (8/h_0)x \cdot h_0 x$, $h_0^2 < 8$) we choose when solving the intermediate problem in the second method (where it appears in the relation $m = n + h_0$). Thus we end up in both cases with

$$\square = \frac{1}{4}\left[\frac{8}{h_0} + h_0\right]^2 x^2, \qquad \square' = \frac{1}{4}\left[\frac{8}{h_0} - h_0\right]^2 x^2.$$

[48] Remember that the application of such methods taken from Book II is never performed, the text giving merely the numerical results.

[49] Taking $h = 2$ would give the previous solution (see the *remark* below).

Both approaches are reducible to the identity

$$\left(\frac{p+q}{2}\right)^2 - \left(\frac{p-q}{2}\right)^2 = p \cdot q,$$

on which some resolutions found in Book I were also based (see p. 236).

**Problem IV,35.**
$$\begin{cases} b^2 + a^3 = \square, \\ b^2 - a^3 = \square'. \end{cases}$$

We put $a = x$, $b = 2x$, so

$$\begin{cases} 4x^2 + x^3 = \square, \\ 4x^2 - x^3 = \square'. \end{cases}$$

Taking $\square = (mx)^2$, $\square' = (nx)^2$, we have

$$4x^2 + x^3 = m^2x^2 \quad \text{and} \quad 4x^2 - x^3 = n^2x^2;$$

therefore

$$x = m^2 - 4 = 4 - n^2.$$

Hence $m^2 + n^2 = 8 = 2^2 + 2^2$, which can be solved using II,9:

$$(y + 2)^2 + (2 - hy)^2 = 2^2 + 2^2,$$

$$y^2 + 4y + 2^2 + 2^2 - 4hy + h^2y^2 = 2^2 + 2^2,$$

so $$y^2(1 + h^2) = 4y(h - 1) \quad \text{and} \quad y = \frac{4(h-1)}{1+h^2} \quad (h > 1).$$

For $h = 2$: $$y = \tfrac{4}{5}, \quad y + 2 = \tfrac{14}{5}, \quad 2 - hy = \tfrac{2}{5}.$$

Thus (since $m > n$)

$$m^2 = (\tfrac{14}{5})^2 = \tfrac{196}{25} = 7\tfrac{21}{25}, \qquad n^2 = (\tfrac{2}{5})^2 = \tfrac{4}{25}.$$

Both equalizations give: $x = \tfrac{96}{25} = 3\tfrac{21}{25}.$

So $$a^3 = \left(\frac{96}{25}\right)^3 = \frac{884{,}736}{15{,}625}, \quad b^2 = \left(\frac{192}{25}\right)^2 = \frac{36{,}864}{625},$$

$$\square = \frac{1{,}806{,}336}{15{,}625} = \left(\frac{1344}{125}\right)^2, \quad \square' = \frac{36{,}864}{15{,}625} = \left(\frac{192}{125}\right)^2.$$

Al-Karajī (*Extrait*, V,36) has the very same problem, solved with comparable prolixity.

---

The group of problems IV,34–35, that is to say the systems

$$\begin{cases} a^3 + b^2 = \square, \\ a^3 - b^2 = \square', \end{cases} \qquad \begin{cases} b^2 + a^3 = \square, \\ b^2 - a^3 = \square', \end{cases}$$

are the first representatives of problems involving two magnitudes such that the one, both increased and diminished by the other, gives a square.[50] We shall encounter two similar groups in Book IV, namely

IV,40–41: $\begin{cases}(b^2)^2 + a^3 = \square, \\ (b^2)^2 - a^3 = \square',\end{cases}$ $\begin{cases}a^3 + (b^2)^2 = \square, \\ a^3 - (b^2)^2 = \square',\end{cases}$

IV,42,a–b: $\begin{cases}(a^3)^3 + (b^2)^2 = \square, \\ (a^3)^3 - (b^2)^2 = \square',\end{cases}$ $\begin{cases}(b^2)^2 + (a^3)^3 = \square, \\ (b^2)^2 - (a^3)^3 = \square',\end{cases}$

which differ from the present group only in the higher powers.

---

**Problem IV,36.** $\begin{cases}a^3 + k \cdot a^2 = \square, \\ a^3 - l \cdot a^2 = \square',\end{cases}$ $k = 4,$ $l = 5.$

We put $a = x$, so we have:

$$\begin{cases}x^3 + 4x^2 = \square, \\ x^3 - 5x^2 = \square'.\end{cases}$$

Taking $\square = (mx)^2$, $\square' = (nx)^2$, we have

$$x = m^2 - 4 = n^2 + 5.$$

Hence $m^2 - n^2 = 9$; an obvious solution is $m^2 = 25, n^2 = 16$ (corresponding to the value 1 of the parameter $h$ in the application of II,10). Hence

$$x = 21.$$

So $\quad a^3 = 21^3 = 9261, \quad \square = 11{,}025 = 105^2, \quad \square' = 7056 = 84^2.$

Al-Karajī (*Extrait*, V,37) has the same values, and he asserts the possibility of the treatment by the double-equation method. He gives the outline of the resolution by the latter method for a similar problem in his *Badīʿ* (fol. 119ᵛ).

• There then follows in our text a remark asserting that, inverting the rôles of the two multipliers, that is, considering the system

$$\begin{cases}x^3 + 5x^2 = \square, \\ x^3 - 4x^2 = \square',\end{cases}$$

a solution would be $x = 20$, hence $x^3 = 8000$, leading to

$$\square = 10{,}000 = 100^2, \qquad \square' = 6400 = 80^2.$$

This remark, expressed like a corollary,[51] is rather odd, since the inversion involved is of limited interest (we shall equate $x^3 + 5x^2$, instead of $x^3 + 4x^2$,

---

[50] Actually, II,30 is of the same type, but it is solved differently.

[51] It is introduced by *istabāna*, probably rendering something like (ἐκ δὴ τούτου) φανερόν; see index, *bāna* (X), p. 436.

to $25x^2$); also, no explanation is given. It is possible that Diophantus wished to point out that the solution of the inverted case is easily obtainable since we are led to the same intermediate problem, without any new condition.[52] Still, some explanation would be desirable. A more likely possibility is that this remark is an addition by a scholiast—perhaps the same one who made a new case in IV,14 by interchanging the multipliers (see IV,14,b and p. 191). Whatever its origin, this remark was written no later than the major commentary, since there is a verification of the solution.

N.B. If ever a general statement were to have been made, it should have asserted that if $x_1$ is a solution of the system

$$\begin{cases} x^{2n+1} + kx^{2n} = \square_1, \\ x^{2n+1} - lx^{2n} = \square_1', \end{cases}$$

then $x_2 = x_1 + k - l$ is a solution of the system

$$\begin{cases} x^{2n+1} + lx^{2n} = \square_2, \\ x^{2n+1} - kx^{2n} = \square_2', \end{cases}$$

as one readily sees by considering the corresponding linear systems.

**Problem IV,37.** $$\begin{cases} a^3 + k \cdot a^2 = \square, & k = 10, \\ a^3 + l \cdot a^2 = \square', & l = 5. \end{cases}$$

We put $a = x$, so

$$\begin{cases} x^3 + 10x^2 = \square, \\ x^3 + \ \ 5x^2 = \square'. \end{cases}$$

Taking $\square = (mx)^2$, $\square' = (nx)^2$, we shall have

$$x = m^2 - 10 = n^2 - 5.$$

Thus $m^2 - n^2 = 5$, again soluble by II,10, keeping in mind that $n^2$ must be larger than 5, as stated in the text. Putting $m = n + h$, we obtain $2nh + h^2 = 5$, or

$$n = \frac{5 - h^2}{2h}.$$

Taking $h = \frac{1}{3}$, we have

$$n = \tfrac{44}{6} = 7\tfrac{1}{3}, \qquad n^2 = 53\tfrac{7}{9}, \quad \text{and} \quad m^2 = 58\tfrac{7}{9} = (7\tfrac{2}{3})^2.$$

Thus $$x = 48\tfrac{7}{9}.$$

[52] Whereas the inversion of the signs of the coefficients in IV,37 modifies a condition for the intermediate problem: see IV,38.

So $$a^3 = \left(\frac{439}{9}\right)^3 = \frac{84{,}604{,}519}{9\cdot 9\cdot 9}, \quad \Box = \frac{917{,}544{,}681}{9\cdot 9\cdot 9\cdot 9} = \left(\frac{30{,}291}{9\cdot 9}\right)^2,$$

$$\Box' = \frac{839{,}492{,}676}{9\cdot 9\cdot 9\cdot 9} = \left(\frac{28{,}974}{9\cdot 9}\right)^2.$$

Al-Karajī (*Extrait*, V,38) has the very same problem, with the condition for $n^2$, and gives $a^3$ the same form. He again asserts the possibility of solving by the method of the double-equation.

**Problem IV,38.** $$\begin{cases} a^3 - l\cdot a^2 = \Box, & l = 5, \\ a^3 - k\cdot a^2 = \Box', & k = 10. \end{cases}$$

We put $a = x$, so

$$\begin{cases} x^3 - 5x^2 = \Box, \\ x^3 - 10x^2 = \Box'. \end{cases}$$

Taking $\Box = (mx)^2$, $\Box' = (nx)^2$, we arrive at $x = m^2 + 5 = n^2 + 10$; hence $m^2 - n^2 = 5$, as before, but now without any condition for $n^2$, as asserted in the text. An obvious solution is $m^2 = 9$, $n^2 = 4$, giving $x = 14$. Hence

$$a^3 = 14^3 = 2744, \quad \Box = 1764 = 42^2, \quad \Box' = 784 = 28^2.$$

Al-Karajī (*Extrait*, V,39) has the very same problem (but does not point out the absence of a condition for $n^2$). An example of this type and one of the following type are also formulated in the *Badīʿ* (fol. 119ᵛ).

**Problem IV,39.** $$\begin{cases} k\cdot a^2 - a^3 = \Box, & k = 7, \\ l\cdot a^2 - a^3 = \Box', & l = 3. \end{cases}$$

We put $a = x$, so that we have the system

$$\begin{cases} 7x^2 - x^3 = \Box, \\ 3x^2 - x^3 = \Box'. \end{cases}$$

Taking $\Box = (mx)^2$, $\Box' = (nx)^2$, we arrive at $x = 7 - m^2 = 3 - n^2$; so $m^2 - n^2 = 4$, with the stated condition $n^2 < 3$.

A solution is readily obtained by dividing by 4 the known relation $25 - 9 = 16$. One obtains the same result from II,10:

$$m^2 - n^2 = (n+h)^2 - n^2 = 4, \quad \text{hence} \quad n = \frac{4 - h^2}{2h}.$$

With $h = 1$:

$$m^2 = (\tfrac{5}{2})^2 = 6\tfrac{1}{4}, \qquad n^2 = (\tfrac{3}{2})^2 = 2\tfrac{1}{4}, \quad \text{and} \quad x = \tfrac{3}{4}.$$

Thus $$a^3 = \frac{27}{8\cdot 8}, \quad \Box = \frac{225}{8\cdot 8} = \left(\frac{15}{8}\right)^2, \quad \Box' = \frac{81}{8\cdot 8} = \left(\frac{9}{8}\right)^2.$$

Al-Karajī (*Extrait*, V,40) has the very same problem and states the condition for $n$.

---

The group formed by IV,36–39 consists of the following systems:

**IV,36:** $\begin{cases} a^3 + ka^2 = \square, \\ a^3 - la^2 = \square', \end{cases}$

which leads, with $\square = m^2a^2$, $\square' = n^2a^2$, to $m^2 - n^2 = k + l$.

**IV,37:** $\begin{cases} a^3 + ka^2 = \square, \\ a^3 + la^2 = \square', \end{cases}$

giving $m^2 - n^2 = k - l$ with the auxiliary condition $n^2 > l$. The linear system to which the above one is reduced by division by $a^2$ was solved in II,11.

**IV,38:** $\begin{cases} a^3 - la^2 = \square, \\ a^3 - ka^2 = \square', \end{cases}$

leads to $m^2 - n^2 = k - l$. The corresponding linear system is II,13.

**IV,39:** $\begin{cases} ka^2 - a^3 = \square, \\ la^2 - a^3 = \square', \end{cases}$

hence $m^2 - n^2 = k - l$ with the auxiliary condition $n^2 < l$. Here again, there is an equivalent linear problem earlier in the *Arithmetica*, namely II,12.

All these problems are thus reducible to II,10, and all are also soluble by the method of the double-equation.[53] The only remaining forms of this kind (in which $a^3$ no longer occurs with the same sign in the two equations), namely

$$\begin{cases} ka^2 + a^3 = \square, \\ la^2 - a^3 = \square', \end{cases} \quad \text{and} \quad \begin{cases} ka^2 - a^3 = \square, \\ a^3 - la^2 = \square', \end{cases}$$

lead to problem II,9.

---

**Problem IV,40.** $\begin{cases} (b^2)^2 + a^3 = \square, \\ (b^2)^2 - a^3 = \square'. \end{cases}$

We put $b = 2x$, and, say, $a = 4x$.[54] The system is then

$$\begin{cases} 16x^4 + 64x^3 = \square, \\ 16x^4 - 64x^3 = \square'. \end{cases}$$

---

[53] If the solution found by the method of II,10 depends on a value $h_0$ of the parameter, one will obtain the same solution by the method of the double-equation whilst using the separation

$$\square - \square' \equiv px^2 = \frac{p}{h_0} x \cdot h_0 x \quad \text{(see p. 207).}$$

[54] Putting simply $a = x$, as previously, would result in a less convenient value for $x$.

With $\square = (mx^2)^2$, $\square' = (nx^2)^2$, we have

$$x = \frac{64}{m^2 - 16} = \frac{64}{16 - n^2}.$$

Thus $\quad m^2 - 16 = 16 - n^2, \quad \text{or} \quad m^2 + n^2 = 32 = 4^2 + 4^2.$

As in IV,35, we apply II,9:

$$(y + 4)^2 + (4 - hy)^2 = 4^2 + 4^2,$$

$$y^2 + 8y + 4^2 + 4^2 - 8hy + h^2y^2 = 4^2 + 4^2,$$

$$y = \frac{8(h - 1)}{h^2 + 1} \qquad (h > 1).$$

An obvious choice is $h = 2$, which gives $y = \frac{8}{5}$; therefore

$$(y + 4)^2 = (\tfrac{28}{5})^2 = \tfrac{784}{25} = 31\tfrac{9}{25} = m^2, \qquad (4 - hy)^2 = (\tfrac{4}{5})^2 = \tfrac{16}{25} = n^2,$$

which are the values given by Diophantus.

Hence $$x = \frac{64}{15\frac{9}{25}} = 4\tfrac{1}{6}.$$

So $$a^3 = \left(16\,\frac{2}{3}\right)^3 = 4629 + \frac{5}{9} + \frac{2}{3}\cdot\frac{1}{9}\left[= 4629\,\frac{17}{27}\right],$$

$$(b^2)^2 = \left(\left(8\,\frac{1}{3}\right)^2\right)^2 = \left(69\,\frac{4}{9}\right)^2 = 4822 + \frac{4}{9} + \frac{7}{9\cdot 9}\left[= 4822\,\frac{43}{81}\right],$$

$$\square = 9452 + \frac{1}{9} + \frac{4}{9\cdot 9}\left[= 9452\,\frac{13}{81}\right] = \left(97\,\frac{2}{9}\right)^2,$$

$$\square' = 192 + \frac{8}{9} + \frac{1}{9\cdot 9}\left[= 192\,\frac{73}{81}\right] = \left(13\,\frac{8}{9}\right)^2.$$

Al-Karajī (*Extrait*, V,41) has the same numerical values; but he gives the results in the form $x = \frac{25}{6}$, $a^3 = 1{,}000{,}000/216$, $b^2 = 2500/36$.[55]

**Problem IV,41.** $$\begin{cases} a^3 + (b^2)^2 = \square, \\ a^3 - (b^2)^2 = \square'. \end{cases}$$

As observed in the translation, this problem might well have been part of the preceding one originally, for the relations of $a$ and $b$ to $x$ are considered as known and are not initially stated, as is usually done.[56]

$$\begin{cases} 64x^3 + 16x^4 = \square, \\ 64x^3 - 16x^4 = \square'. \end{cases}$$

[55] In none of these problems does he give the values of $\square$ and $\square'$, as already mentioned (see above, under IV,27).

[56] As already said, there is some arbitrariness in the subdivision into problems (cf. p. 62, n. 33).

Putting $\square = (mx^2)^2$, $\square' = (nx^2)^2$, we have

$$x = \frac{64}{m^2 - 16} = \frac{64}{n^2 + 16}.$$

Hence $m^2 - n^2 = 32$, soluble by II,10; Diophantus takes the obvious solution $m^2 = 36$, $n^2 = 4$.

So $$x = \tfrac{64}{20} = 3\tfrac{1}{5}$$

and

$$a^3 = \left(12\frac{4}{5}\right)^3 = 2097\frac{95}{625},$$

$$(b^2)^2 = \left(\left(6\frac{2}{5}\right)^2\right)^2 = \left(40 + \frac{4}{5} + \frac{4}{5\cdot 5}\right)^2 \left[= \left(40\frac{24}{25}\right)^2\right] = 1677\frac{451}{625},$$

$$\square = 3774\frac{546}{625} = \left(61\frac{11}{25}\right)^2, \quad \square' = 419\frac{269}{625} = \left(20\frac{12}{25}\right)^2.$$

Al-Karajī (*Extrait*, V,42) has the same values. Here too, his text differs in giving the results in the form $a^3 = 262{,}144/125$, $b^2 = 1024/5\cdot 5$.

The *Faḫrī* then ends with the repetition of a problem already treated (cf. p. 60).

---

**Problem IV,42.** $$\begin{cases}(a^3)^3 + (b^2)^2 = \square,\\ |(a^3)^3 - (b^2)^2| = \square'.\end{cases}$$

We put, say, $a = 2x$, so that $(a^3)^3 = (8x^3)^3 = 512x^9$, and, say, $b = 4x^2$,[57] so that $(b^2)^2 = (16x^4)^2 = 256x^8$.

(a) $$\begin{cases}(a^3)^3 + (b^2)^2 = \square,\\ (a^3)^3 - (b^2)^2 = \square'.\end{cases}$$

We already know, the text says, a solution to this problem (from IV,29–30; cf. p. 203); hence, only the method for finding a solution will be recalled.

$$\begin{cases}512x^9 + 256x^8 = \square,\\ 512x^9 - 256x^8 = \square'.\end{cases}$$

1°. Method of the double-equation.

$$\square - \square' = 512x^8 = px^4\cdot qx^4,$$

then $$512x^9 \pm 256x^8 = \{\tfrac{1}{2}(px^4 \pm qx^4)\}^2 = \{\tfrac{1}{2}(p \pm q)\}^2\cdot x^8.$$

[57] Thus $b = a^2$, so that the system takes the simple form $|a^9 \pm a^8| =$ square, that is, $|a \pm 1| =$ square.

So $$512x^9 = \{\tfrac{1}{2}(p \pm q)\}^2 x^8 \mp 256x^8$$

whence $$x = \frac{\{\tfrac{1}{2}(p \pm q)\}^2 \mp 256}{512}.$$

One should then proceed with the synthesis of the problem.

2°. Search for an identical equation for the proposed pair of equations. Putting $\square = (mx^4)^2$, $\square' = (nx^4)^2$, we shall arrive at

$$512x = m^2 - 256 \quad \text{and} \quad 512x = n^2 + 256;$$

$x$ will have the same value in both cases if $m^2$ and $n^2$ fulfil

$$m^2 - 256 = n^2 + 256,$$

that is, if $m^2 - n^2 = 512$ (soluble by II,10 or simply by multiplication of the solution 36, 4 found in IV,41 by 16). We shall then reconstruct the problem, solve the (single) resulting equation for $x$, and afterwards perform the synthesis.

3°. Initial simplification of the proposed system.

The equations may immediately be reduced to linear ones by dividing by the even power of the left sides (taking some quadratic factor as coefficient of the said power[58]).

So, dividing the system

$$\begin{cases} 512x^9 + 256x^8 = \square, \\ 512x^9 - 256x^8 = \square', \end{cases}$$

e.g., by $16x^8$, we have

$$\begin{cases} 32x + 16 = \square_1, \\ 32x - 16 = \square'_1, \end{cases}$$

which is the new system to be solved. Thus (in the previous manner), one will seek $u$ satisfying

$$\begin{cases} u + 16 = \square_1, \\ u - 16 = \square'_1, \end{cases}$$

and the required $x$ will be equal to $u/32$. We shall then make the synthesis of the problem.

• Then follows the remark that this procedure is applicable to "most" of the systems of two simultaneous equations seen before. The problems excluded are no doubt IV,40 and 41, which, since the even power of $x$ is not the lower

[58] The essence of the resolution is of course not affected by the choice.

one, are not reducible to systems linear in $x$ solely by a division (cf. p. 227, n. 4).

(b) $\begin{cases}(b^2)^2 + (a^3)^3 = \square, \\ (b^2)^2 - (a^3)^3 = \square',\end{cases}$

that is,

$$\begin{cases}256x^8 + 512x^9 = \square, \\ 256x^8 - 512x^9 = \square'.\end{cases}$$

We put $\square = (mx^4)^2$, $\square' = (nx^4)^2$, hence

$$x = \frac{m^2 - 256}{512} = \frac{256 - n^2}{512}.$$

So $m^2 + n^2 = 256 + 256 = 16^2 + 16^2$. One can obtain the solutions given by Diophantus by using II,9 and by taking for the parameter $h$ ($h > 1$) the value 2, or by multiplying the solutions found in IV,40, namely $m_1^2 = 31\frac{9}{25}$, $n_1^2 = \frac{16}{25}$, by 16. Thus the solutions

$$m^2 = 501\tfrac{19}{25} = (22\tfrac{2}{5})^2, \qquad n^2 = 10\tfrac{6}{25} = (3\tfrac{1}{5})^2.$$

Hence $$x = \tfrac{12}{25}.$$

So

$$(a^3)^3 = \left(\left(\frac{24}{25}\right)^3\right)^3 = \left(\frac{13{,}824}{25^3}\right)^3 = \frac{2{,}641{,}807{,}540{,}224}{(25^3)^3} = \frac{105{,}672{,}301{,}608\frac{24}{25}}{(625^2)^2},$$

$$(b^2)^2 = \left(\left(\frac{576}{625}\right)^2\right)^2 = \left(\frac{331{,}776}{625^2}\right)^2 = \frac{110{,}075{,}314{,}176}{(625^2)^2},$$

$$\square = \frac{215{,}747{,}615{,}784\frac{24}{25}}{(625^2)^2} = \left(\frac{464{,}486\frac{2}{5}}{625^2}\right)^2,$$

$$\square' = \frac{4{,}403{,}012{,}567\frac{1}{25}}{(625^2)^2} = \left(\frac{66{,}355\frac{1}{5}}{625^2}\right)^2.$$

---

**Problem IV,43.** $\begin{cases}(a^3)^3 + k\cdot(b^2)^2 = \square, & k = 1\frac{1}{4}, \\ (a^3)^3 - l\cdot(b^2)^2 = \square', & l = \frac{3}{4}.\end{cases}$

We put $a = x$ and, say, $b = 2x^2$; hence

$$\begin{cases}x^9 + 1\frac{1}{4}(16x^8) = x^9 + 20x^8 = \square, \\ x^9 - \frac{3}{4}(16x^8) = x^9 - 12x^8 = \square'.\end{cases}$$

We put $\square = (mx^4)^2$, $\square' = (nx^4)^2$, so

$$x = m^2 - 20 = n^2 + 12, \quad \text{and} \quad m^2 - n^2 = 32;$$

an obvious solution is:

$$m^2 = 36, \qquad n^2 = 4 \quad \text{(as in IV,41).}$$

Then

$$x = 16,$$

and

$$(a^3)^3 = (16^3)^3 = 4096^3 = 68{,}719{,}476{,}736,$$
$$(b^2)^2 = (512^2)^2 = 262{,}144^2 = 68{,}719{,}476{,}736.$$

Hence $(a^3)^3$ is a square, with $(a^3)^3 = (b^2)^2$; so

$$\square = (b^2)^2 + 1\tfrac{1}{4}(b^2)^2 = 2\tfrac{1}{4}(b^2)^2 = (1\tfrac{1}{2}b^2)^2,$$
$$\square' = (b^2)^2 - \tfrac{3}{4}(b^2)^2 = \tfrac{1}{4}(b^2)^2 = (\tfrac{1}{2}b^2)^2.$$

**Problem IV,44.**

(a) $\begin{cases}(a^3)^3 + k\cdot(b^2)^2 = \square, \\ (a^3)^3 + l\cdot(b^2)^2 = \square'.\end{cases}$ (b) $\begin{cases}(a^3)^3 - l\cdot(b^2)^2 = \square, \\ (a^3)^3 - k\cdot(b^2)^2 = \square'.\end{cases}$

(c) $\begin{cases}k\cdot(b^2)^2 - (a^3)^3 = \square, \\ l\cdot(b^2)^2 - (a^3)^3 = \square'.\end{cases}$

We take: $k = 8$, $l = 3$.

(a) Putting $a = x$, $b = 2x^2$, we have:

$$\begin{cases}x^9 + 128x^8 = \square, \\ x^9 + \phantom{1}48x^8 = \square'.\end{cases}$$

Dividing the two expressions by $x^8$ (cf. IV,42,a,3°), we obtain the system:

$$\begin{cases}x + 128 = \square_1, \\ x + \phantom{1}48 = \square_1'.\end{cases}$$

Diophantus immediately gives the value of $x$, which is easily obtainable. Putting $\square_1 = m^2$, $\square_1' = n^2$, we have $x = m^2 - 128 = n^2 - 48$, hence $m^2 - n^2 = 80$ with the condition $n^2 > 48$. One may apply II,10, and obtain $m^2 = 144$, $n^2 = 64$, choosing 4 among the range of allowed values of the parameter; one may also derive the solution from the one seen in IV,38 by multiplying it by 16.

So $x = 16$, and $(a^3)^3$ and $(b^2)^2$ are equal to the $(a^3)^3$ and $(b^2)^2$ found in IV,43. Thus, here too, $(a^3)^3 = (b^2)^2$, and

$$\square = 9\cdot(b^2)^2 = (3b^2)^2, \quad \square' = 4\cdot(b^2)^2 = (2b^2)^2.$$

(b) Putting again $a = x$, $b = 2x^2$, we have:

$$\begin{cases}x^9 - \phantom{1}48x^8 = \square, \\ x^9 - 128x^8 = \square'.\end{cases}$$

or:

$$\begin{cases}x - \phantom{1}48 = \square_1, \\ x - 128 = \square_1'.\end{cases}$$

The solution, directly given in the text, is obtainable in the usual way: putting $\square_1 = m^2$, $\square'_1 = n^2$, we are again led to $m^2 - n^2 = 80$, but this time without a condition. We can still use, though, the same $m^2$, $n^2$ as in part (a), thus obtaining Diophantus' value

$$x = 192.^{59}$$

In the second half of problem IV,44,b, Diophantus constructs a new solution. But, whether certain elements of the reasoning might now be missing because the text was damaged or whether Diophantus himself was sparing in his comments, the reason for and the significance of the computations do not appear from the text, and therefore require some elucidation.

Let us denote by the index "1" the solutions found in IV,44,a (i.e., $a_1 = x_1 = 16$, $b_1 = 2x_1^2 = 512$) and by the index "2" the solutions just found for IV,44,b ($a_2 = x_2, b_2 = 2x_2^2$, with the value $x_2 = 192$). The reasoning seems to be essentially the following: we form the ratios $a_2^3 : a_1^3$ and $b_2^2 : b_1^2$ and reduce the resulting fractions; since, then, the new denominators ($a_1'^3 = 1$, $b_1'^2 = 1$) happen to be another solution of IV,44,a, the numerators ($a_2'^3 = 12^3$, $b_2'^2 = 144^2$) will satisfy IV,44,b.

Let us see how this deduction is applicable. The two systems found in IV,44,a and b are the following:

$$\text{(I)} \quad \begin{cases} (a_1^3)^3 + k(b_1^2)^2 = \square_1, \\ (a_1^3)^3 + l(b_1^2)^2 = \square'_1, \end{cases}$$

and

$$\text{(II)} \quad \begin{cases} (a_2^3)^3 - l(b_2^2)^2 = \square_2, \\ (a_2^3)^3 - k(b_2^2)^2 = \square'_2, \end{cases}$$

with the same values for $k$ and $l$ in the two systems.

Let us put in both cases $a_i = x_i$, $b_i = qx_i^2$—with the same (otherwise arbitrary) $q$—, as we have been accustomed to do in order to obtain consecutive powers.

System (I) becomes

$$\begin{cases} x_1^8\{x_1 + kq^4\} = \square_1, \\ x_1^8\{x_1 + lq^4\} = \square'_1, \end{cases}$$

and solving this amounts to searching for the solutions of

$$\begin{cases} r_1 + k = \text{square}, \\ r_1 + l = \text{square}, \end{cases}$$

with $r_1 = x_1/q^4$. The resolution, performed in the usual way, leads to

$$r_1(h) = \frac{(k-l)^2 - 2(k+l)h^2 + h^4}{4h^2},$$

[59] The smallest integral solution, corresponding to $h = 8$, is $x = 129$. The value 192, however, being divisible by 16, is more convenient for the subsequent reasoning.

whence the solutions $x_1 = r_1(h) \cdot q^4$, where $h$ is a rational number, chosen so as to have $r_1$ positive, and $q$ is fixed by the initial supposition $b_1 = q \cdot x_1^2$. Thus, we have the following solutions of system (I): $a_1 = r_1 \cdot q^4$, $b_1 = r_1^2 \cdot q^9$.

System (II) becomes

$$\begin{cases} x_2^8\{x_2 - lq^4\} = \square_2, \\ x_2^8\{x_2 - kq^4\} = \square_2', \end{cases}$$

and will be solved if we solve

$$\begin{cases} r_2 - l = \text{square}, \\ r_2 - k = \text{square}, \end{cases}$$

where $r_2 = x_2/q^4$. The resolution gives in this case

$$r_2(h) = \frac{(k-l)^2 + 2(k+l)h^2 + h^4}{4h^2},$$

whence the solutions $x_2 = r_2(h) \cdot q^4$, where $h$ is any rational number, while $q$ has the value attributed to it in the initial choice $b_2 = q \cdot x_2^2$. Thus, we have the following solutions of system (II): $a_2 = r_2 \cdot q^4$, $b_2 = r_2^2 \cdot q^9$.

We are now able to see the basis of Diophantus' computations. After having formed the ratio of $a_2$ (resp. $b_2$) to $a_1$ (resp. $b_1$), he reduces each ratio by removing the factor common to its two terms; this amounts to dropping the multiplicative quantities $q^4$ and $q^9$ appearing in the $a_i$'s and the $b_i$'s, respectively.[60]

We see that Diophantus' solutions then have the form $a_i' = r_i$, $b_i' = r_i^2$. In other words, his new sets of solutions correspond to the initial choice $b_i' = x_i^2 = a_i^2$ in replacement of the original one $b_i = qx_i^2 = qa_i^2$ $(q = 2)$.

As to the values of his two $r$'s, namely $r_1 = 1$ and $r_2 = 12$, they are obtained by taking $h = 1$ in the relations $r_1(h)$ and $r_2(h)$.[61] Hence Diophantus' sets of solutions:

$$(q = 2) \qquad a_1 = (q^4 =)\ 16, \qquad b_1 = (q^9 =)\ 512,$$
$$a_2 = 12 \cdot 16 = 192, \qquad (b_2 = 12^2 \cdot 512),$$

and

$$(q = 1) \qquad a_1' = 1, \qquad b_1' = 1,$$
$$a_2' = 12, \qquad b_2' = 144.$$

---

[60] The inference of lines 1569–73—i.e., concluding from the set of relations $(a_2^3)^3 : (a_1^3)^3 = (12^3)^3 : 1$, $(b_2^2)^2 : (b_1^2)^2 = (144^2)^2 : 1$, $(a_1^3)^3 = (b_1^2)^2$, and $1 =$ square cube, that $[(a_2^3)^3 = (12^3)^3$ and $(b_2^2)^2 = (144^2)^2$, i.e.$]$ $a_2 = 12$, $b_2 = 144$—, if not Diophantine, may be an interpolation accounted for by the absence of a reasoning or the corruption of the original reasoning.

[61] In fact, the value $r_1 = 1$, i.e., $x_1 = q^4$, is obvious since the given $k, l$ are both of the form $t^2 - 1$, $t$ rational.

Whence the following, (smallest) integral solution of IV,44,b given at the end:

$$a'_2 = 12, \quad b'_2 = 144,$$
$$((a'_2)^3)^3 = 1728^3 = 5{,}159{,}780{,}352,$$
$$((b'_2)^2)^2 = (144^2)^2 = 20{,}736^2 = 429{,}981{,}696,$$
$$\square = 3{,}869{,}835{,}264 = 62{,}208^2, \quad \square' = 1{,}719{,}926{,}784 = 41{,}472^2.$$

(c) $\begin{cases} 128x^8 - x^9 = \square, \\ 48x^8 - x^9 = \square', \end{cases}$

or:
$$\begin{cases} 128 - x = \square_1, \\ 48 - x = \square'_1. \end{cases}$$

The given solution is easily obtained: putting as usual $\square_1 = m^2, \square'_1 = n^2$, we are led to $128 - m^2 = 48 - n^2$, or $m^2 - n^2 = 80$ with $n^2 < 48$ (which excludes the solutions of part (a)).

With $m = n + h$:

$$n = \frac{80 - h^2}{2h} \quad (h \text{ integral: } 5 \leq h \leq 8 \text{ so that } 0 < n < \sqrt{48}).$$

Taking $h = 8$, we have $n = 1$, and then $x = 47$. So

$$(a^3)^3 = (47^3)^3 = 103{,}823^3\ [=1{,}119{,}130{,}473{,}102{,}767],$$
$$(b^2)^2 = (4418^2)^2 = 19{,}518{,}724^2\ [=380{,}980{,}586{,}588{,}176],$$
$$\square[=1{,}928{,}714{,}219{,}602{,}641] = 43{,}917{,}129^2,$$
$$\square'[=23{,}811{,}286{,}661{,}761] = 4{,}879{,}681^2.$$

The numbers in brackets, presumably because of their size, were not computed by the author of the major commentary. In the last two cases, he limited himself to the calculation of $\sqrt{\square} = \sqrt{47^8 \cdot 81}$ and $\sqrt{\square'} = \sqrt{47^8}$.

---

The last group of Book IV consists thus of the following problems:

**IV,43:** $\begin{cases} (a^3)^3 + k(b^2)^2 = \square, \\ (a^3)^3 - l(b^2)^2 = \square', \end{cases}$

**IV,44,a:** $\begin{cases} (a^3)^3 + k(b^2)^2 = \square, \\ (a^3)^3 + l(b^2)^2 = \square', \end{cases}$

**IV,44,b:** $\begin{cases} (a^3)^3 - l(b^2)^2 = \square, \\ (a^3)^3 - k(b^2)^2 = \square', \end{cases}$

**IV,44,c:** $\begin{cases} k(b^2)^2 - (a^3)^3 = \square, \\ l(b^2)^2 - (a^3)^3 = \square'. \end{cases}$

Putting $a = x, b = qx^2$, we have $(a^3)^3 = x^9$ and $(b^2)^2 = q^4x^8$, and we take accordingly $\square = m^2x^8$, $\square' = n^2x^8$. Thus, these problems end up being analogous in form to the ones in IV,36–39, from which they differ principally by a factor $x^6$, so that IV,44, in particular, is ultimately a derivative of II,11–13 (cf. p. 212).

One may remark that, as in the group IV,36–39, the two combinations not reducible to II,10, namely

$$\begin{cases} k(b^2)^2 + (a^3)^3 = \square, \\ l(b^2)^2 - (a^3)^3 = \square', \end{cases} \quad \text{and} \quad \begin{cases} k(b^2)^2 - (a^3)^3 = \square, \\ (a^3)^3 - l(b^2)^2 = \square', \end{cases}$$

are not treated.

---

The problems involving two equations, numbered IV,34 to 44, result in one of the following systems:[62]

$$\text{(Ia)} \quad \begin{cases} px^{2j\pm 1} + kx^{2j} = \square, \\ px^{2j\pm 1} + lx^{2j} = \square', \end{cases} \quad \text{(nos. 36–39, 43–44)}$$

of which a particular case is:

$$\text{(Ib)} \quad \begin{cases} px^{2j\pm 1} + \alpha^2 x^{2j} = \square, \\ px^{2j\pm 1} - \alpha^2 x^{2j} = \square', \end{cases} \quad \text{(nos. 34, 41, 42a)}$$

and

$$\text{(II)} \quad \begin{cases} \alpha^2 x^{2j} + px^{2j\pm 1} = \square, \\ \alpha^2 x^{2j} - px^{2j\pm 1} = \square'. \end{cases} \quad \text{(nos. 35, 40, 42b)}$$

The first set (Ia, b) is soluble by the (direct) method of the double-equation as well as by the method leading to an intermediate problem, of the form II,10. In the first approach, one puts $(k - l)x^{2j} = [(k - l)/h]x^j \cdot hx^j$ (or $2\alpha^2 x^{2j} = (2\alpha^2/h)x^j \cdot hx^j$) for the decomposition into factors and, in the second approach, one takes $\square = m^2x^{2j}$, $\square' = n^2x^{2j}$, thus obtaining $k - l$ (or $2\alpha^2$) $= m^2 - n^2 \equiv (n + h)^2 - n^2$ (II,10). The parameter $h$ is subject to a limitation (other than the one for $n > 0$) only in problems 37, 39 and (correspondingly) 44a and c, a limitation necessary in order to have $x > 0$.

The second type, (II), is solved in Diophantus' three examples by the second method, which, this time, leads to the intermediate problem $m^2 + n^2 = 2\alpha^2$ (II,9).[63]

---

The first six problems of Book V, reducible to the form

$$\begin{cases} \alpha^2 x^{2j} + kx^{2j\pm 1} = \square, \\ \alpha^2 x^{2j} + lx^{2j\pm 1} = \square', \end{cases} \quad (k, l \text{ positive or negative}),$$

[62] $k$, $l$ positive or negative in (Ia).

[63] In applying the double-equation method, the intermediate problem of representing $4\alpha^2$ as the sum of two squares (II,8) arises in connection with the condition which the parameter must satisfy for $x$ rational.

are obviously related to the above systems. Since these problems have nothing in common with the remaining ones of Book V,[64] their genuineness, or, at the very least, their placement, seems questionable.

The first possibility which comes to mind is that they are interpolated, as are those in the opening sections of the two following Books. But this is difficult to maintain if one considers V,1–6 from a mathematical point of view. For, unlike those problems in the *Arithmetica* which are certainly interpolated, these display real originality and require a notable degree of mathematical proficiency. Unless one admits the possibility of some isolated contributions having been made by a commentator as skilled as Diophantus himself—for which possibility no definite proof can be offered—,[65] one must consider V,1–6 as genuinely Diophantine.

If, on the other hand, we suppose that the present placement is not accidental (as would be the case had the title of Book V, for some reason, slipped back six problems), we must look at the possibility of Diophantus' having deliberately put V,1–6 where they are. He may have done so, motivated by considerations of distribution; for the addition of these six problems to those of Book IV would have made that Book disproportionately long—longer than the three following Books combined—with the risk of its assuming an overwhelming quantitative importance, thrusting the next three Books into the background, a bad policy indeed for a text-book. In any event, such a displacement is all the more acceptable in that V,1–6 involve quite a different type of intermediate problem.

Another, rather arbitrary but somewhat appealing hypothesis is the one which Tannery formulated in order to explain the presence in Book III of four problems (III,1–4) evidently correlated to those at the end of Book II (see p. 467). Tannery suggested that the progressive edition of the different Books might have led Diophantus to add cases omitted in Book II to the beginning of the subsequent installment (*Perte de sept livres*, p. 199 = *Mém. sc.*, II, p. 82).[66] Such a cause for displacement in fact fits our case much better than it does III,1–4. For the beginning of Book III as we have it falls within a very coherent group of problems, thus making it surprising that the group would be broken up, whereas V,1–6, while presenting a similar outward form, do in fact represent a new case, inasmuch as the method of resolution is different.

---

[64] The two remaining groups are quite characteristically dependent upon algebraical identities.

[65] Variations in style, for example, are unreliable indications: they occur even within the Greek text of Diophantus.

[66] That such a progressive production of Books belonging to a single treatise existed in ancient times is attested by the prefaces in Apollonius' *Conica* (see in particular the one to Book I) and in Archimedes' *De Sphaera et Cylindro*.

Tannery speaks of the progressive *edition* of the Books, although he certainly does not mean edition in the (ancient) sense of delivering a copy to an editor and his copyists. No doubt, in the above examples, the treatises were produced serially, the completed parts being sent to friends or colleagues, while the true editing work did not take place until the whole treatise was completed.

# Book V

**Problem V,1.**
$$\begin{cases}(b^2)^2 + k \cdot a^3 = \square, & k = 4,\\ (b^2)^2 - l \cdot a^3 = \square', & l = 3.\end{cases}$$

We put $b = x$, hence the system

$$\begin{cases}x^4 + 4a^3 = \square,\\ x^4 - 3a^3 = \square'.\end{cases}$$

Taking $a^3 = r \cdot x^4$, with $r$ to be determined, we have:

$$\begin{cases}x^4 + 4rx^4 = \square,\\ x^4 - 3rx^4 = \square'.\end{cases}$$

The text then simply states the intermediate problem and gives its numerical results. The full reasoning can be reconstructed as follows.[1]

Admitting $\square = m^2x^4$, $\square' = n^2x^4$, we have

$$r = \frac{m^2 - 1}{4} = \frac{1 - n^2}{3},$$

and therefore:

$$\frac{m^2 - 1}{1 - n^2} = \frac{4}{3}.$$

Let us consider generally:

$$\frac{m_1^2 - p_1^2}{p_1^2 - n_1^2} = \frac{4}{3}.$$

[1] We have added the intermediate computations in the next two problems as well.

From *Arithmetica* II,19, we know how to solve such a problem:

Since $m_1^2 > p_1^2 > n_1^2$, let us put

$$n_1^2 = y^2, \qquad p_1^2 = (y+1)^2, \quad \text{and} \quad m_1^2 = (y+h)^2.$$

Now, $m_1^2 = p_1^2 + \frac{4}{3}(p_1^2 - n_1^2)$, so that $y^2 + 2hy + h^2 = y^2 + 2y + 1 + \frac{8}{3}y + \frac{4}{3}$. Thus

$$y(\tfrac{14}{3} - 2h) = h^2 - \tfrac{7}{3} \quad \text{and} \quad y = \frac{h^2 - \frac{7}{3}}{\frac{14}{3} - 2h}.$$

We may take any suitable $h$ (that is, such that $\sqrt{\frac{7}{3}} < h < \frac{7}{3}$). Let us take $h = 2$; then $y = \frac{5}{2}$.

So, $m_1^2 = \frac{81}{4}$, $p_1^2 = \frac{49}{4}$, $n_1^2 = \frac{25}{4}$; and multiplying these by any square gives a new set of solutions to our problem (in particular, with 4 as multiplier, the integral set $m_2^2 = 81$, $p_2^2 = 49$, $n_2^2 = 25$).

The solution we are seeking in the original problem is fixed by the condition that the coefficient of $x^4$ be unity. Thus our solution will be $\frac{81}{49}$ $(=m^2)$, 1, $\frac{25}{49}$ $(=n^2)$. Hence

$$4rx^4 = (\tfrac{81}{49} - \tfrac{49}{49})x^4 = \tfrac{32}{49}x^4 \quad \text{and} \quad a^3 \equiv rx^4 = \tfrac{8}{49}x^4.$$

We are now left to find a cube in a given ratio to a fourth power. Taking $a = qx$, say $a = 2x$, we have $8x^3 = \frac{8}{49}x^4$, whence

$$x = 49.$$

So

$$a^3 = 98^3 = 941{,}192, \quad (b^2)^2 = (49^2)^2 = 2401^2 = 5{,}764{,}801,$$

$$\square = 9{,}529{,}569 = 3087^2, \quad \square' = 2{,}941{,}225 = 1715^2.$$

**Problem V,2.** $$\begin{cases} (b^2)^2 + k \cdot a^3 = \square, & k = 12, \\ (b^2)^2 + l \cdot a^3 = \square', & l = 5. \end{cases}$$

We put $b = x$, so

$$\begin{cases} x^4 + 12a^3 = \square, \\ x^4 + \phantom{1}5a^3 = \square'. \end{cases}$$

Taking $a^3 = r \cdot x^4$:

$$\begin{cases} x^4 + 12rx^4 = \square =, \text{ say, } m^2x^4, \\ x^4 + \phantom{1}5rx^4 = \square' =, \text{ say, } n^2x^4, \end{cases}$$

hence

$$r = \frac{m^2 - 1}{12} = \frac{n^2 - 1}{5}, \qquad \frac{m^2 - 1}{n^2 - 1} = \frac{12}{5},$$

and (*Elem.*, V,17)

$$\frac{m^2 - n^2}{n^2 - 1} = \frac{7}{5}.$$

We consider generally:

$$\frac{m_1^2 - n_1^2}{n_1^2 - p_1^2} = \frac{7}{5};$$

hence $m_1^2 = n_1^2 + \frac{7}{5}(n_1^2 - p_1^2)$. With $m_1^2 = (y + h)^2$, $n_1^2 = (y + 1)^2$, $p_1^2 = y^2$ $(m_1 > n_1 > p_1)$, we obtain

$$y = \frac{h^2 - \frac{12}{5}}{\frac{24}{5} - 2h} \quad \text{where} \quad \sqrt{\tfrac{12}{5}} < h < \tfrac{12}{5}.$$

We choose $h = 2$, so that $y = 2$, and $m_1^2 = 16$, $n_1^2 = 9$, $p_1^2 = 4$. The norm $p^2 = 1$ gives the desired solution: $m^2 = \frac{16}{4} = 4$, $n^2 = \frac{9}{4}$. Therefore $12rx^4 = (m^2 - 1)x^4 = 3x^4$ and $a^3 \equiv rx^4 = \frac{1}{4}x^4$.

We shall now determine $x$ by putting, say, $a = 2x$; then $8x^3 = \frac{1}{4}x^4$, and

$$x = 32.$$

So

$$a^3 = 64^3 = 262{,}144, \quad (b^2)^2 = (32^2)^2 = 1024^2 = 1{,}048{,}576,$$

$$\square = 4{,}194{,}304 = 2048^2, \quad \square' = 2{,}359{,}296 = 1536^2.$$

**Problem V,3.** $$\begin{cases}(b^2)^2 - l \cdot a^3 = \square, & l = 7, \\ (b^2)^2 - k \cdot a^3 = \square', & k = 12.\end{cases}$$

We put $b = x$, so

$$\begin{cases}x^4 - 7a^3 = \square, \\ x^4 - 12a^3 = \square'.\end{cases}$$

Taking $a^3 = r \cdot x^4$:

$$\begin{cases}x^4 - 7rx^4 = \square =, \text{ say, } m^2x^4, \\ x^4 - 12rx^4 = \square' =, \text{ say, } n^2x^4.\end{cases}$$

Then

$$r = \frac{1 - m^2}{7} = \frac{1 - n^2}{12}, \qquad \frac{1 - n^2}{1 - m^2} = \frac{12}{7};$$

hence (*Elem.*, V,17 and 7, porism)

$$\frac{1 - m^2}{m^2 - n^2} = \frac{7}{5}.^2$$

---

[2] These intermediate, formal transformations can in fact be avoided by ordering the squares with regard to their respective magnitudes: since $x^4 > \square > \square'$, we have immediately

$$\frac{x^4 - \square}{\square - \square'} = \frac{1 - m^2}{m^2 - n^2} = \frac{7}{5}.$$

Similarly in the two previous problems.

Let us seek generally $p_1^2$, $m_1^2$, $n_1^2$ such that

$$\frac{p_1^2 - m_1^2}{m_1^2 - n_1^2} = \frac{7}{5};$$

but we know the solution from the preceding problem: $p_1^2 = 16$, $m_1^2 = 9$, $n_1^2 = 4$. Thus, with the norm $p^2 = 1$: $m^2 = \frac{9}{16}$, $n^2 = \frac{4}{16}$. Hence $a^3 \equiv rx^4 = \frac{1}{16}x^4$, and putting $a = \frac{1}{2}x$ gives $x = 2$.

Thus

$$a^3 = 1, \quad (b^2)^2 = (2^2)^2 = 16, \quad [\square = 9, \square' = 4].$$

---

Let us recapitulate the method used by Diophantus in problems V,1–3. The system

$$\begin{cases} b^4 + ka^3 = \square, \\ b^4 + \ la^3 = \square', \end{cases} \qquad (k, l \text{ positive or negative})$$

can be transformed, assuming that $a^3 = r \cdot b^4$, into[3]

$$\begin{cases} 1 + kr = \square_1 = m^2, \\ 1 + \ lr = \square_1' = n^2, \end{cases}$$

which, since $r = (m^2 - 1)/k = (n^2 - 1)/l$, leads to the intermediate problem of finding $m^2$, $n^2$ fulfilling

$$\frac{m^2 - 1}{n^2 - 1} = \frac{k}{l}.$$

(a) If $k > 0$, $l < 0$:

$$m^2 > 1 > n^2$$

and

$$\frac{m^2 - 1}{1 - n^2} = \frac{k}{|l|} \quad (\text{V,1}).$$

(b) If $k > l > 0$:

$$m^2 > n^2 > 1;$$

then

$$\frac{m^2 - n^2}{n^2 - 1} = \frac{k - l}{l} \quad (\text{V,2}).$$

(c) If $0 > l > k$:

$$1 > n^2 > m^2;$$

---

[3] This new system is in fact never stated explicitly. The laconicism of the text is striking.

hence

$$\frac{1-n^2}{n^2-m^2}=\frac{|l|}{|k|-|l|} \quad (\text{V},3).$$

The proportion being so ordered, we seek the solution using II,19, taking $p^2$ instead of unity and putting $(y+h)^2$ for the largest square, $(y+1)^2$ for the middle, and $y^2$ for the smallest. The known square, namely 1, allows us to find the particular, required solution. $m^2$, $n^2$ being known, $r$ is known, and two numbers remain to be found, $a^3$ and $b^4$, in the ratio $r$. This is easily done by putting $b = x$ and $a = qx$, which gives

$$x=\frac{q^3}{r}.$$

Observe first that (in our explicit representation) the introduction of the ratio into the problem leads to a linear system for $r$, and is thus equivalent to Lagrange's transformation of the system

$$\begin{cases} A_1x^2+B_1x = \text{square}, \\ A_2x^2+B_2x = \text{square}, \end{cases}$$

into a linear one by dividing by $x^2$ and putting $y = 1/x$ (see his *Add. à l'Alg. d'Euler*, VI,62).[4]

Further, the reduction to problem II,19 allows us to treat systems of the type

$$\begin{cases} A_1x+C^2=\square, \\ A_2x+C^2=\square', \end{cases} \qquad (A_1, A_2 \gtrless 0),^5$$

or, more generally,

$$\begin{cases} A_1x+C_1^2=\square, \\ A_2x+C_2^2=\square', \end{cases}$$

which system is reducible to the previous one by multiplying the equations by $C_2^2$ and $C_1^2$, respectively.[6] We shall encounter the three aspects of the former system, with $C^2 \neq 1$, in the coming problems V,4–6.

*Lexicological remark*

The size of the ratio of $a^3$ to $b^4$ is not arbitrary, but depends on the values of the magnitudes $k$ and $l$ settled in the ἔκθεσις; that is, knowing the values of

[4] The transformation $y = 1/x$ is not performed in the *Arithmetica* (cf. Heath, *Dioph.*, p. 87, n. 1 and *supra*, pp. 215–216).

[5] An equivalent problem is II,16; there, the equations are fulfilled successively. This same principle is applied in VII,8–10 (where, this time, the square is considered unknown).

[6] The system $\begin{cases} 10x+9=\square, \\ 5x+4=\square', \end{cases}$ occurring earlier (in III,15), is solved by the method of the double-equation; this cannot be used systematically for the above, general, system (see pp. 231–232).

$k$ and $l$ allows us to assign a numerical value to the ratio, a ratio thus said to be "given".[7] The expression "given ratio" (line 1626) must thus be a faithful translation of δεδομένος λόγος. But, another expression is used (lines 1631–32, 1663–64, 1669–70, 1696–97, 1702): $a^3$ is said to be "given in ratio to $x^4$", which is tantamount to saying that, as in the previous instance, the ratio borne by $a^3$ to $x^4$ is obtainable from the data of the problem. This wording is interesting: the Arabic *mafrūḍ* (or *maᶜlūm*) *al-nisbata* surely renders the Greek δεδομένος (τῷ) λόγῳ; and, although such an association of δεδομένος with the words μέγεθος, εἶδος, and θέσις is quite common,[8] the expression δεδομένος λόγῳ is otherwise found in only one Greek author, Proclus—and has thus been considered to be a form scarcely employed (if at all elsewhere).[9] The example Proclus gave is that of an angle, which is *given in ratio* to some other angle when it is the double, triple, etc. of the latter. He also observes that the angle would be *given in magnitude* were it, e.g., the third of a right angle.

This meaning of "given in ratio" fits in our text well. For we first determine the *ratio* which a quantity, $p$, must have to a square, $q^2$, in order to fulfil the general conditions of the problem ($q^2$, increased or diminished by $kp$, $lp$, must result in a square); then, we determine the *magnitudes* of the actual unknowns $a^3$, $b^4$ from the known ratio by setting a proportion between their sides.

---

The group of problems V,4–6 differs from the previous one by the replacement of $a^3$ by $(a^3)^3$.[10] Setting this time at the outset $a = x$, $b = qx^2$, we shall end up with the *odd* power of $x$ being the higher of the two consecutive powers, so that problems V,4–6—unlike V,1–3—can be directly reduced, by an initial division, to problems linear in $x$ (see above, IV,42,a,3°).

**Problem V,4.**
$$\begin{cases}(b^2)^2 + k\cdot(a^3)^3 = \square, & k = 5,\\ (b^2)^2 - l\cdot(a^3)^3 = \square', & l = 3.\end{cases}$$

---

[7] This use of *given* to mean "potentially given" or "numerically determinable" is extensively employed in Euclid's *Data*, and Marinus of Neapolis, who discusses at length the various interpretations of the word "given", chooses finally to define it as γνώριμον ἅμα καὶ πόριμον, "knowable *and* determinable" (cf. *Euclidis opera*, VI (=*Data* c. comm. Marini), pp. 250,4–8 and 252,3–11; on πόριμον, cf. also p. 240,9–10). Observe, however, that there is in our case not just *one* acceptable ratio, since its numerical value depends on a parameter $h$ which is—within certain limits—optional.

[8] *Vide*, e.g., Euclid, *Data* (=*Opera*, VI), deff.; D.G., I, pp. 402,13; 404,15.

[9] *Comment. in Eucl.* (Friedlein), p. 205,13–14 (=note on I,1); p. 277,12–14 (=note on I,9)—or Steck's transl., pp. 310 and 359. Heath's evaluation of this passage (*Euclid's Elements*, I, pp. 132–33) must thus be modified.

[10] The resulting form, by the way, also amounts to interchanging $(a^3)^3$ and $(b^2)^2$ in IV,43–44,b.

We put $a = x$, $b = 2x^2$, so

$$\begin{cases} 16x^8 + 5x^9 = \square, \\ 16x^8 - 3x^9 = \square'. \end{cases}$$

Dividing by the square $x^8$:

$$\begin{cases} 16 + 5x = \square_1, \\ 16 - 3x = \square'_1. \end{cases}$$

Now, for any square $u^2$, we have:

$$\begin{cases} u^2 + 5 \cdot \dfrac{u^2}{4} = 2\tfrac{1}{4}u^2 = \text{square}, \\[2ex] u^2 - 3 \cdot \dfrac{u^2}{4} = \tfrac{1}{4}u^2 = \text{square}. \end{cases}$$

This set of identities, either because of its banality or because it emerges incidentally from IV,43, is simply stated in the text.[11]

Taking 16 for $u^2$, $x$ will be equal to $u^2/4 = 4$. So

$$(a^3)^3 = (4^3)^3 = 64^3 = 262{,}144, \quad (b^2)^2 = (32^2)^2 = 1024^2 = 1{,}048{,}576,$$

$$\square = 2{,}359{,}296 = 1536^2, \quad \square' = 262{,}144 = 512^2.$$

**Problem V,5.** $$\begin{cases} (b^2)^2 + k \cdot (a^3)^3 = \square, & k = 12, \\ (b^2)^2 + l \cdot (a^3)^3 = \square', & l = 5. \end{cases}$$

We put $a = x$, $b = 2x^2$, so

$$\begin{cases} 16x^8 + 12x^9 = \square, \\ 16x^8 + 5x^9 = \square'. \end{cases}$$

[11] It could also be obtained as before: putting $x = r \cdot 16$, we shall have

$$\begin{cases} 16(1 + 5r) = \square_1 =, \text{ say, } m^2 \cdot 16, \\ 16(1 - 3r) = \square'_1 =, \text{ say, } n^2 \cdot 16, \end{cases} \quad \text{with} \quad m^2 > 1 > n^2.$$

Hence $$\frac{m^2 - 1}{1 - n^2} = \frac{5}{3} \quad \text{or, generally:} \quad \frac{m_1^2 - p_1^2}{p_1^2 - n_1^2} = \frac{5}{3};$$

setting $m_1^2 = (y + h)^2$, $p_1^2 = (y + 1)^2$, $n_1^2 = y^2$ leads to

$$y = \frac{h^2 - \frac{8}{3}}{\frac{16}{3} - 2h}, \qquad \sqrt{\tfrac{8}{3}} < h < \tfrac{8}{3}.$$

$h = 2$: $$y = 1,$$

so $$m_1^2 = 9, \quad p_1^2 = 4, \quad n_1^2 = 1.$$

Norm: $p^2 = 1$; then $m^2 = \frac{9}{4}$, $n^2 = \frac{1}{4}$, so $r = \frac{1}{4}$. Thus the above identities.

Dividing by $x^8$:

$$\begin{cases} 16 + 12x = \square_1, \\ 16 + \ \ 5x = \square_1'. \end{cases}$$

Now, for any square $u^2$, we have the identities (seen in V,2):

$$\begin{cases} u^2 + 12\dfrac{u^2}{4} = \text{square}, \\[2ex] u^2 + \ \ 5\dfrac{u^2}{4} = \text{square}. \end{cases}$$

Putting 16 for $u^2$ gives here again $x = 4$. So

$(b^2)^2$, $(a^3)^3$ as in V,4, $\quad \square = 4{,}194{,}304 = 2048^2, \quad \square'$ as $\square$ in V,4.

**Remark.** We could have deduced from V,2, without any computation, the above solution.

**Problem V,6.** $\quad \begin{cases} (b^2)^2 - l \cdot (a^3)^3 = \square, & l = 4, \\ (b^2)^2 - k \cdot (a^3)^3 = \square', & k = 7. \end{cases}$

We put $a = x$, $b = 3x^2$.[12] So:

$$\begin{cases} 81x^8 - 4x^9 = \square, \\ 81x^8 - 7x^9 = \square', \end{cases} \quad \text{or} \quad \begin{cases} 81 - 4x = \square_1, \\ 81 - 7x = \square_1'. \end{cases}$$

Thus, the text says, we have to seek which given[13] (fractional) quantity, $r$, of a square, $p^2$, will fulfil

$$\begin{cases} p^2 - 4 \cdot rp^2 = \text{square} \equiv m^2, \\ p^2 - 7 \cdot rp^2 = \text{square} \equiv n^2. \end{cases}$$

The text states simply that the answer may be obtained "in the previous manner". Indeed,

$$rp^2 = \frac{p^2 - m^2}{4} = \frac{p^2 - n^2}{7},$$

whence

$$\frac{p^2 - m^2}{m^2 - n^2} = \frac{4}{3}.$$

[12] Putting, as usual, $b = 2x^2$ would lead to the inconvenient solution $x = \frac{128}{81}$.

[13] *mafrūd*, in the sense of "potentially given" (above, p. 228, n. 7).

But a solution of this is known from V,1: $p^2 = 81$, $m^2 = 49$, $n^2 = 25$. This fits our case ($p^2 = 81$), and we have immediately $r = \frac{8}{81}$ and $x = rp^2 = 8$. So

$$(a^3)^3 = (8^3)^3 = 512^3 = 134{,}217{,}728,$$

$$(b^2)^2 = (192^2)^2 = 36{,}864^2 = 1{,}358{,}954{,}496,$$

$$\square = 822{,}083{,}584 = 28{,}672^2, \quad \square' = 419{,}430{,}400 = 20{,}480^2.$$

---

The method used in solving problems V,4–6 does not differ essentially from the one used in V,1–3. The two powers are now $(a^3)^3$ and $(b^2)^2$, and we put $a = x$, $b = qx^2$; the resulting system

$$\begin{cases} q^4x^8 + kx^9 = \square, \\ q^4x^8 + lx^9 = \square', \end{cases} \qquad (k, l \text{ positive or negative}),$$

can then be reduced to a system linear in $x$, and, by taking $x = r \cdot q^4$, to the form

$$\begin{cases} 1 + kr = \square_1, \\ 1 + lr = \square'_1, \end{cases}$$

to be solved as previously, by II,19.[14]

The three related cases, where $(b^2)^2$ appears in a subtraction, that is (with $k, l > 0$),

$$\begin{cases} k(a^3)^3 - (b^2)^2 = \square, \\ l(a^3)^3 - (b^2)^2 = \square', \end{cases} \quad \begin{cases} (b^2)^2 - l(a^3)^3 = \square, \\ k(a^3)^3 - (b^2)^2 = \square', \end{cases} \quad \begin{cases} (b^2)^2 + l(a^3)^3 = \square, \\ k(a^3)^3 - (b^2)^2 = \square', \end{cases}$$

are not soluble similarly,[15] and are therefore considered neither here nor (*mutatis mutandis*) in the previous group.

**Remark.** One might rightly ask whether problems V,1–6 are soluble by the method of the double-equation, that is, whether the system to which they are reducible,

$$\begin{cases} 1 + kr = \square \equiv m^2, \\ 1 + lr = \square' \equiv n^2, \end{cases} \qquad (k, l \text{ positive or negative}),$$

is soluble in the said way.

One sees immediately that in

$$4 + 4kr = \left[\frac{(k-l)r}{h} + h\right]^2 = \left(\frac{k-l}{h}\right)^2 \cdot r^2 + 2(k-l)r + h^2,$$

---

[14] Or else in the way II,16 was solved.

[15] The three squares cannot be put in order of magnitude.

with $h$ rational, the only possible way of arriving at a linear equation is to take $h = 2$, which gives

$$r = \frac{8(k + l)}{(k - l)^2}.$$

Thus we observe:

1°. that the choice of the parameter $h$ is imposed;

2°. that, since $k + l$ can be negative, the solution obtained will not always be acceptable (as, e.g., in V,3 and 6).

This second characteristic in particular would have prevented Diophantus from using here the double-equation method. The first limitation is not really a restriction new to us, since it has already been encountered, though not explicitly pointed out, in some problems of Book III (cf. nos. 13, 15, 17 and 18).

N.B. Besides the example of Book III already mentioned (p. 227, n. 6), there is, this time in a subsequent Book (problem "IV",39), a system of a type similar to the one under consideration, namely

$$\begin{cases} 8x + 4 = \square, \\ 6x + 4 = \square', \end{cases}$$

which Diophantus first attempts to treat by the double-equation method. After having obtained a solution not fulfilling the initial conditions, he reformulates the problem in the manner of II,19. The initial conditions, however, do not allow a mechanical application of II,19 as do V,1–6.

---

**Problem V,7.** $\begin{cases} a + b = k, \\ a^3 + b^3 = l, \end{cases}$ $\quad k = 20,\ l = 2240.$

*Condition*: $(4l - k^3)/3k$ = square, or $(4l - k^3)\frac{3}{4}k$ = square.

In this problem and the next one are given two conditions, which are in fact equivalent since they merely differ by a square factor $\frac{9}{4}k^2$. The second one was probably some marginal addition which was integrated into the text. This would explain why they are joined by an "and" and not given as alternative (surely the scholiast did at least recognize their equivalence).[16]

The problem represented by the condition is constructible (see p. 192), as stated in the text.

[16] He almost certainly obtained his condition by eliminating $a$ and considering the discriminant of the resulting second-degree equation (form $(B/2)^2 + AC = sq.$ for an equation $Ax^2 + Bx = C$; type II on p. 76).

Putting $2x$ as the difference of the two numbers (cf. problem I,27), we shall have:

$$a = \frac{k}{2} + x, \qquad b = \frac{k}{2} - x,$$

which fulfil the condition of the first equation.

Then are given in the text—or, rather, explained at length—the two relations

$$(u + v)^3 = u^3 + 3u^2v + 3uv^2 + v^3,$$
$$(u - v)^3 = u^3 - 3u^2v + 3uv^2 - v^3,$$

which will play an essential rôle in problems V,7–16, and which are encountered here for the first time (remember that cubes do not appear before Book IV); but the excessively lengthy explanations can hardly go back to Diophantus himself. So

$$a^3 = \frac{k^3}{8} + \tfrac{3}{4}k^2x + \tfrac{3}{2}kx^2 + x^3, \qquad b^3 = \frac{k^3}{8} - \tfrac{3}{4}k^2x + \tfrac{3}{2}kx^2 - x^3;$$

then
$$a^3 + b^3 = \frac{k^3}{4} + 3kx^2 = l$$

and
$$x = \frac{1}{2}\sqrt{\frac{4l - k^3}{3k}}.$$

This gives the condition.

For $k = 20$, $l = 2240$: $\qquad x = 2,$

so $\qquad a = 12, \quad a^3 = 1728, \quad b = 8, \quad b^3 = 512.$

**Remark.** The same problem, but without stated condition and with $k = 10$, $l = 370$, is found in the Greek (so-called) fourth Book as first problem. It is obviously an interpolation, as is the next problem, "IV",2, solved by the commentator on the model of V,8. We have already pointed out the significance of these interpolations as an argument for the authenticity of the Arabic Books (p. 4).

**Problem V,8.** $\begin{cases} a - b = k, & k = 10, \\ a^3 - b^3 = l, & l = 2170. \end{cases}$

*Condition*: $(4l - k^3)/3k$ = square, or $(4l - k^3)\frac{3}{4}k$ = square. The two conditions are equivalent (cf. V,7).

Putting $2x$ as the sum of $a$ and $b$, we shall have

$$a = x + \frac{k}{2}, \qquad b = x - \frac{k}{2}.$$

So

$$a^3 - b^3 = x^3 + \tfrac{3}{2}x^2k + \tfrac{3}{4}xk^2 + \frac{k^3}{8} - x^3 + \tfrac{3}{2}x^2k - \tfrac{3}{4}xk^2 + \frac{k^3}{8}$$

$$= \frac{k^3}{4} + 3kx^2 = l.$$

Hence $$x = \frac{1}{2}\sqrt{\frac{4l - k^3}{3k}}.$$

This gives the condition.

For $k = 10, l = 2170$: $x = 8$;

hence $a = 13, \quad a^3 = 2197, \quad b = 3, \quad b^3 = 27.$

The problem "IV",2 does not give any diorism, and has $k = 6, l = 504$.

**Problem V,9.** $\begin{cases} a + b = k, & k = 20, \\ a^3 + b^3 = l(a - b)^2, & l = 140. \end{cases}$

*Condition*: $k^3(l - \tfrac{3}{4}k) =$ square.
Putting $a - b = 2x$, we have:

$$a = \frac{k}{2} + x, \qquad b = \frac{k}{2} - x.$$

So $$a^3 + b^3 = \frac{k^3}{4} + 3kx^2 = l(a - b)^2 = 4lx^2.$$

Hence $$x = \sqrt{\frac{k^3}{4(4l - 3k)}} = \frac{1}{4}\sqrt{\frac{k^3}{l - \frac{3}{4}k}}.$$

This gives the condition.

For $k = 20, l = 140$: $x = 2$;

hence $a = 12, \quad [a^3 = 1728], \quad b = 8, \quad [b^3 = 512].$[17]

**Problem V,10.** $\begin{cases} a - b = k, & k = 10, \\ a^3 - b^3 = l(a + b)^2, & l = 8\tfrac{1}{8}. \end{cases}$

*Condition*: $k^3(l - \tfrac{3}{4}k) =$ square.

Putting $a + b = 2x$, we have:

$$a = x + \frac{k}{2}, \qquad b = x - \frac{k}{2}.$$

---

[17] $a^3$ and $b^3$ have already been computed in V,7.

So $$a^3 - b^3 = 3x^2k + \frac{k^3}{4} = l(a+b)^2 = 4lx^2.$$

Hence $$x = \sqrt{\frac{k^3}{4(4l - 3k)}} = \frac{1}{4}\sqrt{\frac{k^3}{l - \frac{3}{4}k}}.$$

This gives the condition.

For $k = 10, l = 8\frac{1}{8}$: $$x = 10;$$

hence $$a = 15, \quad a^3 = 3375, \quad b = 5, \quad b^3 = 125.$$

**Problem V,11.** $$\begin{cases} a - b = k, & k = 4, \\ a^3 + b^3 = l(a+b), & l = 28. \end{cases}$$

*Condition*: $l - \frac{3}{4}k^2 =$ square.

Putting $a + b = 2x$, we have:

$$a = x + \frac{k}{2}, \qquad b = x - \frac{k}{2}.$$

So $$a^3 + b^3 = 2x^3 + \tfrac{3}{2}k^2x = l(a+b) = 2lx.$$

Hence $$x = \sqrt{l - \tfrac{3}{4}k^2}.$$

This gives the condition.

For $k = 4, l = 28$: $$x = 4,$$

so $$a = 6, \quad a^3 = 216, \quad b = 2, \quad b^3 = 8.$$

**Problem V,12.** $$\begin{cases} a + b = k, & k = 8, \\ a^3 - b^3 = l(a-b), & l = 52. \end{cases}$$

*Condition*: $l - \frac{3}{4}k^2 =$ square.

Putting $a - b = 2x$, we have

$$a = \frac{k}{2} + x, \qquad b = \frac{k}{2} - x.$$

So $$a^3 - b^3 = \tfrac{3}{2}k^2x + 2x^3 = l(a-b) = 2lx.$$

Hence $$x = \sqrt{l - \tfrac{3}{4}k^2}.$$

This gives the condition.

For $k = 8, l = 52$: $$x = 2,$$

so $$a = 6, \quad a^3 = 216, \quad b = 2, \quad b^3 = 8.$$

The set of problems V,7–12 is reminiscent of three elementary cases treated by Diophantus in Book I, namely:

$$(\text{I},27)\ \begin{cases} a+b=k, \\ a\cdot b\ \ =l. \end{cases} \qquad (\text{I},30)\ \begin{cases} a-b=k, \\ a\cdot b\ \ =l. \end{cases} \qquad (\text{I},28)\ \begin{cases} a+b\ \ \ =k, \\ a^2+b^2=l. \end{cases}$$

In both sets, the conditions lead to constructible problems, and in all of the propositions one takes as unknown the sum or the difference of $a$ and $b$, depending on whether their difference or their sum is given.

($\alpha$) Note first that the three above-mentioned problems from Book I did not originate with Diophantus. The first two had already been solved more than two millenia before by Sumerian mathematicians, who, as noted by Vogel (in his *Zur Berechnung d. quadr. Gl. bei den Bab.*), based their resolution on the identity

$$\left(\frac{a-b}{2}\right)^2 + ab = \left(\frac{a+b}{2}\right)^2 \quad (\text{comp. } \textit{Elem.},\text{II},5).$$

Thus, if the product of the unknowns is given as well as their sum, the calculator will endeavour to find their difference by using the above formula, and inversely; the unknowns themselves will then be obtained from the relations

$$a = \tfrac{1}{2}\{(a+b)+(a-b)\},$$
$$b = \tfrac{1}{2}\{(a+b)-(a-b)\}.^{18}$$

Diophantus reproduces, in a more algebraical form, this way of solving; in particular, his diorisms are immediately evident from the identity given above (see p. 192).

He certainly follows an archaic tradition also when he solves the third system,

$$\begin{cases} a+b\ \ \ =k, \\ a^2+b^2=l, \end{cases}$$

by using the identity

$$2(a^2+b^2) = (a+b)^2 + (a-b)^2 \quad (\text{comp. } \textit{Elem.}, \text{II},9)$$

which directly gives the necessary condition of resolution $2l - k^2 =$ square. The Mesopotamian approach relies on the equivalent formula

$$\frac{a^2+b^2}{2} = \left(\frac{a+b}{2}\right)^2 + \left(\frac{a-b}{2}\right)^2,$$

[18] Several examples in Thureau-Dangin's *Textes mathématiques babyloniens* and Neugebauer's *Mathematische Keilschrift-Texte*.

which is also used for solving the system (omitted in the *Arithmetica*)

$$\begin{cases} a - b = k, \\ a^2 + b^2 = l, \end{cases}$$

(cf. the tablet BM 13901, problems 8 and 9).

**Remark.** One should not infer, on the basis of the existence of some types of problems found in both Mesopotamian mathematics and the *Arithmetica*, that Diophantus was a follower of some "Mesopotamian tradition". For, firstly, these common problems are limited to the elementary Book I (the resolution of $a^2 - b^2 = k$ by means of tables—or its construction from them—in BM 85194,4, merely has its formulation in common with Diophantus' II,10). Secondly, it is certainly true that a great part of the *Arithmetica* is ultimately based on elementary identities which were, by virtue of their very simplicity, also known to the Mesopotamians, but it was precisely Diophantus' (and his Greek forerunners') genius which allowed him to derive from these few identities a great number of algebraical problems, often reaching—especially in the later Books—a high level of difficulty.

($\beta$) The above discussion leads one to wonder whether some identity does not lie behind V,7–12 as well. This happens to be the case, for Diophantus doubtless developed his problems from the identity

$$4(a^3 \pm b^3) = 3(a \pm b)(a \mp b)^2 + (a \pm b)^3,$$

various forms of which readily yield the diorisms for the considered cases.

Thus, with

$$\frac{4(a^3 + b^3) - (a + b)^3}{3(a + b)} = (a - b)^2,$$

$$\frac{4(a^3 - b^3) - (a - b)^3}{3(a - b)} = (a + b)^2,$$

one associates V,7–8; next

$$\frac{(a + b)^3}{4\dfrac{a^3 + b^3}{(a - b)^2} - 3(a + b)} = (a - b)^2,$$

$$\frac{(a - b)^3}{4\dfrac{a^3 - b^3}{(a + b)^2} - 3(a - b)} = (a + b)^2,$$

are the basis of V,9–10, while the relations

$$4\,\frac{a^3+b^3}{a+b} - 3(a-b)^2 = (a+b)^2,$$

and

$$4\,\frac{a^3-b^3}{a-b} - 3(a+b)^2 = (a-b)^2$$

underlie the last two problems V,11–12.

Lastly, observe that each problem of the group V,7–12 gives the condition for the rationality of $x$, but neglects to give the condition for the positivity of the smaller required number. Thus, in V,7, where $b = k/2 - x$, we must have

$$\frac{k}{2} > x = \frac{1}{2}\sqrt{\frac{4l-k^3}{3k}}, \quad \text{so} \quad k^2 > \frac{4l-k^3}{3k},$$

hence $k^3 > l$. In V,8 (where $b = x - k/2$), we shall have accordingly $l > k^3$.
We find similarly for the other problems:

$$l > k \quad (\text{V},9),$$
$$k > l \quad (\text{V},10),$$
$$l > k^2 \quad (\text{V},11),$$
$$k^2 > l \quad (\text{V},12).$$

Note that the same omission occurs in I,28:[19] we have $b = k/2 - x$, with $x = \frac{1}{2}\sqrt{2l - k^2}$, so that $b$ will be positive only if $k^2 > l$.

---

**Problem V,13.**
$$\begin{cases} k \cdot a^2 + l = u + v, & k = 9, \quad l = 30, \\ u + a^3 = \square, \\ v + a^3 = \square'. \end{cases}$$

We put $a = x$, so:
$$\begin{cases} kx^2 + l = u + v, \\ u + x^3 = \square, \\ v + x^3 = \square'. \end{cases}$$

Taking $\square = (x+p)^3$, $\square' = (x+q)^3$, we have:

$$u = (x+p)^3 - x^3, \qquad v = (x+q)^3 - x^3.$$

Hence:
$$\begin{aligned} kx^2 + l &= 3x^2p + 3xp^2 + p^3 + 3x^2q + 3xq^2 + q^3 \\ &= 3x^2(p+q) + 3x(p^2+q^2) + (p^3+q^3). \end{aligned}$$

---

[19] I,27 and 30 do not require such a condition.

In order to be left with one term equal to another, we make the $x^2$'s disappear:

*Condition I*: $p + q = k/3$.

We have then

$$x = \frac{l - (p^3 + q^3)}{3(p^2 + q^2)}$$

which must be positive. Thus:

*Condition II*: $p^3 + q^3 < l$.

For given $k = 9$, $l = 30$, we must choose $p$, $q$ such that

$$p^3 + q^3 < 30$$

and

$$p + q = 3.$$

An obvious pair is $p = 2$, $q = 1$;[20] so,

$$x = 1\tfrac{2}{5}.$$

Hence

$$a^3 = (\tfrac{7}{5})^3 = 2\tfrac{93}{125}, \quad u = 36\tfrac{70}{125}, \quad v = 11\tfrac{10}{125}, \quad \square = 39\tfrac{38}{125} = (3\tfrac{2}{5})^3,$$
$$\square' = 13\tfrac{103}{125} = (2\tfrac{2}{5})^3.$$

In order that $x$ be rational or, rather, in order to obtain a linear equation, Diophantus imposed $p + q = k/3$. This determines the sum $p + q$ for given $k$. The other condition to be observed in the choosing of the values of $p$ and $q$ was given by the condition of positivity for $x$ and took the form $p^3 + q^3 < l$.

Thus arises the question whether it is possible, for any given $k$, $l$, to obtain a suitable pair $p, q > 0$. The answer is given by Diophantus in a remark at the end of the problem; this remark would have been a diorism were it not limited to the resolution which he presents (see below).

Let us consider, as in the case of constructible problems, that $k$ is the first assigned value. The condition for positive $x$ can be written as

$$0 < l - (p^3 + q^3) = l - p^3 - \left(\frac{k}{3} - p\right)^3 = l - \left(\frac{k}{3}\right)^3 + \frac{k^2}{3}p - kp^2 \equiv l - f(p).$$

Since $f(p)$ has its minimal value for $p_0 = k/6$, namely $f(p_0) = \frac{1}{4}(k/3)^3$, the smallest possible $l$ will have to be greater than that value; thus Diophantus' limitation $4l > (k/3)^3$.

Since, on the other hand, the maxima of $f(p)$ are found at the limits of the interval $[0, k/3]$, namely $f(0) = f(k/3) = (k/3)^3$, any pair $p, q > 0$ with

[20] In fact, any $p, q > 0$ with $p + q = 3$ satisfy $p^3 + q^3 < 30$ (see what follows).

$p + q = k/3$ will be acceptable for $l \geq (k/3)^3$. But if $l$ is taken within the limits

$$\frac{1}{4}\left(\frac{k}{3}\right)^3 < l < \left(\frac{k}{3}\right)^3,$$

we shall have to find values $p, q$ satisfying the pair of equations

$$\begin{cases} p + q = \dfrac{k}{3}, \\ p^3 + q^3 = l' < l. \end{cases}$$

We know how to solve this from V,7.[21] And it is from the condition underlying the resolution of this problem, namely

$$\frac{4l' - \left(\dfrac{k}{3}\right)^3}{k} = \text{square},$$

that Diophantus must have inferred his condition $4l(>4l') > (k/3)^3$.

**Remark.** Diophantus' specification that $4l > (k/3)^3$ must be fulfilled when we use "this" treatment seems to indicate that he had the existence of another possibility in mind. Indeed, in

$$kx^2 + l = 3x^2(p + q) + 3x(p^2 + q^2) + (p^3 + q^3)$$

we may consider eliminating the units instead of the $x^2$ by putting

$$l = p^3 + q^3,$$

with $k/3 > p + q$ in order that

$$x = \frac{p^2 + q^2}{\dfrac{k}{3} - (p + q)}$$

be positive. If we keep the same $p = 2$, $q = 1$ as above, we have a solution, e.g., for $l = 9$, $k = 10$; here then the specified condition does not apply.

**Problem V,14.** $\begin{cases} k \cdot a^2 - l = u + v, & k = 9, l = 26, \\ a^3 - u = \square, \\ a^3 - v = \square'. \end{cases}$

We put $a = x$, so $\begin{cases} kx^2 - l = u + v, \\ x^3 - u = \square, \\ x^3 - v = \square'. \end{cases}$

[21] $l'$ must be chosen so as to give rational values for $p, q$ (cf. formula below).

Taking $\square = (x - p)^3$, $\square' = (x - q)^3$, we have

$$u = x^3 - (x - p)^3, \qquad v = x^3 - (x - q)^3.$$

Hence: $$kx^2 - l = 3x^2p - 3xp^2 + p^3 + 3x^2q - 3xq^2 + q^3$$
$$= 3x^2(p + q) - 3x(p^2 + q^2) + (p^3 + q^3).$$

The terms in $x^2$ will vanish with the

*Condition*: $k/3 = p + q$.

Then: $$x = \frac{l + p^3 + q^3}{3(p^2 + q^2)}.$$

This is greater than 0 for any $l, p, q > 0$. Hence, as observed in the text, there is only the above condition.

Taking $k = 9, l = 26$, we shall choose $p, q$ with $p + q = 3$, say $p = 2, q = 1$.
Hence $$x = 2\tfrac{1}{3}, \quad a^3 = x^3 = 12\tfrac{19}{27}, \quad u = 12\tfrac{18}{27}, \quad v = 10\tfrac{9}{27},$$

$$\square = \tfrac{1}{27} = (\tfrac{1}{3})^3, \quad \square' = 2\tfrac{10}{27} = (1\tfrac{1}{3})^3.$$

**Problem V,15.** $$\begin{cases} k \cdot a^2 - l = u + v, & k = 9, l = 18, \\ a^3 + u = \square, \\ a^3 - v = \square'. \end{cases}$$

We put $a = x$, so we have

$$\begin{cases} kx^2 - l = u + v, \\ x^3 + u = \square, \\ x^3 - v = \square'. \end{cases}$$

Taking $\square = (x + p)^3$, $\square' = (x - q)^3$ $(q > p)$, we have

$$u = (x + p)^3 - x^3, \qquad v = x^3 - (x - q)^3.$$

Hence $$kx^2 - l = 3x^2p + 3xp^2 + p^3 + 3x^2q - 3xq^2 + q^3$$
$$= 3x^2(p + q) - 3x(q^2 - p^2) + (p^3 + q^3).$$

The terms in $x^2$ will vanish with the

*Condition*: $k/3 = p + q$.

Then: $$x = \frac{l + p^3 + q^3}{3(q^2 - p^2)}.$$

For $k = 9$, $l = 18$, and choosing $p = 1$, $q = 2$, we have

$$x = 3, \quad a^3 = x^3 = 27, \quad u = 37, \quad v = 26, \quad \square = 64, \quad \square' = 1.$$

**Problem V,16.** $$\begin{cases} k \cdot a^2 - l = u + v, & k = 9, l = 16, \\ a^3 - u = \square, \\ v - a^3 = \square'. \end{cases}$$

We put $a = x$, so we have

$$\begin{cases} kx^2 - l = u + v, \\ x^3 - u = \square, \\ v - x^3 = \square'. \end{cases}$$

Taking $\square = (x - p)^3$, $\square' = (q - x)^3$, we have

$$u = x^3 - (x - p)^3, \qquad v = x^3 + (q - x)^3.$$

Hence: $$\begin{aligned} kx^2 - l &= 3x^2p - 3xp^2 + p^3 + q^3 - 3q^2x + 3qx^2 \\ &= 3x^2(p + q) - 3x(p^2 + q^2) + (p^3 + q^3). \end{aligned}$$

The terms in $x^2$ will vanish with the

*Condition*: $k/3 = p + q$.

Then $$x = \frac{l + p^3 + q^3}{3(p^2 + q^2)}.$$

For $k = 9$, $l = 16$, and $p = 1$, $q = 2$:

$$x = 1\tfrac{2}{3}, \quad a^3 = x^3 = 4\tfrac{17}{27}, \quad u = 4\tfrac{1}{3}, \quad v = 4\tfrac{2}{3}, \quad \square = \tfrac{8}{27}, \quad \square' = \tfrac{1}{27}.$$

---

The group of problems V,13–16 is constructed, as is the previous one, from an identity, namely

$$(x + p)^3 + (x + q)^3 = 2x^3 + 3x^2(p + q) + 3x(p^2 + q^2) + p^3 + q^3,$$

in which we allow the signs to vary—keeping in mind, however, that for the resulting problem only those combinations which, for $p, q > 0$, lead to positive values of $x, u, v, \square, \square'$, are admissible.

The possible different combinations are then the following:

1°. $(x + p)^3 + (x + q)^3$, leading to

$$\underbrace{3x^2(p + q)}_{kx^2} + \underbrace{3x(p^2 + q^2) + p^3 + q^3}_{+l} = \underbrace{(x + p)^3 - x^3}_{u} + \underbrace{(x + q)^3 - x^3}_{v},$$

which is problem V,13, that is,

$$\begin{cases} kx^2 + l = u + v, \\ u + x^3 = \square, \\ v + x^3 = \square'. \end{cases}$$

2°. $-(x - p)^3 - (x - q)^3$, giving

$$\underbrace{3x^2(p + q)}_{kx^2} \underbrace{- 3x(p^2 + q^2) + p^3 + q^3}_{-l} = \underbrace{-(x - p)^3 + x^3}_{u} \underbrace{- (x - q)^3 + x^3}_{v},$$

which is problem V,14, that is,

$$\begin{cases} kx^2 - l = u + v, \\ x^3 - u = \square, \\ x^3 - v = \square'. \end{cases}$$

3°. $(x + p)^3 - (x - q)^3$, giving

$$\underbrace{3x^2(p + q)}_{kx^2} + \underbrace{3x(p^2 - q^2) + p^3 + q^3}_{-l} = \underbrace{(x + p)^3 - x^3}_{u} \underbrace{- (x - q)^3 + x^3}_{v},$$

which is problem V,15, that is,

$$\begin{cases} kx^2 - l = u + v, \\ x^3 + u = \square, \\ x^3 - v = \square'. \end{cases}$$

4°. $-(x - p)^3 + (q - x)^3$, resulting in

$$\underbrace{3x^2(p + q)}_{kx^2} \underbrace{- 3x(p^2 + q^2) + p^3 + q^3}_{-l} = \underbrace{-(x - p)^3 + x^3}_{u} + \underbrace{(q - x)^3 + x^3}_{v},$$

which is problem V,16, that is,

$$\begin{cases} kx^2 - l = u + v, \\ x^3 - u = \square, \\ v - x^3 = \square'. \end{cases}$$

Note that the basic expressions in 2° and 4°, although algebraically identical, lead to different problems, since $\square$, $\square'$ will be positive if in one case $x > p, q$ and in the other $p < x < q$. These conditions, as well as that of V,15 ($x > q$), one can fulfil, holding the usual numerical values for the pair $p$, $q$, by choosing a suitable $l$.[22]

The two following combinations which lead also to $k = 3(p + q)$ were not considered by Diophantus:

5°. $(x + p)^3 + (q - x)^3$, corresponding to the problem

$$\begin{cases} kx^2 - l = u + v, \\ u + x^3 = \square, \\ v - x^3 = \square'; \end{cases}$$

6°. $(p - x)^3 + (q - x)^3$, giving the problem

$$\begin{cases} kx^2 - l = u + v, \\ u - x^3 = \square, \\ v - x^3 = \square'. \end{cases}$$

The expressions for $x$ are the same as in problems V,15 and 14, respectively. But, in accordance with the previous considerations, $l$ will have to be chosen smaller than 9 and 6, respectively.

[22] Thus the choices $l > 21$ in V,14, $l > 9$ in V,15, and $6 < l < 21$ in V,16.

# Book VI

**Problem VI,1.** $\begin{cases}(a^3)^2 + (b^2)^2 = \Box, \\ a = mb, \qquad m = 2.\end{cases}$

Putting $b = x$: $\quad 64x^6 + x^4 = \Box[=(nx^3)^2]$.[1]

Hence $n^2 - 64 =$ square (say, $p^2$).

The solution is obvious:

$$n^2 = 100, \quad \left[p^2 = \frac{1}{x^2} = 36\right].$$

The problem is then reconstructed, giving

$$x = \frac{1}{6} = b, \quad (b^2)^2 = \left(\frac{1}{36}\right)^2 = \frac{36}{46{,}656},$$

$$(a^3)^2 = \left(\left(\frac{2}{6}\right)^3\right)^2 = \left(\frac{8}{216}\right)^2 = \frac{64}{46{,}656}, \quad \Box = \frac{100}{46{,}656} = \left(\frac{10}{216}\right)^2.$$

This problem is the first of the large set of interpolated problems which occupies almost half of Book VI. VI,1 repeats the method seen in IV,25, the only differences being that we impose here $a = 2b$ instead of putting $b = 2a$ and set the indeterminate square proportional to $x^6$ instead of to $x^4$.[2]

**Problem VI,2.** $\begin{cases}(a^3)^2 - (b^2)^2 = \Box, \\ a = mb, \qquad m = 2.\end{cases}$

Putting $b = x$: $\quad 64x^6 - x^4 = \Box[=(nx^3)^2]$.

Hence $64 - n^2 =$ square (say, $p^2$).

[1] See the remark on p. 106, n. 55.

[2] The word "our" in "what has been shown previously in our treatise" (line 2181) may go back to the translator. In any event, I do not consider its presence as any proof of the genuineness of the problem.

The solution given, obtained either by multiplying $36 + 64 = 100$ by $\frac{64}{100}$ or by taking $h = 2$ in the application of II,8, is

$$n^2[=(\tfrac{32}{5})^2] = 40\tfrac{24}{25}, \qquad [p^2 = (\tfrac{24}{5})^2 = 23\tfrac{1}{25}].$$

Thus we find $\quad x = \dfrac{5}{24} = b, \quad (b^2)^2 = \left(\dfrac{25}{576}\right)^2 = \dfrac{5625}{2{,}985{,}984},$

$$(a^3)^2 = \left(\left(\frac{5}{12}\right)^3\right)^2 = \left(\frac{125}{1728}\right)^2 = \frac{15{,}625}{2{,}985{,}984}, \quad \Box = \frac{10{,}000}{2{,}985{,}984} = \left(\frac{100}{1728}\right)^2.$$

This (interpolated) problem stems from IV,26,a, from which it differs by the imposed condition $a = 2b$, and, again, by taking the indeterminate square proportional to $x^6$; hence, the intermediate problem is II,8 here instead of II,10.

**Problem VI,3.** $\quad \begin{cases} (b^2)^2 - (a^3)^2 = \Box, \\ a = mb, \qquad m = 2. \end{cases}$

Putting $b = x$, we have

$$x^4 - 64x^6 = \Box[=(nx^3)^2].$$

Considering $n^2 + 64 = p^2$, we know from VI,1 the (simplest) solution $n^2 = 36$. The reconstruction of the problem gives

$$x^2 = \tfrac{1}{100}.$$

So

$$x = \frac{1}{10} = b, \quad (b^2)^2 = \frac{100}{1{,}000{,}000},$$

$$(a^3)^2 = \left(\left(\frac{2}{10}\right)^3\right)^2 = \left(\frac{8}{1000}\right)^2 = \frac{64}{1{,}000{,}000}, \quad \Box = \frac{36}{1{,}000{,}000} = \left(\frac{6}{1000}\right)^2.$$

This problem is the last of the first group of interpolations, i.e., those originating from IV,25–26; it corresponds to IV,26,b. The particular attention bestowed on IV,25–26 perhaps arose from their being the first problems in Book IV leading to the basic methods taught in the group II,8–10.

---

**Problem VI,4.** $\quad \begin{cases} (a^3)^2 + a^3 \cdot b^2 = \Box, \\ a = mb, \qquad m = 5. \end{cases}$

Putting $b = x$, we have

$$15{,}625x^6 + 125x^5 = \Box[=(nx^3)^2].$$

This leads us to consider $n^2 = 15{,}625 + p$, where $p$ is a (simple) number. Hence we may choose for $n^2$ the next (higher) integral square after $15{,}625 = 125^2$, that is, $126^2$; this amounts to putting for $p$ the number $2 \cdot 125 + 1 = 251$.[3]

Thus (as $p = 125/x$), $\qquad x = \frac{125}{251}$.

Then

$$b^2 = \left(\frac{125}{251}\right)^2 = \frac{15{,}625}{63{,}001}, \quad a^3 = \left(\frac{625}{251}\right)^3 = \frac{244{,}140{,}625}{15{,}813{,}251}.$$

The text has here 2,563,001 instead of 15,813,251. There can be no doubt that somebody read as 2,563,001 what was supposed to represent "251 times 63,001". Let us try to trace the origin of this error.

($\alpha$) The confusion cannot have arisen in Arabic times, for all numbers in our text are written in words. Thus, the reading of the Arabic text, "two thousands of thousands and five hundreds sixty-three thousands and one" is unmistakably the translation of what appeared as 2,563,001 in the Greek exemplar.

($\beta$) This must also have been the number read by whoever (probably a Greek: cf. p. 64) added the final statements, since it is repeated in the conclusion.

($\gamma$) From consideration of the Arabic Books, we gather that the author of the major commentary did not leave any results in product-form, except, for practical reasons, a few denominators consisting of identical factors (cf. p. 40). Hence, at the time the commentary was made, "251 times 63001" must already have appeared as "2563001".

But, an explanation for the origin of the corruption itself should still be found. It is conceivable that the error passed through the following steps:

$$\frac{625^3}{251 \cdot 63001}{}^{4} \qquad \frac{625^3}{2563001} \qquad \frac{244140625}{2563001},$$

the first one going back to the author of the problem (interpolated: see below) and the last one to the author of the major commentary. We may tentatively explain the intermediate step by supposing the following: the archetype, omitting a factor, had (in uncials) $\overset{\text{Ϲ}}{\text{M}}$,ΓΑ,[5] and, in the margin, the

---

[3] A so-called γνώμων-number; see, e.g., Aristotle, *Physica*, III.4,203 a 13–15 and Heath, *Math. in Arist.*, pp. 101–2.

[4] Perhaps also $\frac{125^2 \cdot 15625}{251 \cdot 63001}$, or something of that kind.

[5] The numbers placed over the M being "the orthodox way of writing tens of thousands", according to Heath, *Hist. of Gr. Math.*, I, p. 39.

addition CNAEΠIM; this was taken by the next copyist to mean $\overset{\text{CNAC}}{\text{M}}$,ΓA, and the then meaningless A was thus carelessly dismissed.[6]

N.B. The final phrase of the problem, by which the verification, i.e., the computation of the indeterminate square is eluded, should not be taken as an "easy-way-out" explanation left by the author of the major commentary after he failed to obtain a square result. It is unreasonable to expect him to have carried such a lengthy computation further: earlier in the treatise he has left uncomputed numbers of magnitude comparable to those of the above $(a^3)^2$ and $a^3b^2$ (cf. pp. 49–50).

As already observed, VI,4 was inspired by a problem formulated earlier in a corollary (to IV,33: Cor. $2^a$; see pp. 205–206), the difference being merely that we impose here the ratio of $a$ to $b$ and take unity as the value of the given number.

**Problem VI,5.**
$$\begin{cases}(b^2)^2 + a^3 \cdot b^2 = \square, \\ a = b.\end{cases}$$

We have then
$$a^4 + a^5 = \square.$$

The method is clear. With $a = x$:
$$x^4 + x^5 = \square =, \text{ say, } (2x^2)^2;$$
then
$$x^4 + x^5 = 4x^4, \quad \text{and} \quad x = 3.$$
So
$$a^3 = 27, \quad b^2(=a^2) = 9, \quad \square = 324 = 18^2.$$

The origin of this problem is IV,33, Cor. $1^a$ (with $k = 1$), which means that VI,5 and VI,4 have the same source; hence, despite the even greater banality (equating $a$ to $b$), this (and the next two problems) may well go back to the same commentator who added VI,4.

**Problem VI,6.**
$$\begin{cases}a^3 \cdot b^2 - (a^3)^2 = \square, \\ a = b.\end{cases}$$

We have then
$$a^5 - a^6 = \square, \quad \text{or, with } a = x\text{:} \quad x^5 - x^6 = \square.$$

Putting $\square = (x^3)^2$:
$$x^5 - x^6 = x^6, \quad \text{and} \quad x = \tfrac{1}{2}.$$

---

[6] There is, of course, some arbitrariness in assuming this last point; but, again, the whole explanation is no more than tentative.

Note that the combination $\overset{\text{A}}{\text{M}}$ meaning μυριάδες ἁπλαῖ, used in manuscripts (see Pappus, *Collectio* (Hultsch), III, ii, p. 130; Rome, *Comm.* (1–2), p. 397), cannot have played a rôle here if our attempt at reconstruction is tenable (see previous note).

So $$b^2 = \tfrac{1}{4}, \quad a^3 = \tfrac{1}{8}, \quad \square = \tfrac{1}{64} = (\tfrac{1}{8})^2.$$

The origin of this problem, complementary to VI,4, is IV,33, Cor. 2[c]. Note the unimaginative choice of unity as the numerical factor in $\square$.

**Problem VI,7.** $$\begin{cases} a^3 \cdot b^2 - (b^2)^2 = \square, \\ a = b. \end{cases}$$

We have then, with $a = x$:

$$x^5 - x^4 = \square.$$

Putting $\square = (x^2)^2$:

$$x^5 - x^4 = x^4, \quad \text{and} \quad x = 2.$$

So $$b^2 = 4, \quad a^3 = 8, \quad \square = 16.$$

This problem, closely related to VI,5, corresponds to IV,33, Cor. 1[c].

---

**Problem VI,8.** $$a^3 \cdot b^2 + \sqrt{a^3 \cdot b^2} = \square.$$

We put $a^3 = 64$ and $b^2 = x^2$; hence

$$64x^2 + 8x = \square.$$

We put $\square = (nx)^2$, with $n > 8$, say $\square = (10x)^2$. Then:

$$8x = 36x^2 \quad \text{and} \quad x = \tfrac{2}{9}.$$

So $$b^2 = \tfrac{4}{81} \quad \text{and} \quad \square = \tfrac{400}{81} = (\tfrac{20}{9})^2.$$

This problem (and consequently the two following ones) I also consider to be interpolated. They are presumably variations on the preceding ones, the single power being replaced by the term $\sqrt{a^3 b^2}$ ($a^3$ must therefore be a sixth power).

**Problem VI,9.** $$a^3 \cdot b^2 - \sqrt{a^3 \cdot b^2} = \square.$$

We put $a^3 = 64$ and $b^2 = x^2$, so that

$$64x^2 - 8x = \square.$$

We put $\square = (nx)^2$, with $n < 8$, say $\square = (7x)^2$. Then:

$$15x^2 = 8x \quad \text{and} \quad x = \tfrac{8}{15}.$$

So $$b^2 = \tfrac{64}{225}, \quad \square = \tfrac{3136}{225} = (\tfrac{56}{15})^2.$$

**Problem VI,10.** $$\sqrt{a^3 \cdot b^2} - a^3 \cdot b^2 = \square.$$

We put $a^3 = 64$, $b^2 = x^2$, so that

$$8x - 64x^2 = \square.$$

We put $\square = (nx)^2$, say $(4x)^2$, hence

$$8x = 80x^2 \quad \text{and} \quad x = \tfrac{1}{10}.$$

So

$$b^2 = \tfrac{1}{100}, \quad \square = \tfrac{16}{100} = (\tfrac{4}{10})^2.$$

---

**Problem VI,11.** $(a^3)^2 + a^3 = \square.$

We put $a = x$, so we have

$$x^6 + x^3 = \square.$$

I have added at this point a necessary condition, which, however, may not have been in the original version: we shall put $\square = (nx^3)^2$, with $n$ such that $n^2 - 1 =$ cube.

The value $n = 3$ fits, hence $\square = (3x^3)^2$; then

$$x^3 = 8x^6, \quad x^3 = \tfrac{1}{8}, \quad a = x = \tfrac{1}{2} \quad \text{and} \quad \square = \tfrac{9}{64} = (\tfrac{3}{8})^2.$$

This problem is odd. I am convinced that it must be an interpolation, possibly originating from VI,4 (with $b = 1$) or from VI,8 (with $b^2 = a^3$).

I have, as indicated above, added the condition, for the text clearly requires some emendation. But a subsequent remark that, with the choice $\square = (3x^3)^2$, "the problem will be soluble and the treatment will not be impossible", tends to suggest at first that the missing passage might have contained something more than the simple exposition of a condition. But what could the content of the missing passage have been? There is indeed little to say: the solution to the problem $n^2 - 1 =$ cube which the text gives is rather obvious, and is, furthermore, the only one.[7] The discussion might have consisted in trying to put first $\square = n^2$, and then $\square = n^2x^2$ (or $n^2x^4$), the conclusion being that these trials are fruitless.[8]

Since, however, such a discussion is not in the style of the interpolated problems (and serves hardly any purpose), one is inclined to wonder whether the whole passage, starting at the beginning of line 2416 and ending in line 2418, is not an interpolation. But if so, the problem in its original form would have had little if any value because of the absence of any allusion to the particularity of its solution. Given the level of the previous interpolations, this is far from impossible.

With the next problem, we return to somewhat more solid ground, and to what must once have been the beginning of Book VI.

---

[7] $n = \pm 1$ and $n = 0$ are of course out of consideration.

[8] The *min ʿadad* of line 2414 could account for the trial $\sqrt{\square} = n$; perhaps for $\sqrt{\square} = nx$ were there ever a mistranslation stemming from the ambiguity of ἐν ἀριθμοῖς (see p. 67, n. 40).

**Problem VI,12.**
$$\begin{cases} a^2 + \dfrac{a^2}{b^2} = \square, \\ b^2 + \dfrac{a^2}{b^2} = \square', \end{cases} \qquad \text{with } a > b.$$

We put $b = x$, so we have

$$\begin{cases} (1) & a^2 + \dfrac{a^2}{x^2} = \square, \\ (2) & x^2 + \dfrac{a^2}{x^2} = \square'. \end{cases}$$

Let us consider the second equation; if we put

$$\frac{a^2}{x^2} = \frac{9}{16}x^2$$

we shall have:
$$x^2 + \frac{a^2}{x^2} = \frac{25}{16}x^2,$$

so that the said equation will be identically satisfied. Equation (1) becomes:

$$a^2 + \tfrac{9}{16}x^2 = \square;$$

but, we have from above: $a^2 = \tfrac{9}{16}x^4$,

so
$$\tfrac{9}{16}x^4 + \tfrac{9}{16}x^2 = \square[=, \text{say}, m^2x^4].^9$$

Hence $m^2 - \tfrac{9}{16}[= \tfrac{9}{16} \cdot 1/x^2] =$, say, $p^2$, so that we have to solve

$$m^2 - p^2 = \tfrac{9}{16}.$$

The (restored) text states the condition $p^2 < \frac{81}{256}$. For, since we want $a$ to be larger than $b$, we must have

$$\frac{a^2}{b^2} = \frac{a^2}{x^2} > 1, \quad \text{or} \quad \tfrac{9}{16}x^2 > 1, \quad \text{hence} \quad \frac{1}{x^2} < \frac{9}{16}$$

and
$$p^2 \equiv \frac{9}{16} \cdot \frac{1}{x^2} < \frac{81}{256}.$$

The solution is given immediately; it could be obtained by using II,10:

$$m^2 - p^2 = \tfrac{9}{16} = (p + h)^2 - p^2, \quad \text{so} \quad p = \frac{\tfrac{9}{16} - h^2}{2h} \qquad (h < \tfrac{3}{4}).$$

For $h = \frac{1}{2}$: $p = \frac{5}{16}$, $p^2 = \frac{25}{256}$, acceptable value, and $m^2 = \frac{169}{256}$. The problem is then reconstructed, and we obtain:

$$x^2[=(\tfrac{12}{5})^2] = \tfrac{144}{25} = 5\tfrac{19}{25}.$$

---

[9] See the remark on p. 106, n. 55.

So

$$b^2 = \frac{144}{25}, \quad a^2\left[= \left(\frac{108}{25}\right)^2\right] = \frac{11{,}664}{625}, \quad \square = \frac{13{,}689}{625} = \left(\frac{117}{25}\right)^2,$$

$$\square' = \frac{5625}{625} = \left(\frac{75}{25}\right)^2.$$

**Remark.** As already pointed out in the translation, the manuscript incorrectly gives the condition for $p^2$ as $p^2 < 1$. There is, however, little doubt that the original text had $p^2 < \frac{81}{256}$. For the missing part corresponds to a homoeoteleuton in Arabic, and it is difficult to imagine Diophantus' having drawn attention to a condition without having established it.

**Problem VI,13.**

$$\begin{cases} a^2 - \dfrac{a^2}{b^2} = \square, \\ b^2 - \dfrac{a^2}{b^2} = \square', \end{cases} \qquad \text{with } a > b.$$

We put $b = x$, so that we have

$$\begin{cases} (1) & a^2 - \dfrac{a^2}{x^2} = \square, \\ (2) & x^2 - \dfrac{a^2}{x^2} = \square'. \end{cases}$$

Let us consider the second equation, which we shall, as before, satisfy identically. Choosing the usual decomposition $1 = \frac{16}{25} + \frac{9}{25}$, we have

$$x^2 = \tfrac{16}{25}x^2 + \tfrac{9}{25}x^2,$$

so that

$$x^2 - \tfrac{9}{25}x^2 = \tfrac{16}{25}x^2;$$

take:

$$\frac{a^2}{x^2} = \frac{9}{25}x^2, \quad \text{hence} \quad a^2 = \frac{9}{25}x^4.$$

This into (1):

$$\tfrac{9}{25}x^4 - \tfrac{9}{25}x^2 = \square \; [=, \text{say}, m^2x^4].$$

Thus $\frac{9}{25} - m^2 = [\frac{9}{25}(1/x^2) =]$, say, $p^2$, or $p^2 + m^2 = \frac{9}{25}$, the solution of which is immediately given in the text. We can obtain it by applying II,8:

$$\tfrac{9}{25} = y^2 + (hy - \tfrac{3}{5})^2 = y^2 + h^2y^2 - \tfrac{6}{5}hy + \tfrac{9}{25},$$

hence

$$y = \frac{\frac{6}{5}h}{h^2 + 1} \qquad (h \neq 1).$$

With $h = 2$:

$$y = \tfrac{12}{25}, \qquad y^2 = \tfrac{144}{625},$$

and

$$hy - \tfrac{3}{5} = \tfrac{9}{25}, \qquad (hy - \tfrac{3}{5})^2 = \tfrac{81}{625}.$$

With $m^2 = \frac{81}{625}$, thus $p^2 = \frac{144}{625}$, we reconstruct the problem and obtain

$$x^2 = \tfrac{25}{16}.$$

Then: $\quad b^2 = \frac{25}{16}, \quad a^2 = \frac{225}{256}, \quad \Box' = 1, \quad \Box = \frac{81}{256} = (\frac{9}{16})^2.$

But $b^2 = \frac{25}{16} = \frac{400}{256}$ is larger than $a^2$ (and hence $\Box' > \Box$), which is contrary to the requirement.

Diophantus says that (despite this unfortunate outcome) he has reproduced the treatment because it is correct. Let us interrupt the resolution of the problem at this point in order to consider how well founded Diophantus' assertion is—assuming that it is his.

At the very beginning, when we used the identity $1 = \frac{16}{25} + \frac{9}{25}$ to satisfy the second given equation, we had the possibility of putting

(A) either $$\frac{a^2}{x^2} = \frac{9}{25}x^2,$$

(B) or $$\frac{a^2}{x^2} = \frac{16}{25}x^2.$$

Since we want $a > b = x$, case (A) is subject to the condition $x > \frac{5}{3}$ and case (B) to $x > \frac{5}{4}$.

Choosing possibility (A), as does Diophantus, leads us next to fulfil the first equation; hence

$$\tfrac{9}{25}x^4 - \tfrac{9}{25}x^2 = \Box \equiv m^2x^4,$$

thus the condition $m^2 + p^2 = \frac{9}{25}$ where $p^2 = \frac{9}{25} \cdot 1/x^2$. Since (by II,8, with $h = 2$) $\frac{81}{625} + \frac{144}{625} = \frac{9}{25}$, we have the choice of taking $p^2 = \frac{81}{625}$ or $p^2 = \frac{144}{625}$; the first choice gives $x = \frac{5}{3}$ and the second one $x = \frac{5}{4}$, neither of which satisfies the condition $x > \frac{5}{3}$ encountered above. Diophantus chose the second value, $p^2 = \frac{144}{625}$, thus obtaining $b > a$ (the first giving $b = a$).

Hence, holding to possibility (A) compels us to use other solutions of $m^2 + p^2 = \frac{9}{25}$, that is, to use other values of the parameter $h$ in applying II,8 (e.g., $h$ integral $\geq 4$). But we may also choose possibility (B), that is,

$$\frac{a^2}{x^2} = \frac{16}{25}x^2, \quad \text{with} \quad x > \tfrac{5}{4}.$$

The new form taken by the first equation is then

$$\tfrac{16}{25}x^4 - \tfrac{16}{25}x^2 = \Box \equiv m^2x^4,$$

or $$m^2 + p^2 = \tfrac{16}{25} \qquad \left(p^2 = \frac{16}{25} \cdot \frac{1}{x^2}\right),$$

the simplest solution of which is (using II,8, with $h = 2$, or multiplying the above solution by $\frac{16}{9}$):

$$\tfrac{256}{625} + \tfrac{144}{625} = \tfrac{16}{25}.$$

Since we want $x > \frac{5}{4}$, hence $1/x^2 < \frac{16}{25}$, we must have $p^2 < \frac{256}{625}$, so that we shall take

$$p^2 = \tfrac{144}{625}, \qquad m^2 = \tfrac{256}{625};$$

hence $x = \frac{5}{3}$, a value already found in case (A), but which is now acceptable since the inferior limit for $x$ is lower.

In the second part of the problem, Diophantus reaches the solution which we have just calculated, without making any attempt to discover the source of his error in the first method. His second and shorter method consists in fulfilling identically the *first* equation, that is

$$a^2\left(1 - \frac{1}{b^2}\right) = \frac{a^2}{b^2}(b^2 - 1) = \text{square},$$

which amounts to solving

$$b^2 - 1 = \text{square};$$

an obvious solution is $b^2 = \frac{25}{9}$.

It is surprising that Diophantus chose to consider first the less convenient equation linking the terms $b^2$ and $a^2/b^2$ in VI,12 and, especially, in VI,13, where it apparently led him into confusion. But, as we shall see, this is not the only baffling element in Book VI.

Taking thus the value $b = \frac{5}{3}$ leads to

$$a^2 - \frac{a^2}{b^2} = a^2 - \frac{9}{25}a^2 = \frac{16}{25}a^2.$$

The remaining unknown $a = x$ is then determined from the second equation,

$$b^2 - \frac{a^2}{b^2} = \frac{25}{9} - \frac{9}{25}x^2 = \square',$$

by putting $\square' = (\frac{5}{3} - hx)^2$; $h$ is taken to be equal to $\frac{6}{5}$.[10] This gives $x = 2\frac{2}{9}$.

So

$$a^2 = (2\tfrac{2}{9})^2 = \tfrac{400}{81}, \quad b^2 = (\tfrac{5}{3})^2 = \tfrac{25}{9}, \quad \square = \tfrac{256}{81} = (\tfrac{16}{9})^2, \quad \square' = 1.$$

**Problem VI,14.**

$$\begin{cases} (1) \quad \dfrac{a^2}{b^2} - b^2 = \square, \\ (2) \quad \dfrac{a^2}{b^2} - a^2 = \square', \end{cases} \qquad \text{with } a > b.$$

Putting $b = \frac{4}{5}$, the second equation gives:

$$\tfrac{25}{16}a^2 - a^2 = \tfrac{9}{16}a^2$$

and is thus identically satisfied.

[10] $\frac{1}{5} < h < \frac{9}{5}$ must hold in order that $a = x > b = \frac{5}{3}$.

The said value in (1) gives, with $a = x$:

$$\tfrac{25}{16}x^2 - \tfrac{16}{25} = \square =, \text{ say, } (1\tfrac{1}{4}x - 2)^2.$$[11]

So, $$5x = 4\tfrac{16}{25}, \quad \text{and} \quad x = \tfrac{116}{125}.$$

Then:

$$b^2 = \frac{16}{25}, \quad a^2 = \left(\frac{116}{125}\right)^2 = \frac{13{,}456}{15{,}625}, \quad \square = \frac{11{,}025}{15{,}625} = \left(\frac{105}{125}\right)^2,$$

$$\square' = \frac{7569}{15{,}625} = \left(\frac{87}{125}\right)^2.$$

---

**Problem VI,15.** $$\begin{cases} (1) & a^2 + (a^2 - b^2) = \square, \\ (2) & b^2 + (a^2 - b^2) = \square', \end{cases} \quad \text{with } a > b.$$

We put $x$ for the larger number, $a$.

(2) is satisfied identically;[12] (1) will be satisfied if we put

$$a^2 - b^2 = 2x + 1.$$

We must now make $b^2 = x^2 - 2x - 1$ a square; this is done by setting it equal to $(x - 2)^2$.[13]

Thus we obtain $$x = 2\tfrac{1}{2}.$$

So $$a^2 = (2\tfrac{1}{2})^2 = 6\tfrac{1}{4}, \quad b^2 = \tfrac{1}{4}, \quad \square = 12\tfrac{1}{4} = (3\tfrac{1}{2})^2, \quad \square'(\equiv a^2) = 6\tfrac{1}{4}.$$

**Problem VI,16.** $$\begin{cases} (1) & a^2 - (a^2 - b^2) = \square, \\ (2) & b^2 - (a^2 - b^2) = \square', \end{cases} \quad \text{with } a > b.$$

We put $x$ for the larger number, $a$.

(1) is satisfied identically. Putting $a^2 - b^2 = 2x - 1$, we shall have

$$b^2 = x^2 - 2x + 1 = (x - 1)^2,$$

which is a square, smaller than $a^2$ for $x > \tfrac{1}{2}$.

There remains the fulfilment of (2).

$$\begin{aligned} b^2 - (a^2 - b^2) &= x^2 - 2x + 1 - (2x - 1) \\ &= x^2 - 4x + 2 = \text{square} =, \text{ say, } (x - 4)^2 \\ &= x^2 - 8x + 16, \end{aligned}$$

hence $$x = 3\tfrac{1}{2}.$$

So $$a^2 = (3\tfrac{1}{2})^2 = 12\tfrac{1}{4}, \quad b^2 = 6\tfrac{1}{4}, \quad \square(\equiv b^2) = 6\tfrac{1}{4}, \quad \square' = \tfrac{1}{4}.$$

---

[11] Generally, $(1\tfrac{1}{4}x - h)^2$, with any (rational and positive) value for $h$ save those comprised between $\tfrac{4}{5}$ and $\tfrac{8}{5}$ (inclusive) in order that $a$ be larger than $b$.

[12] There is no allusion to this in the text, and the second equation is simply *ignored* in the resolution (the verification of its fulfilment no doubt goes back to the author of the major commentary; cf. p. 69). This is also true of the first equation in the next proposition.

[13] Or, generally, $(x - h)^2$, with $h > 1$ for $x > 0$ (thus $a > b$).

## General Remarks on Book VI: Comparative Weakness and Presumed Purpose

**A.** Of the four extant Arabic Books, Book VI is undoubtedly the weakest. Some of its (genuine) propositions are treated awkwardly while several others are of limited interest—so much so that it is surprising to see material of this level at this stage of the *Arithmetica*, which, in theory, runs progressively from more simple to more difficult, or at least does not regress. The following examples illustrate the Book's weakness.

($\alpha$) Propositions 17 and, particularly, 22 are so unimaginative as to be hardly less trivial than interpolated propositions.

($\beta$) VI,15–16 are also elementary; but what distinguishes them is that, in both cases, one of the equations is *useless* since it is identically satisfied as it stands.

($\gamma$) The less simple treatments are found in VI,12–13. But their (relative) difficulty originates rather with the clumsy approach which Diophantus deliberately chose[16] than with any intrinsic difficulty; furthermore, this approach is applied unskilfully in one instance (VI,13, first part) and, moreover, Diophantus appears to have been unable to trace the origin of its failure.

In addition to these fundamental weaknesses, there are some details which betray a certain carelessness: equations which could be simplified through the elimination of a quadratic factor are not (cf. problems 12, 13, 18, 19), whereas they generally are in the extant Greek Books;[17] an important condition for solving the final equation in VI,22,1° is not mentioned.[18]

**B.** Since it is not the form, the external aspect, which links the genuine problems of Book VI to one another,[19] we must seek a common trait in their treatments: after all, Diophantus must have had some reason for putting together these apparently disparate problems. Examination of the genuine propositions VI,12–23 shows that the following elements are used in their resolutions:

(a) some elementary identities—but of a type not at all particular to Book VI;

(b) the simplest solution of the Pythagorean equation, $9 + 16 = 25$ (also multiplied by $\frac{1}{16}$ and $\frac{1}{25}$); it plays an essential rôle in many of the resolutions, but is not used throughout;

[16] Remember that the whole group VI,12–14 is soluble in two ways, depending on which of the two proposed equations one chooses to satisfy first.

[17] See the complete reductions in II,29 and III,16, and the partial ones in II,28 and "IV",31,1°.

[18] The case of the numerical limits for the choice of the parameters occurring in the final equations (some of which we have indicated in the commentary) is different: they are not given in the text, but they are not given regularly in the Greek Books either.

[19] The occurrence of square powers of the unknowns only is not a satisfactory argument.

(c) the methods taught in II,8 and II,10; they are found only in VI,12 and VI,13 (first part: the "awkward" approach);

(d) the resolution of an equation of the type $Ax^2 + Bx + C = \square$ where either $A$ or $C$ is a square (not nil), performed by setting for $\square$ a suitable trinomial, a resolution well known from Book II (cf. p. 7); except in VI,12, in the first part of VI,13,[20] and in VI,22, all the genuine problems end with the resolution of an equation of this type. Such an equation had not been encountered in any of the two previous Books' problems, despite their frequent reduction to methods from Book II.

Perhaps the *raison d'être* of Book VI was to familiarize the reader with the use of such an equation—even if it had often been met with in Books II and III. At all events, I do not see any other justification for Book VI's presence in the *Arithmetica*—assuming that we are entitled or obliged to justify its presence.

---

[20] That is, those parts of the group VI,12–14 in which the "awkward" approach mentioned above is employed.

# Book VII

## The Introduction

The introduction to Book VII consists of a single sentence, which is to all appearances genuine—although the elucidation of its meaning poses some difficulty. In it the following three points are made.

I. *There will be many problems in the present* (*seventh*) *Book.*

The use of the word "many" here is odd since, not counting the interpolated problems in Book VII, there are only twelve. Any supposition that part of Book VII might have been lost being purely conjectural and without any positive evidence to support it, we can do no more than question the appropriateness of the word "many".

II. *The type* (*jins*) *of the coming problems will not depart from the type of problems seen previously in the fourth and fifth Books—even if they are different in species/appearance* (*nau*ᶜ).

The words *jins* and *nau*ᶜ most probably correspond to the Greek words γένος and εἶδος. For, γένος and *jins* (which stems from the former via Syriac) are natural correspondents, while *nau*ᶜ is the usual translation (in particular in our Arabic Books) of εἶδος. Now, in common language, εἶδος refers to the (exterior) aspect, the form which is seen, while, in philosophy, εἶδος is a sub-kind, a species, of the kind (*genus* = γένος). We must understand then, from point II above, that the problems of Book VII have some fundamental trait in common with those of Books IV and V, but that they differ either externally (in form) or internally (in treatment).

(α) It is easy to see that the difference alluded to cannot be external; for the principal varieties of problems of Book VII are no more—or no less—different in aspect from those of Books IV–V than are the problems of any

Book from those of any other Book. Thus, there is no justification for establishing a comparison based on external features.

We must consider then that the difference between the problems of Book VII and those of Books IV and V is an internal one and related to the resolution. The shortness of Book VII makes it seem likely that this "difference in species" may take the form of some peculiarity shared by the (noninterpolated) problems VII,7–18, thus justifying their placement in this Book; for, at first sight, some of the problems might belong to other Books (for example VII,11–15 to Book II or III, and VII,16–18 to Book VI).

In point of fact, a certain connection does come to light. Considering first VII,7–15, we observe that the solution set of the indeterminate system of the second degree (either proposed, in VII,11–15, or to which VII,7–10 are reduced) is *determined up to an arbitrary quadratic factor*, and thus the unknown which is raised to an even power may be given or arbitrarily chosen. And it is this that links the outwardly quite different set VII,16–18 to the previous problems, since, for the same reason, we are entitled to choose *a priori* the numerical value of one of the unknowns, as Diophantus does at the beginning of the resolutions.

($\beta$) Thus it is possible to show a "difference in species" by bringing out the specific character of Book VII. But that still leaves the question of what Books IV, V, and VII can have in common that is *not* found in Book VI, which would explain why Book VI was not mentioned by Diophantus together with the other three. Diophantus certainly does not mean that, after the relatively mediocre set of Book VI, he will present more interesting propositions. Nor does he mean to say that he will return to more classical types, for the greater and more characteristic part of Book V (V,7–16) is certainly not classical in comparison with the problems of the previous Book or Books. On the whole, consideration of the form and treatment in the Arabic Books does not show any more cohesion between Books IV, V, and VII than fundamental difference(s) between them (or any one of them) and Book VI.

The last part of the introduction speaks of the educational rôle of Book VII.

III. *The problems of Book VII are aimed at increasing "experience and skill"*.

Observe that the Arabic words rendered here by "experience and skill" are precisely those found in the preface to Book IV, so that they undoubtedly correspond to the same Greek words (word in the case of an ἓν διὰ δυοῖν). Since that part of the introduction to Book IV is surely genuine (Diophantine), this parallelism of expression strongly speaks for the genuineness of the introduction to Book VII. The introductions to Books IV and VII are thus linked by a common point—the reader will once again encounter problems leading him to greater dexterity in problem-solving—just as are the introductions to Books I and IV (see pp. 175–176).

The introduction to Book VII would take on greater significance if we could understand point II to refer to Books IV, V, *and* VI, that is, if we could supplement the extant text with *wa'l-sādis* ("and the sixth"), words conceivably omitted by a careless Arabic copyist.[1] The trait or "type" (*jins*) shared by the problems of Books IV to VII would then be their resolution by means of methods taught in the first Greek Books, and of these methods only, the acknowledged purpose of the four Books being thus to enlarge their field of application and to increase the readers' "experience and skill". In this respect, the later Greek Books differ from the preceding ones, for there we find problems requiring other techniques, as, for example, in Book "IV", where we learn how to remodel the initial hypotheses after obtaining an irrational solution (a procedure also used in Books "V" and "VI"), or in Book "V", where we learn the quite elaborate technique of the παρισότητος ἀγωγή.

The addition of *wa'l-sādis* thus has the advantage of giving, on the whole, a fairly plausible explanation of the meaning of the introduction to Book VII. But, though minor insofar as the establishment of the critical text is concerned, this addition has a major impact on the sense of the passage, and it is for this reason that we have chosen not to alter the text in line 2924.

---

As already mentioned, we must go through a certain number of interpolations before reaching the problems of the original Book VII.

**Problem VII,1.**

$$\begin{cases} a^3 \cdot b^3 \cdot c^3 = \square, \\ a = mb, \qquad\qquad m = 2. \\ b = mc, \end{cases}$$

We put $c = x$. Then, $c^3 = x^3$, $b^3 = (2x)^3 = 8x^3$, $a^3 = (4x)^3 = 64x^3$, and the problem is reduced to

$$512x^9[= 2^9x^9] = \square \equiv (nx^4)^2.$$

The text chooses a coefficient $n^2$ leading to a simple value for $x$, namely $32^2 = 1024\ (=2^{10})$, giving

$$x = 2.$$

So $c^3 - 2^3 = 8$, $b^3 = 4^3 = 64$, $a^3 = 8^3 = 512$, $\square = 262{,}144 = 512^2$.

**Problem VII,2.** $(a^2)^3 \cdot (b^2)^3 \cdot (c^2)^3 = \square^2$.

Since $64 = (2^2)^3$, if we put $(a^2)^3 = \frac{1}{64}$ and $(b^2)^3 = 64$, only the condition

$$(c^2)^3 = \square^2 \quad \text{will remain;}$$

the text does this and, with $c = x$, the problem becomes

$$x^6 = \square^2, \quad \text{or} \quad x^3 = \square.$$

---

[1] The word *ḫāmis* and the word *sādis* (which supposedly followed it), though not easily confusable, are somewhat similar. Or could the omission go back to Greek times (confusion between E and Ϛ)?

Putting $\square = 4x^2$, we have $x = 4.$

So $(c^2)^3 = 4096,$ and also $\square^2 = 4096 = 64^2 = (8^2)^2.$

**Remark.** A solution could also be obtained by squaring the results of the preceding problem.

---

**Problem VII,3.** $(a^2)^2 = a_1^3 + a_2^3 + a_3^3.$

We put $a^2 = x^2$, hence $x^4 = a_1^3 + a_2^3 + a_3^3.$

Assuming

$$a_1^3 = x^3, \quad a_2^3 = (2x)^3 = 8x^3, \quad a_3^3 = (4x)^3 = 64x^3,$$

we obtain $73x^3 = x^4,$ or $x = 73.$

Then:

$$(a^2)^2 = (73^2)^2 = 5329^2 = 28{,}398{,}241, \quad a_1^3 = 73^3 = 389{,}017,$$

$$a_2^3 = [146^3 =]3{,}112{,}136, \quad a_3^3 = [292^3 =]24{,}897{,}088.$$

**Remark.** Perhaps the author of the problem had in mind the further condition $a_2 = 2a_1, a_3 = 2a_2$, as in VII,1. For he could have put $a_3 = 3a_1$, which leads to the more convenient solution $x = 36$.

**Problem VII,4.** $(a^2)^3 = a_1^2 + a_2^2 + a_3^2.$

We put $a^2 = x^2$, hence

$$(a^2)^3 = x^6 = a_1^2 + a_2^2 + a_3^2.$$

Taking $a_1^2 = u_1^2x^4$, $a_2^2 = u_2^2x^4$, $a_3^2 = u_3^2x^4$, our problem is reduced to an equation for the coefficients:

$$u_1^2 + u_2^2 + u_3^2 = \square.$$

The solution, directly given in the text, is easily obtainable (see p. 255). We can use the method of III,5 and take first $u_1^2 = 1$, $u_2^2 = 4$; thus $u_3^2 + 5 = \square \equiv (u_3 + m)^2$, yielding, with $m = \frac{5}{3}$, the value $u_3^2 = \frac{4}{9}$.[2]

Hence (after reconstruction of the problem in the text)

$$x^2(=u_1^2 + u_2^2 + u_3^2) = \tfrac{49}{9}.$$

So $$(a^2)^3 = \left(\frac{49}{9}\right)^3 = \frac{117{,}649}{729},$$

$$\left[a_1^2 = u_1^2x^4 = \frac{21{,}609}{729}, \quad a_2^2 = u_2^2x^4 = \frac{86{,}436}{729}, \quad a_3^2 = u_3^2x^4 = \frac{9604}{729}\right].$$

[2] In fact, the solution was probably obtained from the triple {4, 9, 36} (occurring in III,5) by division.

The three figures given in brackets do not appear in the text, which instead computes the magnitudes $u_1^2 x^2$, $u_2^2 x^2$, and $u_3^2 x^2$. We have attributed this error to the author of the major commentary and not to the author of the problem (cf. p. 64, no. 7), even though it is surprising that the latter would not have computed these parts, which are, together with $(a^2)^3$, required magnitudes.[3] Perhaps the text became corrupted at some stage and the values were recomputed by the author of the major commentary in the above way.

---

**Problem VII,5.** $\quad (a^3)^3 \cdot b^3 + (a^3)^3 \cdot c^2 = \square.$

We put $(a^3)^3 = (2^3)^3 = 512$; then

$$512b^3 + 512c^2 = \square.$$

The author of the problem simply takes $b = x, c = x$, thus obtaining

$$512x^3 + 512x^2 = \square \equiv (nx)^2 \quad (n^2 > 512, \text{ not stated in the text}).$$

For $n = 64$: $\quad 512x^3 + 512x^2 = 4096x^2.$

Hence $\quad x = 7,$

and $\quad b^3 = x^3 = 343, \quad c^2 = x^2 = 49, \quad \square = 200{,}704 = 448^2.$

This problem is not only odd in form, but also in treatment, as is seen in particular in the oversimplification of setting the two unknowns $b$ and $c$ equal (cf. VI,5–7). Thus, VII,5 has all the characteristics of an interpolation, and, as interpolations generally appear in groups, one's suspicions aroused by considering the initial problems of Book VII are strengthened. For, of these, only VII,4 (not considering the miscomputations) would deserve to figure among the problems of the *Arithmetica.*

---

**Problem VII,6.** $\quad \begin{cases} \dfrac{a^2 \cdot b^2}{a^2 + b^2} = r, & r, \text{ given ratio,} \\ a^2 + b^2 = \square. \end{cases}$

*Condition*: The number belonging to the given ratio must be a square.

Thus the system

$$\begin{matrix} (1) \\ (2) \end{matrix} \quad \begin{cases} a^2 \cdot b^2 = k^2(a^2 + b^2), & k^2 = 9, \\ a^2 + b^2 = \square. \end{cases}$$

---

[3] Diophantus himself, whose text is throughout characterized by conciseness and brevity, regularly gives the values of the magnitudes actually required, except in some problems of Book "V" which have abbreviated resolutions. An exception is III,10 (possibly the three required numbers are not given there because two of them happen, unfortunately, to have the same value).

If we put $\square = x^2$, with $a^2 = \frac{16}{25}x^2$ and $b^2 = \frac{9}{25}x^2$, (2) will be satisfied identically. Then, (1) becomes

$$\tfrac{144}{625}x^4 = 9x^2,$$

hence
$$x^2 = 39\tfrac{1}{16}.$$

So $\quad a^2 = \frac{16}{25}x^2 = 25, \quad b^2 = \frac{9}{25}x^2 = 14\frac{1}{16}, \quad \square = 39\frac{1}{16} = (6\frac{1}{4})^2.$

This problem probably arose from a scholiast's considerations about the lemma given in VI,23: one is reminded of the expression

$$\frac{k^2(a^2+b^2)}{a^2\cdot b^2} \quad \text{with } a^2+b^2 = \text{square},$$

obtained by verifying the said lemma. The agreement in some numerical values confirms this impression.

Thus, VII,6 must also belong to the group of interpolated problems in Book VII. Note that it is the only one with a visible origin: except for VII,2 and 4, which are related to their immediate predecessors, no link can be made with previous problems.[4] A scholiast might here have simply tried to devise problems of his own.

We now come to the problems apparently belonging to the original Book VII.[5]

---

**Problem VII,7.**
$$\begin{cases} (a^3)^2 = a_1 + a_2 + a_3, \\ a_1 + a_2 = \square, \\ a_2 + a_3 = \square', \\ a_3 + a_1 = \square''. \end{cases}$$

1°. Taking $a = x$, we shall have

$$x^6 = a_1 + a_2 + a_3$$

under the said conditions for the $a_i$'s.

Putting $a_i = u_i \cdot x^4$, the problem is reduced to finding three numbers $u_1, u_2, u_3$ such that their sum and the sum of any two are squares. This has been solved in III,6 with the solution

$$u_1 = 80, \qquad u_2 = 320, \qquad u_3 = 41.$$

We have then:
$$x^2 = 441$$

[4] A link with subsequent problems (say VII,(3–)4 with VII,7) is not altogether evident and would be unusual.

[5] A clear subdivision into groups is in fact noticeable only from VII,8 on. Although some suspicion might be raised about problem 7, there is no serious reason for considering it to be interpolated (see below).

(the problem is reconstructed in the text in order to yield the solution). So

$$a_1 = u_1x^4 = 80 \cdot 194{,}481 = 15{,}558{,}480, \quad a_2 = u_2x^4 = 62{,}233{,}920,$$

$$a_3 = u_3x^4 = 7{,}973{,}721,$$

$$(a^3)^2 = a_1 + a_2 + a_3 = 85{,}766{,}121 = 9261^2 = (21^3)^2,$$

$$a_1 + a_2 = 77{,}792{,}400 = 8820^2, \quad a_2 + a_3 = 70{,}207{,}641 = 8379^2,$$

$$a_3 + a_1 = 23{,}532{,}201 = 4851^2.$$

2°. Another method (which is easier, the text says).

We put $(a^3)^2 = 64 = (2^3)^2$, to be divided as above. From III,6, we have: $320 + 80 + 41 = 441$, with the required properties for the parts.

But the number to be divided is 64; hence, we shall multiply each part by 64 and divide the result by the square 441. We obtain:

$$a_1 = \frac{320 \cdot 64}{441} = \frac{20{,}480}{441}, \quad a_2 = \frac{80 \cdot 64}{441} = \frac{5120}{441},$$

$$a_3 = \frac{41 \cdot 64}{441} = \frac{2624}{441};$$

$$a_1 + a_2 = \frac{25{,}600}{441} = \left(\frac{160}{21}\right)^2, \quad a_2 + a_3 = \frac{7744}{441} = \left(\frac{88}{21}\right)^2,$$

$$a_3 + a_1 = \frac{23{,}104}{441} = \left(\frac{152}{21}\right)^2.$$

This second resolution introduces the method to be used in problems VII,8–11 and 15, which is as follows. The value of the principal unknown $a^2$ (or $a^6$) being initially imposed or chosen (say $a_0^2$ or $a_0^6$), one disregards this numerical condition and solves the similar system of the second degree obtained by replacing the fixed $a^2$ (or $a^6$) by a *required* quantity $u^2$. Let the value found be $u_0^2$. Since the solution of this intermediate system is in all these problems determined up to a quadratic factor, we shall obtain the solution to the original problem by multiplying all the magnitudes found in solving the intermediate problem by $a_0^2/u_0^2$ (or $a_0^6/u_0^2$). As noted earlier (p. 262), this possibility of fixing *a priori* the value of an unknown occurring in an even power was probably meant to be the distinguishing characteristic of Book VII.

Two singularities of the text are noticeable in this alternative resolution. Firstly, the formulation of the problem is misleading, as was the case in VII,3.[6] Secondly, the reference to III,6 is curiously repeated as if it had not been mentioned at all in the first part. Now, if, rendered suspicious, we were to suppose that the alternative resolution was added by some (early) commentator inspired by the following propositions, the genuineness of the *whole* problem would be doubtful, since its link to the following problems

[6] The repetition itself of the formulation at the beginning of an ἄλλως is not unusual: this occurs in the Greek text as well (see in I,21; III,15; "IV",28 and 31).

lies principally in the method employed in the second resolution.[7] However, we do not consider this to be the case, and take the whole problem VII,7 to be genuine, even though this gives rise to a new question—one which cannot be dismissed by assuming a commentator's inadequate reworking (see p. 274, *remark*).

---

**Problem VII,8.**

$$\begin{cases}(a^3)^2 + 2b = \square, \\ (a^3)^2 + \ \ b = \square'.\end{cases}$$

Putting $(a^3)^2 = 64 = (2^3)^2$, we have the new system

$$\begin{cases}64 + 2b = \square, \\ 64 + \ \ b = \square'.\end{cases}$$

Let us consider the general problem:

$$\begin{aligned}&(1)\\&(2)\end{aligned}\quad\begin{cases}u^2 + 2v = \square_1, \\ u^2 + \ \ v = \square_1'.\end{cases}$$

If we put $u^2 = x^2$ and $v = 2x + 1$, (2) will be identically satisfied. Then, (1) gives

$$x^2 + 4x + 2 = \text{square} =, \text{say}, (x - 2)^2,$$

so $$x^2 + 4x + 2 = x^2 - 4x + 4, \quad \text{and} \quad x = \tfrac{1}{4}.$$

Hence $$u^2 = x^2 = \tfrac{1}{16}, \qquad v = 1\tfrac{1}{2}.$$

Since any $u^2t^2$, $vt^2$, $t$ rational, is also a solution, $u_1^2 = 1$, $v_1 = 24$ is an integral solution of the considered pair of equations, and the solution we were looking for will result from its multiplication by 64:[8]

$$(a^3)^2 = 64, \quad b = 64 \cdot 24 = 1536, \quad \text{giving} \quad \square = 3136 = 56^2,$$
$$\square' = 1600 = 40^2.$$

**Problem VII,9.**

$$\begin{cases}(a^3)^2 - \ \ b = \square, \\ (a^3)^2 - 2b = \square'.\end{cases}$$

We put $(a^3)^2 = 64$, thus obtaining

$$\begin{cases}64 - \ \ b = \square, \\ 64 - 2b = \square'.\end{cases}$$

Let us consider the general system:

$$\begin{aligned}&(1)\\&(2)\end{aligned}\quad\begin{cases}u^2 - \ \ v = \square_1, \\ u^2 - 2v = \square_1'.\end{cases}$$

---

[7] The fact that the characteristic of the Book appears only in the *alternative* (easier) method is not a conclusive argument against the genuineness of the whole problem. Assuming that our opinion about the essential rôle of the equation $Ax^2 + Bx + C = \square$ in Book VI is correct, we have a similar example: the equation occurs for the first time in an alternative resolution of the group VI,12–14.

[8] In this and the next two problems, when establishing the required solution from the intermediate one, the text first gives the smallest integral solution of the auxiliary system.

We shall satisfy (1) identically by putting $u^2 = x^2$ and $v = 2x - 1$. Then (2) gives:

$$x^2 - 4x + 2 = \square'_1 =, \text{ say, } (x-3)^2.$$

So $$x^2 - 4x + 2 = x^2 - 6x + 9;$$

hence $$x = 3\tfrac{1}{2} \quad \text{and} \quad u^2 = x^2 = 12\tfrac{1}{4}, \quad v = 6.$$

Therefore, the smallest integral solution of the same system will be $u_1^2 = 49$, $v_1 = 24$.

But we had $(a^3)^2 = 64$; thus, we multiply the latter pair of elements by $\frac{64}{49}$ and obtain as the required solution:

$$(a^3)^2 = 64, \quad b = 24 \cdot \tfrac{64}{49} = \tfrac{1536}{49}, \quad \text{giving} \quad \square = \tfrac{1600}{49} = (\tfrac{40}{7})^2,$$
$$\square' = \tfrac{64}{49} = (\tfrac{8}{7})^2.$$

**Problem VII,10.** $$\begin{cases}(a^3)^2 + b = \square, \\ (a^3)^2 - b = \square'.\end{cases}$$

After putting $(a^3)^2 = 64$, we consider as previously the general system:

$$\begin{matrix}(1) \\ (2)\end{matrix} \quad \begin{cases} u^2 + v = \square_1, \\ u^2 - v = \square'_1.\end{cases}$$[9]

Diophantus chooses to satisfy (2) identically by taking

$$u^2 = x^2 \quad \text{and} \quad v = 2x - 1,$$

which gives for (1):

$$x^2 + 2x - 1 = \square_1 =, \text{ say, } (x-3)^2;$$

hence $$x^2 + 2x - 1 = x^2 - 6x + 9, \quad \text{and} \quad x = 1\tfrac{1}{4},$$

so that $$x^2 = u^2 = \tfrac{25}{16} \quad \text{and} \quad v = \tfrac{24}{16}.$$

Thus $u_1^2 = 25$, $v_1 = 24$ will be an integral solution.

But we assumed $(a^3)^2$ to be 64; the required solution is then

$$(a^3)^2 = 64, \quad b = 24 \cdot \tfrac{64}{25} = \tfrac{1536}{25}, \quad \text{giving} \quad \square = \tfrac{3136}{25} = (\tfrac{56}{5})^2,$$
$$\square' = \tfrac{64}{25} = (\tfrac{8}{5})^2.$$

---

The principle of the resolution used by Diophantus in the group VII,8–10 is the following.

Given a system

$$\begin{cases}(a^3)^2 + kb = \square, \\ (a^3)^2 + \ lb = \square',\end{cases} \qquad (k, l \text{ positive or negative})$$

---

[9] This might remind one of the problem of congruent numbers (cf. p. 83); but $v$ is not imposed in our case.

we examine

$$\begin{cases} u^2 + kv = \square_1, \\ u^2 + lv = \square_1'. \end{cases}$$

Taking $kv = 2mu + m^2$, $m$ arbitrary, the first equation is satisfied, and one has to fulfil now

$$u^2 + \frac{2ml}{k}u + \frac{l}{k}m^2 = \square_1',$$

which is easy to do since $u^2$ occurs with the positive sign.

The solution being $u_0$, $v_0$, we can now construct the solution of the given system for any arbitrary $a$. If, e.g., $a = 2$, thus $(a^3)^2 = 64$, the value of $b$ is obtained by multiplying $v_0$ by $64/u_0^2$.

**Remarks.** 1°. Diophantus' three cases are, in fact, not as general. They lead to intermediate systems involving three squares in an arithmetical progression, and the progression underlying his solutions is always the simplest one, $\{1, 25, 49\}$.

2°. The system resulting from the choice $(a^3)^2 = 64$ (or from any other), namely

$$\begin{cases} 64 + kb = \square, \\ 64 + lb = \square', \end{cases}$$

could be solved directly, i.e., without using the intermediate system, as in problem II,16 (see also p. 227).

---

**Problem VII,11.** $\begin{cases} a^2 = a_1 + a_2, \\ a^2 + a_1 = \square, \\ a^2 - a_2 = \square'.[10] \end{cases}$ $a^2 = 25$, given.

We seek some square fulfilling the equations of the problem, say $u^2$:

$$\begin{aligned} &(1) \\ &(2) \\ &(3) \end{aligned} \begin{cases} u^2 = u_1 + u_2, \\ u^2 + u_1 = \square_1, \\ u^2 - u_2 = \square_1'. \end{cases}$$

With $u^2 = x^2$ and $u_1 = 2x + 1$, (2) will be identically satisfied, and, if we put $u_2 = 2x - 1$, (3) will be identically satisfied. Then, (1) gives:

$$x^2 = (2x + 1) + (2x - 1) = 4x; \quad \text{hence} \quad x = 4,$$

$$x^2 = u^2 = 16, \quad \text{and} \quad u_1 = 9, \quad u_2 = 7.$$

But $a^2 = 25$ and $u^2 = 16$; so we shall multiply each magnitude by $a^2/u^2 = \frac{25}{16}$.

[10] Thus $a_1 = \square'$.

Hence:

$$a_1 = \tfrac{25}{16} \cdot u_1 = \tfrac{225}{16}, \quad a_2 = \tfrac{25}{16} \cdot u_2 = \tfrac{175}{16}, \quad \text{and} \quad \square = \tfrac{625}{16} = (\tfrac{25}{4})^2,$$

$$\square' = \tfrac{225}{16} = (\tfrac{15}{4})^2.$$

- Diophantus then simply states the impossibility of solving:

$$\begin{cases} a^2 = a_1 + a_2, \\ a^2 + a_1 = \square, \\ a^2 + a_2 = \square'. \end{cases}$$

Indeed, adding the last two conditions gives

$$2a^2 + (a_1 + a_2) = 3a^2 = \square + \square',$$

or

$$3 = \frac{\square}{a^2} + \frac{\square'}{a^2}.$$

But 3 (of the form $4n + 3$) cannot be represented as the sum of two squares.

**Remarks.** 1°. The only explicit allusion we have in the *Arithmetica* to a specific number not being representable as a sum of two squares is in "VI",14 (the number being 15). But it is apparent from the (reconstruction of the) diorism of "V",9 that Diophantus knew the general condition, concerning all integers and not only the ones of the form $4n + 3$.

2°. We arrive in "IV",32 at the system

$$\begin{cases} 8 - \phantom{3}x = \square, \\ 8 - 3x = \square', \end{cases}$$

which is not solvable rationally (οὐ ῥητόν ἐστι), the text says, "διὰ τὸ μὴ εἶναι τοὺς ϛ πρὸς ἀλλήλους λόγον ἔχοντας ὃν τετράγωνος ἀριθμὸς πρὸς τετράγωνον ἀριθμόν" (D.G., I, p. 270,5–6). Now, the reason given (that the ratio of the coefficients of $x$ is not a ratio of a square to a square) is wrong;[11]

[11] The system

$$\begin{cases} 8 - \phantom{2}x = \square, \\ 8 - 2x = \square', \end{cases}$$

which has the same characteristic, can be satisfied (by, e.g., $x = 3\tfrac{1601}{2401}$). In general, the condition given by Diophantus (which he satisfies in his subsequent reworking) is merely sufficient for obtaining a solution: the system

$$\begin{cases} A_1 x + B_1 = \square, \\ A_2 x + B_2 = \square', \end{cases}$$

reduces to the equation

$$\frac{A_1}{A_2} y^2 + \left(\frac{A_2 B_1 - A_1 B_2}{A_2}\right) = \square$$

(see p. 79), and, by the said condition, to the simple form $\alpha^2 y^2 + \gamma = \square$.

the real reason for the insolubility of the above system results, here too, from the impossibility of three being the sum of two squares: introducing $x = 8 - \square$ into the second equation gives $8 - 3(8 - \square) = 3\square - 16 = \square'$, hence

$$3 = \frac{16}{\square} + \frac{\square'}{\square}.$$

**Problem VII,12.**
$$\begin{cases} a^2 = a_1 + a_2, & a^2 = 25, \text{ given.} \\ a^2 - a_1 = \square, \\ a^2 - a_2 = \square'.^{12} \end{cases}$$

If we divide $a^2$ into two square parts, the conditions of the problem will be fulfilled; the way to do this has already been shown (II,8). A solution is

$$a_1 = 9, \quad a_2 = 16.$$

---

Three variations on the same basic problem compose the group VII,11–12: a square number is divided into two parts, subject to one of the following conditions:

(1) that the addition of one of the parts to the square and the subtraction of the other part from it result in a square;
(2) that the addition of each part to the square give a square;
(3) that the subtraction of each part from the square give a square.

In the two coming groups the problem is extended to a greater number of parts: VII,13–14 divide the square into three parts,[13] while VII,15 treats (partially) the case of four parts.

---

**Problem VII,13.**
$$\begin{array}{ll} (1) & \\ (2) & \\ (3) & \\ (4) & \end{array} \begin{cases} a^2 = a_1 + a_2 + a_3, & a^2 = 25, \text{ given.} \\ a^2 + a_1 = \square, \\ a^2 + a_2 = \square', \\ a^2 + a_3 = \square''. \end{cases}$$

Adding (2), (3), and (4), we obtain

$$3a^2 + a_1 + a_2 + a_3 = 4a^2 = \square + \square' + \square'',$$

where each of the squares on the right side is larger than $a^2$. Thus we end up with the problem of dividing a known square, $4a^2 = 100$, into three squares, each of which must be larger than $a^2 = 25$. Since the reader knows "how to

---

[12] Thus $a_2 = \square$, $a_1 = \square'$.

[13] Because of the odd number of parts, there is no "mixed" case as found in VII,11 and in VII,15.

divide any square number into square parts",[14] Diophantus (as usual) does not perform the computations. The results given in the text are:

$$\square = 36, \quad \square' = (\tfrac{168}{29})^2 = 33\tfrac{471}{841}, \quad \square'' = (\tfrac{160}{29})^2 = 30\tfrac{370}{841}.^{15}$$

By subtraction, we obtain the solution we were looking for:

$$a_1 = \square - 25 = 11, \quad a_2 = \square' - 25 = 8\tfrac{471}{841}, \quad a_3 = \square'' - 25 = 5\tfrac{370}{841},$$

with $a_1 + a_2 + a_3 = 25$.

**Problem VII,14.**

$$\begin{cases} (1) & a^2 = a_1 + a_2 + a_3, \qquad a^2 = 25, \text{ given.} \\ (2) & a^2 - a_1 = \square, \\ (3) & a^2 - a_2 = \square', \\ (4) & a^2 - a_3 = \square''. \end{cases}$$

Adding (2), (3) and (4), we obtain

$$3a^2 - (a_1 + a_2 + a_3) = 2a^2 = \square + \square' + \square'',$$

where each of the squares on the right side is smaller than $a^2$. Thus we end up with the problem of dividing the double of a known square, $2a^2 = 50$, into

---

[14] An allusion to II,8—which may be repeatedly applied.

[15] One can obtain Diophantus' results in the following way. One first chooses (or computes by II,8) $\square = 36$, which fulfils the condition $\square > 25$. Then

$$64 = \square' + \square'', \qquad \square', \square'' > 25.$$

By II,8:
$$64 = y^2 + \left(8 - \frac{p}{q}y\right)^2 = y^2 + 64 - 16\frac{p}{q}y + \frac{p^2}{q^2}y^2,$$

hence
$$y^2\left(1 + \frac{p^2}{q^2}\right) = 16\frac{p}{q}y, \quad \text{or} \quad y = \frac{16\frac{p}{q}}{1 + \frac{p^2}{q^2}} = \frac{16pq}{p^2 + q^2}, \qquad p, q > 0.$$

Since $64 - 36 = 28$, if the side of one of the two squares lies between 5 and 6, then the side of the other will be larger than 5. Thus, let us impose the condition

$$5 < 8 - \frac{p}{q}y < 6,$$

that is to say,
$$2 < \frac{p}{q}y < 3, \quad \text{or} \quad 2 < \frac{16p^2}{p^2 + q^2} < 3.$$

So
$$16p^2 > 2p^2 + 2q^2, \quad \text{or} \quad q^2 < 7p^2,$$

and
$$16p^2 < 3p^2 + 3q^2, \quad \text{or} \quad q^2 > 4\tfrac{1}{3}p^2.$$

Let us take $p = 2$; then $17\frac{1}{3} < q^2 < 28$. An obvious choice is $q^2 = 25$, giving

$$y = \frac{16pq}{p^2 + q^2} = \frac{160}{29}, \qquad 8 - \frac{p}{q}y = 8 - \frac{64}{29} = \frac{168}{29}.$$

Thus the values given by Diophantus.

three squares, each of which is smaller than the given square $a^2$. The results are directly given since the reader, the text says, knows "how to divide a number into square parts":[16]

$$\square = 16, \quad \square' = (\tfrac{61}{13})^2 = 22\tfrac{3}{169}, \quad \square'' = (\tfrac{45}{13})^2 = 11\tfrac{166}{169},$$

corresponding to the solution

$$a_1 = 25 - \square = 9, \quad a_2 = 25 - \square' = 2\tfrac{166}{169}, \quad a_3 = 25 - \square'' = 13\tfrac{3}{169},$$

with $a_1 + a_2 + a_3 = 25$.

**Remark.** Another form of this problem (without the numerical condition) is

$$\begin{cases} a^2 = a_1 + a_2 + a_3, \\ a_2 + a_3 = \square, \\ a_3 + a_1 = \square', \\ a_1 + a_2 = \square''. \end{cases}$$

[16] This allusion is no doubt to II,9. In order to solve the above problem, we may first choose a square smaller than 25 and such that the subtraction of it from 50 gives a result which can be represented as the sum of two squares; then we shall be able to apply II,9. Or, we may represent $2a^2$ as the sum of two different squares (by II,9) and then divide the larger one into two suitable squares (by II,8). Diophantus apparently used the first approach. Taking $\square = 16$, we have:

$$\square' + \square'' = 34 = 25 + 9.$$

Applying II,9:

$$34 = \left(5 - \frac{p}{q}y\right)^2 + (3 + y)^2 = 25 - 10\frac{p}{q}y + \frac{p^2}{q^2}y^2 + 9 + 6y + y^2,$$

so

$$y = \frac{10\frac{p}{q} - 6}{1 + \frac{p^2}{q^2}}.$$

Now, we must have $3 < 3 + y < 5$, that is to say $0 < y < 2$. This implies

1°. $10\frac{p}{q} > 6$, or $\frac{p}{q} > \frac{3}{5}$

2°. $5\frac{p}{q} - 3 < 1 + \frac{p^2}{q^2}$, $\frac{p^2}{q^2} - 5\frac{p}{q} + 4 > 0$, then $(p/q - \frac{5}{2})^2 > \frac{9}{4}$ and therefore either $p/q > 4$ or $p/q < 1$.

Hence the limitation: $\frac{3}{5} < p/q < 1$ (or $p/q > 4$). Taking as previously $p = 2$, we have

$$2 < q < 3\tfrac{1}{3} \quad (\text{or} \quad 0 < q < \tfrac{1}{2}).$$

An immediate choice is $q = 3$. Thus $p/q = \frac{2}{3}$, $y = \frac{6}{13}$, so that $5 - (p/q)y = \frac{61}{13}$ and $3 + y = \frac{45}{13}$. Hence the values given by Diophantus.

N.B. We could also solve this problem by the more advanced technique of the method of approximation to limits which appears later on, in "V",9–14. The same could have been done for the previous intermediate problem.

Now, the problem in this form has been solved by Diophantus in III,6 (see also VII,7) with the solution $441 = 320 + 80 + 41$; thus we could, as in VII,7, make use of this solution, the further condition $a^2 = 25$ merely necessitating its multiplication by $\frac{25}{441}$.

One wonders why Diophantus did not allude to III,6 in this (surely genuine) problem. Did he not realize that the system had already been solved in another form? This seems to be the case, although the proximity of problem VII,7 should have reminded him of III,6.[17] Perhaps also, since he is believed, like Euclid and Apollonius, to have borrowed from knowledge developed by his predecessors,[18] he merely reproduced this problem as he found it, together with related ones (cf. the group VII,11–15).

---

**Problem VII,15.** $$\begin{cases} a^2 = a_1 + a_2 + a_3 + a_4, \qquad a^2 = 25, \text{ given.} \\ a^2 + a_1 = \square, \\ a^2 + a_2 = \square', \\ a^2 - a_3 = \square'', \\ a^2 - a_4 = \square'''. \end{cases}$$

We shall try to fulfil the equations for some square $u^2$, such that:

$$\begin{array}{ll} (1) & \\ (2) & \\ (3) & \\ (4) & \\ (5) & \end{array} \begin{cases} u^2 = u_1 + u_2 + u_3 + u_4, \\ u^2 + u_1 = \square_1, \\ u^2 + u_2 = \square_2, \\ u^2 - u_3 = \square_3, \\ u^2 - u_4 = \square_4. \end{cases}$$

Let $u^2 = x^2$ be the unknown.

Putting $u_1 = 2x + 1$, (2) will be satisfied identically; (3) will be fulfilled by taking, e.g., $u_2 = 4x + 4$, (4) by taking $u_3 = 2x - 1$, (5) by taking $u_4 = 4x - 4$.

Then, $\quad u_1 + u_2 = 6x + 5, \quad$ and $\quad u_3 + u_4 = 6x - 5;$

hence $\quad x^2 = u^2 = u_1 + u_2 + u_3 + u_4 = 12x, \quad$ and $\quad x = 12.$

So $\quad u^2 = 144, \quad u_1 = 25, \quad u_2 = 52, \quad u_3 = 23, \quad u_4 = 44.$

Since $a^2 : u^2 = 25 : 144$, we shall multiply each of the above results by $\frac{25}{144}$. We obtain:

$$a^2 = 25, \quad a_1 = \tfrac{625}{144}, \quad a_2 = \tfrac{1300}{144}, \quad a_3 = \tfrac{575}{144}, \quad a_4 = \tfrac{1100}{144}.$$

(The values of $\square$, $\square'$, $\square''$, $\square'''$ are not given in this problem).

---

[17] Assuming that VII,7 is genuine (cf. pp. 267–268).

[18] See, e.g., Heath, *Dioph.*, p. 124.

- The text has then the remark that one similarly solves the system:

$$\begin{cases} a^2 = \sum_{k=1}^{8} a_k, & a^2 \text{ given square,} \\ a^2 + a_i = \square_i, & i = 1, \ldots, 4, \\ a^2 - a_j = \square_j, & j = 5, \ldots, 8. \end{cases}$$

Indeed, if, considering $u$'s instead of $a$'s, one puts $u^2 = x^2$ and

$$u_1 = 2x + 1, \quad u_2 = 4x + 4, \quad u_3 = 6x + 9, \quad u_4 = 8x + 16,$$

$$u_5 = 2x - 1, \quad u_6 = 4x - 4, \quad u_7 = 6x - 9, \quad u_8 = 8x - 16,$$

one obtains

$$x^2 = \sum_{k=1}^{8} u_k = 40x, \quad \text{and} \quad x = 40, \quad x^2 = u^2 = 1600;$$

the parts are then:

$$u_1 = 81, \quad u_2 = 164, \quad u_3 = 249, \quad u_4 = 336,$$

$$u_5 = 79, \quad u_6 = 156, \quad u_7 = 231, \quad u_8 = 304,$$

to be multiplied by $a^2/u^2 = a^2/1600$ in order to have the required parts $a_k$.

This (very simple, but elegant) method is generally valid for an even number $2n$ of parts, of which $n$ are additive and $n$ subtractive.

Let $u_1, \ldots, u_n$ be the additive parts and $u_{-1}, \ldots, u_{-n}$ the parts to be subtracted.

Putting

$$\begin{aligned} u_m &= 2mu + m^2, \\ u_{-m} &= 2mu - m^2, \qquad m = 1, \ldots, n, \\ (u_0 &= 0), \end{aligned}$$

we shall have:

$$u^2 = \sum_{m=-n}^{+n} u_m = \sum_{m=1}^{n} (u_m + u_{-m}) = \sum_{m=1}^{n} 4mu = 4u\frac{n(n+1)}{2},$$

hence

$$u = 2n(n + 1).$$

We have seen in this problem and in the remark following it the cases $n = 2$ and $n = 4$; the case $n = 1$ has been treated in VII,11.

**Remark.** The first part $u_1$ (hence also $a_1$) is always a square for, since $u = 2n(n + 1)$, we have

$$u_1 = 2u + 1 = 2[2n(n + 1)] + 1 = (2n + 1)^2.$$

The problems

$$\begin{cases} a^2 = \sum_{k=1}^{4} a_k = 25, \\ a^2 + a_i = \square_i, \qquad i = 1, \ldots, 4 \end{cases} \qquad \begin{cases} a^2 = \sum_{k=1}^{4} a_k = 25, \\ a^2 - a_i = \square_i, \qquad i = 1, \ldots, 4 \end{cases}$$

which might be expected to appear here (cf. p. 272) are soluble in a similar way: the first one amounts to dividing 125 into four squares, each of which is larger than 25, and the second one amounts to dividing 75 into four squares smaller than 25. If one does not wish to take the (tedious) way of iterating the elementary methods of Book II, one may assume one suitable square, say 36 in the first case and 16 in the second one, and then apply to the remainders $89 = 49 + 36 + 4$ and $59 = 49 + 9 + 1$ the παρισότητος ἀγωγή in the manner explained in the later Book "V" (cf. "V",11).

The last three problems of Book VII have quite a different form.

---

**Problem VII,16.** $\begin{cases} a^2 - b^2 = \square, \\ b^2 - c^2 = \square', \\ a^2 : b^2 = b^2 : c^2. \end{cases}$ (hence $c^2 < b^2 < a^2$).

**Lemma.** If $a^2 : b^2 = b^2 : c^2$ and $b^2 - c^2 = $ square, then also $a^2 - b^2 = $ square.

Indeed,

$$\frac{a^2}{b^2} = \frac{b^2}{c^2} \quad \text{implies (\textit{Elem.}, V,17) that} \quad \frac{a^2 - b^2}{b^2} = \frac{b^2 - c^2}{c^2},$$

so that $a^2 - b^2$ is a square if $b^2 - c^2$ is (*Elem.*, VIII,24).

This lemma is a particular case of the one given later on, in VII,18.

We put $c^2 = 1$[19] and (since $a^2 = \alpha^2 b^2 = \alpha^4 c^2$) we put $a^2 = x^4$; then

$$b^2 = \sqrt{a^2 c^2} = x^2.$$

Now, $\quad b^2 - c^2 = x^2 - 1 = \square' =$, say, $(x - 2)^2$;

hence $\quad x^2 - 1 = x^2 - 4x + 4 \quad$ and $\quad x = 1\frac{1}{4}$.

By the above lemma, the remaining condition $a^2 - b^2 = $ square is fulfilled.

So $c^2 = 1$, $\; b^2 = x^2 = \frac{25}{16}$, $\; a^2 = x^4 = \frac{625}{256}$, $\; \square' = \frac{9}{16}$, $\; \square = \frac{225}{256} = (\frac{15}{16})^2$.

**Remark.** A solution to this problem is obtainable from any Pythagorean triplet $h^2 = p^2 + q^2$, taking $a^2 = h^4/p^2$, $b^2 = h^2$, $c^2 = p^2$ (in our case, $h^2 = p^2 + q^2$ is $\frac{25}{16} = \frac{16}{16} + \frac{9}{16}$).

---

[19] This is not a restriction, since any square multiple of the solution will also be a solution. The same holds for the next two problems.

**Problem VII,17.** $$\begin{cases} a^2 + b^2 + c^2 + d^2 = \square, \\ a^2 : b^2 = c^2 : d^2. \end{cases}$$

We put $d^2 = 1$ and $a^2 = 16x^2$. Since $a^2:b^2 = c^2:d^2$, taking $c^2 = m^2x^2$ will leave for $b^2$ a certain number of units: $b^2 = 16/m^2$.

The other condition being

$$a^2 + b^2 + c^2 + d^2 = \square = (16 + m^2)x^2 + (b^2 + 1),$$

we shall take $m^2 = 9$ in order to have an equation of the form $\alpha^2x^2 + \gamma = \square$.

Thus $b^2 = \frac{16}{9}$, and the equation becomes

$$25x^2 + 2\tfrac{7}{9} = \square.$$

We put $\square = (5x + \frac{1}{3})^2$;[20] hence

$$x = \tfrac{8}{10}.$$[21]

So $$a^2 = 16x^2 = \tfrac{1024}{100} = (\tfrac{32}{10})^2 [= (\tfrac{16}{5})^2], \quad b^2 = 1\tfrac{7}{9},$$

$$c^2 = 9x^2 = \frac{576}{100} = \left(\frac{24}{10}\right)^2 \left[ = \left(\frac{12}{5}\right)^2 \right], \quad d^2 = 1,$$

$$\square = \frac{16{,}900}{900} = \left(\frac{130}{30}\right)^2 \left[ = \left(\frac{13}{3}\right)^2 \right].$$

**Problem VII,18.** $$\begin{cases} (1) & a^2 - b^2 = \square, \\ (2) & b^2 - c^2 = \square', \\ (3) & c^2 - d^2 = \square'', \\ (4) & a^2 : b^2 = c^2 : d^2. \end{cases} \qquad (\text{hence } d^2 < c^2 < b^2 < a^2).$$

**Lemma.** If $c^2 - d^2 =$ square and $a^2:b^2 = c^2:d^2$, then $a^2 - b^2 =$ square.

One verifies this lemma as above (problem 16). Thus, equations (2), (3), and (4) remain to be fulfilled. We choose $d^2 = 9$. Then, (3) will be satisfied if we put $c^2 = 25$.

Taking $a^2 = x^2$, (4) gives:

$$9x^2 = 25b^2, \quad \text{or} \quad b^2 = \tfrac{9}{25}x^2.$$

The only remaining condition is (2), which yields the equation:

$$\tfrac{9}{25}x^2 - 25 = \square' =, \text{say}, (\tfrac{3}{5}x - 1)^2 = \tfrac{9}{25}x^2 - \tfrac{6}{5}x + 1;$$

hence $$x = \tfrac{130}{6} [= \tfrac{65}{3}].$$

---

[20] Generally, $(5x + h)^2$ with $h^2 < 2\frac{7}{9}$ (or the reverse for negative $h$).

[21] The Greek text also gives, sometimes, a resulting fractional value in an unsimplified form; see, e.g., problems II,12; II,22; II,34; III,1; III,13. The value of $x$ is given in an unsimplified form in the next problem also.

So

$$a^2 = x^2 = \frac{16{,}900}{36}\left[=\frac{4225}{9}\right], \quad b^2 = \frac{9}{25}x^2 = \frac{6084}{36}\,[=169], \quad c^2 = 25,$$

$$d^2 = 9, \quad \square = \frac{10{,}816}{36} = \left(\frac{104}{6}\right)^2\left[=\left(\frac{52}{3}\right)^2\right], \quad \square' = \frac{5184}{36} = \left(\frac{72}{6}\right)^2 [= 12^2],$$

$$\square'' = 16.$$

**Remark.** This problem is again (see VII,16) soluble using any Pythagorean triplet $h^2 = p^2 + q^2$, where (by II,8) $p^2 = r^2 + s^2$: we shall put $a^2 = h^2p^2/r^2$, $b^2 = h^2$, $c^2 = p^2$ and $d^2 = r^2$.

---

VII,16–18 form thus the last group of problems of our Arabic Books. Comparing VII,16 with the pair VII,17–18, one might expect Diophantus also to have treated

$$\begin{cases} a^2 + b^2 + c^2 = \square, \\ a^2 : b^2 = b^2 : c^2. \end{cases}$$

But this problem has no rational solution, as was suspected at least by late Arabic times (see Nesselmann, *Beha-eddin's Essenz*, pp. 56 and 72).

# Part Four

# Text

We have already discussed in Part One the policies followed by us for the establishment of the Arabic text (see §§4,7, and 11). Thus we need only point out a few editorial procedures concerning the Arabic text and the critical apparatus.

Square brackets, [ ], are used to enclose interpolations (cf. §5), while angle brackets, ⟨ ⟩, enclose our additions to the manuscript's text. The Arabian numerals on the left denote the pages of the manuscript, and the numerals on the right number the lines.

The critical notes are numbered, the corresponding numerals appearing in parentheses after the indication of the line(s) in the text to which the notes refer. For explanations concerning the more notable errors or emendations, the reader is again referred to Part One (§§3–7 and 10–11)[1].

N.B. The few occurrences of a *kāf* without its upper stroke have not been pointed out: they are textually without relevance since this deficient *kāf* cannot be mistaken for a *lām*.

[1] See also the references at the beginning of the General Index.

**المقالة الرابعة من كتاب ذيوفنطس الاسكندرانى فى المربّعات والمكعّبات نقله من اللغة اليونانيّة الى اللغة العربيّة قسطا بن لوقا البعلبكّى وهذا خطّ محمّد بن ابى بكر بن حاكير المنجّم وكتب فى سنة خمس وتسعين وخمس مائة هجريّة**

بسم الله الرحمن الرحيم

المقالة الرابعة من كتاب ذيوفنطس فى المربّعات والمكعّبات

امّا اذ قد اتيتُ فيما تقدّم من القول فى المسائل العدديّة على كثير من المسائل التى إنتهينا فيها بعد الجبر والمقابلة الى نوع واحد يعادل نوعًا واحدًا ما كان منها من نوعى العدد الخطّى والسطحى وايضًا ما كان مزدوجًا منها وجعلتُ ذلك على مراتب يُمكن المتعلّمين حفظُها وتحصيلُ معانيها فانّى ارى ايضًا لئلّا يفوتك شىء ممّا يُمكن عملُه من هذه الصناعة ان اكتب لك فيما يتلو ايضًا كثيرًا من مسائل هذا الفنّ ما يكون منها من نوع العدد الذى يُسمّى الجرمى وايضًا ما كان منه مركّبًا مع النوعين الاوّلين واسلك فيه ذلك المسلك واجعلك فيه ترقى فيه من درجة الى درجة ومن فنّ الى فنّ ليكون ذلك دربة وعادة فانّك متى عرفتَ ما رسمتُ امكنك الجواب فى كثير من المسائل التى لم ارسمها اذ كنتُ قد رسمت لك كيف المسلك فى وجود اكثر المسائل ووصفت لك من كلّ نوع منها مثالاً

فاقول ان كلّ مربّع يُضرب فى ضلعه فانّه يكون كعبًا فمتى قسمتُ الكعب على المال خرج منه ضلع الكعب وان قُسم على شىء [وهو جذر ذلك المال]

---

4 (1): حاكير in codice, حاگير conjecturâ auctoris indicis codicum manuscriptorum bibliothecae mausolei Meschedae.

9 (2): إنتهينا : اسناها in cod.

13 (3): يتلو : ىتلوا in codice, sc. cum *alif* quod apud grammaticos otiosum vocatur.

15 (4): فيه post واجعلك forsan delendum.

(5): ترقى : مُرقي in cod.

20 (6): قسمتُ conjeci, ىسمت in codice. Similiter habet ضربت (lin. 22) neque *ḍammah* neque *fatḥah* in cod.

21 (7): Verba وهو جذر ذلك المال et similia infra (usque ad lineam 51) interpolatori arabico tribuo.

خرج منه مال فان ضربت الكعب فى شى ء خرج منه مثل الذى يخرج من ضرب المال فى مثله وهو يُسمّى مال مال فان قُسم مال المال على كعب خرج منه شى ء[وهو جذر المال] فان قُسم على مال خرج منه مال فان قُسم على شى ء[وهو جذر المال] خرج منه كعب فان ضُرب مال المال فى شى ء[وهو جذر المال] خرج منه مثل الذى من ضرب الكعب فى المال وهو يُسمّى مال كعب وان قُسم مال الكعب على شى ء[وهو جذر المال] خرج منه مال مال وان قُسم على مال خرج منه كعب وان قُسم على كعب خرج منه مال وان قُسم على مال مال / خرج منه شى ء[وهو جذر المال] وان ٢ ضُرب مال الكعب فى شى ء خرج منه مثل الذى من ضرب الكعب فى مثله ومن ضرب المال فى مال مال وهو يُسمّى كعب كعب فان قُسم كعب الكعب على شى ء[وهو جذر المال] خرج منه مال كعب فان قُسم على مال خرج منه مال مال فان قُسم على كعب خرج منه كعب فان قُسم على مال مال خرج منه مال فان قُسم على مال كعب خرج منه شى ء[وهو جذر المال]

وينبغى ان اذا تناهى بنا العملُ بعد الجبر والمقابلة[اعنى بالجبر زيادة ما كان ناقصًا على كلتى الناحيتين وبالمقابلة إلقاء ما كان متساويًا من كلتى الناحيتين] الى نوع واحد من هذه الانواع التى قد وصفنا تضعيف بعضها ببعض وقسمة بعضها على بعض يعادل نوعًا آخر ان نقسم الجميع على واحد من اقعد الناحيتين حتّى يخرج لنا نوع واحد يعادل عددًا

آ نريد ان نجد عددين مكعّبين يكون الجميع منهما عددًا مربّعًا

فنفرض ضلع المكعّب الاصغر شيئًا واحدًا ليكون مكعّبه كعبًا واحدًا ونفرض ضلع المكعّب الاعظم كم شئنا من الاشياء فنفرضه شيئين فيكون المكعّب الاعظم ثمنية كعاب وجملتهما تسعة كعاب فنحتاج ان يكون ذلك يعادل مربّعًا فنعمل المربّع من ضلع كم شئنا من الاشياء فنعمله من ضلع ستّة اشياء حتّى يكون ستّة وثلثين مالاً فاذًا التسعة كعاب

---

22 (8): الكعب : المكعّب in cod.

35-37 (9): Interpretationem vocabulorum جبر et مقابلة, quam a quodam arabico lectore additam fuisse suspicor, seclusi.

41 (10): Problematum numeratio (per litteras), quam per lineam supra scriptam significavi, hic et ubique in codice atramento rubro notatur.

44 (11): وجملتهما : وجملتها in cod.

تعادل ستّة وثلثين مالاً فلان الناحية التى فيها الاموال اقعد من الناحية الاخرى فانّا نقسم الجميع على مال واحد فالتسعة الكعاب اذا قُسمت على مال واحد كان منها تسعة اشياء [وهى تسعة جذور للمال] وامّا الستّة والثلثون مالاً ⟨فانّها⟩ اذا قُسمت على مال واحد خرج منه عدد وهو ستّة وثلثون احداً فاذاً التسعة الاشياء [التى هى الجذور] تعادل ستّة / وثلثين احداً فالشىء الواحد مساوٍ لأربعة آحاد ولانّا ٣
فرضنا المكعّب ⟨الاصفر⟩ من ضلع شىء يكون ضلعه اربعة فيكون المكعّب الاصفر اربعة وستّين ولانّا فرضنا المكعّب الاعظم من ضلع شيئين يكون ⟨ضلعه⟩ ثمنية آحاد فيكون المكعّب الاعظم خمس مائة واثنى عشر وجملة المكعّبين خمسمائة وستّة وسبعون وهو مربّع من ضلع اربعة وعشرين

فقد وجدنا عددين مكعّبين وجميعهما مربّع ⟨و⟩ الاصفر اربعة وستّون والاعظم خمس مائة واثنا عشر وذلك ما اردنا ان نبيّن

ب̅ نريد ان نجد عددين مكعّبين يكون تفاضلهما عدداً مربّعاً

فنفرض المكعّب الاصفر من ضلع شىء واحد فيكون كعباً واحداً ونفرض ضلع الاعظم كم اردنا من الاشياء فلنفرضه من ضلع شيئين حتّى يكون المكعّب الاعظم ثمنية كعاب وتفاضلهما سبعة كعاب وهى تعدل عدداً مربّعاً فلنفرض ضلع المربّع سبعة اشياء حتّى يكون ⟨المربّع⟩ تسعة واربعين مالاً فاذاً السبعة الكعاب تعدل تسعة واربعين مالاً والناحية التى فيها الاموال اقعد الناحيتين فنقسم الجميع على مال واحد فيخرج لنا سبعة اشياء تعادل تسعة واربعين احداً فالشىء الواحد يعدل سبعة آحاد ومن اجل انّا فرضنا المكعّب الاصفر من ضلع شىء واحد يكون

---

50 (12): والثلثون : والثلثين in cod.
(13): فانّها addidi.
51 (14): احداً : واحدا in cod.
52 (15): مساوٍ : مُساوي in cod.
53 (16): الاصفر addidi.
55 (17): ضلعه addidi.
(18): واثنى : واثنا in cod.
57 (19): و (primum): من in cod.
(20): و (secundum) addidi.
63 (21): المربّع addidi.
65 (22): فيها : منها in cod.
(23): Post اقعد addit codex من.

ثلثمائة وثلثة واربعين ويكون ضلع الاعظم من اجل انّه من شيئين اربعة عشر فيكون المكعّب الاعظم ألفين وسبعمائة واربعة واربعين وتفاضلهما ألفان واربعمائة وواحد وهو مربّع ضلعه تسعة واربعون

فقد وجدنا عددين مكعّبين تفاضلهما عدد مربّع وذلك ما اردنا ان نبيّن

جـ نريد ان نجد عددين مربّعين ويكون جميعهما عدداً مكعّباً /

فنفرض المربّع الاصغر مالاً ونفرض المربّع الاعظم اربعة اموال ويكون ٤ جملة المربّعين خمسة اموال ونحتاج ان يكون ذلك يعادل عدداً مكعّباً فلنعمله من ضلع كم شئنا ⟨من الاشياء⟩ فلنفرضه ايضاً من ⟨ضلع⟩ شيء واحد حتّى يكون كعباً واحداً فاذاً الخمسة الاموال تعادل كعباً واحداً فلان الناحية التى فيها الاموال اقعد الناحيتين فانّا نقسم الجميع على مال واحد فيكون شيء واحد يعدل خمسة آحاد فلانّا وضعنا المربّع الاصغر مالاً والمال لانّه انّما يكون من ضرب الشيء فى مثله والشيء خرج لنا خمسة آحاد يكون المال خمسة وعشرين احداً ولانّا جعلنا المربّع الاعظم اربعة اموال يكون مائة وجملة المربّعين مائة وخمسة وعشرون وهو عدد مكعّب وضلعه خمسة آحاد

فقد وجدنا عددين مربّعين وجمعهما عدد مكعّب وهو مائة وخمسة وعشرون وذلك ما اردنا ان نبيّن

د نريد ان نجد عددين مربّعين يكون تفاضلهما عدداً مكعّباً

فنفرض ضلع المربّع الاصغر شيئاً وضلع الاعظم كم شئنا من الاشياء فليكن ضلع الآخر خمسة اشياء حتّى يكون المربّع الاعظم خمسة وعشرين

---

68 (24): من اجل انّه من ; fortasse scribendum est فُرِض ante aut pro من altero.

74 (25): المربّع (prius): العدد in cod.

76 (26): Per homoeoteleuton (ut ita dicam) omissum addidi. Vide etiam adn. 50.

(27): ضلع addidi.

80 (28): Post والمال addit codex خمسة وعشرون.

81 (29): خمسة وعشرُون : خمسة وعشرين in cod.

(30): Pro المربّع praebet codex المال . Vide adn. 44,178;286.

84 (31): Pro جمع in وجمعهما fortasse subjiciendum est جميع , quo interpres in hoc textu uti solet.

87 (32): الاشياء : الاحاد in cod.

مالاً ويكون الاصغر مالاً واحداً وتفاضلهما اربعة وعشرون مالاً وذلك يعادل مكعّباً فلنفرض المكعّب من ضلع كم شئنا من الاشياء فلنفرضه من ضلع شيئين فيكون اربعة وعشرون مالاً تعادل ثمنية كعاب لان المكعّب الذى يكون من الشيئين ثمنية كعاب فنقسم ايضاً الجميع على مال واحد فيكون ثمنية اشياء تعادل اربعة وعشرين احداً فيكون الشىء الواحد ثلثة آحاد ولانّا فرضنا المربّع الاصغر من ضلع شىء واحد وفرضنا ضلع المربّع الاعظم من خمسة اشياء يكون ضلع الاصغر ثلثة ويكون ضلع الاعظم خمسة عشر والمربّع الاصغر تسعة والمربّع / الاعظم مائتان وخمسة وعشرون ٥ وتفاضلهما مائتان وستّة عشر وهى عدد مكعّب من ضلع ستّة آحاد

فقد وجدنا عددين مربّعين وتفاضلهما عدد مكعّب وهما مائتان وخمسة وعشرون وتسعة وذلك ما اردنا ان نبيّن

هـ نريد ان نجد عددين مربّعين يُحيطان بعدد مكعّب

فنفرض الاصغر مالاً والاعظم من ضلع كم شئنا من الاشياء فنفرضه من ضلع شيئين فيكون المربّع الاعظم اربعة اموال والذى يحيطان به اربعة اموال اموال وهى تعادل عدداً مكعّباً فنفرض المكعّب من ضلع شيئين حتّى يكون ثمنية كعاب فاذاً اربعة اموال اموال تعادل ثمنية كعاب فنقسم الجميع على كعب فيكون ثمنية آحاد تعادل اربعة اشياء ⌈لان ثمنية كعاب اذا قُسمت على كعب خرج منها ثمنية آحاد [ولان الواحد فى كعب كعبٌ فاذا قُسم الكعب على الكعب خرج منه واحد] واذا قُسمت اربعة اموال اموال على كعب خرج منه اربعة اشياء فاذاً اربعة اشياء تعادل ثمنية آحاد⌉ فالشىء الواحد يعادل اثنين ولانّا جعلنا المربّع الاصغر مالاً يكون اربعة آحاد لان المال يكون من ضرب الشىء فى مثله ولانّا جعلنا المربّع الاعظم اربعة اموال يكون ستّة عشر والعدد الذى يُحيط به هذان المربّعان اربعة وستّون وهى مكعّب ضلعه اربعة آحاد

---

90 (33): مكعّباً : كعبا in cod.

(34): المكعّب : الكعب in cod.

105 (35): Post آحاد addit codex verba لان الواحد (vide lin. 106, in fine).

105-109 (36): Quin uncis inclusa verba lin. 106-107 interpolata sint, haud dubium est; cetera autem seclusa verba haud genuina esse opinari licet.

112 (37): ضلعه pro ضلعهما codicis substitui. Sed vide adn. 224,434.

فقد وجدنا عددين مربّعين يحيطان بعدد مكعّب وهما الاربعة والستّة عشر وذلك ما اردنا ان نبيّن

وَ نريد ان نجد عددين مربّعاً ومكعّباً يحيطان بعدد مربّع

فنفرض ضلع المربّع ما شئنا من الاشياء فنفرضه شيئاً واحداً حتّى يكون المربّع مالاً واحداً ونفرض ضلع المكعّب ايضاً ما اردنا من الاشياء / فنفرضه شيئين حتّى يكون ⟨المكعّب⟩ ثمنية كعاب والذى يحيطان به ٦
اعنى المال والثمنية الكعاب ثمنية مال كعاب وذلك يعدل مربّعاً فلانّا ⟨ان⟩ فرضنا ضلع المربّع ⟨من⟩ اشياء يجتمع منها ⟨اموال فيكون اموال كعاب تعادل⟩ اموالاً ونحتاج ان نقسم الناحيتين على الاموال فيكون حينئذٍ كعاب تعادل آحاداً لان كعاب الاموال كما ذكرتُ اذا قُسمت على الاموال خرج منها كعاب فنفرض ضلع المربّع من اموال كم شئنا فنفرضه من اربعة اموال حتّى يكون المربّع ستّة عشر مال مال فاذاً ثمنية كعاب مال تعادل ستّة عشر مال مال فنقسم الجميع على مال مال لانّها اقعد الناحيتين فستّة عشر مال مال اذا قُسمت على مال مال خرج منها ستّة عشر احداً وامّا ثمنية كعاب المال فانّها اذا قُسمت على مال مال خرج منها ثمنية اشياء فاذاً ثمنية اشياء تعادل ستّة عشر احداً فاذاً الشىء اثنان فلانّا فرضنا ضلع المربّع شيئاً يكون المربّع اربعة آحاد ويكون المكعّب لانّا فرضناه من ضلع شيئين اربعة وستّين والعدد الذى يحيطان به اعنى المربّع الذى هو اربعة والمكعّب الذى هو اربعة وستّون مائتان وستّة وخمسون وهى مربّع ضلعه ستّة عشر احداً

فقد وجدنا عددين احدهما مربّع والآخر مكعّب ويحيطان بعدد مربّع وهما الاربعة والاربعة والستّون وذلك ما اردنا ان نبيّن

زَ نريد ان نجد الآن عددين احدهما مربّع والآخر مكعّب ويحيطان بعدد مكعّب

---

116 (38): عددين : عدداً in cod.

119 (39): المكعّب addidi.

121 (40): ان deest in cod.

(41): من addidi, sed dubitanter.

121-122 (42): Velut per homoeoarcton omissa addidi.

126 (43): لانّها : لانهما in cod.

130 (44): المربّع (posterius): المال in cod.

132 (45): والمكعّب : والكعب in cod.

فنفرض ضلع المربّع شيئاً فيكون المربّع مالاً ونفرض ضلع المكعّب ما شئنا من الاشياء فنفرضه اربعة اشياء حتّى يكون المكعّب اربعة وستّين كعباً والذى يحيطان به هو اربعة وستّون كعب مال وهى تعادل عدداً مكعّباً ولانّا ان فرضنا ضلع المكعّب من اشياء يكون المكعّب / ٧ من كعاب واذا عادلناها بكعاب اموال إحتجنا ان نقسم الجميع على كعب فيخرج لنا حينئذٍ اموال تعادل آحاداً فنحتاج ⟨ان تكون⟩ الآحاد ⟨التى تعادل المال الواحد⟩ مربّعةً لكن ان فرضناه من اموال يكون المكعّب من كعاب كعاب فاذا عادلنا ذلك بكعاب اموال إحتجنا الى ان نقسم ⟨الناحيتين⟩ على كعب مال فيخرج لنا حينئذٍ اشياء تعادل آحاداً فنفرض ضلع المكعّب من مالين فيكون المكعّب ثمنية كعاب كعاب فاذاً ثمنية كعاب كعاب تعادل اربعة وستّين كعب مال فنقسم الجميع على كعب مال لانّها اقعد الناحيتين فيخرج لنا من قَسم ثمنية كعاب الكعب على كعب مال ثمنية اشياء ومن قَسم الاربعة والستّين مال كعب على مال كعب اربعة وستّون احداً فيكون ثمنية اشياء تعادل اربعة وستّين احداً فالشىء الواحد ثمنية آحاد ولانّا فرضنا ضلع المربّع شيئاً يكون هو اربعة وستّين ويكون المكعّب لانّا فرضنا ضلعه اربعة اشياء يكون ضلعه اثنين وثلثين ويكون المكعّب اثنين وثلثين الفاً وسبعمائة وثمنية وستّين فاذا ضربنا ذلك فى المربّع الذى هو اربعة وستّون إجتمع من ذلك عدد مكعّب لان كلّ واحد منهما مكعّب

فقد وجدنا عددين على الشرط الذى شرطنا وذلك ما اردنا ان نجد

ح̄ نريد ان نجد عددين مكعّبين يحيطان بعدد مربّع

فلانّا ان فرضنا فى هذه المسئلة ايضاً ضلع المكعّب الاصغر شيئاً واحداً يكون المكعّب الاصغر كعباً واحداً ⟨وان⟩ فرضنا الاعظم من ضلع كم شئنا ⟨من الاشياء⟩ كانّا فرضنا⟨ه⟩ من ضلع شيئين فيكون المكعّب الاعظم

---

143 (46): ان تكون addidi.

143-144 (47): Verba التى تعادل المال الواحد , quae valde desiderantur, inserui.

145 (48): الناحيتين addidi.

161 (49): وان addidi.

162 (50): Velut per homoeoteleuton omissa addidi.

(51): Pronomen post فرضنا restitui.

ثمنية كعاب والذى يحيطان به هو ثمنية كعاب كعاب ونحتاج ان يكون ذلك مساوياً لمربّع ولا يستقيم ان نفرض ضلع هذا المربّع من اشياء لان مربّع الاشياء اموال واذا عودلت بكعاب كعاب ثمّ قُسمت على اقعد الناحيتين وهى اموال خرج منه مال مال يعادل آحاداً ⟨لكن ان فرضنا ضلع المربّع من اموال يكون المربّع من اموال اموال فاذا عادلنا ذلك بكعاب كعاب إحتجنا الى ان نقسم الناحيتين على مال مال فيخرج لنا حينئذٍ اموال تعادل آحاداً⟩ فنحتاج ان يكون ⟨ما⟩ يعادل المال / الواحد من الآحاد مربّعاً من اجل ذلك نصير الى ان نطلب مربّعاً ٨ وعدداً مكعّباً يحيطان بعدد مربّع [لِما سيبين من سهولة ذلك فى العمل] فنجد مثل ما تقدّم ذكره احد العددين وهو المربّع اربعة والآخر وهو المكعّب اربعة وستّين والذى يحيط به هذان العددان هو مائتان وستّة وخمسون وهو مربّع ضلعه ستّة عشر احداً وذلك ما اردنا ان نجد

ط نريد ان نجد عددين مكعّبين يحيطان بمربّع

فنفرض ضلع المكعّب الاعظم اربعة اشياء وضلع المكعّب الاصغر شيئاً واحداً فيكون المكعّب الاعظم اربعة وستّين كعباً ويكون المكعّب الاصغر كعباً واحداً والذى يحيطان به اربعة وستّون كعب كعب ونحتاج ان يكون ذلك معادلاً لعدد مربّع فنفرض ضلع المربّع من اموال يكون عددها مساوياً لضلع المربّع الذى يكون من ضرب الاربعة والستّين فى الاربعة

---

164 (52): Post من addit codex ضلع.

165 (53): واذا : فاذا in cod.

166 (54): Per dittographiam praebet codex verbum اموال bis; sub altero autem scripsit eadem ni fallor manus طء ad errorem, ut videtur, delendum.

166-169 (55): Hoc, quod ad sensum necessarium est, sed per homoeoteleuton omissum, restitui.

169 (56): ما deest in cod.

171-172 (57): Verba لِما ... العمل interpolata videntur; pro سيبين praebet codex سسين.

173 (58): المكعّب : كعب in cod.

179 (59): وستّون : وستين in cod.

180 (60): يكون : فيكوُن in cod.

الذى هو مائتان وستّة وخمسون من ⟨ضلع ستّة عشر احداً فاذاً نفرض ضلع المربّع من⟩ ستّة عشر مالاً ليكون مربّعه مائتين وستّة وخمسين مال مال فيكون اربعة وستّون كعب كعب تعادل مائتين وستّة وخمسين مال مال فنقسم الجميع على مال مال لانّها اقعد الناحيتين فيكون الاربعة والستّون كعب كعب اذا قُسمت على مال مال يخرج منها اربعة وستّون مالاً واذا قسمنا المائتين والستّة والخمسين مال المال على مال مال خرج منها مائتان وستّة وخمسون احداً فاذاً الاربعة والستّون المال تعادل مائتين وستّة وخمسين احداً فيكون المال الواحد يعادل اربعة آحاد والمال مربّع والاربعة مربّع فضلعاهما متساويان وضلع المال شىء وضلع الاربعة اثنان فاذاً الشىء هو اثنان فلانّا فرضنا ضلع المكعّب الاصغر شيئاً واحداً يكون المكعّب ⟨الاصغر⟩ ثمنية آحاد ولانّا فرضنا ضلع المكعّب الاعظم اربعة اشياء وهى ثمنية آحاد يكون المكعّب الاعظم خمس مائة واثنى عشر فاذا ضربناه فى المكعّب / ٩ الاصغر إجتمع من ذلك العدد الذى يحيطان به وهو اربعة الٰف وستّة وتسعون وهو مربّع وضلعه اربعة وستّون

فقد وجدنا عددين مكعّبين يحيطان بعدد مربّع وهما الثمنية والخمس مائة والاثنا عشر وذلك ما اردنا ان نجد

فان اردنا ان نجد عدداً مكعّباً اذا قسمناه على مكعّب خرج منه عدد مربّع إلتمسنا عدداً مربّعاً اذا ضربناه فى عدد آخر مكعّب ايضاً نلتمسه فيجتمع من الضرب عدد مكعّب فاذا وجدنا ذلك فالمجتمع من ضرب احدهما فى الآخر هو العدد المكعّب الذى اردناه

---

182-183 (61): Verba ضلع...من , fortasse per homoeoteleuton omissa, addidi (de usu hujus من (quod etiam omittere licet) in indice verborum commemoravi; vide enim sub *farada*, 1°, ٧).

184 (62): مايتان in codice, مائتين scripsi.

187 (63): المايتان والسته والحمسون : المائتين والستّة والخمسين in cod.

190 (64): وضلع : مضلع in cod.

191 (65): Post فلانّا addit codex اذا.

192 (66): المكعّب : الكعب in cod.

(67): الاصغر addidi.

195 (68) Pro scriptura pluralis defectiva الف , quam codex ubique praebet, scripturam الٰف substitui.

202 (69): هو العدد المكعّب : العدد المكعب هو in cod.

وكذلك ان اردنا ان نجد عدداً مربّعاً اذا قسمناه على مربّع خرج منه مكعّب نعمله بالعكس ممّا تقدّم وكذلك كلّ ما كان من طريق القَسم إلتمسناه ممّا تقدّم من جنسه لانّهما واحد اذ كانت القسمة انّما هى عكس الضرب

ى نريد ان نجد عدداً مكعّباً اذا زدنا عليه مثل المربّع الذى يكون من ضلعه كم مرّة شئنا فيجتمع منه عدد مربّع

فنفرض المكعّب من ضلع شىء واحد فيكون كعباً واحداً ونفرض المرّات عشراً ونُضيف الى كعب عشرة امثال ⟨مربّع⟩ ضلع المكعّب الذى هو مال فيكون كعباً وعشرة اموال وذلك يعادل مربّعاً فنفرض ذلك المربّع من ضلع اشياء يكون مربّعها اكثر من عشرة اموال لكي تُمكن المقابلة فنفرض ذلك من ضلع اربعة اشياء فيكون المربّع ستّة عشر مالاً فيكون الكعب وعشرة اموال تعدل ستّة عشر مالاً فلنلق العشرة الاموال المشتركة فيبقى ستّة اموال تعادل كعباً فنقسم ذلك على مال فيخرج شىء واحد يعادل ستّة آحاد ويكون الكعب مأتين وستّة عشر ويكون مربّع الضلع ستّة وثلثين وعشرة امثال ذلك ثلثمائة وستّون فنُضيفها الى الكعب فيجتمع من ذلك خمسمائة وستّة وسبعون وهو مربّع من ضلع اربعة وعشرين

فقد وجدنا عدداً / مكعّباً اذا زدنا عليه عشرة امثال المربّع الذى ١٠
يكون من ضلعه صار بعد الزيادة عدداً مربّعاً وهو مائتان وستّة عشر
وضلعه ستّة وذلك ما اردنا ان نجد

يا نريد ان نجد عدداً مكعّباً اذا نقصنا منه مثل المربّع الذى يكون من ضلعه كم مرّة شئنا يبقى منه عدد مربّع

---

207 (70): Pro زدناه علي codicis زدنا عليه substitui.

210 (71): E textu elapsum verbum مربّع restitui.

213 (72): Fortasse suppleatur المربّع post ذلك.

(73): المكعّب : الكعب in cod.

214 (74): فلنلقي : فلنلق in cod.

217 (75): اموال وذلك : امثال ذلك in codice. Vide etiam adn. 366.

220 (76): Verbum عدداً repetivit librarius in initio decimae paginae.

فنفرض المكعّب من ضلع شيء واحد ليكون كعباً واحداً ونفرض المرّات ستّة ونريد ان **يبقى من الكعب بعد نقصان الستّة الاموال مربّع** فنفرض المربّع من ضلع اشياء كم شئنا فنفرضه من ضلع شيئين حتّى يكون مربّعه اربعة اموال فاذاً الكعب الّا ستّة اموال تعادل اربعة اموال فنجبر الكعب بالستّة الاموال ونزيدها على الاربعة الاموال فيكون **كعب واحد يعادل** عشرة اموال فنقسم الجميع على مال فيخرج لنا شيء واحد يعادل عشرة آحاد فلانّا فرضنا ضلع المكعّب شيئاً واحداً يكون ⟨المكعّب⟩ **الفاً ويكون مربّع الضلع مائة وستّةُ امثال المائة ستّمائة والذى يبقى من الالف بعد ⟨نقصان⟩ ستّمائة اربع مائة وهو عدد مربّع ضلعه عشرون**

فقد وجدنا عدداً مكعّباً اذا نقصنا منه مثل مربّع ضلعه ستّة امثال بقى منه عدد مربّع وهو الف وضلعه عشرة

يبَ نريد ان نجد عدداً مكعّباً اذا زدنا عليه مثل المربّع الذى يكون من ضلعه كم مرّة شئنا يكون الذى يجتمع منه عدداً مكعّباً

فنفرض ضلع المكعّب شيئاً واحداً حتّى يكون ⟨المكعّب⟩ **كعباً واحداً ونُضيف اليه المرّات التى نريد وهى على ما فرضنا فيما تقدّم فيكون كعباً** وعشرة اموال وهى تعادل مكعّباً فنعمل المكعّب من ضلع **شيئين حتّى يكون ثمنية كعاب تعادل كعباً واحداً وعشرة اموال فنلقى الكعب المشترك فيبقى عشرة اموال تعادل سبعة كعاب ونقسم ذلك على مال فيكون سبعة اشياء تعادل عشرة آحاد فيكون الشيء الواحد عشرة** اسباع ويكون المكعّب الفاً بالمقدار الذى هو سُبع سُبع / سُبع فاذا ١١

---

226 (77): **نفرض** in codice, **يبقى من** scripsi; vide lin. 233.

(78): **الكعب : المكعّب** in cod.

227 (79): **مربّع : مُربّعا** in cod.

228 (80): **الكعب : المكعب** in cod.

229 (81): **الكعب : المكعب** in cod.

(82): **ونزيده** in codice, **ونزيدها** scripsi.

230 (83): Pro **فيكون كعب واحد يعادل** praebet codex **سكون كعباً واحداً يعادله**. Cf. adn. 330.

232 (84): **المكعّب** addidi.

233 (85): Deficiens **نقصان** restitui.

235 (86): **ضلعه ستّة امثال : ستّة امثال ضلعه** in cod.

239 (87): Verbum **المكعّب** addidi.

اضفنا اليه عشرة امثال المربّع الذى هو مائة سُبع سُبع التى هى سبعة الف سُبع سُبع سُبع لإجتمع من ذلك ثمنية الف سُبع سُبع سُبع و هى مكعّب من ضلع عشرين سُبعاً

فقد وجدنا مكعّباً يظهر فيه الشرط الذى شُرط لنا وهو الف سُبع سُبع سُبع من ضلع عشرة اسباع وذلك ما اردنا ان نجد

يجـ نريد ان نجد عدداً مكعّباً اذا نقصنا منه مثل المربّع الذى يكون من ضلعه كم مرّة شئنا بقى منه عدد مكعّب

فنفرض المكعّب من ضلع شىء حتّى يكون كعباً واحداً ونفرض المرّات سبعة حتّى يكون الذى يبقى كعباً الّا سبعة اموال فيكون ذلك يعادل عدداً مكعّباً فنفرض ضلع المكعّب بعض شىء فنفرضه من ضلع نصف شىء حتّى يكون المكعّب جزءاً من ثمنية اجزاء من كعب فذلك يعادل كعباً الّا سبعة اموال فنجبر ونقابل فيبقى سبعة اثمان كعب تعادل سبعة اموال فنقسم الجميع على مال فيخرج سبعة آحاد تعادل سبعة اثمان شىء فيكون الشىء الواحد ثمنية آحاد ويكون الكعب خمس مائة واثنى عشر فاذا نقصنا منه سبعة امثال الاربعة والستّين بقى منه اربعة وستّون وهو مكعّب

ونعمل ذلك بوجه آخر نجعل ضلع المكعّب الاوّل كم شئنا من الاشياء فنجعله شيئين حتّى يكون المكعّب ثمنية كعاب فيبقى [فضل ما بين الكعب والثمنية كعاب] سبعة كعاب تعادل سبعة امثال المربّع الذى يكون من ضلع المكعّب الاعظم وضلع المكعّب الاعظم شيئان ومربّعه اربعة اموال وسبعة امثال ذلك ثمنية وعشرون مالاً فثمنية وعشرون مالاً تعادل سبعة كعاب فنقسم الجميع على مال فيكون ثمنية وعشرون

---

254 (88): كعباً : كعبٌ in cod.

256 (89): المكعّب : الكعب in cod.

262 (90): Verba ونعمل ذلك بوجه آخر atramento rubro in codice.

263 (91): Suntne verba شيئًا واحدًا ونجعل المكعّب الثانى من اشياء كم شئنا فنجعله post فنجعله supplenda? Sed insuper desideratur الثانى post المكعّب in linea 263 et etiam illo addito locus non sanaretur (vide adn. seq.).

263-264 (92): Verba فضل ... كعاب a quodam arabico lectore addita esse censeo; lacunam enim habet textus ante فيبقى.

267 (93): وعشرون : وعشرين in cod.

احداً تعادل سبعة اشياء فالشىء الواحد يعادل اربعة آحاد ولذلك يكون المكعّب الاصفر اربعة وستّين [وضلعه فُرض شيئاً واحداً] فامّا المكعّب الاعظم فلان ضلعه فُرض من شيئين يكون ضلعه ثمنية آحاد ويكون المكعّب خمسمائة واثنى عشر فقد تبيّن ان المكعّب الآخر ‹و›هو الاعظم يزيد على المكعّب / الاصفر سبعةَ امثال المربّع ‹من› ضلع ١٢ المكعّب الاعظم وذلك هو الشرط الذى شُرط لنا فى هذه المسئلة وذلك ما اردنا ان نجد

يد نريد ان نجد عدداً اذا ضربناه فى عددين مفروضين كان احدهما مكعّباً والآخر مربّعاً

فنفرض العددين خمسة وعشرة ونريد ان نجد عدداً اذا ضربناه فى العشرة كان مكعّباً واذا ضربناه فى الخمسة كان مربّعاً فنفرض العدد المطلوب شيئاً ونضربه فى الخمسة فيكون خمسة اشياء ثمّ نضربه فى العشرة فيكون عشرة اشياء ونريد ان نعدل العشرة الاشياء بعدد مكعّب والخمسة الاشياء بعدد مربّع فنفرض المربّع المعادل لخمسة اشياء من مربّع ضلع المكعّب المعادل للعشرة الاشياء اىّ جزء شئنا او ايّت اجزاء شئنا بعد ان يكون ضلع الجزء مشاركاً لضلع الكلّ اعنى ان يكون الجزء مربّعاً او ان نفرض مربّع ضلع المكعّب من المربّع المعادل للخمسة الاشياء اىّ جزء او ايّت اجزاء بعد ان يكون مربّعاً فلنفرض مربّع ضلع المكعّب من المربّع المعادل للخمسة الاشياء رُبعاً

---

269 (94): المكعّب : الكعب in cod.

(95): Seclusa verba adnotamentum quod e margine in textum irrepsit esse suspicor.

271 (96): واثنى : واثنا in cod.

(97): و ante هو addidi; etiam arbitrari licet, verba هو الاعظم (lin. 271-272) additionem esse lectoris.

272 (98): Deficiens من restitui.

282 (99): المكعّب : الكعب in cod.

283 (100): Post الجزء addit codex perverse او العشرة .

285 (101): او:و in cod.

286 (102): مربّع: مُربّعا in cod.

(103): المكعّب: الكعب in cod.

(104): المربّع: الربع in cod.

(105): رُبعاً : الرُبّع in cod.

‹حتّى› يكون مربّع ضلع المكعّب المعادل للعشرة الاشياء شيئاً واحداً وربُعَ شىء فيكون هذا المربّع الذى هو الشىء الواحد والرُبع شىء اذا ضربناه فى ضلعه كان منه عشرة اشياء فان قسمنا العشرة الاشياء على الشىء والرُبع شىء كان الذى يخرج منه هو ضلع المكعّب المعادل للعشرة الاشياء والذى يخرج من قسمة العشرة الاشياء على الشىء والرُبع ‹شىء› ثمنية آحاد [لان الاشياء اذا ضُربت فى الآحاد كان منها اشياء] فاذاً الثمنية الآحاد هو ضلع المكعّب المعادل للعشرة الاشياء وهو ضلع المربّع المعادل للشىء الواحد والرُبع شىء والمكعّب الذى يكون ‹من ضلع› ثمنية آحاد هو خمس مائة واثنا عشر وذلك يعادل عشرة اشياء فيكون الشىء الواحد احد وخمسين وخُمساً وايضاً مربّع الثمنية آحاد اربعة وستّون وذلك يعادل شيئاً واحداً وربُعَ شىء
فالشىء الواحد اربعة اخماس الاربعة والستّين / وهو احد وخمسون ١٢
وخُمس فاذا ضربنا الاحد والخمسين والخُمس فى العشرة ادّى ‹ذلك الى› خمس مائة واثنى عشر وهو عدد مكعّب واذا ضربناه فى الخمسة كان مائتين وستّة وخمسين وهو مربّع ضلعه ستّة عشر

فقد وجدنا عدداً اذا ضربناه فى العددين المفروضين وهما العشرة والخمسة اجتمع من ضربه فى العشرة عدد مكعّب ومن ضربه فى الخمسة عدد مربّع وهو الذى اردنا ان نجد

---

287 (106): حتّى addidi.

288 (107): فيكون : يكون in cod.

292 (108): شىء addidi.

292-293 (109): Verba: لان الاشياء ... منها اشياء (cod.: منها احاد) scholiastae cuidam attribuenda esse suspicor; imperitiae autem scholiastae indicium est usus verbi آحاد pro اشياء, nisi erranti librario tribuendus sit (cf. adn. 32).

294 (110): والرُبع شىء : والمُربع شىء in cod.

295 (111): Desiderata verba من ضلع addidi.

(112): واثنا : واسى in cod.

296 (113): وخُمساً : وَخُمس in cod.

297 (114): وستّون : وستّين in cod.

298 (115): وخمسون : وخَمسين in cod.

299-300 (116): ذلك الى addidi.

300 (117): واثنى : واثنا in cod.

فان اردنا ان يكون ما يجتمع من ضرب الشىء فى الخمسة هو المكعّب ومن ضربه فى العشرة هو العدد المربّع فانّا ايضاً خمسة اشياء نعادل بها عدداً مكعّباً وعشرةُ اشياء نعادل بها عدداً مربّعاً ونجعل العدد المربّع الذى من ضلع المكعّب المعادل للخمسة الاشياء ⟨من المربّع⟩ المعادل للعشرة الاشياء رُبعاً ايضاً حتّى يكون مربّع ضلع المكعّب المعادل للخمسة الاشياء شيئين ونصف شىء فاذا قسمنا عليه الخمسة الاشياء كان ما يخرج من ذلك ضلع المكعّب المعادل للخمسة الاشياء ولكن اذا قسمنا الخمسة الاشياء على الشيئين ونصف ⟨شىء⟩ فان الذى يخرج ⟨منه⟩ اثنان فاذاً ضلع المكعّب المعادل للخمسة الاشياء اثنان فاذاً المكعّب الذى يكون معادلاً للخمسة الاشياء ثمنية آحاد فيكون الشىء الواحد ثمنية اخماس واحد فاذا ضربناه فى الخمسة اجتمع منه اربعون خُمساً اعنى ثمنية آحاد وهو عدد مكعّب وان ضربناه فى العشرة اجتمع منه ثمنون خُمساً اعنى ستّة عشر احداً وهو مربّع ضلعه اربعة

ولنفرض فى المسئلة الاولى ان المربّع الذى يكون معادلاً للخمسة الاشياء من المربّع الذى يكون من ضلع المكعّب المعادل للعشرة ⟨الاشياء⟩ فى

---

308 (118): من المربّع addidi.

309 (119): المعادل : لمُعادل in cod.

(120): للعشرة : العشره in cod.

(121): رُبعاً : المربّع in cod.

(122): Post مربّع addit codex من.

310 (123): للخمسة : للعشره in cod.

(124): Verbum شىء sequuntur in codice verba فيكون ضلع المكعّب المُعادل للخمسة شئين ونصف شىء.

312 (125): شىء addidi.

313 (126): منه addidi; vide adn. seq.

314 (127): Post يكون addit codex منه.

315 (128): اربعون : اربعين in cod.

317 (129): ثمنون : ثمنين in cod.

(130): اربعة : مُربّع in cod.

318 (131): Post ان prave inserta sunt verba مربع ضلع المكعّب in cod.

319 (132): Desideratum verbum الاشياء addidi.

نسبة الرُبع حتّى يكون مربّع ضلع المكعّب المعادل للعشرة الاشياء عشرين شيئاً فاذا قُسمت العشرة الاشياء على العشرين الشيء كان القسم نصفاً ويكون ذلك هو ضلع المكعّب المعادل للعشرة / الاشياء ١٤ والمكعّب الذى يكون من النصف هو ثُمن واحد فيكون العشرة الاشياء تعادل ثُمنَ واحد فيكون الشيء الواحد جزءاً من ثمنين جزءاً فاذا ضربناه فى الخمسة اجتمع منه خمسة اجزاء من ثمنين اعنى جزءاً واحداً من ستّة عشر وهو مربّع ضلعه رُبع واحد وان ضُرب فى العشرة كان عشرة اجزاء من ثمنين اعنى ثُمنَ واحد وهو مكعّب ضلعه نصف واحد

فان فرضنا فى عكس المسئلة ان المربّع الذى يكون معادلاً للعشرة الاشياء من مربّع ضلع المكعّب المعادل للخمسة الاشياء ⟨فى نسبة الرُبع كان مربّع ضلع المكعّب المعادل للخمسة الاشياء⟩ اربعين شيئاً فاذا قسمنا عليه الخمسة الاشياء خرج جزء من ثمنية اجزاء من واحد فيكون ضلع المكعّب المعادل للخمسة الاشياء ثُمنَ واحد ⟨و⟩ يكون المكعّب جزءاً واحداً من خمس مائة واثنى عشر ويكون الخمسة الاشياء تعادل جزءاً واحداً من خمس مائة واثنى عشر والشيء الواحد يساوى جزءاً واحداً من ألفين وخمس مائة وستّين فاذا ضربناه فى العشرة كان عشرة اجزاء من ألفين وخمس مائة وستّين اعنى جزءاً واحداً من مأتين وستّة وخمسين وهو مربّع من ضلع جزء واحد من ستّة عشر وان ضربناه فى الخمسة كان منه خمسة اجزاء من ألفين وخمس مائة وستّين اعنى جزءاً واحداً من خمس مائة واثنى عشر وهو مكعّب من ضلع ثُمن واحد

---

320 (133): الرُبع : المربّع in cod.

(134): Post الرُبع supra dictum praebet codex والاشيآ (cf. adn. 132).

322 (135): Codex ab initio quartae decimae paginae a manu altera exaratur.

324 (136): جزءاً (posterius): حرو (sc. جزء ) in cod.

325 (137): واحداً : واحد in cod.

328 (138): فرضنا : صرىنا (sc. ضربنا) in cod.

(139): Post ان addit codex ضلع .

329-330 (140): Per homoeoteleuton omissa restitui; interpretis aut graeci commentatoris impropriis verbis نسبة الرُبع (vide lin. 320) usus sum.

332 (141): و addidi.

فقد وجد نا عد داً اذا ضربناه فى كلّ واحد من العشرة والخمسة كان عد داً مربّعاً وعد داً مكعّباً

ونعمل ايضاً بجهة اخرى نفرض المكعّب الذى يجتمع من ضرب العدد المطلوب فى العشرة من ضلع كم شئنا من الاشياء فلنفرضه من ضلع شى ء حتّى يكون كعباً واحداً ويكون العدد المطلوب جزءاً من عشرة اجزاء من كعب فنحتاج ان يكون هذا الجزء اذا ضُرب فى الخمسة اجتمع منه عدد مربّع ولكن اذا ضربنا جزءاً واحداً / من عشرة ١٥
اجزاء من كعب فى خمسة آحاد اجتمع منه خمسة اجزاء ⟨من عشرة اجزاء من كعب⟩ اعنى نصفَ كعب وذلك معادل لعدد مربّع فلنعمل المربّع من ضلع كم شئنا من الاشياء فلنفرضه من ضلع شيئين حتّى يكون اربعة اموال فاذاً نصف الكعب يعادل اربعة اموال فنقسم الجميع على مال فيكون نصف شى ء يعادل اربعة آحاد فيكون الشى ء الواحد ثمنية آحاد ولانّا فرضنا المكعّب الذى يجتمع من ضرب العدد المطلوب فى العشرة من ضلع شى ء واحد يكون ضلعه ثمنية آحاد ويكون المكعّب خمس مائة واثنى عشر فنقسم الخمس مائة والاثنى عشر على عشرة فيخرج من ذلك العدد المطلوب وهو احد وخمسون وخُمس

وان شئنا فرضنا ضلع المربّع الذى يجتمع من ضرب العدد المطلوب فى الخمسة كم شئنا من الاشياء فلنفرضه من ضلع شى ء واحد حتّى يكون مالاً واحداً فاذاً العدد المطلوب جزء واحد من خمسة اجزاء من مال فاذا ضربناه فى العشرة الآحاد اجتمع عشرة اجزاء من خمسة اجزاء من مال اعنى مالين وذلك يعادل عدداً مكعّباً فنفرض المكعّب من ضلع كم شئنا من الاشياء فلنفرضه من ⟨ضلع⟩ شى ء واحد حتّى يكون كعباً واحداً فيكون مالان يعادلان كعباً واحداً فنقسم الجميع على مال فيكون شى ء واحد يعادل احدين ولانّا فرضنا المربّع من ضلع شى ء واحد يكون ضلعه احدين ويكون هو اربعة آحاد فاذاً العدد المطلوب اذا ضربناه فى الخمسة الآحاد اجتمع منه اربعة آحاد

---

343 (142): Verba ونعمل ايضاً بجهة اخرى atramento rubro in cod.

348-349 (143): Verba من عشرة اجزاء من كعب addidi.

349 (144): كعب : مكعب in cod.

351 (145): الكعب : المكعب in cod.

363 (146): ضلع addidi.

366 (147): اربعة آحاد : الاربعه الاحاد in cod.

فاذاً العدد المطلوب هو اربعة / اخماس فاذا ضربناه فى الخمسة ١٦
اجتمع منه عشرون خُمساً ⟨اعنى⟩ اربعة آحاد وهو مربّع واذا ضربناه فى العشرة اجتمع اربعون خُمساً اعنى ثمنية آحاد وهو مكعّب

فقد وجدنا عدداً اذا ضربناه فى العشرة وفى الخمسة اجتمع منهما عدد مربّع وعدد مكعّب

يه نريد ان نجد عدداً اذا ضربناه فى عددين مفروضين اجتمع من ضربه فى احدهما عدد مكعّب ومن ⟨ضربه فى⟩ الآخر المربّع الذى يكون من ضلع ذلك المكعّب

فليكن احد العددين المفروضين اربعة والآخر عشرة ونريد ان نجد عدداً اذا ضربناه فى العشرة كان منه عدد مكعّب وان ضربناه فى اربعة اجتمع منه المربّع الذى يكون من ضلع المكعّب او عكس ذلك فان المأخذ منهما مأخذ واحد وقياسه بالمتقدّم نفرض العدد ⟨المطلوب⟩ شيئاً فيكون المكعّب عشرة اشياء والمربّع الذى يكون من ضلعه اربعة اشياء وان ضلع المكعّب اذا ضُرب فى مثله اجتمع منه اربعة اشياء وكلّ المكعّب عشرة اشياء فلان الاربعة الاشياء اذا ضُربت فى ضلعها اجتمع منه عشرة اشياء فنقسم العشرة الاشياء على الاربعة الاشياء فيكون ضلع المكعّب اثنين ونصفاً ويكون المربّع الذى يكون منه ستّة آحاد ورُبعاً فاذاً الاربعة اشياء مساوية لستّة آحاد ورُبع فنضرب كلّ ذلك فى اربعة لِمكان الجزء الذى هو الرُبع فيكون ستّة عشر شيئاً تعادل خمسة وعشرين احداً فيكون الشىء الواحد خمسة وعشرين جزءاً من ستّة عشر جزءاً

وعلى جهة المأخذ الثانى نجعل المكعّب الذى يكون من ضرب العدد المطلوب ⟨فى العشرة من ضلع كم شئنا من الاشياء فلنفرضه من ضلع شىء واحد حتّى يكون كعباً واحداً فاذاً العدد المطلوب⟩ عُشر الكعب والذى يجتمع من ضربه فى الاربعة اربعةُ اعشار كعب

---

369 (148): اعنى deest in cod.
(149): واذا : فاذا in cod.
374 (150): In uncis seclusa verba addidi.
379 (151): منهما (forsitan فيهما scribendum): سها in cod.
(152): Pro بالعدم codicis بالمتقدّم scripsi.
(153): المطلوب addidi.
384 (154): ونصفاً : وصف in cod.
389-390 (155): Per homoeoteleuton omissa restitui.

١٧ فاذاً اربعة اعشار كعب تعادل مربّع / ضلع المكعّب الذى هو شىء ومربّعه مال واحد فنضرب كلّ الذى معنا فى عشرة لِمكان الاجزاء التى هى الاعشار فيكون اربعة كعاب تعادل عشرة اموال فنقسم الجميع على مال فيكون اربعة اشياء تعادل عشرة آحاد فيكون الشىء يعادل احدين ونصفاً ويكون ضلع ⟨المكعّب ايضاً احدين ونصفاً واربعة امثال العدد المطلوب التى هى تعادل⟩ مربّع ضلع المكعّب ستّة آحاد وربُع ولذلك يكون العدد المطلوب خمسة وعشرين جزءاً من ستّة عشر جزءاً ومن البيّن ان هذا العدد اذا ضُرب فى الاربعة اجتمع منه مائة جزء من ستّة عشر جزءاً وهو مربّع وان ضُرب فى العشرة اجتمع مائتان وخمسون جزءاً من ستّة عشر فمن البيّن ان المأتين والخمسين من ستّة عشر تكون خمسة عشر احداً ونصفاً وثُمناً وهو مكعّب وضلعه اثنان ونصف ومربّعه ستّة آحاد وربُع وكذلك اذا ضُربت الخمسة والعشرون جزءاً من ستّة عشر فى اربعة كانت مائة جزء من ستّة عشر وهى ستّة آحاد وربُع وهو مربّع ضلعه اثنان ونصف

فقد وجدنا عدداً اذا ضربناه فى عددين مفروضين خرج احدُهما عدداً مكعّباً ومن ضربه فى الآخر المربّع الذى يكون من ضلع ذلك المكعّب

فان اردنا ان نجد عددين يكون احدهما من الآخر فى نسبة مفروضة ويكون احدهما عدداً مكعّباً و⟨يكون⟩ العدد الآخر مربّعاً وفرضنا نسبة هى نسبة الثلثة الى الواحد فرضنا اوّلاً عددين يكون

---

396 (156): ونصفاً : وىصف in cod.

396-397 (157): A librario omissam ut opinor lineam restitui.

398 (158): جزءاً : احرا in cod.

401 (159): مائتان وخمسون : ماىں وحمسں in cod.

402 (160): ونصفاً وثمناً : وىصف وىمں in cod.

403 (161): Pro ولذلك codicis وكذلك scripsi, sed dubitanter; fortasse enim interpolata sunt verba hinc usque ad ونصف (lin.405), quae dicta linearum 399-400 partim iterant.

404 (162): والعشرون : والعشرس in cod.

407 (163): Melius omisisset interpres verba من ضربه فى.

409 (164): نسبة : سىه in codice. Vide adn. 215.

410 (165): يكون addidi.

احدهما ثلثة امثال الآخر ثمّ طلبنا بمثل المأخذ الذى تقدّم عدداً
اذا ضربناه فى كلّ واحد من العددين المفروضين اجتمع منهما عدد
مربّع وعدد مكعّب فنكون قد وجدنا عددين فى نسبة الثلثة الامثال
احدهما مكعّب والآخر مربّع / لان كلّ عدد يُضرب فى عددين فان ١٨
الذى يجتمع من الضرب يكون فى نسبة العددين الاوّلين

يو نريد ان نجد عددين اذا ضربناهما فى عدد مفروض كان الذى يجتمع من احدهما عدداً مكعّباً ومن الآخر ضلع ذلك المكعّب

فنفرض العدد عشرةٍ ونريد ان نجد عددين اذا ضربناهما فى عشرة كان الذى يجتمع من ضرب العشرة فى احدهما عدداً مكعّباً واذا ضربنا العشرة فى الآخر كان الذى يجتمع منه ضلع ذلك المكعّب فلنفرض العدد الاوّل كم شئنا من الاشياء فلنفرضه شيئاً واحداً ونضربه فى عشرة فيكون عشرةَ اشياء وهى ضلع المكعّب ولذلك يكون المكعّب الذى يجتمع من ضرب العدد الثانى فى العشرة الفَ كعب ونفرض العدد الثانى كم شئنا من الاموال فلنفرضه ثلثمائة مال ونضربه فى العشرة فيكون ثلثة آلُف مال فاذاً الف كعب تعادل ثلثة آلُف مال فنقسم الجميع على مال واحد فيكون الف شىء تعادل ثلثة آلُف احد فلذلك يكون الشى الواحد ثلثة آحاد ولانّا فرضنا العدد الاوّل شيئاً واحداً يكون ثلثة آحاد ولانّا فرضنا العدد الثانى ثلثمائة مال والمال تسعة يكون ألفين وسبع مائة فاذا ضربنا العدد الثانى فى العشرة كان سبعة وعشرين الفاً واذا ضربنا العدد الاوّل فى العشرة اجتمع منه ثلثون والثلثون هى ضلع المكعّب الذى هو سبعة وعشرون الفاً

فقد وجدنا عددين على الشرط الذى شُرط لنا وهما الثلثة والألفان والسبع مائة وذلك ما اردنا ان نجد

---

412 (166): عدداً : من عدد in cod.

414 (167): Pro فنكون praebet codex فيكون .

415-416 (168): Melius dixisset interpres: لان كلّ عدد يُضرب فى عددين فان اللذين يجتمعان من الضرب (vel من الضربين) يكونان فى نسبة العددين الاوّلين .

418 (169): Post احدهما addit codex منه , quod delevi; etiam ضرب ante احدهما addere licet.

(170): عدداً مكعّباً : عدد مكعب in cod.

433 (171): وهما : وهى in cod.

يز نريد ان نجد عددين مربّعين يكون ضلعاهما فى نسبة مفروضة واذا
ضُرب كلّ واحد منهما فى عدد / مفروض كان المجتمع من احدهما مكعّباً ١٩
ومن الآخر ضلع ذلك المكعّب

وينبغى ان يكون العدد الذى للنسبة المفروضة مع العدد المفروض
يحيطان بعدد مربّع فهذه من التأتّى التى تُدعى المهيّأة

فلتكن النسبة المفروضة نسبة العشرين مثلاً والعدد المفروض خمسة
آحاد فنريد ان نجد عددين مربّعين يكون ضلع احدهما من ضلع الآخر
فى نسبة العشرين مثلاً واذا ضُرب المربّع الاعظم فى خمسة آحاد كان
المجتمع عدداً مكعّباً واذا ضُرب المربّع الاصفر فى الخمسة كان المجتمع
ضلع ذلك المكعّب فنفرض ضلع المربّع الاصفر شيئاً واحداً حتّى يكون ⟨مربّعه⟩
مالاً واحداً فيكون ضلع المربّع الاعظم عشرين شيئاً والمربّع الاعظم اربع مائة
مال فنضرب الاربع مائة المال فى خمسة فتكون الفى مال ونضرب المال
الواحد فى خمسة آحاد فيكون خمسة اموال ولانّه شُرط فى هذه المسئلة
ان يكون الألفا المال مكعّباً من ضلع الخمسة الاموال فنضرب الخمسة
الاموال فى مثلها ثمّ فى مثلها فيكون ذلك مائة وخمسة وعشرين كعب
كعب فاذاً المائة والخمسة والعشرون كعب كعب تعادل الفى مال
فنقسم الجميع على واحد من اقعد الناحيتين اعنى المال فيكون مائة
وخمسة وعشرون مال مال تعادل الفى احد فلذلك يكون المال المال
الواحد يعادل ستّة عشر احداً ومال المال مربّع مربّع الضلعَ وكذلك
الستّة عشر عدد مربّع مربّع الضلعَ وهما متساويان فضلعا ضلعيهما
ايضاً متساويان وضلع ضلع المال المال شىء واحد وضلع ضلع الستّة
عشر احدان فلذلك الشىء الواحد يعادل احدين / ولان المربّع ٢٠
الاصفر عملناه من ضلع شىء واحد يكون اربعة آحاد ولانّا عملنا المربّع
الاعظم من عشرين شيئاً يكون ضلعه اربعين احداً ويكون هو الفاً وستّمائة
واذا ضربنا الألف والستّمائة فى العدد المفروض الذى هو خمسة آحاد
كان من ذلك ثمنية الٰف وهى مكعّب من ضلع عشرين احداً الذى هو

---

439 (172): فهذه من التأتّى التى تُدعى المهيّأة conjecturâ mea; in codice enim invenitur فدره مرالبابى الدى ىدعا المهىاه .

442 (173): واذا : فادا in cod.

443 (174): Post المجتمع (prius) addit codex و .

444 (175): مربّعه addidi.

456 (176): احدان : احدس in cod.

458 (177): الفًا : الف in cod.

ما يجتمع من ضرب المربّع الاصفر الذى قد تبيّن انّه اربعة آحاد فى العدد المفروض الذى هو خمسة آحاد

فقد وجدنا عددين على الشرط الذى شُرط لنا وهما الاربعة والألف والستّمائة وذلك ما اردنا ان نجد

يح نريد ان نجد عددين مكعّبين يكون ضلعاهما فى نسبة مفروضة واذا ضُرب كلّ واحد منهما فى عدد مفروض كان المجتمع من احدهما مربّعاً ومن الآخر ضلع ذلك المربّع

وينبغى ان يكون العدد المفروض مكعّباً

ولتكن النسبة المفروضة نسبة الثلثة الامثال والعدد المفروض ثمنية آحاد ونريد ان نجد عددين مكعّبين يكون ضلع احدهما من ضلع الآخر فى نسبة الثلثة الامثال ويكون مضروب الاعظم منهما فى الثمنية الآحاد عدداً مربّعاً ومضروب الاصفر منهما فى الثمنية الآحاد ضلع ذلك المربّع فنفرض المكعّب الاصفر من ضلع شىء واحد فيكون كعباً واحداً ويكون ضلع المكعّب الاعظم ثلثة اشياء فيكون هو سبعة وعشرين كعباً واذا ضربنا السبعة والعشرين الكعب فى ثمنية آحاد كانت مأتين وستّة عشر كعباً واذا ضربنا الكعب الواحد فى ثمنية آحاد كانت ثمنية كعاب فلان المأتين والستّة العشر الكعب مربّع من ضلع الثمنية / الكعاب الذى هو اربعة وستّون كعب كعب يكون المائتان ٢١
والستّة عشر كعباً تعادل اربعة وستّين كعب كعب فاذا قسمناهما على الكعب الواحد الذى هو فى اقعد الناحيتين كانت مائتان وستّة عشر احداً تعادل اربعة وستّين كعباً فلذلك الكعب الواحد ثلثة آحاد وثلثة اثمان احدٍ والكعب الواحد مكعّب من ضلع شىء واحد والثلثة الآحاد والثلثة الاثمان الاحد مكعّب من ضلع واحد ونصف فاذاً الشىء الواحد يعدل واحداً ونصفاً ولذلك المكعّب الاصفر ثلثة آحاد وثلثة اثمان احد والمكعّب الاعظم ضلعه اربعة آحاد ونصف وهو احد وتسعون وثُمن واذا ضُرب هذا المكعّب الاعظم فى الثمنية الآحاد اجتمع منه سبع مائة وتسعة وعشرون وهو مربّع من ضلع

---

467 (178): المربّع: المال in cod.
478 (179): المائتان: المامس in cod.
479 (180): كعباً: كعب in cod.
(181): قسمناهما : صماها ( صماها paene) in cod.
484 (182): ولذلك: فكذلك in cod.

سبعة وعشرين احداً التى هى ما يجتمع من ضرب المكّعب الاصغر الذى قد تبيّن انّه ثلثة وثلثة اثمان فى العدد المفروض الذى هو ثمنية آحاد

فقد وجدنا عددين على الشرط الذى شُرط لنا وذلك ما اردنا ان نجد

يط نريد ان نجد عدداً اذا ضربناه فى عددين مفروضين اجتمع من ضربه فى احدهما مكّعب ومن ضربه فى الآخر ضلع ذلك المكّعب

وينبغى ان يكون العددان المفروضان يحيطان بعدد مربّع وهذه ايضاً من المسائل المهيّأة

فليكن احد العددين المفروضين خمسة آحاد والآخر عشرين احداً فنفرض العدد المطلوب شيئاً واحداً ونضربه فى الخمسة الآحاد فيكون خمسة اشياء ونضربه ايضاً فى العشرين فيكون عشرين شيئاً فلان العشرين الشىء مكّعب من / ضلع خمسة اشياء وضلع كلّ مكّعب اذا ضُرب فى ٢٢
مربّعه كان منه المكّعب والمكّعب عشرون شيئاً فهى اذاً اذا قُسمت على ضلعها الذى هو خمسة اشياء كان منها مربّع ضلع العشرين الشىء لكن العشرين الشىء اذا قُسمت على الخمسة الاشياء كان منها اربعة آحاد فاذاً الاربعة الآحاد مربّع من ضلع خمسة اشياء فاذاً ضلع الاربعة الذى هو اثنان يعادل خمسة اشياء فالشىء الواحد خُمساً احد فاذا ضربناه فى العشرين كان منه ثمنية آحاد وهى مكّعب من ضلع احدين اللذين هما ما يجتمع من ضرب ⟨العدد المطلوب⟩ الذى قد تبيّن انّه خُمساً واحد فى العدد الآخر المفروض الذى هو خمسة آحاد

فقد وجدنا عدداً اذا ضربناه فى العددين المفروضين اللذين احدهما خمسة آحاد والآخر عشرون احداً اجتمع من ضربه فى العشرين مكّعب ومن ضربه فى الخمسة ضلع ذلك المكّعب وهو خُمساً واحد وذلك ما اردنا ان نجد

ك نريد ان نجد عدداً مكّعباً اذا ضربناه فى عددين مفروضين اجتمع من ضربه فى احدهما مربّع ومن ضربه فى الآخر ضلع ذلك المربّع

---

497 (183): عشرين : عسرون in cod.

507 (184): العدد المطلوب addidi.

وينبغى ان يكون مربّع احد العددين المفروضين يعدّ العدد الآخر بعدد مكعّب

فليكن احد العددين المفروضين خمسة آحاد والعدد الآخر مائتى احد ونريد ان نجد عدداً مكعّباً اذا ضربناه فى المأتين اجتمع منه مربّع واذا ضربناه فى الخمسة الآحاد اجتمع منه ضلع ذلك المربّع فنفرض المكعّب المطلوب من ضلع شىء واحد حتّى يكون كعباً واحداً ونضربه فى المأتين فيكون مائتى كعب و⟨نضربه ايضاً فى الخمسة فيكون⟩ خمسة كعاب / ٢٣ ولان مائتى الكعب هى مربّع من ضلع خمسة كعاب وكلّ مربّع يُقسم على ضلعه يكون الذى يخرج مساوياً لضلعه ومائتا الكعب اذا قُسمت على خمسة كعاب كان منها اربعون احداً فاذاً الخمسة الكعاب تعادل اربعين احداً ولذلك الكعب الواحد يعادل ثمنية آحاد والكعب الواحد مكعّب من ضلع شىء واحد والثمنية مكعّب من ضلع احدين فيكون الشىء ⟨الذى⟩ فرضناه ضلعاً للمكعّب المطلوب احدين ويكون المكعّب ثمنية آحاد فاذا ضربناه فى المأتين اجتمع منه الف وستّمائة واذا ضربناه فى الخمسة اجتمع منه اربعون وهو ضلع ⟨المربّع⟩ الذى هو الف وستّمائة

فقد وجدنا عدداً مكعّباً اذا ضربناه فى العددين المفروضين اللذين هما المائتان والخمسة الآحاد اجتمع من ضربه فى المأتين مربّع ومن ضربه فى الخمسة ضلع ذلك المربّع وهو ثمنية آحاد وذلك ما اردنا ان نجد

كا نريد ان نجد عدداً مربّعاً اذا ضربناه فى عددين مفروضين اجتمع من ضربه فى احدهما مكعّب ومن ضربه فى الآخر ضلع ذلك المكعّب

---

516 (185): Pro الاحر (sc. الاخير) codicis الآخر scripsi.

520 (186): مربّع bis in codice, primum in fine lineae, iterum in initio sequentis.

522 (187): مائتى : ماس in cod.

(188): Verbis نضربه ..... فيكون textum supplevi.

523 (189): خمسة (prius): حمسه in cod.

528 (190): الذى addidi.

(191): المطلوب : المعروص in cod.

531 (192): المربّع addidi.

وينبغى ان يكون العددان المفروضان يحيطان بعدد مربّع ⟨مربّع⟩ الضلعَ

فليكن احد العددين المفروضين اثنين والعدد الآخر اربعين ونصفاً وظاهر ان العدد المسطّح الذى يحيط به هذان العددان هو واحد وثمنون مربّع مربّع الضلعَ ونريد ان نجد عدداً مربّعاً اذا ضربناه فى الاربعين والنصف وفى الاثنين اجتمع من ضربه فى الاربعين والنصف / مكعّب ومن ضربه فى الاثنين ضلع ذلك المكعّب فنفرض ٢٤ المربّع مالاً واحداً ونضربه فى كلى العددين المفروضين فيكون احد المضروبين اربعين مالاً ونصف مال والمضروب الآخر مالين فلان الاربعين المال والنصف المال مكعّب من ضلع المالين وكلّ ⟨مكعّب⟩ يُقسم على ضلعه يكون منه مربّع ذلك الضلع والاربعون المال والنصف مال اذا قُسمت على مالين كان القسم عشرين ورُبعاً فاذاً مربّع المالين يعادل عشرين ورُبعاً وضلع العشرين والرُبع اربعة آحاد ونصف فالمال الواحد يعدل احدين ورُبعاً وهو مربّع من ضلع واحد ونصف واذا ضُرب هذا المربّع الذى هو اثنان ورُبع فى احد العددين المفروضين الذى هو اربعون ونصف كان منه احد وتسعون وثُمن وهو مكعّب من ضلع اربعة آحاد ونصف الذى هو ما يجتمع من ضرب العدد المربّع المطلوب الذى قد تبيّن انّه اثنان ورُبع فى العدد الآخر المفروض وهو اثنان

فقد وجدنا عدداً مربّعاً على الشرط الذى شُرط لنا وهو اثنان ورُبع وذلك ما اردنا ان نجد

---

538 (193): Per haplographiam omissum مربّع restitui.

540 (194): اثنين : ابيان in cod.

547 (195): Deficiens مكعّب restitui.

548 (196): والاربعون : والاربعس in cod.

550 (197): فالمال : والمال in cod.

551 (198): Pro يعدل habet codex بعدل; librarius enim بعدد in بعدل mutavit.

(199): ورُبعاً : وربع in cod.

553 (200): وثُمن : وثما in cod.

554 (201): هو : هى in cod.

556 (202): اثنان : ماسان (sc. مائتان ) in cod.

وانّما احتجنا ان يكون العددان المفروضان على الشرط الذى وصفنا لِما اقول انّا اذا فرضنا المربّع المطلوب مالاً واحداً ثمّ ضربناه فى كلّ واحد من العددين المفروضين كان كلّ واحد من المضروبين اموالاً وأحد هذين المضروبين مكعّب من ضلع الاموال التى هو المضروب الآخر واذا قُسم الذى هو المكعّب منهما على الذى هو الضلع منهما كان الذى يخرج من القسمة عدداً معادلاً لمربّع الاموال التى هى الضلع / ولذلك ينبغى ان يكون العدد الذى يخرج ٢٥
من القسمة مربّعاً ليكون ضلعه عدداً معادلاً للاموال التى هى الضلع فلذلك احتجنا ان يكون العددان المفروضان اذا قُسم احدهما على الآخر كان منه مربّع والعددان اللذان هما كذلك مضروب احدهما فى الآخر ايضاً مربّع ولمّا كان العدد الذى هو ضلع العدد المربّع الذى هو القسم الذى خرج من قسمة احد العددين على الآخر معادلاً للاموال التى عددها مساوٍ للعدد المقسوم عليه من العددين المفروضين احتجنا ان يكون هذا العدد ايضاً اذا قُسم على الاموال المعادلة له كان القسم مربّعاً ليكون المال الواحد يعادل عدداً مربّعاً ولذلك وجب ان يكون احد العددين المفروضين اذا قُسم على الآخر وكان القسم مربّعاً ان يكون ضلعه ايضاً اذا قُسم على العدد المقسوم عليه كان القسم مربّعاً اعنى ايضاً ان يكون مضروب هذا الضلع فى العدد المفروض ⟨الذى هو⟩ المقسوم عليه مربّعاً والعددان اللذان اذا قُسم احدهما على الآخر كان منه مربّع اذا قُسم ضلعه على العدد المقسوم عليه كان منه مربّع هما العددان اللذان اذا ضُرب احدهما فى الآخر كان منهما مربّع مربّع الضلعَ وذلك ما كان ينبغى ان يبيّن

كب نريد ان نجد عدداً مكعّباً اذا ضربناه فى عددين مفروضين كان منهما مكعّب وضلع ذلك المكعّب

وينبغى ان نجد اوّلاً خاصّة العددين المفروضين فنقول انّا اذا فرضنا المكعّب المطلوب كعباً واحداً وضربناه فى كلّ واحد من العددين المفروضين كان كلّ واحد / من المضروبين كعاباً وأحد هذين المضروبين ٢٦

560 (203): Loco انّا praebet codex لما كا (sc. لمّا كنّا).

566 (204): مربع : مربّعاً in cod.

577 (205): الذى هو addidi.

578 (206): والعددان : والعدد in cod.

مكعّب من ضلع المضروب ⟨الآخر⟩ واذا قُسمت الكعاب التى هى المكعّب من المضروبين على الكعاب التى هى الضلع منهما كان الذى يخرج من القسمة عدداً معادلاً لمربّع الكعاب التى هى الضلع ولذلك ينبغى ان يكون العدد الذى يخرج من القسمة مربّعاً ليكون ضلعه معادلاً للكعاب التى هى الضلع فمن ذا نفرض العددين المفروضين ان يكونا اذا قُسم احدهما على الآخر كان القسم مربّعاً وايضاً لمّا كان العدد الذى هو ضلع العدد المربّع الذى خرج من القسمة معادلاً للكعاب التى هى الضلع وعددها مساوٍ للعدد المقسوم عليه من العددين المفروضين ينبغى ان يكون هذا العدد اذا قُسم على الكعاب المعادلة ⟨له⟩ ان يكون القسم مكعّباً ليكون الكعب الواحد يعادل عدداً مكعّباً فخاصّة هذين العددين اذاً على التمام ⟨و⟩ هو انّه اذا قُسم احدهما على الآخر كان القسم مربّعاً واذا قُسم ضلع هذا المربّع على العدد المقسوم عليه كان القسم مكعّباً

ينبغى ان نستخرج هذين العددين فنفرض احدهما احدين ونريد ان نجد العدد الآخر فمن اجل ان احد هذين العددين اذا قُسم على الآخر كان منه مربّع اذا قُسم ضلعه على العدد المقسوم عليه كان القسم مكعّباً ينبغى ان نطلب عدداً اذا قسمناه على الاثنين كان منه مكعّب وهو ستّة آحاد ونصف ورُبع والستّة آحاد والنصف والرُبع هى ضلع المربّع الذى يخرج من قسمة احد العددين على الآخر والمربّع الكائن من الستّة والنصف والرُبع هو خمسة / واربعون ونصف ٢٧ ونصف ثُمن والعدد الذى يكون من قسمته على الاثنين [هذا العدد الذى ذكرنا] هو احد وتسعون وثُمن فاذاً العدد الآخر الذى

---

587 (207): الآخر addidi.

589 (208): الضلع : المطع (sc. المطلع ) in cod.

591 (209): Pro فمن ذا fortasse scribendum فمن اجل ذلك .

(210): قُسم : قسما in cod.

593 (211): معادلاً : معادل in cod.

595 (212): له post المعادله per haplographiam omisit librarius.

597 (213): و dubitanter addidi.

603 (214): ان نطلب : ان نكون نطلب in cod.

606 (215): الستّة : السسه (pro النسبة ?) in cod.

607-608 (216): Verba هذا العدد الذى ذكرنا e margine in textum irrepsisse censeo.

طلبنا هو احد وتسعون وثُمن .ومثل هذا التدبير تُعلم الخواص المذكورة للاعداد المفروضة فى المسائل المتقدّمة ويكون وجودها

فاذاً احد العددين المفروضين اثنان والعدد الآخر احد وتسعون وثُمن ونريد ان نجد عدداً مكعّباً اذا ضربناه فى الواحد والتسعين والثُمن كان منه مكعّب واذا ضربناه فى الاثنين كان منه ضلع ذلك المكعّب فنفرض المكعّب كعباً واحداً ونعمل كما عملنا فى المسائل المتقدّمة فنعلم ان المكعّب المطلوب ثلثة آحاد وثلثة اثمان وهو اذا ضُرب فى الواحد والتسعين والثُمن كان منه مكعّب وهو ثلثمائة وسبعة آحاد وخمسة وثلثون جزءاً من اربعة وستّين جزءاً واذا ضُرب فى الاثنين كان منه ستّة آحاد ونصف ورُبع احد وهو ضلع المكعّب الذى هو ثلثمائة وسبعة آحاد وخمسة وثلثون جزءاً من اربعة وستّين جزءاً

فقد وجدنا عدداً مكعّباً على الشرط الذى شُرط لنا وذلك ما اردنا ان نجد

كجـ نريد ان نجد عددين مربّعين يكون مربّعاهما مجموعين مكعّباً

فنفرض احد المربّعين مالاً واحداً والمربّع الآخر من ضلع كم شئنا من الاشياء فلنفرضه من ضلع شيئين حتّى يكون اربعة اموال ومربّعا هذين المربّعين امّا الاصغر منهما فمال مال وامّا الاعظم فستّة عشر مال مال وجميعهما سبعة عشر مال مال وهو يعادل عدداً مكعّباً فنعمل / المكعّب من ضلع ثلثة اشياء حتّى يكون سبعة وعشرين كعباً ٢٨ فيكون السبعة العشر المال مال تعادل سبعة وعشرين كعباً ولذلك يكون سبعة عشر شيئاً تعادل سبعة وعشرين احداً فاذاً الشىء الواحد سبعة وعشرون جزءاً من سبعة عشر جزءاً من واحد ولانّا فرضنا المربّع الاصغر من ضلع شىء واحد يكون ضلعه سبعة وعشرين جزءاً من سبعة عشر جزءاً ويكون المربّع الاصغر سبع مائة وتسعة وعشرين جزءاً من مأتين وتسعة وثمنين جزءاً من واحد ولانّا فرضنا المربّع الاعظم من ضلع شيئين يكون ضلع المربّع الاعظم اربعة وخمسين جزءاً من سبعة عشر

---

611 (217): فاذاً (vel فاذن): فلاں in cod.

612 (218): ونريد: ويرىد in cod.

614 (219): Pro عملنا praebet codex علمنا; nam librarius initio علم pro عمل scripsit.

جزءاً والمربّع الاعظم ألفان وتسع مائة وستّة عشر جزءاً من مأتين وتسعة وثمنين جزءاً من واحد ومن اجل ذلك ايضاً مربّع المربّع الاصغر يكون خمس مائة ⟨الف وأحد وثلثين الفاً واربع مائة وأحد واربعين جزءاً من ثلثة وثمنين الفاً وخمس مائة⟩ وأحد وعشرين جزءاً من واحد وامّا مربّع المربّع الاعظم فيكون ثمنية الٰف الف وخمس مائة الف وثلثة الٰف وستّة وخمسين جزءاً من ثلثة وثمنين الفاً وخمس مائة وأحد وعشرين جزءاً من واحد وجملة هذين المربّعين تسعة الٰف الف واربعة وثلثون الفاً واربع مائة وسبعة وتسعون جزءاً من ثلثة وثمنين الفاً وخمس مائة وأحد وعشرين جزءاً من واحد الذى هو ايضاً خمس مائة الف وأحد وثلثون الفاً واربع مائة وأحد واربعون جزءاً من اربعة الٰف وتسع مائة وثلثة عشر جزءاً من واحد وهو مكعّب ضلعه احد وثمنون جزءاً من سبعة عشر جزءاً من واحد

فقد وجدنا عددين مربّعين على الشرط الذى شُرط لنا وهما السبع مائة والتسعة والعشرون الجزء والألفان / والتسع المائة والستّة العشر ٢٩ الجزء من مأتين وتسعة وثمنين جزءاً وذلك ما اردنا ان نجد

كد نريد ان نجد عددين مربّعين يكون تفاضل مربّعيهما عدداً مكعّباً

فنفرض المربّع الاصغر من ضلع شىء واحد والمربّع الاعظم من ضلع شيئين حتّى يكون الاصغر مالاً واحداً ويكون الاعظم اربعة اموال ويكون تفاضل مربّعيهما خمسة عشر مال مال وهو مكعّب فلنفرضه من ضلع خمسة اشياء وكلّ مكعّب يُقسم على ضلعه يكون الذى يخرج من القسمة مساوياً لمربّع ضلعه ويكون الخمسة العشر مال مال التى هى مكعّب ضلعه خمسة اشياء اذا قُسمت على ضلعها الذى هو خمسة اشياء كان القِسم ثلثة كعاب فاذاً الثلثة الكعاب هى مربّع ضلعه خمسة اشياء والمربّع الكائن من الخمسة الاشياء خمسة وعشرون مالاً فاذاً الثلثة الكعاب تعادل خمسة وعشرين مالاً واذا قسمناهما على المال الذى هو فى اقعد الناحيتين كان ثلثة اشياء تعادل خمسة وعشرين احداً فالشىء

---

636 (220): ألفان : العس in cod.

638-639 (221): E textu per homoeoteleuton elapsa restitui.

646 (222): وثمنون : وسبون in cod.

650 (223): وثمنين : وسبين in cod.

656 (224): ضلعه (posterius) pro صلعها codicis, hic ut in linea 112, substitui.

الواحد يعادل ثمنية آحاد وثُلثَ ⟨واحد⟩ ولانّا فرضنا المربّع الاصغر من ضلع شيء واحد والمربّع الاعظم من ضلع شيئين يكون ضلع المربّع الاصغر ثمنية آحاد وثُلثَ ⟨واحد⟩ وضلع المربّع الاعظم ستّة عشر احداً وثُلثى واحد ويكون المربّع الاصغر تسعة وستّين واربعة اتساع ويكون المربّع الاعظم مأتين وسبعة وسبعين وسبعة اتساع ومربّع المربّع الاصغر اربعة الف وثمانى مائة واثنان وعشرون وثلثة واربعون جزءاً من واحد وثمنين جزءاً من واحد / ومربّع المربّع الاعظم سبعة ٣٠
وسبعون الفاً ومائة وستّون واربعون جزءاً من احد وثمنين جزءاً من واحد وتفاضل هذين المربّعين اثنان وسبعون الفاً وثلثمائة وسبعة وثلثون وثمنية وسبعون جزءاً من احد وثمنين جزءاً من واحد التى هى ستّة وعشرون جزءاً من سبعة وعشرين جزءاً من واحد وهو مكعّب ضلعه واحد واربعون وثُلثاَ واحد

فقد وجدنا عددين ⟨مربّعين⟩ على الشرط الذى شُرط لنا وهما التسعة والستّون والاربعة الاتساع والمائتان والسبعة والسبعون والسبعة الاتساع وذلك ما اردنا ان نجد

كه نريد ان نجد عددين مربّعاً ومكعّباً يكون مربّعاهما مجموعين مربّعاً

فنفرض المكعّب من ضلع شيء واحد حتّى يكون كعباً واحداً ونفرض المربّع من ضلع كم شئنا من الاشياء فلنفرضه من ضلع شيئين حتّى يكون اربعة اموال ومربّعاهما امّا مربّع المكعّب فكعب كعب واحد وامّا مربّع المربّع فستّة عشر مال مال وجميعهما كعب كعب واحد وستّة عشر مال مال وذلك يعادل عدداً مربّعاً فينبغى ان نجد العدد الذى هو

---

662 (225): واحد addidi.

663 (226): والمربّع : وصلع المربع in cod.

664 (227): واحد, hic ut supra (lin. 662), addidi.

666 (228): وسبعة (posterius): وسعه in cod.

668 (229): وثمنين جزءاً من واحد / ومربّع : وسس حرا س واحد وس / حرو س واحد ومربع in cod.

671 (230): وسبعون : وسعون in cod.

674 (231): مربّعين addidi.

675 (232): والسبعة : والسعه in cod.

681 (233): فكعب : وكعب in cod.

ضلع هذا المربّع فنقول انّا ان فرضنا هذا الضلع اموالاً يكون المربّع المعادل للكعب كعب الواحد والستّة العشر مال المال اموالَ اموال فاذا نقصنا الستّة العشر مال المال المشتركة من كلى الناحيتين بقيت اموال اموال تعادل كعب كعب واذا قسمناهما على مال مال الواحد الذى هو اقعد الناحيتين كان من ذلك مال يعادل عدداً وينبغى ان يكون / هذا العدد مربّعاً لانّه يعادل مالاً واحداً ولكن هذا العدد ٣١ هو زيادة اموال الاموال التى هى عدد مربّع على الستّة عشر فلذلك ينبغى ان يكون عدد اموال الاموال التى هى عدد مربّع يزيد على الستّة العشر عدداً مربّعاً فمن اجل ذلك نصير الى ان نطلب عددين مربّعين يكون تفاضلهما ستّة عشر فنجد المربّع الاعظم خمسة وعشرين والمربّع الاصغر تسعة آحاد فنجعل المربّع المعادل للكعب الكعب والستّة عشر المال المال خمسة وعشرين مال مال الذى هو من ضلع خمسة اموال ونلقى الستّة عشر مال مال المشتركة من كلى الناحيتين فيبقى كعب كعب يعادل تسعة اموال مال ولذلك يكون مال واحد يعادل تسعة آحاد والمال مربّع ⟨من⟩ ضلع شىء واحد والتسعة الآحاد مربّع ضلعه ثلثة آحاد فاذاً الشىء الواحد ثلثة آحاد ولانّا فرضنا المكعّب من ضلع شىء واحد يكون ضلعه ثلثة آحاد ويكون هو سبعة وعشرين ولانّا فرضنا المربّع من ضلع شيئين يكون ضلعه ستّة آحاد ويكون هو ستّة وثلثين ومربّعاهما امّا مربّع السبعة والعشرين فسبع مائة وتسعة وعشرون وامّا مربّع الستّة والثلثين فألف ومائتان وستّة وتسعون وجميعهما ألفان وخمسة وعشرون وهو مربّع ضلعه خمسة واربعون

---

685 (234): Post مال المال addit codex مال.

686 (235): من كلى : على in cod.

688 (236): Fortasse فى aut واحد من post الذى هو addendum, vel scribendum التى هى (cf. lin. 126,148,185).

(237): Pro كان دلك مالا codicis كان من ذلك مال scripsi.

690 (238): التى هى : الدى هو in codice; etiam عدد post زيادة supplere liceret.

696 (239): المشتركة : المسرك in cod. Vide adn. 246,271,283,350,730.

698 (240): من deest in codice (vide enim adn. seq.).

(241): Pro ضلع habet codex صلع, librarius videlicet scribere incepit صلعه.

703 (242): وعشرون : وعسرس in cod.

فقد وجدنا عددين مكعّباً ومربّعاً ومربّعاهما مجموعين مربّع وهما السبعة والعشرون والستّة والثلثون وذلك ما اردنا ان نجد

٣٢ كو نريد ان نجد عددين مكعّباً ومربّعاً يكون تفاضل مر/بّعيهما عدداً مربّعاً

فنفرض المكعّب كعباً واحداً والمربّع اربعة اموال فيكون مربّع المكعّب كعبَ كعب ومربّع المربّع ستّة عشر مال مال ونريد ان يكون تفاضلهما عدداً مربّعاً فلنطلب اوّلاً ان يكون ⟨مربّع⟩ المكعّب يزيد على ⟨مربّع⟩ المربّع عدداً مربّعاً فنقول كعب كعب الّا ستّة عشر مال مال تعادل عدداً مربّعاً ونطلب بمثل الذى قد تقدّم ذكره فى المسئلة المتقدّمة لهذه المسئلة الاموال التى ينبغى ان تُفرض ضلعاً لهذا المربّع فنجدها ثلثة اموال والمربّع الذى يكون منها تسعة اموال اموال فاذاً كعب كعب الّا ستّة عشر مال مال تعادل تسعة اموال مال فنزيد الستّة العشر مال مال مشتركةً على الناحيتين فيكون كعب كعب يعادل خمسة وعشرين مال مال ولذلك يكون مال واحد يعادل خمسة وعشرين احداً والمال مربّع ضلعه شىء واحد والخمسة والعشرون مربّع ضلعه خمسة آحاد فاذاً الشىء الواحد يعادل خمسة آحاد ولانّا فرضنا المكعّب من ضلع شىء واحد يكون ضلعه خمسة آحاد ويكون هو مائة وخمسة وعشرين ولانّا فرضنا المربّع من ضلع شيئين يكون ضلعه عشرة آحاد ويكون هو مائة ومربّعاهما امّا مربّع المائة والخمسة والعشرين فخمسة عشر الفاً وستّمائة وخمسة وعشرون وامّا مربّع المائة فعشرة الف

---

706 (243): مجموعين : محموعان in codice. Necessarius est accusativus (vide lin. 623,677). Errat Nix, qui in commentario editionis (partis) libri V Apollonii, pag. 13, formam ا و ب مجموعان tradit.

712 (244): Verbum مربّع bis in codice omittitur. Vide etiam adn. 250.

716 (245): منها : سهما, ut videtur, in codice; mendum autem prius lapsui calami tribuendum esse opinor.

718 (246): مشتركةً : مسىركا in cod.

(247): Fortasse كلى, quod verbi على simile est (vide adn. 948), post على addendum est; item in linea 835.

720 (248): مربّع (posterius): مربعه in cod.

وتفاضلهما خمسة الف وستّمائة وخمسة وعشرون وهو مربّع ضلعه خمسة وسبعون

فقد وجدنا عددين مكعّباً ومربّعاً وزيادة ⟨مربّع⟩ المكعّب على ⟨مربّع⟩ المربّع عدد مربّع وهما المائة والمائة والخمسة / والعشرون ٣٣

وايضاً فلنطلب ان تكون زيادة مربّع المربّع على مربّع المكعّب مربّعاً ونفرض المكعّب كعباً واحداً والمربّع من ضلع خمسة اشياء حتّى يكون معنا ستّمائة وخمسة وعشرون مال مال الّا كعبَ كعب واحد تعادل عدداً مربّعاً فلنطلب ضلع هذا المربّع فنقول انّا ان فرضنا⟨ه⟩ اموالاً يكون مربّعه المعادل للستّمائة والخمسة والعشرين المال مال الّا كعبَ كعب اموالَ اموال فاذا زدنا كعب الكعب مشتركاً على كلى الناحيتين صارت الستّمائة والخمسة والعشرون مال المال تعادل كعب كعب واموال مال ثمّ بعد ⟨المقابلة و⟩ القسمة بقى مال يعادل عدداً فينبغى ان يكون هذا العدد مربّعاً لكن هذا العدد هو زيادة المضروب من الخمسة والعشرين ⟨فى مثلها⟩ على عدد الاموال التى هى مربّع الضلع المطلوب فاذاً ينبغى ان تُقسم الستّمائة والخمسة والعشرون التى هى عدد مربّع بعددين مربّعين على ما وصفنا فى المقالة الثانية وليكن احد القسمين اربع مائة والآخر مأتين وخمسة وعشرين ونفرض المربّع المعادل للستّمائة والخمسة والعشرين مال مال الّا كعبَ كعب اموالَ اموال مساوية لأحد هذين القسمين فلنفرضه مأتين وخمسة وعشرين مال مال ومن بعد الجبر والمقابلة والقسمة يبقى مال يعادل اربع مائة ولذلك يكون الشىء الذى فرضناه ضلعَ المكعّب

---

727 (249): وسبعون : وسعوں in cod.

728 (250): مربّع bis, ut supra (cf. adn. 244), addidi.

733 (251): Pronomen post فرضنا addidi.

734 (252): Post كعب كعب addit codex verba: ادا ىعادل عددا مربعا فلنطلب صلع هدا المربع مربع وانا ان فرصناه اموالا ىكون مربعه.

735 (253): فاذا زدنا كعب الكعب sic in codice: وادا اردنا الكعب الكعب.

737 (254): المقابلة و addidi.

738 (255): فى : من in cod.

739 (256): فى مثلها addidi.

(257): الاموال (posterius): الامول in codice (cf. adn. 346).

(258): على : هى in cod.

عشرين ويكون هو ثمنية الٰف ويكون ⟨ضلع⟩ المربّع لانّا فرضناه خمسة اشياء مائة ويكون هو عشرة الٰف ومربّعاهما امّا مربّع المكعّب الذى هو ثمنية الٰف فأربعة وستّون الف الف وامّا مربّع المربّع الذى هو عشرة الٰف فمائة الٰف / الٰف وتفاضلهما ستّة وثلثون الف الف وهو مربّع ٣٤ ضلعه ستّة الٰف

فقد وجدنا عددين مكعّباً ومربّعاً ومربّع المربّع يزيد على مربّع المكعّب عدداً مربّعاً وهما العشرة الالٰف والثمنية الالٰف وذلك ما اردنا ان نجد

كزَ نريد ان نجد عددين مكعّباً ومربّعاً يكون مربّع المكعّب مع امثال مفروضة للعدد المربّع عدداً مربّعاً

فنفرض المكعّب ⟨كعباً⟩ واحداً ونضربه فى مثله فيكون كعب كعب واحد ونفرض المربّع من ضلع اموال كم شئنا فلنفرضه من ضلع مالين فيكون هو اربعة اموال مال ولتكن الامثال المفروضة خمساً ونضرب الاربعة الاموال الاموال فى خمسة فتكون عشرين مال مال ونزيدها على الكعب الكعب الواحد فيكون كعبَ كعب وعشرين مال مال وذلك مربّع ولنطلب مربّعين تفاضلهما عشرون وهما ستّة وثلثون وستّة عشر فنجعل المربّع الذى هو كعب كعب وعشرون مال مال يعادل ستّة وثلثين مال مال وننقص العشرين مال المال المشتركة من كلى الناحيتين فيبقى ستّة عشر مال مال تعادل كعب كعب ولنقسم جميع ذلك على مال مال فيكون ستّة عشر تعادل مالاً والستّة عشر مربّع ضلعه اربعة آحاد فاذاً اربعة تعادل ضلع المال الذى هو شىء واحد ولانّا فرضنا المكعّب كعباً واحداً يكون ضلعه اربعة آحاد ويكون هو اربعة وستّين ولانّا فرضنا ضلع المربّع مالين يكون ضلعه اثنين وثلثين ويكون هو الفاً واربعة وعشرين وخمسةُ امثال ذلك خمسة الٰف ومائة وعشرون ونزيد

---

747 (259): Desideratum verbum addidi.

749 (260): فأربعة : وارىعه in cod.

753 (261): الالٰف (prius): الاف in codice; forma آلاف esse vix potest, cum pluralis semper defective notetur (vide adn. 68).

(262): الالٰف (posterius): الاالف in cod.

756 (263): للعدد : العدد in cod.

757 (264): كعباً deest in codice. Vide etiam adn. 294.

769 (265): الفاً : الف in cod.

ذلك على العدد المكعّب فيكون خمسة الف ومائة واربعة وثمنين وذلك عدد مربّع ضلعه اثنان وسبعون

فقد وجدنا عددين مربّعاً ومكعّباً ومربّع المكعّب مع خمسة / امثال ٣٥
العدد المربّع عدد مربّع وهما الاربعة والستّون والألف والاربعة والعشرون وذلك ما اردنا ان نجد

كح نريد ان نجد عددين مكعّباً ومربّعاً يكون مربّع المربّع مع امثال مفروضة للعدد المكعّب عدداً مربّعاً

فلتكن الامثال المفروضة عشرة امثال ونجعل المكعّب كعباً واحداً ونضربه فى العشرة فيكون عشرةَ كعاب ونفرض ضلع المربّع شيئين فيكون ⟨المربّع⟩ اربعة اموال ومربّعه ستّة عشر مال مال ونزيد ذلك على العشرة الكعاب فيكون ذلك ستّة عشر مال مال وعشرة كعاب وذلك يعادل عدداً مربّعاً ونفرض المربّع من ضلع ستّة اموال وكلّ مربّع يُقسم على ضلعه فان الذى يخرج من القسمة يكون معادلاً لضلعه فنقسم الستّة عشر المال المال والعشرة الكعاب على ستّة اموال فتكون مالين وثُلثى مال وشيئاً وثُلثى شىء وذلك يعادل ستّة اموال فننقص المالين والثُلثى مال المشتركة من كلى الناحيتين فيبقى ثلثة اموال وثُلث ⟨مال⟩ تعادل شيئاً وثُلثى شىء ولذلك يكون ثلثة اشياء وثُلث ⟨شىء⟩ تعادل واحداً وثُلثى واحد فيكون الشىء الواحد يعادل نصفاً فلانّا فرضنا المكعّب من ضلع شىء واحد يكون ضلعه نصفاً ويكون المكعّب ثُمناً وعشرة امثاله واحد ورُبع ولانّا فرضنا المربّع من ضلع شيئين يكون ضلعه واحداً ويكون هو ايضاً واحداً فاذا زدنا ذلك على الواحد والرُبع الذى هو عشرة امثال المكعّب كان ذلك عدداً مربّعاً ⟨و⟩ هو اثنان ورُبع وضلعه واحد ونصف

---

776 (266): مكعّباً : مكعب in cod.

777 (267): للعدد : العدد in cod.

779 (268): المربّع addidi.

784 (269): وشيئاً : وسى in cod.

785 (270): فننقص : فسسم in cod.

(271): Pro المسرك codicis المشتركة scripsi.

786 (272): مال addidi.

787 (273): ولذلك : وكذلك in cod.

(274): شىء addidi.

792 (275): و ante هو addidi.

فقد وجدنا عددين مكعّباً ومربّعاً ومربّع المربّع مع عشرة امثال العدد المكعّب عدد مربّع وهما الواحد والثُمن واحد وذلك ما اردنا ان نجد

---

كط نريد ان نجد عددين مكعّباً ومربّعاً يكون مكعّب المكعّب ومربّع المربّع مجموعين / عدداً مربّعاً ٣٦

فنفرض المكعّب كعباً واحداً ⟨حتّى يكون⟩ مكعّبه كعب كعب فى كعب ويُسمّى ذلك كعبَ كعب كعب ونفرض ⟨ضلع⟩ المربّع ما شئنا من الاموال فلنفرضه مالين حتّى يكون المربّع اربعة اموال اموال ويكون مربّعه ستّة عشر مال مال ⟨فى مال⟩ مال وذلك يعادل ستّة عشر كعب كعب مضروبة فى مال ويُسمّى واحدُه كعبَ كعب مال فاذاً كعب كعب كعب مع الستّة العشر كعب كعب مال تعادل عدداً مربّعاً فلنفرض ضلعه ايضاً ما شئنا من الاموال اموال فنفرضه من ضلع ستّة اموال اموال ونضربها فى مثلها فتكون ستّة وثلثين مال مال ⟨فى مال⟩ مال اعنى ستّة وثلثين كعب كعب مال فاذاً الكعب كعب كعب مع الستّة عشر كعب كعب مال تعدل ستّة وثلثين كعب كعب مال فلنلق الستّة عشر كعب كعب مال المشتركة من كلى الناحيتين فيبقى كعب كعب كعب يعادل عشرين كعب كعب مال ونقسم كلّ واحد منهما على واحد من اقعد الناحيتين الذى هو كعب كعب مال فاذاً العشرون كعب كعب مال اذا قُسمت على كعب كعب مال كانت عشرين احداً وانّ الكعب كعب كعب انّما هو يجتمع من ضرب كعب ⟨كعب⟩ فى كعب وهو ايضاً مجتمع من ضرب كعب كعب مال فى شىء فاذاً كعب كعب كعب اذا قُسم على كعب كعب مال كان القسم شيئاً واحداً فاذاً الشىء الواحد يعادل عشرين احداً ولانّا فرضنا ضلع المكعّب شيئاً واحداً يكون ضلع المكعّب عشرين احداً

---

798 (276): حتّى يكون addidi.

799 (277): Deficiens ضلع restitui.

801 (278): Textum secundum lineas 830-831 et 833 supplevi.

804 (279): ستّة bis in cod.

805 (280): فى مال ut supra (cf. adn. 278) addidi.

805-806 (281): كعب كعب مال : كعب كعب كعب مال in cod.

807 (282): فلنلق : فلملعى in cod.

(283): المشتركة : المسرك in cod.

812 (284): Per haplographiam omissum verbum restitui.

ويكون المكعّب ثمنية الٰف ولانّا فرضنا المربّع من ضلع مالين والمال اربع مائة يكون ضلع ⟨المربّع⟩ ثمان مائة ويكون المربّع ستّمائة الف واربعين الفاً ويكون مكعّب المكعّب خمس مائة الف الف الف واثنى عشر الف الف الف ويكون مربّع المربّع اربع مائة الف الف الف وتسعة/
الٰف الف الف وستّمائة الف الف والجميع منهما تسع مائة الف الف الف ٣٧
وأحد وعشرون الف الف الف وستّمائة الف الف وذلك عدد مربّع ضلعه تسع مائة الف وستّون الفاً

فقد وجدنا عددين مكعّباً ومربّعاً ومكعّب المكعّب مع مربّع المربّع مجموعين عدد مربّع وهما ثمنية ⟨الٰف⟩ وستّمائة الف واربعون الفاً وذلك ما اردنا ان نجد

ل نريد ان نجد عددين مكعّباً ومربّعاً تكون زيادة مكعّب المكعّب على مربّع المربّع عدداً مربّعاً

فنفرض المكعّب كعباً واحداً حتّى يكون مكعّبه كعب كعب فى كعب اعنى ⟨الذى⟩ يُسمّى كعب كعب كعب ونفرض المربّع من ضلع مالين حتّى يكون المربّع اربعة اموال اموال ويكون مربّعه ستّة عشر مال مال فى مال مال اعنى ستّة عشر كعب كعب مال فاذاً كعب كعب كعب الّا ستّة عشر كعب كعب مال تعادل عدداً مربّعاً فلنفرض ضلعه مالى مال حتّى يكون مربّعه اربعة اموال مال فى مال مال اعنى اربعة كعاب كعاب مال فاذاً كعب كعب كعب الّا ستّة عشر كعب كعب مال تعادل اربعة كعاب ⟨كعاب⟩ مال فلنزد الستّة العشر الكعب كعب مال مشتركةً على الناحيتين فيكون كعب كعب كعب يعادل عشرين كعب كعب مال ولنقسم الجميع على كعب كعب مال الذى هو واحد من اقعد الناحيتين فيكون

---

816 (285): ولانّا : وادا in cod.

(286): والمال : والمربع in cod.

817 (287): المربّع addidi.

824 (288): محموعان codicis, hic ut alibi (vide adn. 243), correxi.

(289): الٰف addidi.

829 (290): الذى addidi.

835 (291): Verbum كعاب, per haplographiam omissum, addidi.

(292): Loco على forsan على كلى scribendum, ut antea (in adn. 247) notavi.

من بعد القسمة شيء واحد يعادل عشرين احداً فلانّا فرضنا ضلع
المكعّب شيئاً واحداً يكون ضلعه عشرين ويكون هو ثمنية الف ولانّا
فرضنا ضلع المربّع مالين والمال اربع ماثة يكون ضلع المربّع ثماني ماثة
ويكون المربّع ستّمائة الف واربعين الفاً فامّا مكعّب المكعّب فتبيّن انّه
خمس مائة الف الف الف واثنا عشر الف الف / الف وامّا مربّع المربّع ٢٨
فانّه اربع مائة الف الف الف وتسعة الف الف الف وستّمائة الف الف
وتفاضلهما اعنى زيادة مكعّب المكعّب على مربّع المربّع مائة الف الف
الف والفاً الف الف واربع مائة الف الف وهو عدد مربّع ضلعه ثلثمائة
الف وعشرون الفاً وقد تبيّن فى المسئلة المتقدّمة ان جملة هذين
العددين ايضاً عدد مربّع

فقد وجدنا عددين مكعّباً ومربّعاً وزيادة مكعّب المكعّب على مربّع المربّع عدد مربّع وهما الثمنية الالف والستّمائة الف والاربعون الفاً وذلك ما اردنا ان نجد

وبهذا يستبان انّا قد وجدنا ايضاً عددين مكعّباً ومربّعاً ومكعّب المكعّب اذا زيد عليه مربّع المربّع كان المجتمع منه عدداً مربّعاً واذا نُقص من مكعّب المكعّب مربّع المربّع فان الباقى عدد مربّع وهما هذان العددان ايضاً

لا̅ نريد ان نجد عددين مربّعاً ومكعّباً تكون زيادة مربّع المربّع على مكعّب المكعّب عدداً مربّعاً

فنفرض المكعّب ⟨كعباً⟩ واحداً فيكون مكعّبه كعب كعب فى كعب وهو المسمّى كعب كعب كعب ونفرض ضلع المربّع مالين فيكون المربّع اربعة اموال مال ويكون مربّعه ستّة عشر مال مال فى مال ⟨مال⟩ وهو المسمّى كعب كعب مال فاذاً الستّة عشر كعب كعب مال التى هى مربّع العدد المربّع تزيد على كعب كعب ⟨كعب⟩ عدداً مربّعاً فلنفرض ضلع ذلك المربّع مالى مال وكلّ مربّع يُقسم على ضلعه فان الذى يخرج من القسمة يكون مساوياً لضلعه فاذاً الستّة عشر كعب كعب المال الّا كعب الكعب الكعب اذا قسمناها على مالى مال كان الذى يخرج من القسمة يعادل

---

845 (293): والعى : والفا in cod.

857 (294): كعباً ut supra (vide adn. 264) addidi.

859 (295): Deficiens مال addidi.

861 (296): Deficiens كعب addidi.

مالى مال ولكن السّتّة عشر كعب كعب المال الاّ كعب كعب كعب كعب اذا قُسم على مالى مال ⟨امّا⟩ السّتّة عشر كعب كعب / المال فلانّها مجتمعة ٣٩ من ضرب ستّة عشر مال مال فى مال مال فانّها اذا قُسمت على مالى مال كان القسم ثمنية اموال مال وامّا كعب كعب الكعب فلانّه مجتمع من ضرب كعب كعب فى كعب والكعب كعب مجتمع من ضرب مال مال فى مال فان كعب كعب الكعب مجتمع من ضرب مال مال فى مال كعب فاذاً كعب كعب الكعب اذا قُسم على مالى مال كان القسم نصفَ مال كعب فاذاً الذى يخرج من القسمة هو ثمنية اموال مال الاّ نصفَ مال كعب وذلك يعادل مالى مال فنجعل نصف المال ⟨كعب⟩ مشتركاً مزيدَه على كلى الناحيتين فيكون ثمنية اموال مال تعادل مالى مال ونصف مال كعب ولنلق مالى مال المشتركين من كلى الناحيتين فيبقى نصف مال كعب يعادل ستّة اموال ⟨اموال⟩ ومن بعد القسمة بقى نصف شىء يعادل ستّة آحاد فاذاً الشىء الواحد يعادل اثنى عشر احداً ولانّا فرضنا ضلع المكعّب شيئاً واحداً يكون ضلعه اثنى عشر ويكون المكعّب الفاً وسبع مائة وثمنية وعشرين ولانّا فرضنا ضلع المربّع مالين والمال مائة واربعة واربعون لان الشىء اثنا عشر يكون ضلع المربّع مائتين وثمنية وثمنين ويكون المربّع اثنين وثمنين الفاً وتسع مائة واربعة واربعين فامّا مكعّب المكعّب فانّه خمسة الف الف الف ومائة الف الف وتسعة وخمسون الف الف وسبع مائة الف وثمنون الفاً وثلثمائة واثنان وخمسون وامّا مربّع المربّع فيكون ستّة الف الف الف وثمانى مائة الف الف وتسعة وسبعين الف الف وسبع مائة الف وسبعة الف ومائة وستّة وثلثين وزيادة هذا العدد على مكعّب المكعّب تكون الف الف الف وسبع مائة الف الف وتسعة عشر الف الف وتسع مائة الف وستّة وعشرين

---

866 (297): امّا addidi.

(298): فلانّها : فانها in cod.

873 (299): كعب addidi.

(300): مزيدَه pro مربده codicis scripsi, sed de loco dubitans.

876 (301): اموال addidi.

886 (302): مكعّب : كعب in cod.

887 (303): وتسعة : وسبعه, ut videtur, in cod.

(304): وعشرين : وعسرون in cod.

الفاً وسبع / مائة واربعة وثمنين وهو عدد مربّع ضلعه واحد واربعون ٤٠ الفاً واربع مائة واثنان وسبعون

فقد وجدنا عددين على الشرط الذى اردنا وهما الألف والسبع مائة والثمنية والعشرون والاثنان والثمنون الفاً والتسع مائة والاربعة والاربعون وذلك ما اردنا ان نجد

لَب نريد ان نجد عددين مكعّباً ومربّعاً يكون مكعّب المكعّب مع امثال مفروضة لِما يجتمع من ضرب المربّع فى المكعّب عدداً مربّعاً

فلتكن الامثال المفروضة خمساً ونفرض المكعّب كعباً واحداً فيكون مكعّبه كعب كعب كعب ونفرض المربّع ⟨من⟩ ضلع كعبين فيكون المربّع اربعة كعاب كعب ونضرب ذلك فى العدد المكعّب الذى فرضناه كعباً واحداً فيجتمع منه اربعة كعاب كعب كعب وخمسة امثالها عشرون كعب كعب كعب فنزيدها على مكعّب المكعّب فيجتمع منها واحد وعشرون كعب كعب كعب وذلك يعادل عدداً مربّعاً فلنفرض ضلعه سبعة اموال مال فيكون المربّع تسعة واربعين كعب كعب فى مال وذلك يعادل واحداً وعشرين كعب كعب كعب ولنقسم كلّ واحد منهما على كعب كعب مال فيكون واحد وعشرون شيئاً تعادل تسعة واربعين احداً فالشىء الواحد يعادل اثنين وثُلث ⟨واحد⟩ ولانّا فرضنا المكعّب من ضلع شىء واحد يكون ضلع المكعّب اثنين وثُلث واحد ويكون المكعّب من اجل ان ضلعه سبعة اثلاث ثلثمائة وثلثة واربعين جزءاً من سبعة وعشرين جزءاً ولانّا فرضنا ضلع المربّع كعبين يكون ضلع المربّع ستّمائة وستّة وثمنين جزءاً من سبعة وعشرين جزءاً من واحد ويكون المربّع اربع مائة الف وسبعين الفاً وخمس مائة وستّة وتسعين / جزءاً من سبع مائة ٤١ وتسعة وعشرين جزءاً من واحد فامّا مكعّب المكعّب فانّه يكون اربعين الف الف وثلثمائة الف وثلثة وخمسين الفاً وستّمائة وسبعة اجزاء من تسعة عشر الفاً وستّمائة وثلثة وثمنين جزءاً من واحد وامّا ما يجتمع من

---

889 (305): وسبعون : وسبعوں in cod.

896 (306): من addidi.

898 (307): واحداً : واحرا (pro واجزاء) in cod.

(308): كعاب كعب كعب : كعاب كعب كعب in cod.

903 (309): وعشرون : وعسرس in cod.

(310): فالشىء : بالسى in codice, lapsu ut videtur calami.

904 (311): واحد addidi.

ضرب العدد المربّع فى العدد المكعّب فانّه يكون مائة الف الف وأحد وستّين الف الف واربع مائة الف واربعة عشر الفاً واربع مائة وثمنية وعشرين جزءاً من تسعة عشر الفاً وستّمائة وثلثة وثمنين جزءاً من واحد وخمسة امثال ذلك ثمانى مائة الف الف وسبعة الف الف واثنان وسبعون الفاً ومائة واربعون جزءاً واذا زدنا ذلك على مكعّب المكعّب كان المجتمع منهما ثمانى مائة الف الف وسبعة واربعين الف الف واربع مائة الف وخمسة وعشرين الفاً وسبع مائة وسبعة واربعين جزءاً من تسعة عشر الفاً وستّمائة وثلثة وثمنين جزءاً من واحد وذلك ايضاً هو مائتا الف الف واثنان وثمنون الف الف واربع مائة الف وخمسة وسبعون الفاً ومائتان وتسعة واربعون جزءاً من ستّة الف وخمس مائة وأحد وستّين جزءاً من واحد وذلك عدد مربّع ضلعه ستّة عشر الفاً وثمان مائة وسبعة اجزاء من واحد وثمنين جزءاً من واحد

فقد وجدنا عددين على الشرط الذى شرطنا وهما الثلثمائة والثلثة والاربعون الجزء من السبعة والعشرين الجزء من واحد والاربع مائة الف والسبعون الفاً والخمس المائة والستّة والتسعون الجزء من السبع مائة والتسعة والعشرين جزءاً من واحد وذلك ما اردنا ان نجد

---

لج نريد ان نجد عددين مكعّباً ومربّعاً يكون مكعّب المكعّب يزيد على امثال مفروضة لِما يجتمع من ضرب العدد المربّع فى العدد المكعّب عدداً مربّعاً

فلتكن الامثال المفروضة ثلثة امثال ونفرض المكعّب / كعباً واحداً ٤٢
فيكون مكعّبه كعب كعب كعب ونفرض المربّع من ضلع نصف كعب فيكون المربّع رُبعَ كعب كعب ونضرب ذلك فى العدد المكعّب الذى فرضناه كعباً واحداً فيجتمع منه رُبع كعب كعب كعب وثلثة امثال ذلك تكون

---

915 (312): وثمنين : ولس in cod.

916 (313): وسبعة : وسعه in cod.

917 (314): زدنا : اردنا in cod.

918 (315): واربعين : وارىعون in cod.

923 (316): وسبعة : وسه in cod.

924 (317): وثمنين : ولس in cod.

926 (318): والاربعون : والاربعس in cod.

927 (319): والسبعون : والسعس in cod.

933 (320): مكعّبه : مكعب in codice. Cf. adn. 345,819.

ثلثة ارباع كعب كعب كعب وننقصها من مكعّب المكعّب فيبقى رُبع كعب كعب كعب يعادل عدداً مربّعاً فلنفرض ضلعه ما شئنا من اموال الاموال فلنفرضه مال مال واحد فيكون كعب كعب مال يعادل رُبع كعب كعب كعب ومن بعد القسمة يكون رُبع شىء يعادل واحداً فالشىء كلّه يعادل اربعة آحاد فلانّا فرضنا ضلع المكعّب شيئاً واحداً يكون ضلعه اربعة آحاد ويكون المكعّب اربعة وستّين ولانّا فرضنا المربّع من ضلع نصف كعب يكون ضلع المربّع اثنين وثلثين فيكون المربّع الفاً واربعة وعشرين فامّا مكعّب المكعّب فيكون مأتى الف واثنين وستّين الفاً ومائة واربعة واربعين وامّا ما يجتمع من ضرب العدد المربّع فى العدد المكعّب فيكون خمسة وستّين الفاً وخمس مائة وستّة وثلثين وثلثة امثال ⟨ذلك⟩ يكون مائة الف وستّة وتسعين الفاً وستّمائة وثمنية واذا نقصنا ذلك من مكعّب المكعّب بقى خمسة وستّون الفاً وخمس مائة وستّة وثلثون وهو مربّع من ضلع مأتين وستّة وخمسين

فقد وجدنا عددين على الشرط الذى شرطنا وهما الاربعة والستّون والألف والاربعة والعشرون وذلك ما اردنا ان نجد

وعلى مثال ما قد وصفنا نجد ما بقى من اقسام هذا الفنّ من المسائل مثل ان نجد عددين مكعّباً ومربّعاً يكون مربّع المربّع مع امثال مفروضة لِما يجتمع من ضرب العدد المربّع فى العدد المكعّب عدداً مربّعاً
وايضاً مكعّب المربّع مع امثال / مفروضة لِما يجتمع من ضرب العدد ٤٣
المربّع فى العدد المكعّب عدد مربّع وعكس ذلك وما اشبهه

لدَ نريد ان نجد عددين مكعّباً ومربّعاً يكون المكعّب اذا زيد عليه المربّع اجتمع منه عدد مربّع واذا نُقص منه المربّع ايضاً بقى منه عدد مربّع

فنفرض المكعّب كعباً واحداً ونفرض المربّع اربعة اموال فيكون كعب واربعة اموال تعادل عدداً مربّعاً وكعب الّا اربعة اموال تعادل عدداً مربّعاً ايضاً فنعمل فى ذلك بعمل المساواة المثنّاة فنأخذ الفضل بين هذين المربّعين وهو ثمنية اموال ونطلب عددين يكون ضرب احدهما

---

940 (321): فلانّا ; ولا نا (pro ولانّا ?) in cod.

942 (322): Pro نصف المكعب codicis نصف كعب scripsi.

(323): الفاً : الف in cod.

946 (324): ذلك addidi.

955 (325): عدد مربّع : عددا مربعا in cod.

959 (326): الّا اربعة : الاربعه in cod.

فى الآخر ثمنية اموال وهما شيئان واربعة اشياء وتفاضلهما شيئان ونصف الشيئين ⟨شئ⟩ واحد ومربّعه مال واحد وذلك يعادل كعباً الّا اربعة اموال فاذا زدنا الاربعة الاموال مشتركةً على كلى الناحيتين كان الكعب الواحد يعادل خمسة اموال وايضاً ان زدنا الشيئين على الاربعة الاشياء كان ذلك ستّة اشياء ونصفها ثلثة اشياء ومربّع الثلثة الاشياء تسعة اموال وذلك يعادل كعبًا واربعة اموال فنلقى الاربعة الاموال المشتركة من كلى الناحيتين فيبقى كعب واحد يعادل خمسة اموال فقد تساوت المعادلة فى كلى الطرفين وإنتهت فى كلّ واحد منهما الى كعب يعادل خمسة اموال فلنقسم جميع ذلك على مال فيكون شئ واحد يعادل خمسة آحاد ولذلك يكون ضلع المكعّب خمسة آحاد ويكون المكعّب مائة وخمسة وعشرين ويكون ضلع المربّع عشرة آحاد ويكون المربّع مائة وهو اذا زيد على العدد المكعّب اجتمع منهما مائتان وخمسة وعشرون وهو عدد مربّع ضلعه خمسة عشر وان نُقص من العدد المكعّب بقى منه خمسة وعشرون وهو/ مربّع ضلعه ٤٤ خمسة آحاد

ونعمل ذلك ايضا بعمل غير المساواة ⟨المثنّاة⟩ فنقول من اجل ان الكعب والاربعة الاموال تعادل عدداً مربّعاً اذا نحن فرضنا ضلعه اشياء كان المربّع اموالاً تعادل كعباً واربعة اموال واذا نقصنا الاربعة الاموال المشتركة من كلى الناحيتين بقى كعب يعادل اموالاً واذا قسمناهما على مال امّا الكعب فيكون شيئاً وامّا الاموال فتكون عدداً ولذلك يكون العدد الذى فُرض فى المسئلة شيئاً واحداً مساوياً لعدد الاموال الباقية وايضاً من اجل ان الكعب الّا اربعة اموال تعادل

---

963 (327): شئ addidi.

964 (328): الّا اربعة : الاربعه in cod.

(329): كلى ut كلى in codice scriptum. Vide etiam adn. 334, 599.

971 (330): شئ واحد يعادل : سا واحدا بعادله in cod.

977 (331): ونعمل ... المساواة atramento rubro in codice. Litteras autem duas ultimas verbi المساواة scripsit librarius supra lineam, penuriâ videlicet spatii; hac re forsan verbum المثنّاة praetermisit.

981 (332): Post عدداً addit codex verba مساوبه لعدد الاموال.

983 (333): الّا اربعة : الاربعه in cod.

عدداً مربّعاً اذا نحن فرضنا ايضاً ضلعه اشياء كان المربّع اموالاً واذا زدنا الاربعة الاموال مشتركةً على كلى الناحيتين اجتمع كعب يعادل اموالاً ولذلك يكون العدد الذى فُرض فى المسئلة شيئاً مساوياً لعدد الاموال المجتمعة فلذلك ينبغى ان تكون الاموال الباقية فى المعادلة الاوّلة عددها مساوٍ لعدد الاموال المجتمعة فى المعادلة الثانية لكن الاموال الباقية فى المعادلة الاوّلة هى ما بقى من عدد مربّع بعد نقصان اربعة آحاد والاموال المجتمعة فى المعادلة الثانية هى عدد مجتمع من عدد مربّع واربعة آحاد من اجل ذلك نطلب عددين مربّعين اذا نقصنا من اعظمهما اربعة آحاد وزدنا على اصغرهما اربعة آحاد إستويا فاذاً ينبغى ان نطلب عددين مربّعين تفاضلهما ثمنية آحاد وهما اثنا عشر ورُبع وعشرون ورُبع ونجعل المربّع الاعظم المعادل للكعب والاربعة الاموال عشرين مالاً ورُبع مال والمربّع الاصغر المعادل للكعب الّا اربعة اموال اثنى عشر مالاً ورُبعَ مال ففى كلى المعادلتين ننتهى الى كعب واحد يعادل ستّة عشر / مالاً ورُبع مال ٤٥
ولذلك يكون الشىء الواحد يعادل ستّة عشر احداً ورُبع احد ولانّا فرضنا ضلع المكعّب شيئاً واحداً يكون ضلع المكعّب ستّة عشر ورُبعاً ويكون المكعّب اربعة الف ومأتين وواحداً وتسعين ‹و›جزءاً من اربعة وستّين جزءاً من واحد ولانّا فرضنا ضلع المربّع شيئين يكون ضلع المربّع اثنين وثلثين ونصفاً ويكون المربّع الفاً وستّة وخمسين ورُبعاً فاذا زدناه على العدد المكعّب اجتمع منهما خمسة الف وثلثمائة وسبعة واربعون وسبعة عشر جزءاً من اربعة وستّين وهو عدد مربّع ضلعه ثلثة وسبعون وثمن واحد واذا نقصنا‹ه› من العدد المكعّب بقى منه ثلثة الف ومائتان واربعة وثلثون وتسعة واربعون جزءاً من اربعة وستّين جزءاً من واحد وهو مربّع ضلعه ستّة وخمسون وسبعة اثمان

---

985 (334): كلى ut كلى in codice scriptum. Vide adn. 329.
988 (335): عددها مساوٍ : عددها مساو in cod.
(336): لعدد : بعدد in cod.
995 (337): للكعب : الكعب in cod.
996 (338): الّا اربعة : الاربعه in cod.
999 (339): ورُبعاً : وربع in cod.
1000 (340): وواحداً : وواحد in cod.
(341): و addidi.
1005 (342): Affixum pronomen addidi.

فقد وجدنا عددين مكعّباً ومربّعاً والعدد المكعّب اذا زيد عليه العدد المربّع اجتمع منهما عدد مربّع واذا نُقص منه العدد المربّع بقى منه ايضاً عدد مربّع

له نريد ان نجد عددين مكعّباً ومربّعاً ويكون العدد المربّع اذا زيد عليه العدد المكعّب اجتمع منهما عدد مربّع واذا نُقص منه العدد المكعّب بقى منه عدد مربّع

فنفرض المكعّب كعباً واحداً والمربّع اربعة اموال فيكون اربعة اموال وكعب تعادل عدداً مربّعاً واربعة اموال الّا كعباً تعادل عدداً مربّعاً فاذا نحن فرضنا ضلع المربّع المعادل للاربعة الاموال والكعب اشياء كان المربّع اموالاً تعادل اربعة اموال وكعباً فاذا نقصنا الاربعة الاموال المشتركة من كلى الناحيتين بقى كعب يعادل اموالاً ويكون العدد الذى فُرض فى المسئلة ⟨شيئاً⟩ مساوياً لعدد الاموال/ ٤٦ الباقية وايضاً فانّا اذا فرضنا ضلع المربّع المعادل للاربعة الاموال الّا كعباً اشياء كان المربّع اموالاً تعادل اربعة اموال الّا كعباً فاذا زدنا الكعب مشتركاً على كلى الناحيتين صارت اموال وكعب تعادل اربعة اموال فاذا نقصنا الاموال المشتركة من كلى الناحيتين بقى كعب يعادل اموالاً ويكون العدد المفروض فى المسئلة شيئاً مساوياً لعدد الاموال الباقية ايضاً فينبغى ان تكون الاموال الباقية فى المعادلة الاوّلة مساويةً لعدد الاموال الباقية فى المعادلة الثانية لكن الاموال الباقية فى المعادلة الاولى هى عدد مربّع نُقص منه اربعة والاموال الباقية فى المعادلة الثانية هى عدد مربّع نُقص من اربعة فنقول مربّع الّا اربعة آحاد تعادل اربعة آحاد الّا مربّعاً آخرَ واذا زدنا الاربعة الآحاد المستثناة من المربّع الاوّل مشتركةً

---

1014 (343): والمربّع : والمربد mendo scripturae (vix enim legendum والمزيد) in cod.

1016 (344): للاربعة : الاربعه in cod.

(345): والكعب : والكعه in cod.

1018 (346): اموالاً : اموال in cod.

1019 (347): Post المسئلة, sc. loco deficientis verbi شيئاً, habet codex الاولى.

1020 (348): للاربعة : الاربعه in cod.

1029 (349): الّا اربعة : الاربعه in cod.

1030 (350): مشتركةً : مسىركا in cod.

على كلى الناحيتين اجتمع مربّع يعادل ثمنية آحاد الاّ مربّعاً فاذا زدنا المربّع الآخر مشتركاً على كلى الناحيتين اجتمع منه مربّعان وكانا يعادلان ثمنية آحاد لكن الثمنية مؤلّفة من عددين مربّعين متساويين فينبغى اذاً ان نقسم الثمنية بعددين مربّعين آخرين كما بيّنّا في المقالة الثانية فليكن احد القسمين اربعة اجزاء من خمسة وعشرين من واحد والآخر سبعة وواحداً وعشرين جزءاً من خمسة وعشرين جزءاً من واحد ونجعل المربّع المعادل للاربعة الاموال والكعب سبعةَ اموال وواحداً وعشرين جزءاً من خمسة وعشرين جزءاً من مال والمربّع المعادل للاربعة الاموال الاّ كعباً اربعة اجزاء من خمسة وعشرين جزءاً من مال فننتهى / ٤٧ فى كلّ واحدة من المعادلتين بعد الجبر والمقابلة الى ثلثة اموال وواحد وعشرين جزءاً من خمسة وعشرين جزءاً من مال تعادل كعباً واذا قسمناهما على مال كانت ثلثة واربعة اخماس وخُمس خُمس واحد تعادل شيئاً واحداً ولانّا فرضنا المكعّب من ضلع شىء واحد يكون ضلع المكعّب ستّة وتسعين جزءاً من خمسة وعشرين جزءاً ويكون المكعّب ثمانى مائة الف واربعة وثمنين الفاً وسبع مائة وستّة وثلثين جزءاً من خمسة عشر الفاً وستّمائة وخمسة وعشرين جزءاً من واحد ولانّا فرضنا ضلع المربّع شيئين يكون ضلع المربّع مائة واثنين وتسعين جزءاً من خمسة وعشرين جزءاً من واحد ويكون المربّع ستّة وثلثين الفاً وثمانى مائة واربعة وستّين جزءاً من ستّمائة وخمسة وعشرين جزءاً من واحد وهو ايضاً تسع مائة الف وواحد وعشرون الفاً وستّمائة جزء من خمسة عشر الفاً وستّمائة وخمسة وعشرين جزءاً من واحد فاذا زدنا ذلك على العدد

---

1031 (351): اجتمع : بعى (sc. بقى) in codice. Vide etiam adn. 405,424,497.

1033 (352): مؤلّفة : مولعه, ut videtur, in cod.

1037 (353): للاربعة : الاربعه in cod.

1038 (354): وواحداً : وواحد in cod.

1039 (355): للاربعة : الاربعه in cod.

1040 (356): واحدة : واحد in codice; vide etiam adn. 383,394 (in ceteris locis emendate scriptum). Mentionem hujus modi erroris facit M. Simon in editione Anatomicorum Galeni (vide vol.I,p.xlii).

1045 (357): واربعة : واربع in cod.

1047 (358): مائة : ماىس in cod.

1051 (359): وعشرين : وعسرون in cod.

المكعّب اجتمع منهما الف الف وثمانى مائة الف وستّة الف وثلثمائة وستّة وثلثون جزءاً ⟨من خمسة عشر الفاً وستّمائة وخمسة وعشرين جزءاً⟩ وذلك عدد مربّع ضلعه الف وثلثمائة واربعة واربعون جزءاً من مائة وخمسة وعشرين جزءاً من واحد واذا نقصنا منه العدد المكعّب بقى منه ستّة وثلثون الفاً وثمانى مائة واربعة وستّون جزءاً من خمسة عشر الفاً وستّمائة وخمسة وعشرين جزءاً من واحد ⟨وذلك عدد مربّع ضلعه مائة واثنان وتسعون جزءاً من مائة وخمسة وعشرين جزءاً من واحد⟩

فقد وجدنا عددين مكعّباً ومربّعاً والمربّع اذا زيد عليه المكعّب كان منهما عدد مربّع واذا نُقص منه المكعّب بقى منه عدد مربّع وهما الثمانى مائة الف والاربعة والثمنون الفاً والسبع مائة والستّة والثلثون الجزء والتسع المائة الألف والواحد والعشرون الفاً والستّمائة الجزء من خمسة عشر الفاً وستّمائة وخمسة و/عشرين جزءاً من واحد وذلك ما اردنا ٤٨
ان نجد

لو نريد ان نجد عدداً مكعّباً اذا زدنا عليه امثالاً مفروضةً للمربّع الذى يكون من ضلعه اجتمع منه عدد مربّع وان نقصنا منه امثالاً اخرَ مفروضةً للمربّع الذى يكون من ضلعه بقى منه عدد مربّع

ولتكن الامثال الزائدة اربعة امثال والامثال الناقصة خمسة امثال ونريد ان نجد عدداً مكعّباً اذا زدنا عليه اربعة امثال المربّع الذى يكون من ضلعه اجتمع منه عدد مربّع وان نقصنا منه خمسة امثال المربّع الذى يكون من ضلعه بقى منه عدد مربّع فنفرض المكعّب كعباً واحداً ليكون المربّع الذى يكون من ضلعه مالاً واحداً ونطلب عددين مربّعين اذا نقصنا من اعظمهما اربعة آحاد وزدنا على اصغرهما خمسة آحاد إعتدلا وهو ان نطلب عددين مربّعين تفاضلهما تسعة آحاد فنجد احد المربّعين ستّة عشر والمربّع الآخر خمسة وعشرين ونزيد على المكعّب

---

1053 (360): Denominatorem addidi.

1054 (361): واربعون : واربعس in cod.

1057-1058 (362): A librario per homoeoteleuton (ut opinor) omissa verba restitui.

1059 (363): Verba المربّع et المكعّب permutavi.

1072 (364): ونطلب : وبعرض (sc. ونفرض ) in codice. Vide adn. 191, 658-659.

1073 (365): اعظمهما : اعطمها in cod.

اربعة امثال المربّع الذى يكون من ضرب ضلعه فى نفسه فيكون كعباً واحداً واربعة اموال وهو عدد مربّع فلنجعل المربّع المعادل له اموالاً مساويةً للمربّع الاعظم من المربّعين اللذين تفاضلهما تسعة آحاد وهو خمسة وعشرون مالاً ونلقى الاربعة الاموال المشتركة من كلى الناحيتين فيبقى كعب واحد يعادل واحداً وعشرين مالاً وايضاً ننقص من المكعّب خمسة امثال المربّع الذى يكون من ضلعه وهو خمسة اموال فيبقى كعب الّا خمسة اموال وذلك يعادل عدداً مربّعاً فلنجعل المربّع المعادل له اموالاً مساويةً للمربّع الاصغر من المربّعين اللذين تفاضلهما تسعة آحاد وهو ستّة عشر مالاً ونزيد الخمسة الاموال الناقصة من الكعب مشتركةً على كلى الناحيتين / فيكون كعب واحد يعادل واحداً وعشرين ٩ مالاً ففى كلى المعادلتين قد إنتهينا الى كعب واحد يعادل واحداً وعشرين مالاً ولنقسم كلّ واحد منهما على مال فيكون شىء واحد يعادل واحداً وعشرين ولانّا فرضنا ضلع المكعّب شيئاً واحداً يكون ضلع المكعّب واحداً وعشرين ويكون المكعّب تسعة الف ومأتين وواحداً وستّين والمربّع الذى يكون من ضرب ضلعه فى نفسه اربع مائة وواحد واربعون واربعة امثاله ⟨الف و⟩سبع مائة واربعة وستّون فاذا زدناها على العدد المكعّب اجتمع منها احد عشر الفاً وخمسة وعشرون وهو مربّع ضلعه مائة وخمسة آحاد وخمسة امثال مربّع ضلع المكعّب الفان ومائتان وخمسة آحاد فاذا نقصناها من العدد المكعّب بقى منه سبعة الف وستّة وخمسون وهو عدد مربّع ضلعه اربعة وثمنون

فقد وجدنا عدداً مكعّباً اذا زدنا عليه اربعة امثال المربّع الذى يكون من ضلعه اجتمع منه عدد مربّع واذا نقصنا ⟨منه⟩ خمسة امثال المربّع

---

1081 (366): Loco امثال habet codex اموال. Vide adn. 75.

1084 (367): الكعب scripsi, المكعب in cod.

1086 (368): إنتهينا : اسهىا in cod.

1088 (369): واحداً يكون : واحد لىكوں, ut videtur, in cod.

1090 (370): واربعون : وارىعىں in cod.

1091 (371): الف و addidi.

(372): وستّون : وسىں in cod.

1093 (373): الفان ومائتان : العں وماىىں in cod.

1094 (374): سبعة : سىعه in cod.

1097 (375): منه addidi.

الذى يكون من ضلعه بقى منه عدد مربّع وهو تسعة الٰف ومائتان وأحد وستّون وذلك ما اردنا ان نجد

ونستبين ايضاً انّا لو اردنا ان تكون الامثال الزائدة خمساً والامثال الناقصة اربعاً ان ضلع المكعّب يكون عشرين والمكعّب ثمنية الٰف واذا زدنا عليه خمسة امثال المربّع الذى يكون من ضلعه وهو الفان اجتمع منه عشرة الٰف وهو عدد مربّع ضلعه مائة واذا نقصنا منه اربعة امثال المربّع الذى يكون من ضلعه وهو الف وستّمائة بقى منه ستّة الٰف واربع مائة وهو عدد مربّع ضلعه ثمنون

لز نريد ان نجد عدداً مكعّباً اذا ضربنا المربّع الذى يكون / من ٥٠
ضلعه فى عددين مفروضين وزدنا ما يجتمع من كلّ واحد منهما على العدد المكعّب اجتمع منه عدد مربّع

وليكن احد العددين خمسة والآخر عشرة ونريد ان نجد عدداً مكعّباً اذا ضربنا مربّع ضلعه فى خمسة وفى عشرة وزدنا ما يجتمع من ضربه فى كلّ واحد منهما على العدد المكعّب اجتمع منه عدد مربّع فنفرض المكعّب كعباً واحداً ونضرب مربّع ضلعه الذى هو مال واحد فى الخمسة وفى العشرة فيكون خمسة اموال وعشرة اموال ونزيد كلّ واحد منهما على الكعب فيكون كعب وخمسة اموال تعادل عدداً مربّعاً وكعب وعشرة اموال تعادل عدداً مربّعاً فاذا جعلنا ضلع المربّع الذى هو كعب وخمسة اموال اشياء كان مربّعها اموالاً ثمّ اذا نقصنا الخمسة الاموال المشتركة من كلى الناحيتين بقى كعب يعادل اموالاً وظاهر ان العدد المفروض فى هذه المسئلة شيئاً يكون مساوياً لعدد الاموال الباقية وايضاً فانّا اذا جعلنا ضلع المربّع الذى هو كعب وعشرة اموال اشياء كان مربّعها اموالاً فاذا نقصنا العشرة الاموال المشتركة من كلى الناحيتين بقى كعب يعادل اموالاً ولذلك يكون العدد المفروض فى هذا التحليل شيئاً مساوياً لعدد الاموال الباقية فينبغى اذاً ان تكون الاموال الباقية فى المعادلة الاوّلة مساويةً للاموال الباقية فى المعادلة الثانية لكن الاموال الباقية فى المعادلة الاوّلة هو عدد مربّع الّا خمسة آحاد والاموال الباقية فى المعادلة الثانية هو مربّع الّا عشرة آحاد واذاً

1098 (376): تسعة : سعه in cod.

1114 (377): كعب (in utroque loco): كعبا in cod.

1118 (378): شيئاً : سى in cod.

1123 (379): للاموال : الاموال in cod.

1124 & 1125 (380): Pro هو expectandum erat هى.

ينبغى ان نجد عددين مربّعين اذا نقصنا من اعظمهما عشرة آحاد ومن اصغرهما / خمسة آحاد إستويا فنقول مربّع الّا خمسة آحاد يعادل (١ د) مربّعاً آخرَ الّا عشرة آحاد ونزيد العشرة الآحاد مشتركةً على كلى الناحيتين فيجتمع مربّع وخمسة آحاد تعادل مربّعاً فينبغى ان نطلب مربّعين تفاضلهما خمسة آحاد ويكون الاصغر منهما اكثر من خمسة آحاد وليكن المربّع الاصغر ثلثة وخمسين وسبعة اتساع وضلعه سبعة آحاد وثُلث واحد والمربّع الاعظم ثمنية وخمسين وسبعة اتساع وضلعه سبعة آحاد وثُلثا واحد ونجعل المربّع المعادل للكعب والخمسة الاموال ثلثة وخمسين مالاً وسبعة اتساع مال ونجعل المربّع المعادل للكعب والعشرة الاموال ثمنية وخمسين مالاً وسبعة اتساع مال ففى كلّ واحدة من المعادلتين ننتهى الى كعب واحد يعادل ثمنية واربعين مالاً وسبعة اتساع مال فاذا قسمناهما على مال واحد كان شىء واحد يعادل ثمنية واربعين احداً وسبعة اتساع واحد ولانّا فرضنا ضلع المكعّب شيئاً واحداً يكون ضلعه اربع مائة وتسعة وثلثين تُسعاً ويكون المكعّب اربعة وثمنين الف الف وستّمائة الف واربعة الف وخمس مائة وتسعة عشر تُسع تُسع تُسع وهو ايضاً سبع مائة الف الف وواحد وستّون الف الف واربع مائة الف واربعون الفاً وستّمائة وواحد وسبعون تُسع تُسع تُسع تُسع ومربّع ضلع المكعّب مائة الف واثنان وتسعون الفاً وسبع مائة وواحد وعشرون تُسع تُسع وهو ايضاً خمسة عشر الف الف وستّمائة الف وعشرة الف واربع مائة تُسع تُسع تُسع تُسع ⟨و⟩ تُسع تُسع تُسع تُسع واحد واذا ضربنا ذلك فى خمسة آحاد اجتمع منه ثمنية وسبعون الف الف / واثنان وخمسون الفاً وخمسة اتساع تُسع تُسع تُسع واذا زدنا ذلك (٢) على العدد المكعّب اجتمع منهما ثمانى مائة الف الف وتسعة وثلثون الف الف واربع مائة الف واثنان وتسعون الفاً وستّمائة وستّة وسبعون تُسع تُسع تُسع تُسع ⟨وهو مربّع ضلعه ثمنية وعشرون الفاً وتِسع مائة واربعة وسبعون تُسع تُسع⟩ وايضاً اذا ضربنا ⟨مربّع⟩ ضلع المكعّب فى عشرة

1128 (381): مشتركةً : المسركه in cod.

1134 (382): ونجعل : ويحعل in cod.

1135 (383): واحدة : واحد in cod.

1145 (384): و addidi.

1146-1147 (385): Pro الف الف praebet codex الف الف الف.

1150-1151 (386): Per homoeoteleuton omissum addidi.

1151 (387): مربّع addidi.

آحاد اجتمع منه مائة الف الف وستّة وخمسون الف الف ومائة الف واربعة الف وعشرة اتساع تُسع تُسع تُسع واذا زدناه على العدد المكعّب اجتمع منهما تِسع مائة الف الف وسبعة عشر الف الف وخمس مائة الف واربعة واربعون الفاً وستّمائة وواحد وثمنون تُسع تُسع تُسع تُسع وهو مربّع ضلعه ثلثون الفاً ومائتان وواحد وتسعون تُسع تُسع

فقد وجدنا عدداً مكعّباً على الشرط الذى شرطنا [وهما العددان اللذان ذكرنا]

لح نريد ان نجد الآن عدداً مكعّباً اذا ضربنا المربّع الذى يكون من ضلعه فى عددين مفروضين ونقصنا كلّ واحد منهما من العدد المكعّب بقى منه عدد مربّع

وليكن احد العددين خمسة آحاد والآخر عشرة آحاد ونريد ان نجد عدداً مكعّباً اذا ضربنا مربّع ضلعه فى خمسة وفى عشرة ونقصنا ما يجتمع من كلّ واحد منهما من العدد المكعّب بقى منه عدد مربّع وكذلك ايضاً نجعل المكعّب كعباً واحداً ونضرب مربّع ضلعه وهو مال واحد فى خمسة وفى عشرة فيكون خمسة اموال وعشرة اموال وننقص كلّ واحد منهما من العدد المكعّب فيبقى كعب الّا خمسة اموال وكعب الّا عشرة اموال وكلّ واحد منهما يعادل عدداً مربّعاً فامّا الاشياء التى هى ضلع المربّع المعادل للكعب الناقص خمسةَ اموال / فانّه اذا زيد على مربّعها ٥٣
خمسة اموال كان المجتمع من ذلك اموالاً عددها هو العدد المفروض فى المسئلة شيئاً واما الاشياء التى هى ضلع المربّع المعادل للكعب الناقص عشرةَ اموال فانّه اذا زيد على مربّعها عشرة اموال كان المجتمع من ذلك اموالاً عددها هو العدد المفروض فى المسئلة شيئاً فلذلك ينبغى ان نأخذ عددين مربّعين اذا زدنا على اعظمهما خمسة آحاد وزدنا على اصغرهما عشرة آحاد إستويا فنقول مربّع اعظم وخمسة آحاد تعادل مربّعاً صغيراً وعشرة آحاد فننقص الخمسة الآحاد المشتركة من

---

1155 (388): وثمنون : ولمون in cod.

1157-1158 (389): Uncis inclusa verba a quodam stulto lectore (forsan a librario nostro) addita esse censeo.

1170 (390): اموالاً : اموال in cod.

1171 (391): للمربع in codice, المربّع scripsi.

1173 (392): اموالاً : اموال in cod.

1174 (393): نأخذ : melius dixisset نطلب .

كلى الناحيتين فيبقى مربّع صغير وخمسة آحاد تعادل مربّعاً عظيماً فاذاً تفاضل المربّعين خمسة آحاد فلنطلب عددين مربّعين تفاضلهما خمسة آحاد اىّ عددين إتفقا وليكن الاصغر منهما اربعة آحاد والاعظم تسعة آحاد ونجعل المربّع المعادل للكعب الناقص خمسةَ اموال تسعة اموال ونجعل المربّع المعادل للكعب الناقص عشرةَ اموال اربعة اموال فننتهى فى كلّ واحدة من المعادلتين الى كعب يعادل اربعة عشر مالاً ويكون الشىء الواحد يعادل اربعة عشر احداً ولانّا فرضنا ضلع المكعّب شيئاً واحداً يكون ضلعه اربعة عشر ويكون المكعّب ألفين وسبع مائة واربعة واربعين والمربّع الذى يكون من ضلعه مائة وستّة وتسعون واذا ضربناه فى خمسة آحاد خرج تسع مائة وثمنون فاذا نقصناه من العدد المكعّب بقى الف وسبع مائة واربعة وستّون وهو مربّع ضلعه اثنان واربعون وايضاً اذا ضربنا مربّع ضلع المكعّب فى عشرة آحاد اجتمع منه الف وتسع مائة وستّون / واذا نقصناه من ٥٤
العدد المكعّب بقى منه سبع مائة واربعة وثمنون وهو مربّع ضلعه ثمنية وعشرون

فقد وجدنا عدداً مكعّباً على الشرط الذى شرطنا وهو ألفان وسبع مائة واربعة واربعون وذلك ما اردنا ان نجد

لط نريد ان نجد عدداً مكعّباً اذا ضربنا مربّع ضلعه فى عددين مفروضين ونقصنا المكعّب من كلّ واحد منهما بقى من كلّ واحد منهما عدد مربّع

فليكن العددان المفروضان ثلثة وسبعة ونريد ان نجد عدداً مكعّباً اذا ضربنا مربّع ضلعه فى ثلثة وفى سبعة ونقصنا المكعّب من كلّ واحد من المضروبين بقى من كلّ واحد منهما عدد مربّع فلنفرض المكعّب كعباً واحداً ونضرب مربّع ضلعه الذى هو مال واحد فى ثلثة وفى سبعة وننقص الكعب من كلّ واحد منهما فيبقى ثلثة اموال الّا كعباً تعادل مربّعاً وسبعة اموال الّا كعباً تعادل مربّعاً فنجعل ضلع المربّع المعادل للثلثة الاموال الناقصة كعباً اشياء ونضربها فى مثلها فتصير

1182 (394): واحدة : واحد in cod.

1186 (395): وتسعون : وسعس in cod.

1199 (396): من (prius) scripsit librarius (cum puncto) super واحد (lin. 1198).

(397): المضروبين : المعروصس in cod.

اموالاً تعادل ثلثة اموال غير كعب ونزيد الكعب مشتركاً على كلى الناحيتين فيصير اموال وكعب تعادل ثلثة اموال فاذا نقصنا الاموال المشتركة من الثلثة الاموال بقى كعب يعادل اموالاً ويكون الشىء مساوياً لعدد الاموال الباقية وايضاً اذا جعلنا ضلع المربّع المعادل للسبعة الاموال الناقصة كعباً اشياء وضربناها فى مثلها وجبرنا وقابلنا بقى كعب واحد ايضاً يعادل بقيّة السبعة ⟨الاموال⟩ ويكون الشىء ايضاً مساوياً ⟨لبقيّة⟩ السبعة ولذلك يجب ان تكون الاموال الباقية من الثلثة الاموال مساويةً للاموال الباقية من السبعة الاموال لكن / الباقية ٥٥ من الثلثة الاموال هى ثلثة الّا عدداً مربّعاً والباقية من السبعة الاموال هى سبعة الّا عدداً مربّعاً فثلثة الّا عدداً مربّعاً تعادل سبعة الّا عدداً مربّعاً ونزيد كلّ واحد من المربّعين مشتركاً على كلى الناحيتين فيجتمع سبعة ومربّع صغير تعادل ثلثة ومربّعاً عظيماً ونلقى الثلثة المشتركة فيبقى مربّع عظيم يعادل مربّعاً صغيراً واربعة آحاد فلذلك ينبغى ان نطلب عددين مربّعين تفاضلهما اربعة آحاد وليكن اصغرهما اقلّ من ثلثة آحاد وهما اثنان ورُبع وستّة آحاد ورُبع ونجعل المربّع المعادل للثلثة الاموال الناقصة كعباً مالين ورُبعَ مال ونجعل المربّع المعادل للسبعة الاموال الناقصة كعباً ستّة اموال ورُبعَ مال ففى كلّ واحدة من المعادلتين ننتهى الى كعب يعادل ثلثة ارباع مال ولذلك يكون الشىء ثلثة ارباع واحد ويكون الكعب سبعة وعشرين ثُمن ثُمن و⟨يكون⟩ مربّع ضلع المكعّب ستّة وثلثين ثُمن ثُمن واذا ضربناه فى ثلثة اجتمع منه مائة وثمنية اثمان ثُمن واذا نقصنا منها العدد المكعّب بقى

---

1204 (398): Pro الاموال codicis اموال scripsi.

1205 (399): اموال وكعب : اموالا وكعبا in cod.

1209 (400): الاموال addidi.

1210 (401): لبقيّة addidi.

(402): السبعة : لسبعه in codice (sequitur enim مساوياً).

(403): Post hoc لسبعه habet codex verbum اموال, quod supra (aut cum, aut sine articulo) desiderabatur.

1211 (404): للاموال : الاموال in cod.

1215 (405): فيجتمع : فسعى (sc. فيبقى) in cod.

1219 (406): ورُبعَ مال : وربعا in cod.

1220 (407): ورُبعَ مال : وربعا in cod.

1223 (408): يكون addidi.

1224 (409): ثمن bis in cod.

منها احد وثمنون ثُمن ثُمن وهو مربّع ضلعه تسعة اثمان واذا ضربنا ايضاً مربّع ضلع المكعّب وهو ستّة وثلثون ثُمن ثُمن فى سبعة آحاد اجتمع منه مائتان واثنان وخمسون ثُمن ثُمن فاذا نُقص منها العدد المكعّب بقى منها مائتان وخمسة وعشرون ثُمن ثُمن وهو عدد مربّع ضلعه خمسة عشر ثُمناً

فقد وجدنا عدداً مكعّباً على الشرط الذى شرطنا وهو سبعة وعشرون ثُمن ثُمن وذلك ما اردنا ان نجد

مٓ نريد ان نجد عددين مربّعاً ومكعّباً يكون مربّع المربّع اذا زيد عليه المكعّب اجتمع منهما عدد مربّع واذا نُقص منه المكعّب بقى منه/ عدد ٥٦ مربّع

فلنفرض المربّع من ضلع شيئين فيكون المربّع اربعة اموال ويكون مربّعه ستّة عشر مال مال ولنفرض المكعّب من ضلع كم شئنا من الاشياء فلنفرضه من ضلع اربعة اشياء فيكون المكعّب اربعة وستّين كعباً ونزيد هذا المكعّب على ستّة عشر مال مال وننقصه منه فيكون ستّة عشر مال مال واربعة وستّون كعباً تعادل عدداً مربّعاً وستّة عشر مال مال الّا اربعة وستّين كعباً تعادل عدداً مربّعاً ثمّ نأخذ فى طلب شىء يساوى معادلة الطرفين على نحو ما عملنا فيما تقدّم فنقول انّا اذا فرضنا ضلع المربّع المعادل للستّة عشر مال مال والاربعة والستّين الكعب اموالاً كان مربّعها اموالَ اموال تعادل ستّة عشر مال مال واربعة وستّين كعباً فاذا نقصنا الستّة عشر مال مال المشتركة من كلى الناحيتين بقى اربعة وستّون كعباً تعادل اموال اموال فاذا قسمنا جميع ذلك على كعب كانت اشياء تعادل اربعة وستّين احداً ولذلك يكون العدد المفروض فى المسئلة ⟨شيئاً⟩ مساوياً لِما يخرج من قسمة عدد الكعاب التى هى اربعة وستّون على عدد الاموال الاموال الباقية وايضاً فانّا اذا فرضنا ضلع المربّع المعادل للستّة عشر مال مال الّا اربعة وستّين كعباً اموالاً كان مربّعها اموالَ اموال تعادل ستّة عشر مال مال الّا اربعة وستّين

---

1239 (410): الّا اربعة : الا ربعه in cod.

1245 (411): Post كعباً iterantur verba فاذا نقصنا . . . . وستّون كعباً in codice per dittographiam.

1247 (412): شيئاً addidi.

1249 (413): الّا اربعة : الا ربعه in cod.

1250 (414): الّا اربعة : الا ربعه in cod.

كعباً واذا زدنا الكعاب الناقصة مشتركةً على كلى الناحيتين صارت
اموال اموال واربعة وستّون كعباً تعادل ستّة عشر مال مال فاذا نقصنا
اموال الاموال المشتركة بقيت اربعة وستّون كعباً تعادل اموال/اموال ٥٧
واذا قسمنا جميع ذلك على كعب كانت اربعة وستّون احداً تعادل
اشياء ولذلك يكون الشیء هو العدد الذى يخرج من قسمة الاربعة
والستّين على عدد ⟨اموال⟩ الاموال الباقية فلذلك يكون عدد ⟨اموال⟩
الاموال الباقية فى المعادلة الاوّلة مساوياً لعدد ⟨اموال⟩ الاموال الباقية
فى المعادلة الثانية لكن عدد ⟨اموال⟩ الاموال الباقية فى المعادلة
الاوّلة هو عدد مربّع الّا ستّة عشر و⟨اموال⟩ الاموال الباقية فى المعادلة
الثانية هى ستّة عشر الّا عدداً مربّعاً فاذاً عدد مربّع عظيم الّا ستّة عشر
تعادل ستّة عشر الّا عدداً مربّعاً صغيراً فاذا زدنا المربّع الصغير
مشتركاً وزدنا ايضاً الستّة عشر الناقصة من المربّع العظيم مشتركةً على
كلى الناحيتين اجتمع مربّع عظيم ومربّع صغير يعادلان اثنين وثلثين
احداً لكنْ الاثنان والثلثون مؤلّفة من مربّعين متساويين يُمكن اذاً ان
تُقسم بعددين مربّعين مختلفين فلتُقسم وليكن احد المربّعين ستّة عشر
خُمس خُمس والمربّع الآخر واحداً وثلثين وتسعة اخماس خُمس ونجعل
المربّع المعادل للستّة عشر مال مال واربعة وستّين كعباً واحداً وثلثين

---

1251 (415): مشتركةً : المسركه in cod.

1256 (416): اموال bis addidi. Vide etiam adn. 418,419,421 et infra, 437.

1257 (417): مساوياً : مساوه in cod.

(418): اموال addidi.

1258 (419): اموال addidi.

1259 (420): هو : هى in cod.

(421): اموال addidi.

1262 (422): مشتركاً : مساركا (sc. مشاركاً) in codice. Vide etiam adn. 676.

(423): على : مى in cod.

1263 (424): اجتمع : هى in codice, errore pro بقى (vide adn. 451).

(425): يعادلان : سعادل in cod.

1264 (426): Pro مولف codicis مؤلّفة scripsi.

1265 (427): وليكن : ولكں in cod.

مال مال وتسعة اخماس خُمس مال مال والمربّع المعادل للستّة عشر مال مال الّا اربعة وستّين كعباً ستّة عشر خُمس خُمس مال مال ففى كلّ واحدة من المعادلتين ننتهى الى اربعة وستّين كعباً تعادل خمسة عشر مال مال وتسعة اخماس خُمس مال ⟨مال⟩ ونقسم جميع ذلك على كعب فيكون خمسة عشر شيئاً وتسعة اخماس خُمس شىء تعادل اربعة وستّين احداً فالشىء الواحد هو ما يخرج من قسمة الف وستّمائة على ثلثمائة واربعة وثمنين وهو اربعة آحاد وسُدس واحد / ٥٨ فلانّا فرضنا ضلع المربّع شيئين يكون ضلع المربّع ثمنية آحاد وثُلثَ واحد ويكون المربّع تسعة وستّين احداً واربعة اتساع واحد ويكون مربّع المربّع اربعة الف وثمانى مائة واثنين وعشرين واربعة اتساع وسبعة اتساع تُسع ولانّا فرضنا ضلع المكعّب اربعة اشياء يكون ضلع المكعّب ستّة عشر احداً وثُلثى احد ويكون المكعّب اربعة الف وستّمائة وتسعة وعشرين وخمسة اتساع وثُلثى تُسع وهو اذا زيد على العدد المجتمع من ضرب العدد المربّع فى مثله اجتمع منهما تسعة الف واربع مائة واثنان وخمسون وتُسع واربعة اتساع تُسع وهو عدد مربّع ضلعه سبعة وتسعون احداً وتُسعاً واحد واذا نُقص هذا العدد من مربّع العدد المربّع بقى منه مائة واثنان وتسعون وثمنية اتساع وتُسع تُسع وهو عدد مربّع ضلعه ثلثة عشر احداً وثمنية اتساع واحد

فقد وجدنا عددين على الشرط الذى اشترطنا وهما العددان اللذان حدّدنا

ما نريد ان نجد عددين آخرين مكعّباً ومربّعاً يكون العدد المكعّب اذا زيد عليه مربّع المربّع اجتمع منه عدد مربّع واذا نُقص منه مربّع المربّع بقى منه عدد مربّع

---

1268 (428): للستّة : السه in cod.

1269 (429): الّا اربعة : الاربعه in cod.

1271 (430): Per haplographiam omissum مال addidi.

1274 (431): وثمنين : ولبس in cod.

(432): Post ولبس supra dictum praebet codex شيئاً , سا videlicet (cf. adn. 721) cum signo (deletionis?) nescio quali.

1281 (433): مثله : صله in cod.

وعلى مثال ما وصفنا نقول اربعة وستّون كعباً وستّة عشر مال مال تعدل عدداً مربّعاً فاذا جعلنا ضلعها اموالاً كان المربّع اموالَ اموال تعدل اربعة وستّين كعباً وستّة عشر مال مال فاذا نقصنا الستّة العشر المال المال المشتركة من كلى الناحيتين بقى اربعة وستّون كعباً تعادل اموال اموال فاذا قسمناهما على كعب كانت اربعة وستّون احداً تعادل اشياء فالشىء هو ما يخرج من قسمة الاربعة والستّين على عدد اموال الاموال الباقية وايضاً /اربعة وستّون كعباً الّا ستّة عشر مال مال تعادل ٥٩ عدداً مربّعاً فاذا جعلنا ضلعها اموالاً كان مربّعها اموالَ اموال تعادل اربعة وستّين كعباً الّا ستّة عشر مال مال ونزيد الستّة العشر مال مال مشتركةً على كلى الناحيتين فيجتمع اموال اموال تعادل اربعة وستّين كعباً واذا قسمناهما على كعب كانت اشياء تعادل اربعة وستّين احداً ويكون الشىء هاهنا ايضاً هو ما يخرج من قسمة الاربعة والستّين على عدد ⟨اموال⟩ الاموال المجتمعة فينبغى ان يكون عدد ⟨اموال⟩ الاموال الباقية فى المعادلة الاوّلة وهو عدد مربّع الّا ستّة عشر مساوياً لعدد ⟨اموال⟩ الاموال المجتمعة فى المعادلة الثانية وهو عدد مربّع وستّة عشر فاذاً عدد مربّع الّا ستّة عشر تعادل مربّعاً آخرَ وستّة عشر ونزيد الستّة عشر الناقصة مشتركةً على كلى الجهتين فيجتمع مربّع واثنان وثلثون احداً تعادل مربّعاً عظيماً ولذلك نطلب عددين مربّعين تفاضلهما اثنان وثلثون احداً فيكون المربّع الاعظم منهما اكثر من ستّة عشر وليكن المربّع الاصغر اربعة آحاد والمربّع الاعظم ستّة وثلثين ونفرض ⟨المربّع⟩ المعادل للاربعة والستّين الكعب والستّة العشر مال مال ستّة وثلثين مال مال والمربّع المعادل للاربعة والستّين الكعب الّا ستّة عشر مال مال اربعة اموال اموال ففى كلّ واحدة من المعادلتين ننتهى الى اربعة وستّين كعباً تعادل عشرين مال مال ولنقسم كلّ واحد منهما على كعب فيكون عشرون شيئاً

---

1292 (434): Loco ضلعها expectes ضلعه , ad عدد مربّع referendum; item in linea 1298.

1297 (435): للاموال : الاموال in cod.

1300 (436): المسركه : مشتركةً in cod.

1303-1305 (437): اموال ter addidi.

1311 (438): المربّع addidi.

(439): الاربعه : للاربعة in cod.

1315 (440): عسرس : عشرون in cod.

تعادل اربعة وستّين احداً فيكون الشیء الواحد ثلثة آحاد وخُمس واحد ولانّا فرضنا المربّع من ضلع شيئين يكون ⟨ضلعه⟩ ستّة آحاد وخُمسی واحد ويكون / المربّع اربعين احداً واربعة اخماس واحد ٦٠ واربعة اخماس خُمس واحد ويكون مربّع المربّع الفاً وستّمائة وسبعة وسبعين واربع مائة وأحد وخمسين جزءاً من ستّمائة وخمسة وعشرين جزءاً من واحد ولانّا فرضنا ضلع المكعّب اربعة اشياء يكون ضلع المكعّب اثنی عشر واربعة اخماس ويكون المكعّب ألفين وسبعة وتسعين وخمسة وتسعين جزءاً من ستّمائة وخمسة وعشرين جزءاً من واحد وهو اذا ما زيد عليه مربّع المربّع اجتمع منهما ثلثة الُف وسبع مائة واربعة وسبعون وخمس مائة وستّة واربعون جزءاً من ستّمائة وخمسة وعشرين جزءاً من واحد وهو مربّع ضلعه واحد وستّون وأحد عشر جزءاً من خمسة وعشرين جزءاً من واحد واذا ما نُقص من هذا المكعّب مربّع العدد المربّع بقى منه اربع مائة وتسعة عشر احداً ومائتان وتسعة وستّون جزءاً من ستّمائة وخمسة وعشرين جزءاً من واحد وهو عدد مربّع ضلعه عشرون احداً واثنا عشر جزءاً من خمسة وعشرين جزءاً من واحد

فقد وجدنا عددين على الشرط الذی اشترطنا وهما العددان اللذان حدّدنا وذلك ما اردنا ان نجد

مب نريد ان نجد عددين مكعّباً ومربّعاً يكون مكعّب المكعّب ومربّع المربّع جميعهما عدد مربّع وتفاضلهما عدد مربّع

ونفرض ضلع المكعّب ما شئنا من الاشياء ولنفرضه شيئين فيكون ⟨المكعّب⟩ ثمنية كعاب ومكعّب هذا المكعّب خمس مائة واثنا عشر كعب كعب كعب ونفرض المربّع ايضاً من ضلع كم شئنا من الاموال فلنفرضه من ضلع اربعة اموال حتّى يكون المربّع ستّة عشر مال مال ومربّع المربّع مائتان وستّة وخمسون مال مال فی مال مال / اعنی الذی يُسمّی واحدُه ٦١ كعبَ كعب مال

ولنطلب اوّلاً ان يكون مكعّب المكعّب اذا زيد عليه مربّع المربّع كان مربّعاً واذا نُقص منه مربّع المربّع بقی منه مربّع وقد كنّا وجدنا فيما تقدّم

---

1317 (441): ضلعه deest in cod.

1319 (442): الفاً : الف in cod.

1334 (443): عدد مربّع (in utroque loco): عددا مربعا in cod.

1336 (444): المكعّب addidi.

(445): عشر bis in cod.

عددين على هذه الصفة على طريق الاتّفاق من غير ان نقصد لوجودها نريد ان نذكر هاهنا الجهة التى بها يكون وجودهما فنقول خمس مائة واثنا عشر كعب كعب كعب ومائتان وستّة وخمسون كعب كعب مال تعادل عدداً مربّعاً وايضاً خمس مائة واثنا عشر كعب كعب كعب الّا مأتين وستّة وخمسين كعب كعب مال تعادل مربّعاً

فان شئنا عملنا فى ذلك على جهة المساواة المثنّاة وهو ان نأخذ الفضل بين هذين المربّعين وهو خمس مائة واثنا عشر مال مال فى مال مال ونطلب عددين من اموال اموال اذا ضربنا احدهما فى الآخر كان ما يجتمع من الضرب خمس مائة واثنى عشر مال مال فى مال مال فانّا اذا أخذنا نصف جملة العددين وضربنا ما اجتمع فى مثله وقابلنا به المربّع الاعظم الذى هو خمس مائة واثنا عشر كعب كعب كعب ومائتان وستّة وخمسون كعب كعب مال ثمّ أخذنا ⟨نصف⟩ الفضل بين العددين وضربنا ما حصل فى مثله وقابلنا به المربّع الاصغر الذى هو خمس مائة واثنا عشر كعب كعب كعب الّا مأتين وستّة وخمسين مال مال فى مال مال إنتهينا فى كلّ واحدة من المعادلتين الى خمس مائة واثنى عشر كعب كعب كعب تعادل عدداً بِعينه من كعاب كعب مال فاذا قسمنا كلّ واحد منهما على واحد من اقعدهما وهو كعب كعب مال كانت خمس مائة واثنا عشر شيئاً تعادل عدداً ومن قبل ذلك نعلم الشىء واذا علمنا الشىء رجعنا الى الاصول التى كنّا ثبتنا عليها / فركّبنا بعد معرفة ٦٢ الشىء جميع ما فى المسئلة

فان شئنا عملنا بعمل طلب المساواة فى معادلة الطرفين كما قد وصفنا فى المسائل المتقدّمة وهو انّا نقول اذا نحن جعلنا ضلع المربّع الاعظم اموال اموال كان مربّعها كعاب كعاب كعاب مال تعادل المربّع الاعظم فاذا نقصنا كعاب كعاب مال المأتين والستّة والخمسين المشتركة

---

1352 (446): وضربنا ما : وصرساهما in cod.

(447): اجتمع : احمع in cod.

1354 (448): نصف deest in cod.

1357 (449): المعادلتين : المعاىلس (pro المتقابلين (?) in codice; genus grammaticale defectivi verbi praecedens واحدة indicat.

1361 (450): ثبتنا conjeci pro سىا codicis; item in lineis 1409 et 1494 (vide adn. 498).

من كلى الناحيتين بقى خمس مائة واثنا عشر كعب كعب كعب كعب تعادل كعاب كعاب مال فاذا قسمناهما على كعب كعب مال كانت خمس مائة واثنا عشر شيئاً تعادل عدداً ولذلك يكون العدد المساوى لعدد الكعاب كعب مال الباقية اذا قُسم على خمس مائة واثنى عشر كان الذى يخرج هو العدد المفروض فى المسئلة شيئاً وايضاً اذا نحن فرضنا ضلع المربّع الاصغر اموال اموال كان مربّعها كعاب كعب مال تعادل المربّع الاصغر فاذا زدنا كعاب كعب المال المأتين والستّة والخمسين مشتركةً على كلى الجهتين اجتمعت كعاب كعاب مال تعادل خمس مائة واثنى عشر كعب كعب كعب واذا قُسما على واحد من اقعد الناحيتين وهو كعب كعب مال كانت خمس مائة واثنا عشر شيئاً تعادل عدداً ولذلك يكون هذا العدد اذا قُسم على خمس مائة واثنى عشر كان العدد الذى يخرج من القسمة هو ⟨العدد⟩ المفروض فى المسئلة شيئاً فينبغى ان يكون عدد كعاب كعاب المال الباقية فى المعادلة الاوّلة مساوياً لعدد كعاب كعاب المال المجتمعة فى المعادلة الثانية لكن عدد كعاب كعاب المال الباقية فى المعادلة ⟨الاوّلة هو عدد مربّع الّا مأتين وستّة وخمسين وعدد كعاب كعاب المال المجتمعة فى المعادلة⟩ الثانية هو عدد مربّع ومائتان وستّة وخمسون فلذلك ينبغى ان نطلب عددين تفاضلهما ضعف المأتين/ ٦٣ والستّة والخمسين اعنى خمس مائة واثنى عشر فاذا وجدناهما جعلنا الاعظم منهما كعاب كعاب مال وعاد لنا بها المربّع الاعظم وجعلنا الاصغر منهما كعاب كعاب مال وعاد لنا بها المربّع الاصغر فننتهى فى كلّ واحدة من المعادلتين بعد ذلك الى خمس مائة واثنى عشر شيئاً تعادل عدداً واحداً بِعينه ومن قبل ذلك نعلم الشىء الذى نطلب معرفةَ مقدارِه ثمّ نرجع ⟨ونأخذ⟩ فى تركيب المسئلة

---

1367 (451): بقى : هى in cod.

1369 (452): ولذلك : وكذلك in cod.

1375 (453): قُسما : قسمت in cod.

1378 (454): العدد addendum esse censui.

1380 (455): مساوياً : مساويه in cod.

1381-1383 (456): Per homoeoteleuton omissum addidi.

1383 (457): Post وخمسون addit codex مه.

1384 (458): مربّعين post عددين forsan subauditum.

1388 (459): واحداً : واحد in cod.

1389 (460): ونأخذ addidi.

وان شئنا قلنا خمس مائة واثنا عشر كعب كعب كعب ومائتان وستّة وخمسون كعب كعب مال تعادل مربّعاً وخمس مائة واثنا عشر كعب كعب كعب الّا مأتين وستّة وخمسين كعب كعب مال تعادل مربّعاً وكلّ مربّع يُقسم على مربّع فان الذى يخرج من القَسم يكون مربّعاً فاذا قسمنا الخمس المائة والاثنى عشر الكعب كعب كعب والمأتين والستّة والخمسين الكعب كعب مال على مربّع وهو كعب كعب مال او اربعة كعاب كعاب مال او تسعة كعاب كعاب مال او ستّة عشر كعب كعب مال او على ما شئنا من الاعداد المربّعة بعد ان نجعل كلّ واحد منها كعاب كعاب مال امّا كعاب كعاب مال فانّه يخرج من قسمتها على كعاب كعاب مال عدد وامّا كعاب كعب كعب فانّه تخرج اشياء فلنُنزل انّا قسمناهما على ستّة عشر كعب كعب مال فخرج من القسمة اثنان وثلثون شيئاً وستّة عشر احداً وعلى مثل ما قسمنا هذا المربّع فلنقسم المربّع الآخر الذى هو خمس مائة واثنا عشر كعب كعب كعب الّا مأتين وستّة وخمسين كعب كعب مال فيكون اثنين وثلثين شيئاً الّا ستّة عشر احداً فاثنان وثلثون شيئاً وستّة عشر احداً مربّع واثنان وثلثون شيئاً الّا ستّة عشر احداً مربّع فلنطلب عدداً اذا زدنا /عليه عدداً مفروضاً وهو ستّة عشر كان مربّعاً وان نقصنا ٦٤ منه عدداً مفروضاً وهو ستّة عشر كان مربّعاً فاذا وجدنا ذلك العدد قسمناه على اثنين وثلثين فما خرج من القسمة فهو الشىء فاذا عرفناه عُدنا فركّبنا المسئلة على نحو ما كنّا ثبتنا عليه فى تحليلها

وهكذا اكثر المسائل ذوات الطرفين التى تقدّم ذكرها يُمكن عملُها على هذه ⟨الجهة⟩ التى وصفنا

فلنطلب ان يكون مربّع المربّع اذا زيد عليه مكعّب المكعّب كان مربّعاً وان نُقص منه مكعّب المكعّب بقى منه عدد مربّع

وكذلك فلنقل كما قلنا آنفاً مائتان وستّة وخمسون كعب كعب مال وخمس مائة واثنا عشر كعب كعب كعب تعادل مربّعاً ومائتان وستّة وخمسون كعب كعب مال الّا خمس مائة واثنى عشر كعب كعب كعب تعادل مربّعاً ونعمل فى ذلك بطلب مساواة المعادلة فى كلى

---

1396 (461): الكعب كعب مال : الكعب كعب مال مال in cod.

1398-1399 (462): Post منها praebet codex كعب كعب مال, pro quo formam pluralis كعاب كعاب مال scripsi.

1400 (463): عدد : عددا in cod.

1411 (464): الجهة addidi.

الطرفين كما قد وصفنا فيما تقدّم من هذا النوع من المسائل فننتهى الى ان نقسم ضعف المأتين والستّة والخمسين الذى هو عدد مربّع وضعفه خمس مائة واثنا عشر بعددين مربّعين مختلفين فليكن الاصغر من العددين المربّعين عشرة آحاد وستّة اجزاء من خمسة وعشرين جزءًا من واحد وضلعه ثلثة آحاد وخُمس واحد والعدد المربّع الاعظم منهما خمس مائة وواحداً وتسعة عشر جزءًا من خمسة وعشرين جزءًا من واحد وضلعه اثنان وعشرون وخُمساً واحد فاذا جعلنا اصغر هذين المربّعين معادلاً لاصغر المربّعين الاوّلين واعظمهما معادلاً لاعظمهما [المربّعين الاوّلين] فننتهى فى كلّ واحدة من المعادلتين الى خمس مائة واثنى
عشر كعب كعب كعب / تعادل مأتين وخمسة واربعين كعب كعب مال ٦٥
وتسعة عشر جزءًا من خمسة وعشرين جزءًا من كعب كعب مال ولنقسم كلّ ⟨واحد⟩ منهما على كعب كعب مال فيكون خمس مائة واثنا عشر شيئاً تعادل مأتين وخمسة واربعين احداً وتسعة عشر جزءًا من خمسة وعشرين من واحد ولذلك يكون الشىء الواحد اثنى عشر جزءًا من خمسة وعشرين ولانّا فرضنا ضلع المكعّب شيئين يكون ضلع المكعّب اربعة وعشرين جزءًا من خمسة وعشرين جزءًا من واحد ويكون المكعّب ثلثة عشر الفاً وثمانى مائة واربعة وعشرين جزءًا من مكعّب الخمسة والعشرين و⟨يكون⟩ مكعّب هذا المكعّب الفى الف الف الف وستّمائة وواحداً واربعين الف الف الف وثمانى مائة وسبعة الف الف وخمس مائة واربعين الفاً ومأتين واربعة وعشرين جزءًا من مكعّب ⟨مكعّب⟩ الخمسة والعشرين وهو ايضاً يكون مائة

---

1418 (465): من (prius): مى in cod.

1423 (466): وواحداً : وواحد in cod.

1424 (467): جعلنا : حمعا in cod.

1425-1426 (468): Verba المربّعين الاوّلين interpolata esse censeo. A lectore, ut opinor, idcirco scripta sunt, quod explanatio pronominis هما in لاعظمهما desiderabatur.

1428 (469): كعب كعب مال : واحد in cod.

1429 (470): Deficiens واحد restitui.

1434 (471): مكعّب (prius): مربع in codice. Vide adn. 474.

(472): يكون addidi.

1435 (473): وواحداً : وواحد in cod.

1437 (474): مكعّب : مربع in cod.

(475): مكعّب alterum deest in cod.

الف الف الف وخمسة آلف الف الف وستّمائة الف الف واثنين وسبعين الف الف وثلثمائة الف والفاً وستّمائة وثمنية ⟨اجزاء⟩ واربعة وعشرين جزءاً ⟨من خمسة وعشرين جزءاً⟩ من جزء واحد من مربّع مربّع الستّمائة والخمسة والعشرين ولان المربّع فرضنا ضلعه اربعة اموال والمال هو مائة واربعة واربعون جزءاً من ستّمائة وخمسة وعشرين وذلك لان الشيء اثنا عشر جزءاً من خمسة وعشرين يكون ضلع المربّع خمس مائة وستّة وسبعين جزءاً ⟨من ستّمائة وخمسة وعشرين جزءاً⟩ ويكون المربّع ثلثمائة الف وواحداً وثلثين الفاً وسبع مائة وستّة وسبعين جزءاً من مربّع الستّمائة والخمسة والعشرين ويكون مربّع هذا المربّع مائة الف الف الف وعشرة آلف الف الف وخمسة وسبعين الف الف وثلثمائة الف واربعة عشر الفاً ومائة وستّة وسبعين جزءاً من مربّع /مربّع الستّمائة والخمسة ٦٦
والعشرين وهو اذا ما زيد عليه مكعّب العدد المكعّب اجتمع منهما مائتا الف الف الف وخمسة عشر الف الف وسبع مائة الف الف وسبعة واربعون الف الف وستّمائة الف وخمسة عشر الفاً وسبع مائة واربعة وثمنون جزءاً واربعة وعشرون جزءاً من خمسة وعشرين من جزء مربّع مربّع الستّمائة والخمسة والعشرين وهو مربّع ضلعه اربع مائة الف واربعة وستّون الفاً واربع مائة وستّة وثمنون جزءاً وخُمساً جزء من مربّع الستّمائة والخمسة والعشرين واذا ما نُقص من مربّع هذا المربّع مكعّب العدد المكعّب بقي منه اربعة آلف الف الف واربع مائة الف الف وثلثة آلف الف واثنا عشر الفاً وخمس مائة وسبعة وستّون جزءاً

---

1439 (476): والفاً : والف in cod.
(477): Verbo اجزاء textum supplevi (cf. lin. 1452,1457).
1440 (478): Per homoeoteleuton omissum addidi.
1444 (479): Per homoeoteleuton omissum addidi.
1445 (480): وثلثين : وسس in cod.
1449 (481): مكعّب : كعب in codice. Vide adn. 486.
1451 (482): واربعون : واربعس in cod.
1452 (483): وعشرون . . . وثمنون : وعسرس . . . ومسس in cod.
1454 (484): وثمنون : وعسرون in cod.
(485): وخُمساً : وحمسى in cod.
1456 (486): مكعّب : كعب in cod.
1457 (487): وثلثة آلف : ولىماىه in cod.
(488): وستّون : وسس in cod.

وجزء من خمسة وعشرين جزءًا من جزء مربّع مربّع الستّمائة والخمسة والعشرين وهو مربّع ضلعه ستّة وستّون الفاً وثلثمائة وخمسة وخمسون جزءًا وخُمس جزء من مربّع الستّمائة ⟨والخمسة⟩ والعشرين

فقد وجدنا عددين على ما اردنا وهما العددان اللذان حدّدنا وذلك ما اردنا ان نجد

مجـ نريد ان نجد عددين مكعّباً ومربّعاً يكون مكعّب المكعّب اذا زدنا عليه امثالاً مفروضةً لمربّع المربّع اجتمع منهما عدد مربّع وان نقصنا منه امثالاً مفروضةً لمربّع المربّع بقى منه عدد مربّع

فلنفرض المكعّب كعباً واحداً فيكون مكعّبه كعب كعب كعب ونفرض ضلع المربّع ما شئنا من الاموال فلنفرضه من ضلع مالين فيكون اربعةَ اموال اموال ومربّع المربّع ستّة عشر كعب كعب مال ولتكن الامثال المفروضة للزيادة مثلاً وربعَ مثل .والامثال المفروضة للنقصان نصفَ وربعَ مثلٍ ونزيد على مكعّب المكعّب مثلَ وربعَ مثلِ ⟨مربّعِ⟩ المربّع / ٦٧ وهو عشرون كعب كعب مال فيكون كعبَ كعب كعب وعشرين كعب كعب مال وهو يعادل عدداً مربّعاً ولننقص من مكعّب المكعّب نصفَ وربعَ مربّعِ المربّع وهو اثنا عشر كعب كعب مال فيكون كعب كعب كعب الّا اثنى عشر كعب كعب مال تعادل عدداً مربّعاً فاذا فرضنا ضلع المربّع المعادل للكعب كعب كعب والعشرين كعب كعب مال اموالَ اموال كان مربّعها اموال اموال فى اموال اموال وهى التى يُسمّى واحدُها كعبَ كعب مال فاذا عادلنا بها ⟨الكعب⟩ كعب كعب والعشرين كعب كعب مال ثمّ نقصنا العشرين ⟨كعب كعب مال⟩ المشتركة بقى كعب كعب كعب يعادل كعاب كعب مال عددها مساوٍ لمربّع الّا عشرين وذلك هو العدد المفروض فى هذا العمل شيئاً وايضاً فانّا اذا فرضنا ضلع المربّع المعادل للكعب كعب كعب الّا اثنى عشر كعب كعب مال اموال اموال كان مربّعها كعاب كعاب مال فاذا زدنا عليها

---

1460 (489): والخمسة addidi.
1470 (490): Deficiens مربّع addidi.
1471 (491): وعشرين : وعسرون in cod.
1475 (492): للكعب : لكعب in cod.
1477 (493): الكعب addidi.
1478 (494): كعب كعب مال addidi.
1481 (495): الّا اثنى : الاسى in cod.

الاثنى العشر الكعب الكعب المال الناقصة من كعب كعب الكعب وجعلناها زيادةً مشتركةً على الناحيتين كلتيهما اجتمع كعب كعب كعب يعادل كعاب كعاب مال عددها مساوٍ لمربّع واثنى عشر وذلك هو العدد المفروض فى المسئلة شيئاً فاذاً مربّع الآ عشرين تعادل مربّعاً صغيراً واثنى عشر ونزيد العشرين مشتركةً على الناحيتين جميعاً فيكون مربّع صغير واثنان وثلثون تعادل مربّعاً عظيماً فالمربّع الصغير هو اربعة آحاد واذا زيد عليه اثنان وثلثون اجتمع منهما ستّة وثلثون وهو المربّع العظيم فنجعل المربّع المعادل للكعب كعب كعب وعشرين كعب كعب مال ستّة وثلثين كعب كعب مال والمربّع المعادل للمربّع الآخر اربعة كعاب كعاب مال ففى كلّ واحدة من المعادلتين / بعد ٦٨

الجبر والمقابلة والقسمة ننتهى الى شىء يعادل ستّة عشر احداً ونركّب المسئلة على نحو ما كنّا ثبتنا عليه ‹فى› تحليلها وضلع المكعّب جعلناه شيئاً فهو ستّة عشر احداً والمكعّب اربعة الف وستّة وتسعون وضلع المربّع فرضناه مالين والمال مائتان وستّة وخمسون فضلع المربّع خمس مائة واثنا عشر والمربّع مائتا الف واثنان وستّون الفاً ومائة واربعة واربعون فامّا مكعّب المكعّب فيكون ثمنية وستّين الف الف الف وسبع مائة الف الف وتسعة عشر الف الف واربع مائة الف وستّة وسبعين الفاً وسبع مائة وستّة وثلثين وامّا مربّع المربّع فيكون مثل هذا العدد ايضاً فاذاً مكعّب المكعّب مربّع مساوٍ لِما يجتمع من ضرب العدد المربّع فى مثله فهو اذاً اذا زيد عليه مثل مربّع المربّع ومثل رُبعه كان الذى يجتمع من ذلك مثلى مربّع المربّع ومثل رُبعه وذلك عدد مربّع ضلعه مثلُ ونصفُ العدد المربّع وهو ايضاً ان نُقص منه ثلثة ارباع مربّع المربّع كان الباقى منه مثل رُبع مربّع العدد المربّع وذلك مربّع ضلعه نصف العدد المربّع

---

1484 (496): كلتيهما : كلاهما in cod.

(497): اجتمع : ىعى in cod.

1494 (498): ثبتنا : ىىىا (pro بيّنّا?) in cod.

(499): فى addidi.

1496 (500): فضلع : وضلع in cod.

1499 (501): وسبعين : وسىس (sc. وستّين ) in cod.

1501-1502 (502): العدد المربّع فى مثله : العدد فى المربع in cod.

1503 (503): Pro سلس لمربع codicis مثلى مربّع scripsi.

فقد وجدنا عددين على الصفة التى وصفنا وهما العددان اللذان حدّدنا وذلك ما اردنا ان نجد

مد نريد ان نجد عددين مكعّباً ومربّعاً يكون مربّع العدد المربّع اذا ضربناه فى عددين ⟨مفروضين⟩ وزدنا مكعّب المكعّب على كلّ واحد منهما اجتمع من زيادته على كلّ واحد منهما عدد مربّع او ان نقصنا كلّ واحد منهما من مكعّب المكعّب كان الذى يبقى منه عدداً مربّعاً او ان نقصنا من كلّ واحد منهما ⟨مكعّب⟩ المكعّب كان الذى يبقى من كلّ واحد منهما عدداً مربّعاً

وليكن احد العددين المفروضين ثلثة والآخر ثمنية ونريد ان نجد عددين / مكعّباً ومربّعاً ويكون مربّع المربّع اذا ضربناه فى ثلثة وفى ٦٩
ثمنية وزدنا ما يجتمع من كلّ واحد منهما على مكعّب المكعّب اجتمع من زيادة كلّ واحد منهما عدد مربّع او ان نقصنا ما يجتمع من كلّ واحد منهما من مكعّب المكعّب كان الذى يبقى من مكعّب المكعّب بعد نقصان كلّ واحد منهما عدداً مربّعاً او ان نقصنا ⟨مكعّب⟩ المكعّب من كلّ واحد منهما بقى من كلّ واحد منهما عدد مربّع

فلنطلب الاوّل من الثلثة ونفرض ضلع المكعّب شيئاً واحداً فيكون ⟨المكعّب كعباً و⟩ مكعّبه كعب كعب كعب ونفرض المربّع من ضلع مالين فيكون المربّع اربعة اموال اموال ويكون مربّع المربّع ستّة عشر كعب كعب مال فاذا ضربناه فى ثلثة وفى ثمنية اجتمع منهما ثمنية واربعون كعب كعب مال ومائة وثمنية وعشرون كعب كعب مال واذا زدنا كلّ واحد منهما على مكعّب المكعّب كانا كعبَ كعب كعب وثمنية واربعين كعب كعب مال

---

1510 (504): مفروضين addidi.

(505): Pro وردناه على مكعب المكعب codicis وزدنا مكعّب المكعّب على scripsi.

1512 (506): عدداً مربّعاً : عدد مربع in codice. Vide etiam adn. 508, 510.

1513 (507): مكعّب addidi.

1514 (508): عدداً مربّعاً : عدد مربع in cod.

1520 (509): نقصان : النقصان من in cod.

(510): عدداً مربّعاً : عدد مربع in cod.

(511): مكعّب, ut supra (vide adn. 507), addidi.

1523 (512): المكعّب كعباً و addidi.

وكعبَ كعب كعب ومائة وثمنية وعشرين كعب كعب مال وكلّ واحد منهما مربّع وكلّ مربّع يُقسم على مربّع فانّه يكون مربّعاً فلنقسم كلّ واحد منهما على مربّع وليكن ذلك المربّع كعبَ كعب مال فيكون احد القسمين شيئاً وثمنية واربعين احداً وهو يعادل عدداً مربّعاً لانّه يخرج من قسمة مربّع على مربّع ويكون القسم الآخر شيئاً ومائة ⟨وثمنية⟩ وعشرين احداً وهو يعادل عدداً مربّعاً لانّه خرج من قسمة مربّع على مربّع فالشئ هو اذاً اذا زيد عليه ثمنية واربعون اجتمع منه مربّع واذا زيد عليه مائة وثمنية/
وعشرون صار ايضاً مربّعاً فلنطلب عدداً اذا زدناه على هذين ٧٠
العددين صار مع كلّ واحد منهما مربّعاً وهو ستّة عشر احداً فالشئ اذاً ستّة عشر ولانّا فرضنا ضلع المكعّب شيئاً يكون ضلعه ستّة عشر ويكون المكعّب هو العدد المكعّب الذى بيّنّاه فى المسئلة المتقدّمة ويكون مكعّبه ايضاً العدد الذى هو مكعّبه فى المسئلة المتقدّمة وكذلك ايضاً مربّع المربّع مساوٍ لمكعّب المكعّب وهو اذا ضُرب فى ثلثة آحاد وزيد عليه مكعّب المكعّب اجتمع منهما اربعة امثال مربّع المربّع وهو مربّع ضلعه مثلاً العدد المربّع وهو ايضاً اذا ما ضُرب فى ثمنية آحاد وزيد ⟨ذلك⟩ على مكعّب المكعّب اجتمع منهما تسعة امثال مربّع المربّع وهو مربّع ضلعه ثلثة امثال العدد المربّع

فقد وجدنا عددين مكعّباً ومربّعاً اذا ضربنا مربّع المربّع فى ثلثة وفى ثمنية ثمّ زدنا كلّ واحد منهما على مكعّب المكعّب اجتمع من كلّ واحد منهما عدد مربّع وهما امّا المكعّب فأربعة الف وستّة وتسعون وامّا المربّع فمائتا الف واثنان وستّون الفاً ومائة واربعة واربعون

وايضاً فلنطلب الثانى من الثلثة وكذلك اذا جعلنا المكعّب كعباً واحداً والمربّع اربعة اموال مال اجتمع لنا مربّعان احدهما كعب كعب كعب الّا ثمنية واربعين كعب كعب مال والآخر كعب كعب كعب الّا مائة وثمنية وعشرين كعب كعب مال وكلّ مربّع يُقسم على مربّع فان

---

1530 (513): شيئاً : سماه (sc. ستّمائة ) in cod.

1532 (514): وثمنية addidi.

1533 (515): Post مربّع (posterius) iterantur in codice verba ويكون القسم الآخر ....مربّع على مربّع per dittographiam. Animadvertendum est, verbum وثمنية hic etiam deficere.

1535 (516): Pro سا (sc. شيئاً ) codicis عدداً substitui. Cf. adn. 520.

1543 (517): ذلك addidi.

القسم ايضاً يكون مربّعاً وليكن المربّع الذى نقسم عليه كعب كعب كعب الاّ ⟨ثمنية واربعين كعب كعب مال وكعب كعب كعب الاّ مائة وثمنية عشرين كعب كعب مال⟩ كعب كعب مال وهو الكائن من ضرب مال مال فى نفسه فيكون احد القسمين شيئاً الاّ ثمنية واربعين والآخر شيئاً الاّ مائة وثمنية وعشرين وكلّ واحد منهما مربّع فلنطلب عدداً اذا نقصنا منه ثمنية واربعين /بقى منه عدد مربّع واذا نقصنا منه مائة وثمنية ٧١ وعشرين بقى منه ايضاً عدد مربّع وذلك العدد هو المفروض فى عمل المسئلة شيئاً وهو مائة واثنان وتسعون ولان ضلع المكعّب الذى وجدنا فى المسئلة المتقدمة ستّة عشر وهو ضلع المكعّب مائة واثنان وتسعون يكون ضلع هذا المكعّب عند ضلع ذلك المكعّب فى نسبة اثنى عشر مثلاً ويكون هذا المكعّب عند ذلك المكعّب فى نسبة مكعّب الاثنى عشر الى الواحد ولان ضلع المربّع فرضناه مالين وهذا الشىء عند ذلك الشىء الذى فى المسئلة المتقدّمة فى نسبة الاثنى عشر الى الواحد يكون هذا المال عند ذلك المال فى نسبة مربّع الاثنى عشر الى الواحد وكذلك ضلع المربّع عند ضلع ذلك المربّع فامّا المربّع عند المربّع ففى نسبة ⟨مربّع⟩ المائة والاربعة والاربعين الى الواحد فامّا مكعّب هذا المكعّب عند مكعّب ذلك المكعّب ففى نسبة مكعّب المكعّب الكائن من الاثنى عشر الى الواحد وامّا مربّع المربّع عند مربّع ذلك المربّع ففى نسبة مربّع مربّع المائة والاربعة والاربعين الى الواحد ومربّع ذلك المربّع كان مساوياً لمكعّب ذلك المكعّب فاذاً مكعّب ذلك المكعّب مربّع ⟨و⟩الواحد ايضاً مكعّب مربّع فلذلك يكون ضلع هذا المكعّب اثنى عشر ويكون المكعّب الفاً وسبع مائة وثمنية وعشرين ويكون ضلع المربّع مائة واربعة واربعين ويكون المربّع عشرين

---

1553 (518): وليكن : ولكن in cod.

1554-1555 (519): A librario omissam (ut opinor) lineam restitui.

1558 (520): Pro سى (sc. شىء) codicis عدد substitui (vide enim lin. seq.).

1562 (521): نسبة : نفسه (pro نفسه ?) in cod.

1563 (522): نسبة : نفسه in cod.

1568 (523): مربّع addidi.

1573 (524): و post مربّع addidi.

1574 (525): الفاً : الف in cod.

الفاً وسبع مائة وستّة وثلثين ومكعّب هذا المكعّب يكون خمسة الف
الف الف ومائة الف الف وتسعة وخمسين الف الف وسبع مائة الف
وثمنين الفاً وثلثمائة واثنين وخمسين ومربّع هذا المربّع يكون اربع
مائة الف الف وتسعة/وعشرين الف الف وتسع مائة الف وأحد وثمنين ٧٢
الفاً وستّمائة وستّة وتسعين وثلثة امثال مربّع المربّع يكون الف الف الف
ومأتين وتسعة وثمنين الف الف وتسع مائة ⟨الف وخمسة⟩ واربعين الفاً
وثمنية وثمنين واذا نقصناه من مكعّب المكعّب بقى منه ثلثة الف الف
الف وثمانى مائة وتسعة وستّون الف الف وثمانى مائة وخمسة وثلثون
الفاً ومائتان واربعة وستّون وهو مربّع ضلعه اثنان وستّون الفاً ومائتان
وثمنية وثمنيةُ امثال مربّع المربّع ثلثة الف الف الف واربع مائة وتسعة
وثلثون الف الف وثمانى مائة وثلثة وخمسون الفاً وخمس مائة وثمنية
وستّون وهو اذا نُقص من مكعّب المكعّب بقى منه الف الف الف وسبع
مائة وتسعة عشر الف الف وتسع مائة وستّة وعشرون الفاً وسبع مائة
واربعة وثمنون وهو مربّع ضلعه واحد واربعون الفاً واربع مائة واثنان
وسبعون

فقد وجدنا عددين مكعّباً ومربّعاً اذا ضربنا مربّع المربّع فى ثلثة وفى ثمنية ونقصنا كلّ واحد منهما من مكعّب العدد المكعّب بقى منه عدد مربّع وهما العددان اللذان وجدنا

فلنطلب الآن الوجه الباقى من الثلثة الذى حدّدناه ونقول ثمنية
واربعون كعب كعب مال الّا كعب كعب كعب تعادل مربّعاً ومائة
وثمنية وعشرون كعب كعب مال الّا كعب كعب كعب تعادل مربّعاً
ولنقسمهما على كعب كعب مال فيكون احد القسمين ثمنية واربعين احداً
الّا شيئاً والآخر مائة وثمنية وعشرين احداً الّا شيئاً وكلّ واحد منهما
مربّع فلنطلب عدداً اذا نقصناه من الثمنية والاربعين ومن المائة
والثمنية والعشرين بقى من كلّ واحد منهما مربّع وليكن سبعة واربعين
و/ذلك هو العدد المفروض شيئاً فى عمل هذه المسئلة ولانّا فرضنا ٧٣

---

1576 (526): خمسة : ىلىه in cod.

1581 (527): الف وخمسة addidi.

1582 (528): وثمنية وثمنين : وىلىه وىلىس in cod.

1588 (529): وسبع : وسع in cod.

1597 (530): كعب كعب مال : كعب كعب كعب مال in cod.

1600 (531): Post وليكن addit codex ضلعه.

ضلع المكعّب شيئاً يكون ضلعه سبعة واربعين فيكون المكعّب مائة الف وثلثة الٰف وثمانى مائة وثلثة وعشرين ولانّا فرضنا ضلع المربّع مالين والمال ألفان ومائتان وتسعة آحاد يكون ضلع المربّع اربعة الٰف واربع مائة وثمنية عشر ويكون المربّع تسعة عشر الف الف وخمس مائة الف وثمنية عشر الفاً وسبع مائة واربعة وعشرين ومكعّب المكعّب اذا نُقص من ثلثة امثال مربّع هذا المربّع كان الباقى مربّعاً ضلعه اربعة الٰف الف وثمانى مائة الف وتسعة وسبعون الفاً وستّمائة وواحد وثمنون وان نُقص من ثمنية امثال مربّع المربّع بقى منه مربّع ضلعه ثلثة واربعون الف الف وتسع مائة وسبعة عشر الفاً ومائة وتسعة وعشرون

فقد وجدنا عددين مكعّباً ومربّعاً ومربّع المربّع اذا ضُرب فى ثلثة وفى ثمنية ونُقص مكعّب المكعّب من ⟨كلّ⟩ واحد منهما بقى من كلّ واحد منهما عدد مربّع وهما العددان اللذان حدّدنا وذلك ما اردنا ان نجد

تمّ القول الرابع من كتاب ذيوفنطس فى المربّعات والمكعّبات وهو اربع واربعون مسئلة

بسم الله الرحمن الرحيم

المقالة الخامسة من كتاب ذيوفنطس الاسكندرانى فى المسائل العدديّة

ا نريد ان نجد عددين مربّعاً ومكعّباً يكون مربّع المربّع اذا زدنا عليه امثالاً مفروضةً للعدد المكعّب اجتمع منهما عدد مربّع وان نقصنا منه امثالاً أُخرَ مفروضةً للعدد المكعّب بقى منه عدد مربّع

ولتكن الامثال الزائدة اربعاً والناقصة ثلثاً ونريد ان نجد عددين على ما وصفنا فنجعل ضلع المربّع شيئاً واحداً ليكون المربّع مالاً واحداً
ومربّع المربّع مال مال واحد وهو مع اربعة/امثال مكعّب ما تعادل ٢٤
مربّعاً والّا ثلثة امثال ذلك المكعّب ايضاً تعادل مربّعاً فاذاً المكعّب يعادل مقداراً ما لَهُ الى المال مال نسبة مفروضة اذا زيد اربعة

---

1608 (532): وواحد : prius و super posterius in cod.

1609 (533): امثال : اماله in cod.

1612 (534): كلّ addidi.

1616 (535): اربع : اربعه in cod.

1618 (536): Verba hujus lineae atramento rubro in codice.

1619 (537): زدنا : ربد in codice. Cf. adn. 653.

امثاله على المال مال كان مربّعاً واذا نُقص ثلثة امثاله من المال مال بقى منه مربّع فنطلب ثلثة اعداد مربّعة تكون نسبة زيادةِ الاعظم منها على الاوسط الى زيادة الاوسط على الاصغر مثل نسبة الاربعة الى الثلثة ولتكن تلك الاعداد واحداً وثمنين وتسعة واربعين وخمسة وعشرين فاذا جعلنا المال مال تسعة واربعين جزءاً فان المقدار المفروض النسبةَ اليه الذى اذا زيد اربعة امثاله على المال ⟨مال⟩ وهو اثنان وثلثون جزءاً من تسعة واربعين جزءاً كان مربّعاً واذا نُقص من المال مال ثلثة امثاله وهو اربعة وعشرون جزءاً من تسعة واربعين جزءاً كان مربّعاً ثمنيةُ اجزاء من تسعة واربعين جزءاً من المال مال فاذاً المكّعب المطلوب يعادل ثمنية اجزاء من تسعة واربعين جزءاً من مال مال ولنفرض المكّعب من ضلع كم شئنا من الاشياء فلنفرضه من ضلع شيئين فيكون المكّعب ثمنية كعاب فاذاً الثمنية كعاب تعادل ثمنية اجزاء من تسعة واربعين جزءاً من مال مال فلنقسم كلّ واحد منهما على كعب فيكون ثمنية اجزاء من تسعة واربعين جزءاً من شىء تعادل ثمنية آحاد فاذاً الشىء الواحد يعادل تسعة واربعين احداً ولذلك يكون ضلع المربّع تسعة واربعين ويكون المربّع ألفين واربع مائة وواحداً ولان ضلع المكّعب فرضناه شيئين يكون ضلع المكّعب ثمنية وتسعين ويكون المكّعب تسع مائة الف وأحد واربعين الفاً ومائة واثنين وتسعين وامّا مربّع المربّع فانّه يكون خمسة الٰف الف وسبع مائة الف واربعة وستّين الفاً وثمانى مائة وواحداً واذا زيد عليه اربعة/ امثال العدد المكّعب الذى هو ثلثة الٰف الف وسبع مائة الف ٧٥ واربعة وستّون الفاً وسبع مائة وثمنية وستّون اجتمع منهما تسعة الٰف الف وخمس مائة الف وتسعة وعشرون الفاً وخمس مائة وتسعة وستّون وهو مربّع ضلعه ثلثة الٰف وسبعة وثمنون وان نُقص منه ثلثة امثال العدد المكّعب الذى هو الفَا الف وثمانى مائة الف وثلثة وعشرون الفاً وخمس مائة وستّة وسبعون بقى الفَا الف وتسع مائة وأحد واربعون الفاً ومائتان وخمسة وعشرون وهو عدد مربّع ضلعه الف وسبع مائة وخمسة عشر

فقد وجدنا عددين على الشرط الذى اردنا وذلك ما اردنا ان نجد

---

1632 (538): مال addidi. Vide adn. 552.

1636 (539): الكعب : المكّعب in cod.

بَ نريد ان نجد عددين مربّعاً ومكعّباً يكون العدد المكعّب اذا ضربناه فى عددين معلومين وزدنا كلّ واحد منهما على مربّع المربّع اجتمع من كلّ واحد منهما مربّع

ونجعل العددين المعلومين اثني عشر احداً وخمسة آحاد ونفرض ضلع المربّع شيئاً واحداً فيكون المربّع مالاً واحداً ومربّعه مال مال واحد وهو مع اثنى عشر مكعّباً تعادل مربّعاً ومع خمسة امثال ذلك المكعّب ايضاً تعادل مربّعاً ولذلك فلنطلب المقدار المعلوم النسبةَ من المال مال الذى اذا زيد عليه اثنا عشر مثلاً لَهُ كان مربّعاً وان زيد عليه ايضاً خمسة امثاله كان مربّعاً ولذلك نصير الى ان نطلب ثلثة اعداد مربّعة يكون نسبة زيادة الاعظم منها على الاوسط الى زيادة الاوسط على الاصغر كنسبة زيادة الاثنى عشر على الخمسة الى الخمسة اعنى نسبة المثل والخمسى مثل فلتكن تلك الاعداد ستّة عشر وتسعة واربعة فاذا جعلنا المال مال مال اربعة اجزاء فظاهر ان المقدار المعلوم نسبتَه منه الذى اذا زيد عليه خمسة امثاله اعنى خمسة اجزاء كان / مربّعاً واذا زيد عليه اثنا عشر مثلاً له اعنى اثنى عشر جزءاً كان مربّعاً ٧٦ هو رُبع مال مال فاذاً رُبع مال مال يعادل عدداً مكعّباً فلنفرض المكعّب من ضلع شيئين فيكون المكعّب ثمنية كعاب وهو يعادل رُبع مال مال فلنقسم كلّ واحد منهما على كعب فيكون رُبع شىء يعادل ثمنية آحاد ولذلك يكون الشىء يعادل اثنين وثلثين ولذلك يكون ضلع المربّع اثنين وثلثين ويكون المربّع الفاً واربعة وعشرين و‹يكون› مربّع المربّع الف الف وثمنية واربعين الفاً وخمس مائة وستّة وسبعين ولانّا فرضنا ضلع المكعّب شيئين يكون ضلع المكعّب اربعة وستّين ويكون المكعّب ‹مائتى الف و› اثنين وستّين الفاً ومائة واربعة واربعين وهو اذا ضُرب فى اثنى عشر اجتمع منه ثلثة الف الف

---

1662 (540): مكعّباً : مكعب in cod.

(541): خمسة : الخمسه in cod.

1666 (542): على (posterius): الى in cod.

1669 (543): Pro وطاهر codicis فظاهر scripsi.

1671 (544): اثنا : اسى in cod.

1672 (545): رُبع (prius): مربع in cod.

1677 (546): يكون addidi.

1679 (547): مائتى الف و deest in cod.

وماثة وخمسة واربعون الفاً وسبع ماثة وثمنية وعشرون واذا زيد ذلك على مربّع المربّع اجتمع منهما اربعة الف الف ومائة واربعة وتسعون الفاً وثلثمائة واربعة وهو مربّع ضلعه ألفان وثمنية واربعون وايضاً اذا ضُرب العدد المكعّب فى خمسة آحاد اجتمع منه الف الف وثلثمائة وعشرة الف وسبع مائة وعشرون واذا زيد ذلك على مربّع المربّع اجتمع منهما الفَا الف وثلثمائة الف وتسعة وخمسون الفاً ومائتان وستّة وتسعون وهو مربّع ضلعه الف وخمس مائة وستّة وثلثون

فقد وجدنا عددين على الشرط الذى اردناه وهما العددان اللذان حدّدنا

جـ نريد ان نجد عددين آخرين مكعّباً ومربّعاً يكون المكعّب اذا ضربناه فى عددين مفروضين ونقصنا كلّ واحد منهما من مربّع المربّع بقى منه مربّع

وليكن احد العددين المعلومين اثنى عشر والآخر سبعة آحاد
ونفرض ايضاً المربّع مالاً واحداً حتّى يكون مربّع المربّع /ايضاً مال مال ٧٧
واحد فيكون مال مال الّا اثنى عشر مكعّباً تعادل مربّعاً والّا سبعة
مكعّبات ايضاً تعادل مربّعاً ولنطلب المقدار المعلوم النسبةَ من
المال مال الذى اذا نُقص اثنا عشر مثلاً له من المال ⟨مال⟩ بقى منه
مربّع واذا نُقص منه سبعة امثاله ايضاً بقى منه مربّع وهو ان نطلب
ثلثة اعداد مربّعة تكون نسبة زيادة الاعظم منها على الاوسط الى
زيادة الاوسط على الاصغر كنسبة السبعة الى نقصانها من الاثنى عشر
وهى الاعداد التى ذكرناها آنفاً [ستّة عشر وتسعة واربعة] ولذلك
يكون المقدار المعلوم النسبةَ من المال مال الذى حدّدناه هو جزء
من ستّة عشر جزءاً من مال مال فاذاً المكعّب يعادل جزءاً من ستّة

---

1681 (548): واربعس . . . . وعسرس : واربعون . . . وعشرون in cod.

1682 (549): وسعس : وتسعون in cod.

1686 (550): الف : الفَا in cod.

1695 (551): سبع : سبعة in cod.

1697 (552): Deficiens مال addidi.

1701 (553): Valores quaesitorum numerorum, quamvis expectentur, interpolatos esse videntur.

1702 (554): Verba الذى حدّدناه quoque interpolata esse censeo.

1703 (555): الكعب : المكعّب in cod.

عشر جزءًا من مال مال ونفرض ضلع المكعّب نصف شیء فيكون المكعّب ثُمن كعب فاذن ثُمن كعب يعادل جزءًا من ستّة عشر من مال مال ولذلك يكون نصف ثُمن شیء يعادل ثُمنَ واحد فاذاً الشیء الواحد يعادل احدين ولذلك يكون المربّع اربعة آحاد ومربّع المربّع ستّة عشر احداً ولان ضلع المكعّب فرضناه نصفَ شیء يكون ضلع المكعّب واحداً ويكون المكعّب ايضاً واحداً وهو اذا ضُرب فی اثنی عشر وفی سبعة ونُقص كلّ واحد منهما من مربّع المربّع بقی منه مربّع

دَ نريد ان نجد عددين مربّعاً ومكعّباً يكون مربّع المربّع اذا زدنا عليه امثالاً مفروضةً لمكعّب المكعّب اجتمع منهما عدد مربّع ⟨و⟩اذا نقصنا منه امثالاً أُخرَ مفروضةً لمكعّب المكعّب بقی منه ايضاً عدد مربّع

ولتكن الامثال الزائدة خمسة والامثال الناقصة ثلثة ولنفرض المكعّب من ضلع شیء واحد فيكون كعباً واحداً ومكعّبه كعب كعب كعب واحد ونفرض المربّع من ضلع مالين فيكون المربّع اربعة اموال/مال ومربّعه ٨' ستّة عشر كعب كعب مال فاذن ستّة عشر كعب كعب مال وخمسة كعاب كعب كعب تعادل مربّعاً وستّة عشر كعب كعب مال الّا ثلثة كعاب كعب كعب تعادل مربّعاً وكلّ مربّع يُقسم على مربّع فان القسم يكون مربّعاً فلنقسم كلّ واحد من هذين المربّعين على المربّع الذی هو كعب كعب مال واحد فيكون احد القسمين ستّة عشر احداً وخمسة اشياء والآخر ستّة عشر احداً الّا ثلثة اشياء وكلّ واحد منهما مربّع وكلّ عدد مربّع يزاد عليه خمسة امثال رُبعه فانّه يكون مربّعاً ويُنقص منه ثلثة امثال رُبعه فانّه يكون مربّعاً ايضاً فاذاً الشیء الواحد هو رُبع الستّة عشر وهو اربعة آحاد ومن اجل انّا فرضنا ضلع المكعّب شيئاً واحداً يكون ضلع المكعّب اربعة آحاد ويكون المكعّب اربعة وستّين ولانّا فرضنا ضلع المربّع مالين والمال ستّة عشر احداً يكون ضلع المربّع اثنين وثلثين ويكون المربّع الفاً واربعة وعشرين فامّا مربّع المربّع فانّه يكون الف الف وثمنية واربعين الفاً وخمس مائة وستّة وسبعين وامّا مكعّب المكعّب فانّه يكون مائتی الف واثنين وستّين الفاً ومائة واربعة واربعين وهو اذا زيد خمسة امثاله على مربّع المربّع اجتمع منهما الفَا الف وثلثمائة الف وتسعة وخمسون

---

1706 (556): Pro مما واحدا codicis ثُمنَ واحد scripsi.

1712 (557): و omissum restitui.

1732 (558): الفَا : العى in cod.

الفاً ومائتان وستّة وتسعون وهو مربّع ضلعه الف وخمس مائة وستّة وثلثون واذا نُقص ثلثة امثاله من مربّع المربّع بقى منه مائتا الف واثنان وستّون الفاً ومائة واربعة واربعون وهو مربّع ضلعه خمس مائة واثنا عشر احداً

فقد وجدنا عددين على الشرط الذى إشترطنا وهما العددان اللذان حدّدنا

ه̅ نريد ان نجد عددين/مكعّباً ومربّعاً يكون مكعّب المكعّب اذا ٧٩
ضربناه فى عددين مفروضين وزدنا كلّ واحد منهما على مربّع المربّع اجتمع منه عدد مربّع

فنجعل احد العددين المعلومين اثنى عشر والآخر خمسة آحاد ونريد ان نجد عددين على ما وصفنا فنفرض ضلع المكعّب شيئاً واحداً فيكون المكعّب كعباً واحداً ويكون مكعّبه كعب كعب كعب واحد ونفرض ضلع المربّع مالين فيكون المربّع اربعة اموال مال ومربّع المربّع ستّة عشر كعب كعب مال ولذلك يكون ستّة عشر كعب كعب مال واثنا عشر كعب كعب كعب تعادل مربّعاً وستّة عشر كعب كعب مال وخمسة كعاب كعاب كعاب تعادل ⟨مربّعاً⟩ وكلّ مربّع يُقسم على مربّع فان القسم مربّع فلنقسم كلّ واحد منهما على المربّع الذى هو كعب كعب مال ومن اجل ذلك يكون ستّة عشر احداً واثنا عشر شيئاً وستّة عشر احداً وخمسة اشياء كلّ واحد منهما مربّع لكن كلّ مربّع يزاد عليه خمسة امثال رُبعه فانّه يكون مربّعاً واذا زيد عليه ايضاً اثنا عشر مثلاً لرُبعه كان مربّعاً فاذن الشىء الواحد هو رُبع الستّة عشر اعنى اربعة آحاد ولذلك يكون المكعّب اربعة وستّين والمربّع الفاً واربعة وعشرين وظاهر ان مربّع هذا المربّع اذا زيد عليه اثنا عشر مثلاً لمكعّب هذا المكعّب الذى هو ثلثة الف الف ومائة الف وخمسة واربعون الفاً وسبع مائة وثمنية وعشرون يصير اربعةَ الف الف ومائة الف واربعة وتسعين الفاً وثلثمائة واربعة آحاد وهو مربّع ضلعه ألفان وثمنية واربعون وقد

---

1734 (559): مائتا : ماه in cod.

1739 (560): مكعّب : كعب in cod.

1743 (561): فنفرض bis in cod.

1748 (562): Omissum verbum restitui.

1755 (563): لمكعّب : المكعب in cod.

1758 (564): وثمنية واربعون : وارىعه وعسرون (cf. lin. 1754) in cod.

تبيّن فى المسئلة المتقدّمة انّه اذا زيد عليه خمسة امثال مكعّب المكعّب ايضاً يصير مربّعاً

وَ نريد ان نجد عددين مكعّباً ومربّعاً يكون / مكعّب المكعّب اذا ضربناه ٨٠
فى عددين مفروضين ونقصنا كلّ واحد منهما من مربّع المربّع بقى منه مربّع

فليكن العددان المفروضان سبعة واربعة ونعمل المكعّب من ضلع شىء واحد فيكون المكعّب كعباً واحداً ومكعّبه كعب كعب كعب واحد ونعمل المربّع من ضلع ثلثة اموال فيكون المربّع تسعة اموال مال ومربّع المربّع واحداً وثمنين كعب كعب مال ولذلك يكون واحد وثمنون كعب كعب مال الّا سبعة كعاب كعب كعب تعادل مربّعاً والّا اربعة كعاب كعب كعب تعادل ايضاً مربّعاً ولنقسم كلّ واحد منهما على المربّع الذى هو كعب كعب مال واحد فيكون واحد وثمنون الّا سبعة اشياء تعادل مربّعاً وواحد وثمنون الّا اربعة اشياء تعادل ايضاً مربّعاً ولنطلب المقدار المفروض من كلّ مربّع الذى اذا نُقص سبعة امثاله من المربّع بقى مربّع واذا نُقص ايضاً من المربّع اربعة امثاله بقى مربّع وطلب ذلك على النحو الذى قد تقدّم وليكن ذلك المقدار ثمنية اتساع تُسع فان الواحد والثمنين اذا نُقص منها سبعة امثال ثمنية اتساع تُسعها اعنى ستّة وخمسين بقى مربّع وهو خمسة وعشرون واذا نُقص منها اربعة امثال ثمنية اتساع تُسعها اعنى اثنين وثلثين بقى مربّع وهو تسعة واربعون فاذن الشىء الواحد هو ثمنية اتساع تُسع الواحد والثمنين وهو ثمنية آحاد ولانّا فرضنا ضلع المكعّب شيئاً واحداً يكون المكعّب خمس مائة واثنى عشر ولانّا فرضنا ضلع المربّع ثلثة اموال والمال اربعة وستّون يكون ضلع المربّع مائة واثنين وتسعين ويكون المربّع ستّة وثلثين الفاً وثمان مائة واربعة وستّين ومربّع
المربّع يكون الف الف الف وثلثمائة الف / الف وثمنية وخمسين الف ٨١
الف وتسع مائة واربعة وخمسين الفاً واربع مائة وستّة وتسعين وامّا

1767 (565): واحداً : واحد in cod.

1768 (566): والّا اربعة : والا ربعه in cod.

1775 (567): والثمنين : والىمىں in cod.

1780 (568): Pro واحدا scripsit librarius falso ولهدا.

1781-1782 (569): واثنين وتسعين : واىاں وسعوں in cod.

1784 (570): وتسعين : وسعىں in cod.

مكعّب المكعّب فانّه يكون مائة واربعة وثلثين الف الف ومائتى 〈الف〉 وسبعة عشر الفاً وسبع مائة وثمنية وعشرين وهو اذا نُقص سبعة امثاله من مربّع المربّع بقى منه اربع مائة الف الف وتسعة عشر الف الف واربع مائة الف وثلثون الفاً واربع مائة وهو مربّع ضلعه عشرون الفاً واربع مائة وثمنون واذا نُقص اربعة امثاله من مربّع المربّع بقى ثمانى مائة الف الف واثنان وعشرون الف الف وثلثة وثمنون الفاً وخمس مائة واربعة وثمنون وهو مربّع ضلعه ثمنية وعشرون الفاً وستّمائة واثنان وسبعون

فقد وجدنا عددين على الشرط الذى شرطنا وهما امّا المكعّب فخمس مائة واثنا عشر وامّا المربّع فستّة وثلثون الفاً وثمانى مائة واربعة وستّون وذلك ما اردنا ان نجد

ز̄ نريد ان نجد عددين يكون جملتهما وجملة مكعّبيهما مثل عددين مفروضين

وينبغى ان تكون اربعة امثال العدد المفروض لجملة مكعّبى العددين منهما تزيد على مكعّب العدد المفروض لجملة العددين عدداً اذا قُسم على ثلثة امثال العدد المفروض لجملة العددين كان القسم مربّعاً واذا ضُرب فى ثلثة ارباع العدد المفروض لجملة العددين كان مربّعاً وهذه من المسائل المهيّأة

فليكن العدد المفروض لجملة العددين عشرين احداً والعدد 〈المفروض〉 لجملة مكعّبى العددين ألفين ومأتين واربعين ونريد ان نجد عددين يكون جملتهما عشرين احداً وجملة مكعّبيهما ألفين ومأتين واربعين احداً فنجعل تفاضل العددين شيئين فيكون احدهما عشرة آحاد وشيئاً والآخر عشرة آحاد غير شىء ونعمل / من كلّ واحد ٨٢ منهما مكعّباً وينبغى كلّما اردنا ان نعمل مكعّباً من ضلع مركّب من نوعين مختلفين لئلّا تُغلّطنا كثرةُ الانواع ان نأخذ مكعّبى كلّ واحد من النوعين المختلفين ونُضيف اليهما ثلثة امثال ما يجتمع من ضرب

---

1785 (571): وثلثين : وبلبون in cod.

(572): A librario verbum الف omissum (vide enim terminationem praecedentis verbi) restitui.

1786 (573): وثمنية : وىمان in cod.

1803 (574): المفروض addidi.

1804 (575): عشرين : عسره من in cod.

1806 (576): وشيئاً : وسى in cod.

مربّع كلّ واحد منهما فى النوع الآخر فيكون ما يجتمع مركّباً من اربعة انواع وهو المكعّب الكائن من جملة النوعين المختلفين فان كان النوعان احدهما مستثنىً من الآخر فانّا نأخذ مكعّب الاعظم ونُضيف اليه ثلثة امثال ما يجتمع من ضرب مربّع النوع الاصغر فى النوع الاعظم ونلقى منهما مكعّب النوع الاصغر وثلثة امثال ما يجتمع من ضرب ⟨مربّع⟩ النوع الاعظم فى النوع الاصغر فيكون ما يبقى هو المكعّب الكائن من الفضل بين النوعين المختلفين ولذلك صار المكعّب الكائن من ضلع عشرة آحاد وشئ هو ما يجتمع من مكعّب العشرة وهو الف ومن مكعّب الشئ وهو كعب وثلثة امثال ما يجتمع من ضرب العشرة فى مربّع الشئ وهو ثلثون مالاً وثلثة امثال ما يجتمع ايضاً من ضرب الشئ فى مربّع العشرة وهو ثلثمائة شئ فاذاً المكعّب الكائن من عشرة وشئ هو الف احد وكعب واحد وثلثمائة شئ وثلثون مالاً وايضاً فان المكعّب الكائن من ضلع عشرة آحاد غير شئ هو ايضاً مساوٍ لمكعّب العشرة الذى هو الف ولثلثة امثال ما يجتمع من ضرب العشرة فى مربّع الشئ الذى هو مال وذلك ثلثون مالاً الّا ما يجتمع من مكعّب الشئ الذى هو كعب واحد والّا ثلثة امثال ما يجتمع من ضرب الشئ فى مربّع العشرة وذلك ثلثمائة شئ فاذن المكعب الكائن من عشرة آحاد غير شئ هو الف وثلثون مالاً الّا كعباً وثلثمائة شئ وجملة هذين المكعّبين ألفان وستّون مالاً وذلك ان/الكعب ٨٣ والثلثمائة الشئ الناقصة فى احدهما تذهبها الكعبُ والثلثمائة الشئ الزائدة فى المكعّب الآخر فاذن الالفان والستّون مالاً تعادل ألفين ومأتين واربعين احداً ولنلق الالفين التي فى احدى الناحيتين من العدد الذى فى الناحية الاخرى فيبقى ستّون مالاً تعادل مأتين واربعين احداً ولذلك يكون المال الواحد اربعة آحاد وكلّ واحد

---

1810 (577): Post ما يجتمع praebet codex من الصرب.

1811 (578): من : عں in cod.

1812 (579): مستثنىً : مسىسا in cod.

1815 (580): مربّع addidi.

1827 (581): كعباً : كعب in cod.

1829 (582): تذهبها conjeci pro سهمها codicis.

1831 (583): ولنلق : وللعى in cod.

1832 (584): Post فيبقى perverse addita sunt verba مالا فى احدى الىاحىس in cod.

منهما مربّع فضلعاهما ايضاً متساويان لكن ضلع المال هو شىٔ واحد وضلع الاربعة آحاد احدان فاذن الشىٔ الواحد هو احدان ولانّا فرضنا العدد الاعظم من العددين المطلوبين عشرة آحاد وشيئاً يكون هذا العدد اثنى عشر احداً ولانّا فرضنا العدد الاصغر عشرة آحاد غير شىٔ يكون ثمنية آحاد ومكعّب العدد الاعظم الف وسبع مائة وثمنية وعشرون ومكعّب العدد الاصغر خمس مائة واثنا عشر احداً وجملتهما ألفان ومائتان واربعون احداً

فقد وجدنا عددين جملتهما عشرون احداً وجملة مكعّبيهما ألفان ومائتان واربعون احداً وهما اثنا عشر وثمنية آحاد وذلك ما اردنا ان نجد

ح̄ نريد ان نجد عددين يكون تفاضلهما وتفاضل مكعّبيهما مثل عددين مفروضين

وينبغى ان تكون اربعة امثال العدد المفروض لتفاضل المكعّبين تزيد على مكعّب العدد المفروض لتفاضل العددين عدداً اذا قُسم على ثلثة امثال العدد المفروض ⟨لتفاضل العددين⟩ كان مربّعاً واذا ضُرب فى ثلثة ارباع العدد الذى لتفاضل العددين كان مربّعاً

فليكن العدد المفروض لتفاضل العددين عشرة آحاد والعدد المفروض لتفاضل المكعّبين ألفين ومائة وسبعين احداً ونريد /ان نجد ٨٤ عددين تفاضلهما عشرة آحاد وتفاضل مكعّبيهما ألفان ومائة وسبعون احداً فنفرض جملة العددين شيئين فيكون احدهما شيئاً وخمسة آحاد والآخر شيئاً الّا خمسة آحاد وذلك ليكون تفاضلهما عشرة آحاد ونعمل من كلّ واحد منهما مكعّباً فيكون المكعّب الذى ضلعه شىٔ وخمسة آحاد كما وصفنا مساوياً لمكعّب الشىٔ وهو كعب ولمكعّب الخمسة وهو مائة وخمسة وعشرون ولثلثة امثال ما يجتمع من ضرب مربّع

---

1840 (585): و ante مائتان bis in codice, primum in fine lineae, iterum in initio lineae sequentis.

1842 (586): وهما : وهو in cod.

1844 (587): Pro تفاضلهما praebet codex, ut videtur, معاصلهما. Vide etiam adn. 630.

1848 (588): Uncis inclusa verba addidi.

1854 (589): Verba وذلك ليكون تفاضلهما عشرة آحاد forsan a lectore quodam addita.

الشىء فى الخمسة وهو خمسة عشر مالاً ولثلثة امثال ما يجتمع من ضرب مربّع الخمسة فى الشىء وهو خمسة وسبعون شيئاً فاذن المكعّب الكائن من ضلع شىء وخمسة آحاد هو كعب ومائة وخمسة وعشرون وخمسة عشر مالاً وخمسة وسبعون شيئاً ويكون المكعّب الذى ضلعه شىء الّا خمسة آحاد مساوياً لمكعّب الشىء وهو كعب ولثلثة امثال ما يجتمع من ضرب مربّع الخمسة التى هى ناقصة من الشىء فى الشىء وهو خمسة وسبعون شيئاً الّا مكعّب الخمسة الذى هو مائة وخمسة وعشرون والّا ثلثة امثال ما يجتمع من ضرب مربّع الشىء فى الخمسة الآحاد وهو خمسة عشر مالاً فاذن المكعّب الكائن من ضلع شىء الّا خمسة آحاد هو كعب وخمسة وسبعون شيئاً الّا خمسة عشر مالاً والّا مائة وخمسة وعشرين احداً ولنلق هذا المكعّب من المكعّب الاوّل فيكون الباقى مأتين وخمسين وثلثين مالاً وذلك ان الخمسة العشر المال والمائة والخمسة والعشرين الناقصة فى هذا المكعّب بسبب النقصان زائدة ⟨وتزاد على الخمسة العشر المال والمائة والخمسة والعشرين الزائدة⟩ فى المكعّب الآخر وتذهب الكعب والخمسة والسبعون الشىء من كلّ واحد منهما فاذن المائتان /والخمسون والثلثون ٥. المال تعادل ألفين ومائة وسبعين احداً فلنلق المأتين والخمسين المشتركة من كلى الناحيتين فيبقى الف وتسع مائة وعشرون احداً تعادل ثلثين مالاً ولذلك يكون المال اربعة وستّين وكلّ واحد منهما مربّع فضلعاهما متساويان وضلع المال هو شىء واحد وضلع الاربعة

---

1861 (590): المكعّب : الكعب in cod.

1864 (591): مكعّب : كعب in cod.

1866 (592): فاذن : ادں in cod.

(593): Post المكعّب addit codex و.

1868 (594): ولنلق : ولىلعى in cod.

1869 (595): مأتين وخمسين وثلثين : ماىان وحمسون وىلىون in cod.

1871-1872 (596): Pro راىد codicis زائدة scripsi et uncis inclusis verbis temptavi, ut locum sanarem.

1872 (597): الكعب : المكعب in cod.

1874 (598): فلنلق : فلىلعى in cod.

1875 (599): كلى ut كلى in codice scriptum (vide adn. 329).

1876 (600): وستّين : وسىون in cod.

1877 (601): فضلعاهما : وصلعاهما in cod.

والستّين ثمنية آحاد  فاذاً الشيء الواحد هو ثمنية آحاد  ولانّا فرضنا العدد الاعظم شيئاً وخمسة آحاد يكون ثلثة عشر احداً  وكنّا فرضنا العدد الاصغر شيئاً الّا خمسة آحاد  فلذلك يكون الاصغر ثلثة آحاد ومكعّباهما امّا مكعّب الاعظم فألفان ومائة وسبعة وتسعون وامّا مكعّب الاصغر فسبعة وعشرون وتفاضلهما ألفان ومائة وسبعون

فقد وجدنا عددين تفاضلهما عشرة آحاد وتفاضل مكعّبيهما ⟨ألفان و⟩مائة وسبعون  وهما ثلثة عشر وثلثة آحاد  وذلك ما اردنا ان نجد

ط  نريد ان نقسم عدداً مفروضاً بقسمين تكون جملة مكعّبيهما امثالاً مفروضةً لمربّع تفاضلهما

وينبغي ان تكون الامثال المفروضة اكثر من ثلثة ارباع العدد المفروض بعدد يحيط مع مكعّب العدد المفروض بعدد مربّع

فليكن العدد المفروض عشرين والامثال مائة واربعين مثلاً  ونريد ان نقسم العشرين بقسمين تكون جملة مكعّبيهما مائة واربعين مثلاً لمربّع الفضل بينهما  ولنفرض ايضاً تفاضل القسمين شيئين حتّى يكون احد القسمين عشرة آحاد وشيئاً  والآخر عشرة آحاد الّا شيئاً  وتكون جملة مكعّبيهما على نحو ما وصفنا فيما تقدّم ألفي احد وستّين مالاً  ولكن مربّع تفاضل العددين هو اربعة اموال  فاذن/ ألفا احد وستّون مالاً ٨٦
تعادل مائة واربعين مثلاً لاربعة اموال اعني خمس مائة وستّين مالاً ونلقي الستّين المال المشتركة من كلى الناحيتين  فيبقى ألفا احد تعادل خمس مائة مال  ولذلك فان المال الواحد يعادل اربعة آحاد وضلع المال شيء واحد وضلع الاربعة الآحاد احدان  فاذن الشيء الواحد يعادل احدين  ولانّا فرضنا احد القسمين عشرة آحاد

---

1880 (602): Post فلذلك يكون praebet codex verba الاعظم ىلىه عسر احدا و quae delevi, quoniam notam rem inutiliter repetunt.

1881 (603): ومكعّباهما : ومكعىىهما in cod.

(604): فألفان.....وتسعون : مالعس.....وسعىں in cod.

1882 (605): وعشرون : وعسرىں in cod.

1883-1884 (606): ألفان و addidi.

1888 (607): بعدد (prius): لعدد in codice (sequitur enim verbum المفروض).

1891 (608): احد : احدا in cod.

1892 (609): Pro شيئاً (in utroque loco) praebet codex سى .

1895 (610): خمس : حمسه in cod.

وشيئاً يكون اثنى عشر احداً ولانّا فرضنا القسم الآخر عشرة آحاد الّا شيئاً يكون ثمنية آحاد ومكعّب الاثنى عشر اذا زيد عليه مكعّب الثمنية كان ألفين ومأتين واربعين احداً وتفاضل القسمين اربعة آحاد ومربّعها ستّة عشر احداً والالفان والمائتان والاربعون احداً هى مائة واربعون مثلاً للستّة عشر احداً التى هى مربّع الفضل بين القسمين اللذين وجدناهما

فقد قسمنا العشرين بقسمين على ما اردنا والقسم الاعظم اثنا عشر احداً والاصغر ثمنية آحاد وذلك ما اردنا ان نفعل

ى نريد ان نجد عددين يكون تفاضلهما عدداً مفروضاً ويكون تفاضل مكعّبيهما عند مربّع جملتهما فى نسبة مفروضة

وينبغى ان يكون العدد الذى للنسبة المفروضة اكثر من ثلثة ارباع العدد المفروض لتفاضل العددين <بعدد يحيط مع مكعّب العدد المفروض لتفاضل العددين بعدد مربّع>

فليكن العدد المفروض لتفاضل العددين المطلوبين عشرة آحاد والعدد الذى للنسبة المفروضة ثمنية امثال وثُمن مثل ونريد ان نجد عددين يكون تفاضلهما عشرة آحاد ونسبة تفاضل مكعّبيهما الى مربّع جملتهما نسبةَ الثمنية والثمن الى الواحد ونفرض جملتهما شيئين ونجعل احد العددين شيئاً وخمسة آحاد والآخر شيئاً الّا خمسة / ٨٧ آحاد ليكون تفاضلهما عشرة آحاد ونأخذ الفضل بين مكعّبيهما وهو مائتان وخمسون وثلثون مالاً ومربّع جملة العددين اربعة اموال فاذن المائتان والخمسون والثلثون المال تعادل ثمنية امثال وثُمن مثل الاربعة اموال وذلك اثنان وثلثون مالاً ونصف مال فلنلق الثلثين المال المشتركة من كلى الناحيتين فيبقى مائتان وخمسون احداً تعادل مالين ونصف مال ولذلك يكون المال الواحد يعادل مائة ولذلك الشىء يكون عشرة آحاد ولانّا فرضنا احد العددين شيئاً وخمسة آحاد

---

1906 (611): العشرين : العسره ادں in cod.

1911-1912 (612): A librario omissam, ut opinor, lineam restitui.

1914 (613): نجد per dittographiam bis in cod.

1920 (614): Pro للاربعه codicis الاربعة scripsi.

1921 (615): فلنلق : فلبلعى in cod.

1923 (616): ونصف مال : وىصعا in cod.

يكون خمسة عشر احداً ولانّا فرضنا العدد الآخر شيئاً الّا خمسة آحاد يكون خمسة آحاد ومكعّب الخمسة عشر ثلثة الٰف وثلثمائة وخمسة وسبعون ومكعّب الخمسة مائة وخمسة وعشرون وتفاضلهما ثلثة الٰف ومائتان وخمسون ومربّع جملة العددين اربع مائة ونسبة الثلثة الٰف والمأتين والخمسين الى الاربع مائة هى نسبة الثمنية الامثال والثُمن مثل

فقد وجدنا عددين تفاضلهما عشرة آحاد وتفاضل مكعّبيهما ثمنية امثال وثُمن مثل مربّع جملتهما وهما الخمسة العشر الاحد والخمسة الآحاد وذلك ما اردنا ان نجد

يا نريد ان نجد عددين يكون تفاضلهما عدداً مفروضاً وجملة مكعّبيهما عند جملتهما فى نسبة مفروضة

وينبغى ان يكون العدد الذى للنسبة المفروضة يزيد على ثلثة ارباع مربّع العدد المفروض لتفاضل العددين عدداً مربّعاً

فليكن تفاضل العددين اربعة آحاد والعدد الذى للنسبة المفروضة ثمنية وعشرين مثلاً ونريد ان نجد عددين يكون تفاضلهما اربعة آحاد وجملة مكعّبيهما عند جملتهما فى نسبة الثمنية والعشرين مثلاً فنفرض جملة العددين شيئين فيكون / احدهما شيئاً وأحدين والآخر شيئاً الّا ٨٨
احدين فيكون مكعّب الاعظم كعباً وثمنية آحاد وستّة اموال واثنى عشر شيئاً ومكعّب الاصغر كعباً واثنى عشر شيئاً الّا ستّة اموال والّا ثمنية آحاد وجملتهما كعبان واربعة وعشرون شيئاً وذلك ان الستّة الاموال والثمنية الآحاد الناقصة فى مكعّب العدد الاصغر تحيّزها الثمنيةُ الآحاد والستّة الاموال الزائدة فى مكعّب العدد الاعظم فاذن الكعبان والاربعة والعشرون الشىء تعادل ثمنية وعشرين مثلاً لجملة العددين التى هى شيئان وذلك ستّة وخمسون شيئاً ونلقى الاربعة والعشرين الشىء المشتركة من كلى الناحيتين فيبقى كعبان يعدلان

---

1927 (617): وسبعون : وسعس in cod.

(618): وعشرون : وعسرس in cod.

1930 (619): ثمنية : سلىه in cod.

1940 (620): العددين : العدد in cod.

1944 (621): تحيّزها conjeci pro حىرها codicis (vide indicem verborum, sub *ḥāza*).

1946 (622): الكعبان : المكعبان in cod.

1947 (623): التى هى : الدى هو in codice. Vide etiam adn. 625.

اثنين وثلثين شيئاً فنقسم كلّ واحد منهما على شيء واحد فيكون مالان يعدلان اثنين وثلثين احداً فالمال يعادل ستّة عشر احداً والمال مربّع ضلعه شيء واحد والستّة عشر مربّع ضلعه اربعة آحاد فاذن الشيء الواحد يعادل اربعة آحاد ولانّا فرضنا العدد الاعظم شيئاً واحداً واثنين يكون العدد الاعظم ستّة آحاد ولانّا فرضنا العدد الاصغر شيئاً الّا احدين يكون العدد الاصغر احدين ومكعّب الاعظم مائتان وستّة عشر احداً ومكعّب الاصغر ثمنية آحاد ومجموع هذين المكعّبين مائتان واربعة وعشرون احداً وهو ثمنية وعشرون مثلاً لجملة العددين التى هى ثمنية آحاد

فقد وجدنا عددين تفاضلهما اربعة آحاد وجملة مكعّبيهما ثمنية وعشرون مثلاً لجملتهما وهما ستّة آحاد وأحدان وذلك ما اردنا ان نجد

يب نريد ان نقسم عدداً مفروضاً بقسمين يكون تفاضل مكعّبيهما امثالاً مفروضةً لتفاضلهما /

وينبغى ان يكون العدد الذى للنسبة المفروضة يزيد على ثلثة ارباع ٨٩
مربّع العدد المفروض ايضاً عدداً مربّعاً

وليكن العدد المفروض ثمنية آحاد والامثال التى للنسبة المفروضة نسبةَ الاثنين والخمسين مثلاً ونريد ان نقسم الثمنية بعددين يكون تفاضل مكعّبيهما اثنين وخمسين مثلاً لتفاضلهما فنفرض تفاضل العددين شيئين فيكون القسم الاعظم اربعة آحاد وشيئاً والقسم الاصغر اربعة آحاد الّا شيئاً ومكعّب القسم الاعظم اربعة وستّين احداً وكعباً وثمنية واربعين شيئاً واثنى عشر مالاً ومكعّب القسم الاصغر اربعة وستّين احداً واثنى عشر مالاً الّا كعباً والّا ثمنية واربعين شيئاً وتفاضلهما كعبان وستّة وتسعون شيئاً فاذن الكعبان والستّة والتسعون الشيء تعادل اثنين وخمسين مثلاً لتفاضل العددين الذى هو شيئان وذلك مائة شيء واربعة اشياء فنلقى الستّة والتسعين الشيء المشتركة من

---

1955 (624): Post الاصغر iterantur verba شيئاً الّا احدين...ومكعّب الاصغر per dittographiam in cod.

1957 (625): التى هى : الذى هو in cod.

1965-1966 (626): Scribere debuisset interpres والامثال التى للنسبة المفروضة اثنين وخمسين مثلاً.

1970 (627): وكعباً : وكعب in cod.

كلى الناحيتين فيبقى كعبان يعادلان ثمنية اشياء فنقسم كلّ واحد منهما على شيء فيكون مالان يعادلان ثمنية آحاد ولذلك يكون المال الواحد يعادل اربعة آحاد والشيء الواحد يعدل احدين ولانّا فرضنا القسم الاعظم اربعة آحاد وشيئاً يكون القسم الاعظم ستّة آحاد ولانّا فرضنا القسم الاصغر اربعة آحاد الّا شيئاً يكون القسم الاصغر احدين ومكعّب القسم الاعظم مائتان وستّة عشر احداً ومكعّب القسم الاصغر ثمنية آحاد وتفاضلهما مائتان وثمنية آحاد وهو اثنان وخمسون مثلاً لتفاضل القسمين الذى هو اربعة آحاد

فقد قسمنا اذن الثمنية بقسمين تفاضل مكعّبيهما اثنان وخمسون مثلاً لتفاضلهما وهما ستّة واثنان وذلك ما اردنا ان نفعل

يج نريد ان نجد عدداً مكعّباً اذا زدنا على امثال مفروضة / لمربّع ٩٠
ضلعه عدداً مفروضاً كان ذلك مساوياً لجملة عددين كلّ واحد منهما اذا زيد على المكعّب كان مكعّباً

فليكن العدد المفروض ثلثين احداً والامثال المفروضة تسعة امثال ونريد ان نجد عدداً مكعّباً اذا زدنا تسعة امثال المربّع الكائن من ضلعه على ثلثين احداً كان ذلك مساوياً لعددين كلّ واحد منهما اذا زيد على العدد المكعّب صار مكعّباً فلنفرض ضلع المكعّب شيئاً واحداً فيكون المكعّب كعباً واحداً ولنأخذ تسعة امثال المربّع الكائن من ضلعه وهو تسعة اموال ونزيد ذلك على الثلثين فيكون تسعة اموال وثلثين احداً فلان هذه التسعة الاموال والثلثين الاحد هى مساوية لعددين كلّ واحد منهما اذا زيد على المكعّب الذى هو كعب واحد صار مكعّباً فانّا اذا عملنا مكعّبين من ضلعين يكون كلّ واحد منهما شيئاً وآحاداً ما وأخذنا زيادة كلّ واحد منهما على المكعّب وأقمنا الزيادتين مقام العددين وجمعناهما وعادلنا بهما التسعة الاموال والثلثين الاحد اصبنا ما طلبنا لكن تينك الزيادتين هما مركّبتان من اموال واشياء وعدد فينبغى ان تكون تلك الاموال التى فى

---

1982 (628): الذى : اللدس in cod.

1983 (629): Post بقسمين habet codex محلعس (sc. مختلفين ), quod ob inutilitatem delevi.

(630): تغاضل : معاصل in cod.

1997 (631): وأخذنا : فاحدىا in cod.

1999 (632): تينك الزيادتين هما مركّبتان : ىلك الربادات هى مركبه in cod.

الزيادتين مجموعتين تسعة اموال وان يكون العدد الذى معها
اقلّ من ثلثين لكي ننتهى الى عدد يعادل شيئاً ولانّا نريد ان نعمل
〈المكعّبين〉 من ضلعين يكون كلّ واحد منهما شيئاً وعدداً حتّى تكون
جملة الاموال التي فى كلّ واحد من المكعّبين تسعة اموال وتكون الآحاد
اقلّ من ثلثين التي هى العدد المفروض ولكن الاموال الزائدة فى
كلّ واحد من المكعّبين هى ثلثة امثال كلّ واحد من العددين
الزائدين على الشىء فى ضلعى المكعّبين وجملة الآحاد التى فى
المكعّبين هى جملة مكعّبيهما فاذن ينبغى ان تكون جملة العددين
الزائدين على الشىء ثلثةَ آحاد /ليكون ثلثة امثالها عددَ الاموال التى ٩١
هى تسعة وينبغى ان نقسم الثلثة بقسمين تكون جملة مكعّبيهما اقلّ
من ثلثين وهما اثنان وواحد ونعمل احد المكعّبين من ضلع شىء
وأحدين فيكون كعباً وستّة اموال واثنى عشر شيئاً وثمنية آحاد ونعمل
المكعّب الآخر من ضلع شىء وواحد فيكون كعباً وثلثة اموال وثلثة
اشياء وواحداً فلان الستّة الاموال والاثنى عشر الشىء والثمنية الآحاد
اذا زيدت على كعب كان مكعّباً وكذلك الثلثة الاموال والثلثة الاشياء
والواحد فنجعل المجتمع منهما وهو تسعة اموال وخمسة عشر شيئاً
وتسعة آحاد كما ذكرنا وهو المعادل للتسعة الاموال والثلثين الاحد
ونلقى التسعة الاموال المشتركة من كلى الناحيتين فيبقى خمسة عشر
شيئاً وتسعة آحاد تعادل ثلثين احداً فنلقى التسعة الآحاد المشتركة
من كلى الناحيتين فيبقى خمسة عشر شيئاً تعادل واحداً وعشرين احداً
ولذلك يكون الشىء الواحد واحداً وخمسى واحد ولانّا فرضنا ضلع
المكعّب المطلوب شيئاً واحداً يكون سبعة اخماس واحد ويكون المكعّب
احدين وثلثة وتسعين جزءاً من مائة وخمسة وعشرين جزءاً من واحد فامّا
مربّع ضلع المكعّب فهو واحد واربعة وعشرون جزءاً من خمسة وعشرين
من واحد وتسعة امثال ذلك سبعة عشر احداً وستّة عشر جزءاً من

---

2003 (633): المكعّبين addidi.

2005 (634): ولكن : ولىكں in cod.

(635): الزائدة : الرىاده in cod.

2011 (636): ثلثين : عسرں in cod.

(637): وهما : وهى in cod.

2017 (638): De loco dubitare licet; fortasse scribendum معادلا pro كما ذكرنا وهو المعادل (vide lineas 2068,2111-2112,2147).

2024 (639): Loco خمسة وعشرين habet codex: مايه وحمسه وىس.

2025 (640): وستّة : وسبعه in cod.

خمسة وعشرين جزءاً اعنى ثمانين جزءاً من مائة وخمسة وعشرين جزءاً واذا زدنا ذلك على ثلثين كان سبعة واربعين احداً وثمنين جزءاً من مائة وخمسة وعشرين جزءاً وكنّا فرضنا احد قسمى هذا العدد المجتمع ستّة اموال واثنى عشر شيئاً وثمنية آحاد فامّا الستّة اموال فتكون احد عشر احداً وخمسة وتسعين جزءاً من مائة وخمسة وعشرين جزءاً من واحد والاثنا عشر/شيئاً ستّة عشر احداً ومائة جزء من مائة وخمسة وعشرين جزءاً ٩٢ من واحد فاذن كلّ العدد الاوّل ستّة وثلثون احداً وسبعون جزءاً من مائة وخمسة وعشرين جزءاً من واحد ويكون العدد الآخر بقيّة السبعة والاربعين احداً والثمنين جزءاً من مائة وخمسة وعشرين وهو احد عشر احداً وعشرة اجزاء من مائة وخمسة وعشرين واذا زدنا العدد الاوّل من هذين العددين على العدد المكعّب الذى هو احدان وثلثة وتسعون جزءاً من مائة وخمسة وعشرين كانت جملة ذلك تسعة وثلثين احداً وثمنية وثلثين جزءاً من مائة وخمسة وعشرين جزءاً وهو عدد مكعّب ضلعه ثلثة آحاد وخُمساً واحد واذا زدنا العدد الثانى منهما على العدد المكعّب كانت جملة ذلك ثلثة عشر احداً ومائة جزء وثلثة اجزاء من مائة وخمسة وعشرين جزءاً من واحد وذلك عدد مكعّب ضلعه احدان وخُمساً احد

فقد وجدنا عدداً مكعّباً اذا زدنا تسعة امثال مربّع ضلعه على ثلثين احداً كان ذلك مساوياً لعددين كلّ واحد منهما اذا زيد على العدد المكعّب صار مكعّباً وهو المكعّب الذى حدّدناه وذلك ما اردنا ان نجد

ينبغى ان يُعلم ان هذه المسئلة تخرج بهذا العمل متى كان مكعّب ثُلث عدد المرّات اصغرَ من اربعة امثال العدد المفروض

يد نريد ان نجد عدداً مكعّباً اذا نقصنا من امثال مفروضة لمربّع ضلعه عدداً مفروضاً كان ذلك مساوياً لعددين اذا نُقص كلّ واحد منهما من المكعّب بقى منه مكعّب

---

2029 (641): ستّة : سه عسر in cod.

(642): واثنى : واسا in cod.

2031 (643): والاثنا : والاسى in cod.

2038 (644): وثمنية : ولىه in cod.

2044 (645): احداً : احد in cod.

فليكن العدد المفروض ستّة وعشرين والامثال المفروضة تسعة امثال ونريد ان نجد عدداً مكعّباً على ما وصفنا فنفرض ضلع المكعّب شيئاً واحداً فيكون المكعّب كعباً واحداً ونأخذ تسعة امثال مربّع ضلع المكعّب الذى هو مال وذلك تسعة اموال فنلقى منها العدد المفروض فتبقى تسعة اموال / الّا ستّة وعشرين احداً وهى مساوية ٩٣ لعددين كلّ واحد منهما اذا نُقص من المكعّب بقى منه مكعّب وعلى مثل ما وصفنا فى المسئلة المتقدّمة فلنعمل مكعّبين يكون كلّ واحد منهما من ضلع شىء الّا عدداً حتّى يكون مجموع الاموال الناقصة التى فيهما تسعة اموال ولسنا نحتاج فى هذه المسئلة الى ان تكون جملة العددين اللذين فيهما اقلّ من الآحاد المفروضة بل كيف إتّفق ذلك لنا ولنعمل احد المكعّبين من ضلع يكون شيئاً الّا احدين فيكون ذلك المكعّب كعباً واثنى عشر شيئاً الّا ستّة اموال والّا ثمنية آحاد ولنعمل الآخر من ضلع شىء الّا واحداً فيكون المكعّب كعباً وثلثة اشياء الّا ثلثة اموال والّا واحداً فلان الستّة الاموال والثمنية الآحاد والّا اثنى عشر شيئاً اذا نقصناها من المكعّب صار مكعّباً وكذلك الثلثة الاموال والواحد الّا ثلثة اشياء اذا نقصناها ايضاً من المكعّب بقى منه مكعّب فلنجعل مجموع هذين العددين معادلاً للتسعة الاموال والّا ستّة وعشرين احداً لكن مجموعهما هو تسعة اموال وتسعة آحاد الّا خمسة عشر شيئاً فهى اذن تعادل تسعة اموال الّا ستّة وعشرين احداً فلنزد الستّة والعشرين احداً على كلى الناحيتين وكذلك الخمسة العشر الشىء ونلقى التسعة الاموال المشتركة من كلى الناحيتين فيبقى من بعد الجبر والمقابلة خمسة عشر شيئاً تعادل خمسة وثلثين احداً فلذلك يكون الشىء احدين وثُلثَ واحد ولانّا فرضنا ضلع المكعّب شيئاً واحداً يكون ضلع المكعّب احدين وثُلث واحد ويكون المكعّب اثنى عشر احداً وتسعة عشر جزءاً من سبعة وعشرين جزءاً من واحد ومربّع ضلع المكعّب خمسة آحاد واثنا عشر

---

2059 (646): عدداً : عدد in cod.

2062 (647): احد المكعّبين : احدا المكعس in codice, per dittographiam litterae *alif*.

2064 (648): واحداً : واحد in cod.

2066 (649): والّا اثنى : والا سى in cod.

2071 (650): فلنزد : فلىرىد in cod.

(651): Fortasse addendum est مشتركةً post posterius احداً.

جزءاً من سبعة وعشرين جزءاً وتسعة امثال ذلك تسعة واربعون احداً فلنلق منها الستّة والعشرين المفروضة فتبقى ثلثة وعشرون/احداً وكنّا ٩٤
فرضنا احد قسمي هذه الثلثة والعشرين ستّةَ اموال وثمنية آحاد الّا اثنى عشر شيئاً لكن الستّة الاموال هى اثنان وثلثون وثمنية عشر جزءاً من سبعة وعشرين والاثنا عشر الشىء هو ثمنية وعشرون احداً فلذلك يكون العدد الاعظم من العددين اثنى عشر احداً وثمنية عشر جزءاً من سبعة وعشرين من واحد ولذلك يكون العدد الاصغر عشرة آحاد وتسعة اجزاء من سبعة وعشرين جزءاً من واحد لكن العدد الاعظم من هذين العددين اذا نُقص من المكعّب الذى قد بيّنّا انّه اثنا عشر احداً وتسعة عشر جزءاً من سبعة وعشرين جزءاً من واحد بقى منه جزء واحد من سبعة وعشرين جزءاً من واحد وهو مكعّب ضلعه ثُلث واحد واذا نقصنا العدد الاصغر من العدد المكعّب بقى منه احدان وعشرة اجزاء من سبعة وعشرين جزءاً من واحد وهو مكعّب ضلعه واحد وثُلث واحد

فقد وجدنا عدداً مكعّباً على التحديد الذى حددنا وذلك ما اردنا ان نجد

---

يهَ نريد ان نجد عدداً مكعّباً اذا نقصنا من امثال مفروضة لمربّع ضلعه عدداً مفروضاً كان ذلك مساوياً لعددين اذا زيد احدهما على المكعّب كان مكعّباً واذا نُقص الآخر من المكعّب ايضاً بقى منه مكعّب

فلتكن الامثال المفروضة تسعة امثال والعدد المفروض ثمنية عشر احداً ونريد ان نجد عدداً مكعّباً اذا نُقص من تسعة امثال مربّع ضلعه ثمنية عشر احداً كان الباقى من ذلك عددين ⟨اذا زيد⟩ احدهما على المكعّب اجتمع منهما مكعّب واذا نُقص الآخر من المكعّب بقى منه مكعّب فنفرض ضلع المكعّب شيئاً واحداً فيكون المكعّب كعباً واحداً ونأخذ تسعة امثال مربّع ضلعه وهو تسعة اموال فنلقى منه ثمنية عشر احداً ثمّ نعمل مكعّبين يكون ضلع احدهما شيئاً وعدداً ويكون ضلع الآخر شيئاً الّا عدداً حتّى تكون الاموال الزائدة فى احد المكعّبين مع الاموال الناقصة/فى المكعّب الآخر تسعة اموال وليكن ٩٥

---

2079 (652): فلنلق : فلنلعى in cod.

2094 (653): نقصنا : نعص in cod.

2099 (654): Deficientia verba اذا زيد restitui.

المكعّب الاوّل من ضلع ⟨شىء⟩ الّا احدين حتّى يكون كعباً واثنى عشر شيئاً الّا ثمنية آحاد والّا ستّة اموال وليكن المكعّب الآخر من ضلع شىء وواحد ليكون كعباً وثلثة اموال وثلثة اشياء وواحداً فلان الستّة الاموال والثمنية الآحاد الّا اثنى عشر شيئاً اذا نقصنا⟨ها⟩ من المكعّب المطلوب الذى هو كعب واحد صار الباقى مكعّباً والثلثة الاموال والواحد والثلثة اشياء اذا زيدت على المكعّب المطلوب اجتمع منهما مكعّب فلنجعل مجموع هذين العددين معادلاً للتسعة الاموال الّا ثمنية عشر احداً لكن مجموعهما تسعة اموال وتسعة آحاد الّا تسعة اشياء فهى اذن تعادل تسعة اموال الّا ثمنية عشر احداً ولنجبر ذلك ونقابل به فيبقى بعد الجبر والمقابلة سبعة وعشرون احداً تعادل تسعة اشياء فان الشىء الواحد يعادل ثلثة آحاد ولانّا فرضنا ضلع المكعّب شيئاً واحداً يكون ضلع المكعّب ثلثة آحاد ويكون المكعّب سبعة وعشرين ومربّع ضلع المكعّب تسعة آحاد وتسعة امثال ذلك احد وثمنون ولنلق منها العدد المفروض الذى هو ثمنية عشر احداً فتبقى ثلثة وستّون وقد كنّا فرضنا احد العددين ستّة اموال وثمنية آحاد الّا اثنى عشر شيئاً فلذلك يكون ستّة وعشرين احداً والعدد الآخر هو بقيّة ثلثة وستّين وهو سبعة وثلثون احداً والستّة والعشرون اذا نُقصت من المكعّب الذى هو سبعة وعشرون بقى منه واحد وهو مكعّب والسبعة والثلثون اذا زيدت على المكعّب الذى هو سبعة وعشرون اجتمع منها اربعة وستّون وهو مكعّب وضلعه اربعة آحاد

فقد وجدنا مكعّباً على الشرط الذى اشترطنا وذلك ما اردنا ان نجد

يوَ نريد ان نجد عدداً مكعّباً اذا نقصنا من امثال مفروضة لمربّع
ضلعه عدداً مفروضاً كان الباقى من ذلك / مساوياً لعددين اذا ٩٦

2106 (655): شىء addidi.

2108 (656): وواحداً : وواحد in cod.

2109 (657): Pronomen addidi.

(658): المطلوب : المعروض in cod.

2111 (659): المطلوب : المعروض in cod.

2117 (660): سبعة sic in codice: سلعه; primum enim scripsit librarius سا (sc. شيئاً), quod statim correxit.

2119 (661): ولنلق : ولىلعى in cod.

2124 (662): والسبعة : والسعه in cod.

نُقص احدهما من المكعّب بقى منه مكعّب واذا نُقص المكعّب من العدد الآخر بقى منه مكعّب

فلتكن الامثال المفروضة ايضاً تسعة امثال والعدد المفروض ستّة عشر ونريد ان نجد عدداً مكعّباً اذا نقصنا من تسعة امثال مربّع ضلعه ستّة عشر احداً كان ما يبقى من ذلك مساوياً لعددين اذا نُقص احدهما من المكعّب بقى منه مكعّب واذا نُقص المكعّب من العدد الآخر بقى منه مكعّب فنجعل المكعّب ايضاً كعباً واحداً وننقص من تسعة امثال مربّع ضلعه ستّة عشر احداً ونعمل مكعّبين يكون ضلع احدهما شيئاً الّا عدداً وضلع الآخر عدداً الّا شيئاً ولتكن الاموال التى تقع فيهما تسعة اموال فنعمل احد المكعّبين من ضلع شىء الّا واحداً فيكون كعباً وثلثة اشياء الّا ثلثة اموال والّا واحداً ونعمل المكعّب الآخر من ضلع احدين الّا شيئاً فيكون ثمنية آحاد وستّة اموال الّا كعباً والّا اثنى عشر شيئاً فلان الثلثة الاموال والواحد الّا ثلثة اشياء اذا نُقصت من المكعّب بقى منه مكعّب وهو كما قلنا كعب وثلثة اشياء الّا ثلثة اموال والّا واحداً والستّة اموال والثمنية آحاد الّا اثنى عشر شيئاً اذا نُقص منها المكعّب الذى هو كعب واحد بقى مكعّب وهو كما قلنا ايضاً ثمنية آحاد وستّة اموال الّا اثنى عشر شيئاً والّا كعباً فليكن مجموعهما معادلاً للتسعة الاموال الّا ستّة عشر لكن مجموعهما تسعة اموال وتسعة آحاد الّا خمسة عشر شيئاً فهو اذن يعادل تسعة اموال الّا ستّة عشر احداً ولنجبر ذلك ولنقابل به فننتهى بعد الجبر والمقابلة الى خمسة عشر شيئاً تعادل خمسة وعشرين احداً فلذلك يكون الشىء الواحد واحداً وثُلثى واحد وهو ضلع المكعّب ولذلك يكون المكعّب اربعة آحاد وسبعة عشر جزءاً من سبعة وعشرين من واحد فامّا /مربّع ضلع المكعّب فانّه يكون ٩٧
احدين وواحداً وعشرين جزءاً من سبعة وعشرين جزءاً وتسعة امثال ذلك خمسة وعشرون ولنلق منها الستّة عشر فتبقى تسعة آحاد وقد كنّا فرضنا العدد المنقوص من المكعّب من العددين اللذين جملتهما تسعة آحاد ثلثةَ اموال وواحداً الّا ثلثة اشياء والثلثة الاموال هى ثمنية آحاد وثُلث والثلثة اشياء هى خمسة آحاد فاذن هذا العدد

---

2140 (663): واحداً (prius): واحد in cod.

2143 (664): مكعّب : كعب in cod.

2155 (665): ولنلق : ولىلعى in cod.

الذى ذكرنا يكون اربعة آحاد وثُلثَ واحد والعدد الآخر بقيّة التسعة الآحاد وهو اربعة وثُلثَا واحد والاربعة الآحاد والثُلث واحد اذا نُقصت من المكعّب الذى هو اربعة آحاد وسبعة عشر جزءاً من سبعة وعشرين جزءاً بقى ثمنية اجزاء من سبعة وعشرين جزءاً من واحد وهو مكعّب ضلعه ثُلثَا واحد واما العدد الآخر الذى هو اربعة آحاد وثُلثَا واحد فانّه اذا نُقص منه المكعّب بقى منه جزء واحد من سبعة وعشرين جزءاً من واحد وهو مكعّب ضلعه ثُلث واحد

فقد وجدنا عدداً مكعّباً على الشرط الذى اشترطنا وذلك ما اردنا ان نجد

تمّت المقالة الخامسة من كتاب ذيوفنطس فى المسائل العدديّة وهى ستّ عشرة مسئلة

بسم الله الرحمن الرحيم

المقالة السادسة من كتاب ذيوفنطس

ا نريد ان نجد عددين احدهما مكعّب والآخر مربّع يكون ضلعاهما فى نسبة مفروضة اذا جُمع مربّعاهما كان ⟨المجتمع⟩ عدداً مربّعاً

فلتكن النسبة المفروضة نسبة المثلين ونريد ان نجد عددين احدهما مكعّب والآخر مربّع يكون ضلع المكعّب مثلى ضلع المربّع اذا جُمع ⟨مربّعاهما⟩ كان ⟨المجتمع⟩ عدداً مربّعاً فلنجعل ضلع المربّع شيئاً فيكون المربّع مالاً وضلع المكعّب شيئين فيكون المكعّب ثمنية كعاب / ويكون المجتمع من مربّع المكعّب و⟨مربّع⟩ المربّع اربعة وستّين كعب ٩٨ كعب ومال مال ونحتاج ان يكون مربّعاً فلنطلب عدداً مربّعاً اذا نُقص منه اربعة وستّون احداً كان الباقى مربّعاً ووُجِدَ ان ذلك سهل

2162 (666): Pro واحد praebet codex كعب.
2169 (667): عشرة : عسر in cod.
2171 (668): Linea atramento rubro in cod.
2173 (669): المجتمع addidi.
2176 (670): مربّعاهما addidi.
(671): المجتمع addidi.
2178 (672): مربّع addidi.
2180 (673): وستّون : وسىس in cod.

على ما بيّنّا فيما تقدّم من كتابنا وذلك مائة احد فلنجعلها ⟨كعاب كعاب فتكون⟩ مائة كعب كعب ونقابل بها اربعة وستّين كعب كعب ومال مال ونلقى المشتركات فيبقى ستّة وثلثون كعب كعب تعادل مال مال فنقسم الناحيتين على اقعدهما وهو مال مال فيكون واحد يعادل ستّة وثلثين مالاً فالمال جزء من ستّة وثلثين جزءاً من واحد والشيء جزء من ستّة اجزاء من واحد وهو ضلع العدد المربّع وضلع العدد المكعّب مثلاه وهو جزءان من ستّة اجزاء من الواحد والمكعّب ثمنية اجزاء من مأتين وستّة عشر جزءاً من الواحد فاذا اضيف مربّعه وهو اربعة وستّون جزءاً من ستّة واربعين الفاً وستّمائة وستّة وخمسين جزءاً من الواحد الى مربّع العدد المربّع الذى هو ستّة وثلثون جزءاً من ستّة واربعين الفاً وستّمائة وستّة وخمسين جزءاً كان ذلك مائة جزء من ستّة واربعين الفاً وستّمائة وستّة وخمسين وهو عدد مربّع ضلعه عشرة اجزاء من مأتين وستّة عشر جزءاً من الواحد

فقد وجدنا عددين على التحديد الذى حُدّ لنا وهما ثمنية اجزاء من مأتين وستّة عشر جزءاً من الواحد وستّة اجزاء من مأتين وستّة عشر جزءاً من الواحد وذلك ما اردنا ان نجد

بَ نريد ان نجد عددين احدهما مكعّب والآخر مربّع ويكون ضلعاهما في نسبة مفروضة اذا نُقص من ⟨مربّع⟩ المكعّب مربّع المربّع كان الباقى مربّعاً

فلتكن النسبة المفروضة نسبة المثلين ونريد ان نجد عددين احدهما مكعّب والآخر مربّع يكون ضلع المكعّب مثلى ضلع المربّع واذا نُقص من مربّع المكعّب مربّع المربّع / ٩٩ كان الباقى منه مربّعاً فلنجعل ضلع العدد المربّع شيئاً وضلعَ العدد المكعّب شيئين فيكون العدد المربّع مالاً ومربّعه مال مال والعدد المكعّب ثمنية كعاب ومربّعه اربعة وستّون كعب كعب فاذا نقصنا من اربعة وستّين كعب ⟨كعب⟩

---

2181-2182 (674): In uncis seclusa verba dubitanter addidi.

2183 (675): ونلقى : وسعى (sc. ويبقى) in cod.

(676): المشتركات : المساركات in codice. Vide adn. 422.

2187 (677): والمكعّب : والكعب in cod.

2188 (678): اضيف : اصف in cod.

2198 (679): Verbum مربّع addidi.

2205 (680): Per haplographiam deficiens كعب restitui.

مالَ مال يكون الباقى اربعة وستّين كعب كعب الّا مال مال فنحتاج ان يكون عدداً مربّعاً فلنطلب عدداً مربّعاً اذا نقصناه من اربعة وستّين كان الباقى عدداً مربّعاً ووجِدَان ذلك سهل على ما تقدّم بيانه فيكون اربعين احداً واربعة وعشرين جزءاً من خمسة وعشرين جزءاً من الواحد فلنجعلها كعاب كعاب فتكون اربعين كعب كعب واربعة وعشرين جزءاً من خمسة وعشرين جزءاً من كعب كعب فهى تعادل اربعة وستّين كعب كعب الّا مال مال فنقابل بها ونلقى المشتركات فيبقى ثلثة وعشرون كعب كعب وجزء من خمسة وعشرين جزءاً من كعب كعب تعادل مال مال فنقسم الناحيتين على مال مال فيكون واحد يعادل ثلثة وعشرين مالاً وجزءاً من خمسة وعشرين من مال فالمال خمسة وعشرون جزءاً من خمس مائة وستّة وسبعين جزءاً من الواحد والشىء خمسة اجزاء من اربعة وعشرين جزءاً من واحد وكنّا جعلنا ضلع العدد المكعّب شيئين وذلك خمسة اجزاء من اثنى عشر جزءاً من الواحد فالمكعّب مائة وخمسة وعشرون جزءاً من الف وسبع مائة وثمنية وعشرين جزءاً من الواحد ومربّعه خمسة عشر الفاً وستّمائة وخمسة وعشرون جزءاً من الفى الف وتسع مائة الف وخمسة وثمنين الفاً وتسع مائة واربعة وثمنين جزءاً فاذا نقصنا منه مربّع العدد المربّع وهو خمسة الٰف وستّمائة وخمسة وعشرون جزءاً من الفى الف وتسع مائة الٰف وخمسة وثمنين الفاً وتسع مائة واربعة وثمنين كان الباقى منه عشرة الٰف جزء من الفى الف / وتسع مائة الف وخمسة وثمنين الفاً وتسع مائة واربعة وثمنين جزءاً وذلك عدد مربّع وضلعه مائة جزء من الف وسبع مائة وثمنية وعشرين جزءاً

فقد وجدنا عددين على التحديد الذى حُدّ لنا وهما مائة وخمسة وعشرون جزءاً من الف وسبع مائة وثمنية وعشرين جزءاً من الواحد وخمسة وسبعون جزءاً من الف وسبع مائة وثمنية وعشرين جزءاً من الواحد وذلك ما اردنا ان نجد

جـ نريد ان نجد عددين احدهما مكعّب والآخر مربّع ويكون ضلعاهما فى نسبة مفروضة اذا نقصنا مربّع المكعّب من مربّع العدد المربّع كان الباقى منه مربّعاً

فلتكن النسبة المفروضة نسبة المثلين ونريد ان نجد عددين احدهما مكعّب والآخر مربّع ويكون ضلع المكعّب من ضلع المربّع فى نسبة

---

2215 (681): مالاً : مال in cod.

المثلين واذا نُقص مربّع العدد المكعّب من مربّع العدد المربّع كان الباقى منه مربّعاً فانّا نجعل ضلع العدد المربّع شيئاً فيكون ضلع المكعّب شيئين والمكعّب ثمنية كعاب ومربّعه اربعة وستّون كعب كعب وقد كنّا جعلنا ضلع العدد المربّع شيئاً فالمربّع مال ومربّعه مال مال فنلقى منه مربّع المكعّب وهو اربعة وستّون كعب كعب فيبقى مال مال الّا اربعة وستّين كعب كعب فنحتاج ان يكون مربّعاً ونريد ان نطلب عدداً مربّعاً اذا زيد عليه اربعة وستّون كان المجتمع مربّعاً وذلك ستّة وثلثون وضلعه ستّة فنجعل ضلع مال المال الّا اربعة وستّين كعب كعب ستّة كعاب ونضربه فى مثله فيكون ستّة وثلثين كعب كعب فهى تعادل مال مال الّا اربعة وستّين كعب كعب فنجبر ونقابل فيكون مال مال يعادل مائة كعب كعب فنقسم كلّ ذلك على مال مال فيكون واحد يعادل مائة مال فالمال جزء من مائة وهو عُشر عُشر واحد والشىء جزء من عشرة / ١٠١ وهو عُشر واحد وقد كنّا جعلنا المكعّب ⟨من ضلع⟩ شيئين فضلعه جزءان من عشرة ويكون المكعّب ثمنية اجزاء من الف ومربّعه اربعة وستّون جزءاً من الف الف فاذا نقصناه من مربّع العدد المربّع الذى هو مائة جزء من الف الف كان الباقى منه ستّة وثلثين جزءاً من الف الف وهو عدد مربّع وضلعه ستّة اجزاء من الف جزء من الواحد

فقد وجدنا عددين على التحديد الذى حُدّ لنا وهما ثمنية اجزاء من الف جزء وعشرة اجزاء من الف جزء وذلك ما اردنا ان نجد

د نريد ان نجد عددين احدهما مكعّب والآخر مربّع ويكون ضلع المكعّب من ضلع المربّع فى نسبة مفروضة اذا زيد على العدد الذى يحيطان به مربّع المكعّب كان الذى يجتمع مربّعاً

فلتكن النسبة نسبة الخمسة الامثال فنريد ان نجد عددين احدهما مكعّب والآخر مربّع ويكون ضلع المكعّب خمسة امثال ضلع المربّع واذا زيد على العدد الذى يحيطان به مربّع العدد المكعّب كان الذى

2249-2250 (682): من ضلع addidi.

2250 (683): فضلعه : فصله in cod.

2253 (684): وثلثين : ولىلىوں in cod.

(685): وهو : هو supra و in codice (eadem ut videtur manu) scriptum.

2256 (686): نجد : يحد in cod.

يجتمع من ذلك مربّعاً فلنفرض ضلع المربّع شيئاً فيكون المربّع مالاً ويكون ضلع المكعّب خمسة اشياء والمكعّب مائة وخمسة وعشرون كعباً فالعدد الذى يحيطان به مائة وخمسة وعشرون كعباً مضروبة فى مال فنزيد على ذلك مربّع المكعّب [ومربّع المكعّب خمسة عشر الفاً وستّمائة وخمسة وعشرون كعب كعب] فيكون المجتمع من ذلك خمسة عشر الفاً وستّمائة وخمسة وعشرين كعب كعب ومائة وخمسة وعشرين كعباً مضروبة فى مال فنحتاج ان يكون مربّعاً فلنطلب عدداً مربّعاً اذا نقصنا منه خمسة عشر الفاً وستّمائة وخمسة وعشرين كان الباقى منه عدداً يسيراً وليس بنا حاجة الى ان يكون ما يبقى عدداً مربّعاً وذلك العدد خمسة عشر الفاً وثمان مائة وستّة وسبعون وضلعه مائة وستّة وعشرون فلنجعله [ذلك خمسة عشر الفاً وستّمائة/وخمسة وعشرين كعب كعب ١٠٢ ومال مضروب فى مائة وخمسة وعشرين كعباً] مائةَ وستّة وعشرين كعباً ونضربه فى مثله فيكون خمسة عشر الفاً وثمان مائة وستّة وسبعين كعب كعب فهى تعادل خمسة عشر الفاً وستّمائة وخمسة وعشرين كعب كعب ومائة وخمسة وعشرين كعباً مضروبة فى مال فنلقى الخمسة عشر الفاً والستّمائة والخمسة والعشرين كعب كعب المشتركة من الناحيتين فيبقى مائتان وأحد وخمسون كعب كعب تعادل مائة وخمسة وعشرين كعباً مضروبة فى مال فأقسمْ الناحيتين على كعب مضروب فى مال فيكون مائة وخمسة وعشرون احداً تعادل مائتين وأحد وخمسين شيئاً فالشىء مائة وخمسة وعشرون جزءاً من مائتين وأحد وخمسين وهو ضلع المربّع والمربّع خمسة عشر الفاً وستّمائة وخمسة وعشرون جزءاً من مربّع مائتين

---

2263 (687): فيكون pro و codicis substitui.

2266-2267 (688): Verba ومربّع المكعّب ..... كعب كعب interpolata videntur et a lectore quodam ad lacunam textus explendam addita.

2268 (689): Post كعباً praebet codex و.

2270 (690): عدداً يسيراً : عدد سىر in cod.

2272 (691): وسبعون : وسبعس in cod.

2273-2274 (692): Verba uncis inclusa interpolata esse censeo.

2281 (693): وعشرون : وعسرس in cod.

2282 (694): وعشرون : وعسرس in cod.

2283 (695): وعشرون : وعسرس in cod.

وأحد وخمسين وذلك ثلثة وستّون الفاً ‹و›واحد وقد كان ضلع العدد المكعّب خمسة امثال ضلع العدد المربّع وذلك ستّمائة وخمسة وعشرون جزءاً من مأتين وأحد وخمسين ويكون المكعّب مائتى الف الف واربعة واربعين الف الف ومائة الف واربعين الفاً وستّمائة وخمسة وعشرين جزءاً من الفى الف وخمس مائة وثلثة وستّين الفاً وواحد ونكتفى بصحّة عمل هذه المسئلة عن اصحابها

فقد وجدنا عددين على التحديد الذى حُدّ لنا وهما خمسة عشر الفاً وستّمائة وخمسة وعشرون جزءاً من ثلثة وستّين الفاً وواحد ومائتا الف الف واربعة واربعون الف الف ومائة الف واربعون الفاً وستّمائة وخمسة وعشرون جزءاً من الفى الف وخمس مائة وثلثة وستّين الفاً وواحد وذلك ما اردنا ان نجد

ه نريد ان نجد عددين احدهما مكعّب والآخر مربّع ويكون ضلع المكعّب مساوياً لضلع المربّع اذا زيد على العدد الذى يحيطان به مثل مربّع العدد المربّع كان المجتمع من ذلك مربّعاً

فلنجعل ضلع المربّع شيئاً فيكون المربّع / مالاً وايضاً فضلع المكعّب ٠٣
شىء فالمكعّب كعب والعدد الذى يحيطان به هو كعب مضروب فى مال ونزيد عليه مربّع العدد المربّع وهو مال مال فيكون كعباً مضروباً فى مال ومالَ مال وهو يعادل عدداً مربّعاً فلنفرض ضلعه ماليين فيكون اربعة اموال مال تعادل مال مال ومالاً مضروباً فى كعب فنلقى المال مال المشترك فيبقى مال مضروب فى كعب يعادل ثلثة اموال مال فنقسم كلّ ما معنا على مال مال فيكون شىء يعادل ثلثة آحاد وهو ضلع المربّع والمربّع تسعة آحاد وايضاً فقد كان ضلع المكعّب مساوياً لضلع المربّع فهو ثلثة آحاد والعدد المكعّب سبعة وعشرون احداً والعدد الذى يحيطان به هو ما يكون من ضرب تسعة فى سبعة وعشرين وذلك مائتان وثلثة واربعون فاذا زيد عليه مربّع العدد المربّع وهو واحد وثمنون كان المجتمع من ذلك ثلثمائة واربعة وعشرين وذلك عدد مربّع ضلعه ثمنية عشر

فقد وجدنا عددين على التحديد الذى حُدّ لنا وذلك ما اردنا ان نجد

---

2284 (696): و addidi.

2304 (697): شىء : سا in cod.

وَ نريد ان نجد عددين احدهما مربّع والآخر مكعّب ويكون ضلع المكعّب مساوياً لضلع المربّع واذا نقصنا من العدد الذى يحيطان به مربّعَ العدد المكعّب كان الباقى عدداً مربّعاً

فلنفرض ضلع العدد المربّع شيئاً فيكون العدد المربّع مالاً وايضاً فان ضلع العدد المكعّب شىء فالعدد المكعّب هو كعب ونحتاج ان ننقص مربّع كعب من العدد الذى يحيط بـه كعبٌ ومالٌ ولكن العدد الذى يحيط به كعب ومال هو كعب مضروب فى مال فاذا نقصنا منه مربّع العدد المكعّب وهو كعب كعب كان الباقى كعباً مضروباً فى مال الّا كعب كعب فنحتاج ان يكون مربّعاً فلنجعل ضلعه كعباً واحداً ونضربه فى مثله فيكون كعب كعب فهو يعادل كعباً مضروباً فى مال الّا كعب كعب فلنزد كعب كعب على الناحيتين ونقسم/الناحيتين ٤٠ على اقعد الناحيتين وذلك كعب مضروب فى مال فيكون شيئان يعادلان واحداً فالشىء الواحد يعادل نصفاً وقد كنّا جعلنا ضلع المربّع شيئاً فيكون المربّع جزءاً من اربعة وهو رُبع واحد وايضاً فضلع المكعّب نصف واحد فالمكعّب ثُمن واحد ولان العدد الذى يحيطان به جزء من اثنين وثلثين اذا أُلقى منه مربّع المكعّب وهو جزء من اربعة وستّين جزءاً يكون الباقى جزءاً من اربعة وستّين وهو مربّع وضلعه جزء من ثمنية

فقد وجدنا عددين على التحديد الذى حُدّ لنا وهما رُبع واحد وثُمن واحد وذلك ما اردنا ان نجد

زَ نريد ان نجد عددين احدهما مكعّب والآخر مربّع يكون ضلعاهما متساويين اذا نُقص من العدد الذى يحيطان به مثل مربّع العدد المربّع كان الباقى منه مربّعاً

فلنفرض ضلع العدد المربّع شيئاً فيكون العدد المربّع مالاً ولمّا كان ضلع المكعّب مساوياً لضلع العدد المربّع وجب ان يكون العدد المكعّب كعباً فالعدد الذى يحيطان به كعب مضروب فى مال ولكنّا اذا القينا منه مربّع العدد المربّع وهو مال مال كان الباقى كعباً مضروباً فى مال ⟨الّا مال مال⟩ فنحتاج ان يكون مربّعاً فلنفرض له ضلعاً يكون

---

2324 (698): اقعد : معد in cod.

2326 (699): فضلع : وصلع in cod.

2327 (700): فالمكعّب : فالكعب in cod.

2340 (701): Per homoeoteleuton omissum addidi.

مالاً ونضربه فى مثله فيكون مالَ مال يعادل كعباً مضروباً فى مال منقوصاً منه مالَ مال فنجبر ونقابل فيكون الشى ء اثنين وكنّا فرضنا ضلع المربّع شيئاً فهو اثنان والمربّع اربعة آحاد وايضاً فان ضلع المكعّب اثنان والمكعّب ثمنية ولمّا كان المربّع اربعة والمكعّب ثمنية كان العدد الذى يحيطان به اثنين ⟨وثلثين⟩ احداً فاذا نقصنا منه مثل مربّع العدد المربّع كان الباقى ستّة عشر احداً وذلك عدد مربّع وضلعه اربعة

فقد وجدنا عددين على التحديد الذى حُدّ لنا وهما ثمنية آحاد اربعة آحاد وذلك ما اردنا ان نجد

حَ نريد ان نجد عددين احدهما مكعّب والآخر مربّع اذا زيد على العدد الذى/يحيطان به مثل ضلعه كان المجتمع من ذلك مربّعاً ٠٥

فانّا نفرض العدد المكعّب اربعة وستّين احداً والعدد المربّع مالاً فيكون العدد الذى يحيطان به اربعة وستّين مالاً ولكنّا اذا زدنا عليه مثل ضلعه وهو ثمنية اشياء كان المجتمع اربعة وستّين مالاً وثمنية اشياء فنحتاج الى ان يكون مربّعاً فلنفرض ضلعه ما شئنا من الاشياء بعد ان يكون اكثر من ثمنية ⟨اشياء⟩ فكأنّا جعلنا(ه) عشرة اشياء وضربناه فى مثله فكان مائة مال فهى تعادل اربعة وستّين مالاً وثمنية اشياء فنلقى اربعة وستّين مالاً من الناحيتين فيبقى ستّة وثلثون مالاً تعادل ثمنية اشياء فنقسم ستّة وثلثين مالاً على شى ء فتكون ستّة وثلثين شيئاً ونقسم ثمنية اشياء على شى ء فتكون ثمنية آحاد فالثمنية الآحاد تعادل الستّة والثلثين الشى ء فالشى ء جزءان من تسعة وكنّا فرضنا ضلع المربّع شيئاً فالمربّع ⟨اربعة⟩ اجزاء من واحد وثمنين جزءاً من الواحد وهو العدد المربّع والعدد المكعّب اربعة وستّون احداً والعدد الذى يحيطان به مائتان وستّة وخمسون جزءاً من واحد وثمنين

2345 (702): وثلثين deest in cod.

2348-2349 (703): Ab hac propositione usque ad finem codicis dantur in conclusionibus problematum valores quaesitorum numerorum plerumque asyndetôs.

2355 (704): فنحتاج : محعل (sc. فنجعل) in cod.

2356 (705): اشياء addidi.

(706): Pronomen addidi.

2362 (707): اربعة deest in cod.

2364 (708): Loco به praebet codex سع .

جزءاً من الواحد فاذا زدنا عليه ضلعه وهو ستّة عشر جزءاً من تسعة اعنى مائة واربعة واربعين جزءاً من واحد وثمنين فكان المجتمع من ذلك اربع مائة جزء من واحد وثمنين وهو عدد مربّع وضلعه عشرون جزءاً من تسعة

فقد وجدنا عددين على التحديد الذى حُدّ لنا وهما اربعة وستّون احداً اربعة اجزاء من واحد وثمنين جزءاً من واحد وذلك ما اردنا ان نجد

---

ط نريد ان نجد عددين احدهما مكعّب والآخر مربّع ولكنهما يحيطان بعدد اذا نُقص منه ضلعه كان الباقى منه مربّعاً

فانّا نفرض العدد المكعّب اربعة وستّين احداً والعدد المربّع مالاً فالعدد الذى يحيطان به اربعة وستّون مالاً ولكنّا اذا نقصنا منه ضلعه تبقى اربعة وستّون مالاً الّا ثمنية اشياء فنحتاج /ان تكون مربّعاً ١٠٦
فلنفرض ضلعه ما شئنا من الاشياء ولكن يكون اقلّ من ثمنية اشياء فكأنّا فرضناه سبعة اشياء وضربناها فى مثلها فكانت تسعة واربعين مالاً فهى تعادل اربعة وستّين مالاً الّا ثمنية اشياء فنجبر ونقابل فيكون خمسة عشر مالاً تعادل ثمنية اشياء فنقسم ذلك ⟨على⟩ شىء فيكون خمسة عشر شيئاً تعادل ثمنية آحاد فالشىء ثمنية اجزاء من خمسة عشر جزءاً من الواحد وكنّا فرضنا ضلع العدد المربّع شيئاً فالعدد المربّع اربعة وستّون جزءاً من مأتين وخمسة وعشرين جزءاً من الواحد ولمّا كان العدد ⟨المكعّب اربعة وستّين احداً كان العدد⟩ الذى يحيطان به اربعة الف وستّة وتسعين جزءاً من مأتين وخمسة وعشرين جزءاً فاذا نقصنا منه ضلعه وهو اربعة وستّون جزءاً من خمسة عشر اعنى تسع مائة وستّين جزءاً من مأتين وخمسة وعشرين كان الباقى ثلثة الف ومائة وستّة وثلثين جزءاً من مأتين وخمسة وعشرين وهو عدد مربّع وضلعه ستّة وخمسون جزءاً من خمسة عشر

---

2366 (709): واربعة : واربع in cod.
(710): فكان : وكان in cod.
2367 (711): جزء : حرا in cod.
2378 (712): وضربناها : وصرساها in cod.
2380 (713): Deficiens على restitui.
2384 (714): Per homoeoteleuton omissum addidi.
2389 (715): خمسة عشر : سه عسر in cod.

فقد وجدنا عددين على التحديد الذى حُدّ لنا وهما اربعة وستّون احداً وثمنية اجزاء من خمسة عشر جزءاً من الواحد [اعنى اربعة وستّين جزءاً من مأتين وخمسة وعشرين] وذلك ما اردنا ان نجد

ى نريد ان نجد عددين احدهما مكعّب والآخر مربّع اذا نُقص العدد الذى يحيطان به من ضلعه كان الباقى عدداً مربّعاً

فلنفرض العدد المكعّب اربعة وستّين احداً والعدد المربّع مالاً فالعدد الذى يحيطان به اربعة وستّون مالاً فاذا نقصنا اربعة وستّين مالاً من ضلع اربعة وستّين مالاً وهو ثمنية اشياء كان الباقى ثمنية اشياء الّا اربعة وستّين مالاً ولكنّا نحتاج ان يكون مربّعاً فنجعل ضلعه كم شئنا من الاشياء فكأنّا جعلناه اربعة اشياء فيكون ستّة عشر مالاً تعادل ثمنية اشياء الّا اربعة وستّين مالاً فنجبر ونقابل فيكون ثمنية اشياء / تعادل ثمنين مالاً فنقسم الناحيتين على شىء فيكون ثمنية آحاد تعادل ثمنين شيئاً فالشىء جزء من عشرة وكنّا فرضنا ضلع المربّع شيئاً فيكون المربّع جزءاً من مائة جزء من الواحد والعدد المكعّب اربعة وستّون احداً فالعدد الذى يحيطان به اربعة وستّون جزءاً من مائة جزء ولكنّا اذا نقصناه من ضلعه وهو ثمنية اجزاء من عشرة اعنى ثمنين جزءاً من مائة كان الباقى ستّة عشر جزءاً من مائة وهو عدد مربّع ضلعه اربعة اجزاء من عشرة ٠٧

فقد وجدنا عددين على التحديد الذى حُدّ لنا وهما اربعة وستّون احداً وجزء من مائة جزء من الواحد وذلك ما اردنا ان نجد

يا نريد ان نجد عدداً مكعّباً اذا زدناه على مربّعه كان الذى يجتمع عدداً مربّعاً

فانّا نفرض العدد المكعّب من ضلع يكون شيئاً فيكون العدد المكعّب كعباً ولكن اذا زدناه على مربّعه وهو كعب كعب يكون المجتمع كعب كعب

---

2391-2392 (716): In uncis seclusa verba manifeste e margine in textum irrepserunt.

2403 (717): وستّون : وسىس in cod.

2405 (718): اجزاء : ادا in cod.

2408 (719): وستّون : وسىس in cod.

2409 (720): احداً : حرا (sc. جزءاً) in cod.

2412 (721): Pro شيئاً scripsit librarius سىا ; scriptura vero سا uti solet.

وكعباً فنحتاج ان يكون مربّعاً فلنفرض ضلعه من عدد 〈كعاب اذا نقصنا من مربّعها كعب كعب واحد كان الباقى مكعّباً وهو〉 يكون ثلثة كعاب فيكون اذا نقصنا من مربّع الثلثة كعاب كعبَ كعب كان الباقى منه ثمنية كعاب كعب وهو عدد مكعّب فاذا قابلنا به عدداً مكعّباً خرجت المسئلة ولم يستحل العمل فلنضرب الثلثة الكعاب فى مثلها فتكون تسعةَ كعاب كعب فهى تعادل كعب كعب وكعباً فنلقى الكعب كعب المشترك فيبقى ثمنية كعاب كعب تعادل كعباً فنقسم الناحيتين على كعب فيخرج ثمنية كعاب تعادل واحداً فالكعب ثُمن واحد وهو جزء من ثمنية فاذا زدنا عليه مربّعه وهو جزء من اربعة وستّين جزءاً من الواحد كان المجتمع تسعة اجزاء من اربعة وستّين جزءاً من الواحد وهو عدد مربّع وضلعه ثلثة اجزاء من ثمنية

فقد وجدنا عدداً على التحديد الذى حُدّ لنا وهو جزء من ثمنية اجزاء من الواحد وذلك ما اردنا ان نجد

يب / نريد ان نجد عددين مربّعين تكون قسمة الاعظم منهما على ٠٨
الاصفر اذا زيدت على الاعظم كان المجتمع مربّعاً وان زيدت ايضاً على الاصفر كان 〈المجتمع〉 مربّعاً

فلنفرض العدد الاصفر مالاً ونجعل قسمة الاعظم على الاصفر نصفَ مال ونصف ثُمن مال فيكون اذا زدناه على مال كان المجتمع مربّعاً ويكون العدد الاعظم نصف مال مال ونصف ثُمن مال مال فاذا زدنا عليه نصف مال ونصف ثُمن مال يكون نصفَ مال مال ونصف ثُمن مال مال ونصف مال ونصف ثُمن مال ولكن نحتاج ان يكون عدداً مربّعاً فلنطلب عدداً مربّعاً اذا نقصنا منه نصفاً ونصفَ ثُمن كان الباقى منه عدداً مربّعاً ولنتعمد ان يكون المربّع الذى يبقى اقلّ من واحد 〈وثمنين جزءاً من مأتين وستّة وخمسين جزءاً من الواحد〉 ووِجْدَ ان ذلك سهل على ما تقدّم بيانه فى المقالة الثانية فيكون ذلك مائة وتسعة وستّين جزءاً من

---

2414 (722): وكعباً : وكعب in cod.

2414-2415 (723): In uncis seclusa verba addidi, sed locum, ut opinor, non sanavi.

2416 (724): فيكون : نكون in cod.

2428 (725): زيدت (prius): زيد in cod.

2429 (726): المجتمع addidi.

2436-2437 (727): Velut per homoeoteleuton omissum addidi.

مأتين وستّة وخمسين جزءاً من الواحد وضلعه ثلثة عشر جزءاً من ستّة عشر جزءاً من الواحد وقد علمنا انّا متى نقصنا من مائة وتسعة وستّين جزءاً من مأتين وستّة وخمسين جزءاً من الواحد نصفَ واحد ونصف ثُمن واحد اعنى مائة واربعة واربعين جزءاً من مأتين وستّة وخمسين جزءاً كان الباقى خمسة وعشرين جزءاً من مأتين وستّة وخمسين جزءاً من الواحد وذلك عدد مربّع ضلعه خمسة اجزاء من ستّة عشر جزءاً فلنفرض ضلع نصفِ مال مال ونصف ثُمن مال مال ونصف مال ونصف ثُمن مال ثلثة عشر جزءاً من ستّة عشر جزءاً من مال ونضربه فى مثله فيكون مائة وتسعة وستّين جزءاً من مأتين وستّة وخمسين جزءاً من مال مال فهو يعادل نصف مال مال ونصف ثُمن مال مال ونصف مال ونصف ثُمن مال فلنلق النصف مال مال والنصف ثُمن مال مال المشتركة فيبقى خمسة وعشرون جزءاً من مأتين وستّة وخمسين جزءاً /من مال مال تعادل نصف مال ونصف ثُمن مال ولنضرب كلّ ما معنا فى عشرة وستّة اجزاء من خمسة وعشرين فيكون مال مال يعادل خمسة اموال وتسعة عشر جزءاً من خمسة وعشرين جزءاً من مال فنقسم الناحيتين على مال فيخرج مال يعادل خمسة آحاد وتسعة عشر جزءاً من خمسة وعشرين جزءاً من الواحد وقد كنّا فرضنا العدد الاصغر مالاً فهو خمسة آحاد وتسعة عشر جزءاً من خمسة وعشرين جزءاً من الواحد فلنضرب ذلك فى خمسة وعشرين فيكون مائة واربعة واربعين جزءاً من خمسة وعشرين جزءاً ولمّا فرضنا العدد الاعظم نصف مال مال ونصف ثُمن مال مال علمنا انّه احد عشر الف جزء وستّمائة واربعة وستّون جزءاً من ستّمائة وخمسة وعشرين جزءاً من الواحد فلنجعل المائة والاربعة والاربعين الجزء من خمسة وعشرين جزءاً التى هى المربّع الاصغر اجزاء من ستّمائة وخمسة وعشرين وذلك ان نضربها فى خمسة وعشرين فيكون المربّع الاصغر ثلثة الف وستّمائة جزء من ستّمائة وخمسة وعشرين وقسمة المربّع الاعظم على المربّع الاصغر ثلثة آحاد وستّة اجزاء من خمسة وعشرين من الواحد فلنجعل ذلك اجزاء من ستّمائة وخمسة ٠٩

---

2440 (728): Post جزءاً iterantur verba من ستّة عشر جزءاً in codice per dittographiam.

2448 (729): فلنلق : فلىلعى in cod.

2449 (730): المشتركة : المسرك in codice, quod dubitanter correxi.

2451 (731): كلّ ما : كلما in cod.

2460 (732): Post المربّع praebet codex و.

2461 (733): Post ستّمائة praebet codex وحمسه bis.

وعشرين فيكون الفى جزء وخمسة وعشرين جزءاً من ستّمائة وخمسة وعشرين فاذا زدناه على المربّع الاعظم الذى هو احد عشر الفاً وستّمائة واربعة وستّون جزءاً من ستّمائة وخمسة وعشرين فيكون ثلثة عشر الفاً وستّمائة وتسعة وثمنين جزءاً من ستّمائة وخمسة وعشرين من الواحد وذلك عدد مربّع وضلعه مائة وسبعة عشر جزءاً من خمسة وعشرين / جزءاً وايضاً فلنزد ١٠
الالفين والخمسة والعشرين الجزء من ستّمائة وخمسة وعشرين ⟨على المربّع الاصغر الذى هو ثلثة الف وستّمائة جزء من ستّمائة وخمسة وعشرين فيكون المجتمع خمسة الف وستّمائة وخمسة وعشرين جزءاً من ستّمائة وخمسة وعشرين⟩ وذلك عدد مربّع ضلعه خمسة وسبعون جزءاً من خمسة وعشرين

فقد وجدنا عددين على التحديد الذى حُدّ لنا وهما احد عشر الف جزء وستّمائة جزء واربعة وستّون جزءاً من ستّمائة وخمسة وعشرين جزءاً من الواحد ثلثة الف جزء وستّمائة جزء من ستّمائة وخمسة وعشرين جزءاً من الواحد وذلك ما اردنا ان نجد

يج نريد ان نجد عددين مربّعين يكون ما يخرج من قسمة الاعظم منهما على الاصغر اذا نُقص من كلّ واحد منهما كان الباقى مربّعاً

فلنفرض ضلع المربّع الاصغر شيئاً فيكون المربّع الاصغر مالاً ونجعل ما يخرج من قسمة المربّع الاعظم على المربّع الاصغر الذى هو مال ما اذا نقصنا⟨ه⟩ من مال كان الباقى مربّعاً ونحتاج ايضاً ان يكون الذى ننقصه من المال مربّعاً فلنقسم المال بقسمين مربّعين وذلك ستّة عشر جزءاً من خمسة وعشرين جزءاً من مال وتسعة اجزاء من خمسة وعشرين جزءاً من مال فلنجعل التسعة الاجزاء من الخمسة والعشرين جزءاً من مال ما يخرج من قسمة المربّع الاعظم على مال فنضرب التسعة الاجزاء من خمسة وعشرين جزءاً من مال فى مال فتكون تسعة اجزاء من خمسة وعشرين جزءاً من مال مال وذلك العدد الاعظم ومن البيّن انّا اذا نقصنا ما يخرج من قسمة العدد الاعظم الذى هو تسعة اجزاء من خمسة

---

2465 (734): Verba فيكون الفى جزء وخمسة وعشرين iterantur in codice per dittographiam.

2469 (735): وسبعة : وسعه in cod.

2470-2473 (736): Per homoeoteleuton omissa restitui.

2475 (737): جزء (posterius): حروا (sc. جزءاً ) in cod.

2482 (738): Affixum pronomen addidi.

2488 (739): انّا eadem (ut videtur) manu supra lineam in cod.

وعشرين جزءاً من مال مال على العدد الاصغر الذى هو مال كان الذى يخرج من القَسم تسعة اجزاء من خمسة وعشرين جزءاً من مال فاذا نقصناه من العدد الاصغر اعنى مالاً كان الباقى منه ستّة عشر جزءاً من خمسة وعشرين جزءاً من/مال وهو مربّع ضلعه اربعه اخماس شىء ولكن نريد ١١ ان ننقص تسعة اجزاء من خمسة وعشرين جزءاً من مال من العدد الاعظم وهو تسعة اجزاء من خمسة وعشرين جزءاً من مال مال فيبقى عدد مربّع ولكنّا اذا نقصنا تسعة اجزاء من خمسة وعشرين جزءاً ⟨من مال من تسعة اجزاء من خمسة وعشرين جزءاً⟩ من مال مال يكون الباقى تسعة اجزاء من خمسة وعشرين جزءاً من مال مال الّا تسعة اجزاء من خمسة وعشرين جزءاً من مال ⟨وذلك⟩ يعادل عدداً مربّعاً فلنطلب عدداً مربّعاً اذا نقصناه من تسعة اجزاء من خمسة وعشرين كان الباقى عدداً مربّعاً وذلك واحد وثمنون جزءاً من ستّمائة وخمسة وعشرين جزءاً من الواحد ⟨فاذا نقصناه من تسعة اجزاء من خمسة وعشرين اعنى مأتين وخمسة وعشرين جزءاً من ستّمائة وخمسة وعشرين جزءاً من الواحد⟩ كان الباقى مائة واربعة واربعين جزءاً من ستّمائة وخمسة وعشرين وهو عدد مربّع ضلعه اثنا عشر جزءاً من خمسة وعشرين جزءاً من الواحد فاذ قد صرنا الى ما طلبنا فلنجعل جذر تسعة اجزاء من خمسة وعشرين جزءاً من مال مال الّا تسعة اجزاء من خمسة وعشرين جزءاً من مال ⟨تسعةَ اجزاء من خمسة وعشرين جزءاً من مال⟩ ونضربه فى مثله فيكون واحداً وثمنين جزءاً من ستّمائة وخمسة وعشرين جزءاً من مال مال فهو يعادل تسعة اجزاء من خمسة وعشرين جزءاً من مال مال الّا تسعة اجزاء من خمسة وعشرين جزءاً من مال اعنى مأتين وخمسة وعشرين جزءاً من ستّمائة وخمسة وعشرين جزءاً من مال مال الّا مأتين وخمسة وعشرين جزءاً من ستّمائة وخمسة وعشرين جزءاً من مال فنجبر ونقابل ونلقى المشتركات فيبقى مائة واربعة

---

2496-2497 (740): Per homoeoteleuton omissa addidi.

2499 (741): وذلك deest in cod.

2501 (742): جزءاً (prius): احرا in cod.

2501-2503 (743): Deficiens per homoeoteleuton addidi.

2507-2508 (744): Per homoeoteleuton omissum restitui.

2508 (745): واحداً : واحد in cod.

2511 (746): Post مال praebet codex verba اعنى ماىس وحمسه وعسرس حرا مں مال .

واربعون جزءًا من ستّمائة وخمسة وعشرين جزءًا من مال مال تعادل مأتين
وخمسة وعشرين جزءًا ⟨من ستّمائة وخمسة وعشرين جزءًا⟩ من مال فأُقسمْ
الناحيتين على مال فيخرج مائة واربعة واربعون جزءًا /من ستّمائة وخمسة ١٢
وعشرين جزءًا من مال تعادل مأتين وخمسة وعشرين جزءًا من ستّمائة
وخمسة وعشرين جزءًا من واحد والمال يعادل واحداً وأحد وثمنين جزءًا
من مائة واربعة واربعين جزءًا من الواحد اعنى واحداً وتسعة اجزاء من
ستّة عشر جزءًا وقد كنّا فرضنا المربّع الاصفر مالاً فهو خمسة وعشرون
جزءًا من ستّة عشر جزءًا من الواحد والعدد الاعظم تسعة اجزاء من
خمسة وعشرين جزءًا من مربّع العدد الاصفر وذلك مائتان وخمسة
وعشرون جزءًا من مأتين وستّة وخمسين جزءًا من الواحد وقسمة العدد
الاعظم الذى هو مائتان وخمسة وعشرون جزءًا من مأتين وستّة وخمسين
جزءًا من واحد على العدد الاصفر وهو خمسة وعشرون جزءًا من ستّة
عشر اعنى اربع مائة جزء من مأتين وستّة وخمسين جزءًا من واحد كان
الذى يخرج نصفَ واحد ونصف ثُمن واحد اعنى مائة واربعة واربعين
جزءًا من مائتين وستّة وخمسين جزءًا فاذا نقصناه من احد المربّعين
اعنى من الاربع مائة جزء من مأتين وستّة وخمسين جزءًا من الواحد كان
الباقى مأتين وستّة وخمسين جزءًا من مأتين وستّة وخمسين جزءًا اعنى
واحداً وهو مربّع ضلعه واحد فاذا نقصنا ايضاً ما خرج من القَسم وهو
مائة واربعة واربعون جزءًا من مأتين وستّة وخمسين جزءًا من الواحد من
المربّع الذى هو مائتان وخمسة وعشرون جزءًا من مأتين وستّة وخمسين
كان الباقى واحداً وثمنين جزءًا من مأتين وستّة وخمسين وهو مربّع
وضلعه تسعة اجزاء من ستّة عشر

فقد وجدنا عددين على التحديد الذى حُدّ لنا وهما اربع مائة جزء
من مأتين وستّة وخمسين جزءًا /من واحد مائتان وخمسة وعشرون جزءًا ١٣
من مأتين وستّة وخمسين جزءًا من الواحد وذلك ما اردنا ان نجد

وقد كان غرضُنا فى هذه المسئلة ان يكون العدد المقسوم هو العدد
الاكبر فانتهى بنا العمل الى ان يكون العدد الاكبر هو المقسوم عليه

---

2515 (747): Per homoeoteleuton omissum addidi.

2516 (748): واربعون : واربعس in cod.

2522-2523 (749): مائتان وخمسة وعشرون : ماىس وحمسه وعسرس in cod.

2534 (750): من : مس in cod.

ولمّا كان عملُنا صحيحاً لا ريب فيه اثبتناه ونعمل هذه المسئلة عملاً ثانياً يؤدّى الى ما قصدنا اليه من قسمة المربّع الاعظم على المربّع الاصغر وليكن عملاً هو اقرب من العمل المتقدّم

فنفرض المربّع الاصغر من ضلع يكون واحداً وثلثى واحد فيكون المربّع احدين وسبعة اتساع احد ونفرض المربّع الاعظم من ضلع يكون شيئاً فيكون المربّع الاعظم مالاً ونقسم المربّع الاعظم وهو مال على المربّع الاصغر وهو احدان وسبعة اتساع احد فيخرج القِسم تسعةَ اجزاء من خمسة وعشرين جزءاً من مال فاذا نقصناه من المربّع الاعظم وهو مال كان الباقى منه ستّة عشر جزءاً من خمسة وعشرين جزءاً من مال وهو عدد مربّع ضلعه اربعة اخماس شىء وايضاً فننقص ما خرج من القَسم وهو تسعة اجزاء من خمسة وعشرين جزءاً من مال من المربّع الاصغر وهو احدان وسبعة اتساع احد فيبقى منه احدان وسبعة اتساع احد الّا تسعة اجزاء من خمسة وعشرين جزءاً من مال فنحتاج ان يكون مربّعاً فنفرض ضلعه واحداً وثلثى واحد الّا شيئاً وخُمسَ شىء ونضربه فى مثله فيكون احدين وسبعة اتساع احد ومالاً وأحد عشر جزءاً من خمسة وعشرين جزءاً من مال الّا اربعة اشياء فهو يعادل احدين وسبعة اتساع احد الّا تسعة اجزاء من خمسة وعشرين جزءاً من مال فأُجبرْ / ١٤ ، كلّ ناحية من الناحيتين بما نُقص منها وزدْ مثله على الناحية الاخرى وأُلقِ المتشابهات المشتركات فيبقى مال واربعة اخماس مال تعادل اربعة اشياء فأقسمْ كلى الناحيتين على شىء فيكون شىء واربعة اخماس شىء تعادل اربعة آحاد فالشىء الواحد احدان وتُسعان وقد كان ضلع المربّع الاعظم شيئاً فضلعه احدان وتُسعان والمربّع الاعظم اربع مائة جزء من واحد وثمنين جزءاً من واحد فاذا قسمناه على المربّع الاصغر الذى هو احدان وسبعة اتساع احد اعنى مأتين وخمسة وعشرين جزءاً من واحد وثمنين جزءاً من الواحد كان الذى يخرج من القَسم واحداً وسبعة اتساع اعنى مائة واربعة واربعين جزءاً من واحد وثمنين جزءاً فاذا نقصناه من المربّع الاعظم الذى هو اربع مائة جزء من واحد وثمنين جزءاً يكون الباقى مأتين وستّة وخمسين جزءاً ⟨من واحد وثمنين جزءاً⟩ وهو عدد مربّع ضلعه ستّة عشر جزءاً من تسعة واذا

---

2541 (751): لا ريب فيه اثبتناه conjecturâ mea; lectio enim codicis talis est: لاں سـ فىه اسىاه.

2568-2569 (752): Denominatorem addidi.

نقصناه من المربّع الاصغر الذى هو مائتان وخمسة وعشرون جزءاً من واحد وثمنين جزءاً يكون الباقى واحداً وثمنين جزءاً من واحد وثمنين اعنى واحداً وهو مربّع ضلعه واحد

فقد وجدنا عددين على التحديد الذى حُدّ لنا وهما اربع مائة جزء من واحد وثمنين جزءاً من الواحد مائتان وخمسة وعشرون جزءاً من احد وثمنين جزءاً من الواحد وذلك ما اردنا ان نجد

يد نريد ان نجد عددين مربّعين اذا قُسم الاعظم منهما على الاصفر يخرج من القَسم ما اذا نقصنا منه المربّع الاعظم كان الباقى مربّعاً وان نقصنا منه ايضاً المربّع الاصغر كان الباقى منه مربّعاً

فلنفرض ضلع المربّع الاعظم شيئاً فيكون المربّع الاعظم مالاً ونفرض ايضاً ضلع / المربّع الاصغر اربعة اخماس واحد فيكون المربّع الاصغر ١١٥ ستّة عشر جزءاً من خمسة وعشرين جزءاً من واحد ومن البيّن انّا اذا قسمنا المربّع الاعظم وهو مال على المربّع الاصغر وهو ستّة عشر جزءاً من خمسة وعشرين جزءاً من واحد كان الذى يخرج من القَسم مالاً وتسعة اجزاء من ستّة عشر جزءاً من مال واذا نقصنا منه المربّع الاعظم وهو مال يكون الباقى تسعة اجزاء من ستّة عشر جزءاً من مال وهو مربّع ضلعه ثلثة ارباع شىء ولكنّا نلقى منه المربّع الاصغر وهو ستّة عشر جزءاً من خمسة وعشرين جزءاً من الواحد فتبقى خمسة وعشرون جزءاً من ستّة عشر جزءاً من مال الّا ستّة عشر جزءاً من خمسة وعشرين جزءاً من واحد فنحتاج ان تكون مربّعاً فلنفرض له ضلعاً يكون شيئاً ورُبعَ شىء الّا احدين ونضربه في مثله فيكون خمسة وعشرين جزءاً من ستّة عشر جزءاً من مال واربعة آحاد الّا خمسة اشياء فذلك يعادل خمسة وعشرين جزءاً من ستّة عشر جزءاً من مال الّا ستّة عشر جزءاً من خمسة وعشرين جزءاً من واحد فنجبر كلّ واحدة من الناحيتين بما نُقص منها ونزيد مثله على الناحية الاخرى ونلقى المشتركات فيبقى خمسة اشياء تعادل اربعة آحاد وستّة عشر جزءاً من خمسة وعشرين جزءاً من واحد فالشىء الواحد خُمس اربعة آحاد وستّة عشر من خمسة وعشرين جزءاً من واحد وذلك مائة وستّة عشر جزءاً ⟨من⟩ مائة وخمسة وعشرين جزءاً من الواحد وكنّا فرضنا ضلع المربّع

---

2571 (753): واحداً وثمنين : واحد وسوں in cod.

2586 (754): Pro وكنا codicis ولكنّا scripsi.

2597 (755): من addidi.

(756): وخمسة وعشرين : وحمسه عسر in cod.

الاعظم شيئاً فضلعه مائة وستّة عشر جزءاً من مائة وخمسة وعشرين جزءاً من الواحد فالمربّع اذاً ثلثة عشر الفاً واربع مائة وستّة وخمسون جزءاً من خمسة عشر الفاً وستّمائة وخمسة وعشرين فاذا قسمناه على المربّع الاصغر/ وهو ستّة عشر جزءاً من خمسة وعشرين اعنى عشرة الٰف جزء من خمسة عشر ١٦ الفاً وستّمائة وخمسة وعشرين فيكون واحداً وثلثة الٰف جزء واربع مائة وستّة وخمسين جزءاً من عشرة الٰف اعنى واحداً وعشرين الفاً وخمسة وعشرين جزءاً من خمسة عشر الفاً وستّمائة وخمسة وعشرين فاذا نقصنا منه المربّع الاعظم وهو ثلثة عشر الفاً واربع مائة وستّة وخمسون جزءاً كان الباقى منه سبعة الٰف وخمس مائة وتسعة وستّين جزءاً من خمسة عشر الفاً وستّمائة وخمسة وعشرين وذلك عدد مربّع ضلعه سبعة وثمنون جزءاً من مائة وخمسة وعشرين جزءاً وايضاً فانّا اذا نقصنا منه العدد الاصغر وهو عشرة الٰف جزء كان الباقى احد عشر الف جزء وخمسة وعشرين جزءاً من خمسة عشر الفاً وستّمائة وخمسة وعشرين وذلك عدد مربّع ضلعه مائة جزء وخمسة اجزاء من مائة وخمسة وعشرين جزءاً من الواحد

فقد وجدنا عددين على التحديد الذى حُدّ لنا وهما ثلثة عشر الفاً واربع مائة وستّة وخمسون جزءاً من خمسة عشر الفاً وستّمائة وخمسة وعشرين جزءاً من الواحد عشرة الٰف جزء من خمسة عشر الفاً وستّمائة وخمسة وعشرين جزءاً من الواحد وذلك ما اردنا ان نجد

يهَ نريد ان نجد عددين مربّعين اذا زيد فضل الاعظم منهما على الاصغر على كلّ واحد منهما كان المجتمع من ذلك مربّعاً

فلنفرض ضلع المربّع الاعظم شيئاً فيكون المربّع الاعظم مالاً ونجعل زيادته على المربّع الاصغر شيئين وواحداً فيكون العدد الاصغر مالاً الّا شيئين وواحداً وبيّن انّا اذا زدنا زيادة الاعظم من العددين ⟨على الاصغر⟩ وذلك شيئان وواحد على الاصغر وهو مال / الّا شيئين ١٧

---

2599 (757): وخمسون : وحمسس in cod.

2605 (758): وخمسون : وحمسس in cod.

2607-2608 (759): Loco جزءاً من praebet codex errore nescio quo سىا (pro شيئاً ?).

2612 (760): وهما : وهو in cod.

2619 (761): وواحداً : وواحد in cod.

2620 (762): اذا زدنا : ادردىا (juncta, ut videtur).

2620-2621 (763): Omissa verba على الاصغر restitui.

وواحداً كان المجتمع من ذلك مالاً وهو العدد المربّع الاعظم وهو مربّع ولكنّا اذا زدنا زيادة المربّع الاعظم على المربّع الاصغر وهي شيئان وواحد على المربّع الاعظم وهو مال كان مالاً وشيئين وواحداً وذلك عدد مربّع وضلعه شیء وواحد فنحتاج ان يكون العدد الاصغر الذى هو مال الّا شيئين وواحداً مربّعاً فلنفرض ضلعه شيئاً الّا احدين ونضربه فى مثله فيكون مالاً واربعة آحاد الّا اربعة اشياء فذلك يعادل مالاً الّا شيئين وواحداً فنزيد ⟨اربعة اشياء⟩ على مال واربعة آحاد الّا اربعة اشياء فتكون مالاً واربعة آحاد ونزيد ايضاً اربعة اشياء على مال الّا شيئين وواحداً فتكون مالاً وشيئين الّا واحداً فنزيد واحداً على الناحيتين جميعاً ونلقى المشترك ليبقى نوع واحد يعادل نوعاً ⟨واحداً⟩ فيبقى خمسة آحاد تعادل شيئين فالشیء الواحد احدان ونصف وكنّا فرضنا ضلع المربّع الاعظم شيئاً فضلعه اثنان ونصف والمربّع الاعظم ستّة آحاد ورُبع فاذا القينا منه شيئين وواحداً وذلك ستّة آحاد كان الباقى منه رُبع واحد وهو المربّع الاصغر ومن البيّن ان زيادة المربّع الاعظم وهو ستّة آحاد ورُبع واحد على المربّع الاصغر وهو رُبع واحد هى ستّة آحاد فاذا زيدت على المربّع الاعظم كان المجتمع من ذلك اثنى عشر ورُبعاً وذلك عدد مربّع ضلعه ثلثة ونصف وان زيدت ايضاً على المربّع الاصغر كان المجتمع من ذلك ستّة ورُبعاً وذلك عدد مربّع وضلعه اثنان ونصف

فقد وجدنا عددين على التحديد الذى حُدّ لنا وهما ستّة آحاد ورُبع واحد ورُبع واحد وذلك ما اردنا ان نجد

يو نريد ان نجد عددين مربّعين تكون زيادة الاعظم منهما على الاصغر اذا نُقصت من الاعظم كان الباقى /منه عدداً مربّعاً وان نُقصت ايضاً من ١٨ الاصغر كان الباقى عدداً مربّعاً

فلنفرض ضلع المربّع الاعظم شيئاً فيكون المربّع الاعظم مالاً ونجعل زيادته على المربّع الاصغر شيئين الّا واحداً فيكون المربّع الاصغر مالاً

2622 (764): وواحداً : وواحد in cod.

2623 (765): وهي : وهو in cod.

2624 (766): وواحداً : وواحد in cod.

2628 (767): اربعة اشياء addidi.

2632 (768): واحداً addidi.

2635 (769): رُبع : مربع in cod.

وواحداً الّا شيئين وبيّن انّا اذا نقصنا زيادة المربّع الاعظم على المربّع الاصفر وهى شيئان الّا واحداً من المربّع الاعظم وهو مال ان الباقى مال وواحد الّا شيئين وهو المربّع الاصفر وهو ⟨مربّع⟩ ولكن اذا نقصنا زيادة المربّع الاعظم على المربّع الاصفر وهى شيئان الّا واحداً من المربّع الاصفر وهو مال وواحد الّا شيئين كان الباقى منه مالاً وأحدين الّا اربعة اشياء فنحتاج ان يكون مربّعاً فلنفرض ضلعه شيئاً الّا اربعة آحاد ونضربه فى مثله فيكون مالاً وستّة عشر احداً الّا ثمنية اشياء فهو يعادل مالاً وأحدين الّا اربعة اشياء فنزيد ثمنية اشياء على الناحيتين جميعاً ونلقى المال والاثنين المشتركة فيبقى اربعة اشياء تعادل اربعة عشر احداً فالشىء الواحد ثلثة آحاد ونصف وقد كنّا فرضنا ضلع المربّع الاعظم شيئاً فضلعه ثلثة آحاد ونصف ومربّعه اثنا عشر احداً ورُبع واحد فنلقى منه مثل جذريه الّا واحداً وذلك ستّة آحاد فيبقى منه ستّة آحاد ورُبع واحد وذلك العدد هو المربّع الاصفر وزيادة المربّع الاعظم على المربّع الاصفر ستّة آحاد فان نقصنا ستّة آحاد من المربّع الاعظم كان الباقى منه ستّة ورُبعاً وذلك عدد مربّع ضلعه احدان ونصف وان نقصنا ايضاً الستّة الآحاد من المربّع الاصفر كان الباقى رُبعَ واحد وذلك عدد مربّع وضلعه نصف واحد

فقد وجدنا عددين على التحديد الذى حُدّ لنا وهما اثنا عشر
احداً ورُبع واحد / ستّة آحاد ورُبع واحد وذلك ما اردنا ان نجد ١٩

يز نريد ان نجد ثلثة اعداد مربّعة اذا جُمعت كان المجتمع منها مربّعاً ويكون الاوّل من هذه الاعداد مساوياً لضلع الثانى والثانى مساوياً لضلع الثالث

فلنفرض الاوّل مالاً فيكون الثانى مالَ مال [لانّه مربّع المال] [والمال مساوٍ لضلع الثانى] ويكون الثالث مال مال مال مال [وهو مثل مربّع الثانى

---

2650 (770): Loco مال وواحد الّا شيئين وهو المربّع الاصفر وهو ⟨مربّع⟩ habet codex verba وهو المربع الاصعر وهو مال وواحد الا سىس . Forsan sunt وهو المربّع الاصفر و interpolata; vide casum similem in adn. 927.

2670-2671 (771): Verba والمال مساوٍ لضلع الثانى (et fortasse etiam praecedentia ab لانّه) interpolata esse videntur. Pro مساوٍ praebet codcx مساوى , pro الثانى autem المال .

[والثانى ضلعه] وتجتمع الاعداد الثلثة فتكون مالَ مال مال مال ومالَ مال ومالاً ونحتاج ان تكون عدداً مربّعاً فلنفرض له ضلعاً يكون مال مال ونصفَ واحد ونضربه فى مثله فيكون مال مال مال ومال مال ورُبع واحد فهو يعادل مال مال مال ومال مال ومالاً فنلقى المتشابهات المشتركات فيبقى مال يعادل رُبع واحد وكنّا فرضنا العدد الاوّل من الثلثة الاعداد مالاً فهو رُبع واحد وهو مساوٍ لضلع الثانى والثانى نصف ثُمن واحد والثانى ايضاً مساوٍ لضلع الثالث والثالث جزء من مأتين وستّة وخمسين جزءاً من واحد وتُجمع هذه الاعداد الثلثة فتكون واحداً وثمنين جزءاً من مأتين وستّة وخمسين جزءاً من الواحد وهو عدد مربّع وضلعه تسعة اجزاء من ستّة عشر

فقد وجدنا ثلثة اعداد على التحديد الذى حُدّ لنا وهى رُبع واحد نصف ثُمن واحد جزء من مأتين وستّة وخمسين جزءاً من واحد وذلك ما اردنا ان نجد

يح نريد ان نجد ثلثة اعداد مربّعة اذا ضاعفنا العدد الاوّل بالعدد الثانى ثمّ ما اجتمع بالعدد الثالث وزدنا على ما بلغ ذلك العدد المركّب من جملة الثلثة الاعداد كان المجتمع من ذلك عدداً مربّعاً

فلنفرض العدد الاوّل واحداً والثانى /نصفَ واحد ونصفَ ثُمن واحد ٢٠
والثالث مالاً ثمّ نضاعف الاوّل وهو واحد بالثانى وهو تسعة اجزاء من ستّة عشر فيكون تسعة اجزاء من ستّة عشر جزءاً من الواحد فنضاعفه بالعدد الثالث وهو مال فيكون تسعة اجزاء من ستّة عشر جزءاً من مال فنزيد عليه العدد المركّب من جملة الثلثة الاعداد وذلك مال وخمسة وعشرون جزءاً من ستّة عشر جزءاً من واحد فيكون المجتمع من ذلك خمسة وعشرين جزءاً من ستّة عشر جزءاً من مال وخمسة وعشرين جزءاً من ستّة عشر جزءاً من الواحد فنحتاج ان يكون ذلك عدداً مربّعاً فلنفرض له ضلعاً

---

2672 (772): Verba والثانى ضلعه (forsan etiam praecedentia ab وهو in linea 2671) interpolata esse videntur.

2673 (773): ومالاً : ومال in cod.

(774): له : لنا in cod.

2675 (775): ومالاً : ومال in cod.

2675-2676 (776): Melius dixisset interpres المتشابهين المشتركين .

2692 (777): وذلك : ودلك in codice; مدلك primo scriptum statim correxit librarius.

يكون شيئاً ورُبع شيء وربُع واحد ونضرب ذلك في مثله فيكون خمسة وعشرين جزءاً من ستّة عشر جزءاً من ⟨مال وعشرة اجزاء من ستّة عشر جزءاً من⟩ شيء وجزءاً من ستّة عشر جزءاً من واحد فهو يعادل خمسة وعشرين جزءاً من ستّة عشر جزءاً من مال وخمسة وعشرين جزءاً من ستّة عشر جزءاً من واحد فنلقى المتشابهات المشتركات فيبقى اربعة وعشرون جزءاً من ستّة عشر جزءاً من واحد تعادل عشرة اجزاء من ستّة عشر جزءاً من شيء فالشيء الكامل يعادل احدين وخُمسي واحد وقد كنّا فرضنا ضلع العدد الثالث شيئاً فضلعه اثنان وخُمساً احد والعدد الثالث مائة واربعة واربعون جزءاً من خمسة وعشرين جزءاً من الواحد والعدد الاوّل واحد كما كنّا فرضناه والعدد الثاني تسعة اجزاء من ستّة عشر جزءاً من واحد كما كنّا فرضناه فنضاعف العدد الاوّل بالعدد الثاني ثمّ ما اجتمع بالعدد الثالث فيكون واحداً وثمنين جزءاً من خمسة وعشرين جزءاً من الواحد اعنى الفاً ومأتين وستّة وتسعين جزءاً من اربع مائة جزء من الواحد فنزيد عليه العدد المركّب من الثلثة الاعداد وهو مائة واربعة واربعون / جزءاً من خمسة وعشرين جزءاً وواحد وتسعة اجزاء من ١٢١ ستّة عشر جزءاً من الواحد اعنى ألفين وتسع مائة وتسعة وعشرين جزءاً من اربع مائة فيكون المجتمع اربعة ألف جزء ومائتي جزء وخمسة وعشرين جزءاً من اربع مائة جزء من الواحد وذلك عدد مربّع وضلعه خمسة وستّون جزءاً من عشرين جزءاً من الواحد

فقد وجدنا ثلثة اعداد على التحديد الذي حُدّ لنا وهي مائة واربعة واربعون جزءاً من خمسة وعشرين جزءاً من الواحد وواحد وتسعة اجزاء من ستّة عشر جزءاً من الواحد وذلك ما اردنا ان نجد

يط نريد ان نجد ثلثة اعداد مربّعة اذا ضوعف الاوّل بالثاني وما اجتمع بالثالث ونُقص ممّا بلغ العدد المركّب من جملة الثلثة الاعداد كان الباقي من ذلك مربّعاً

2696 (778): ورُبع شيء : وربعا in cod.

2697-2698 (779): Per homoeoteleuton omissa verba restitui.

2705 (780): الثاني : الثالث in cod.

2708 (781): الفاً : الف in cod.

2716 (782): وواحد و addidit librarius in margine et locum lacunae lineamento curvato significavit.

فلنفرض العدد الاوّل واحداً والثانى واحداً وتسعة اجزاء من ستّة عشر جزءاً والثالث مالاً ونضاعف الاوّل بالثانى وما اجتمع بالثالث فيكون مالاً وتسعة اجزاء من ستّة عشر جزءاً من مال فننقص منه العدد المركّب من جملة الثلثة الاعداد وذلك مال وأحدان وتسعة اجزاء من ستّة عشر جزءاً من واحد فيكون الباقى تسعة اجزاء من ستّة عشر جزءاً من مال الّا احدين وتسعة اجزاء من ستّة عشر جزءاً من واحد فنحتاج الى ان يكون مربّعاً فنفرضه من ضلع ثلثة ارباع شىء الّا رُبعَ واحد ونضربه فى مثله فيكون تسعةَ اجزاء من ستّة عشر جزءاً من مال وجزءاً من ستّة عشر جزءاً من واحد الّا ستّة اجزاء من ستّة عشر جزءاً من شىء فذلك يعادل تسعة اجزاء من ستّة عشر جزءاً من مال الّا احدين وتسعة اجزاء من ستّة عشر جزءاً من واحد فنزيد على الناحيتين جميعاً احدين وتسعة اجزاء من ستّة عشر جزءاً من واحد وستّة اجزاء من ستّة عشر جزءاً من / شىء فيكون بعد الزيادة تسعة اجزاء من ستّة عشر جزءاً من مال ٢٢
وستّة اجزاء من ستّة عشر جزءاً من شىء تعادل تسعة اجزاء من ستّة عشر جزءاً من مال وأحدين وعشرة اجزاء من ستّة عشر جزءاً من واحد فنلقى التسعة الاجزاء ⟨من ستّة عشر جزءاً من⟩ المال المشتركة من الناحيتين فيبقى اثنان واربعون جزءاً من ستّة عشر جزءاً من واحد تعادل ستّة اجزاء من ستّة عشر جزءاً من شىء فالشىء يعادل سبعة آحاد وقد كنّا فرضنا ضلع المربّع الثالث شيئاً فهو سبعة آحاد والمربّع الثالث تسعة واربعون احداً والمربّع الاوّل واحد على ما كنّا فرضناه والمربّع الثانى واحد وتسعة اجزاء من ستّة عشر جزءاً من واحد كما كنّا فرضناه فنضاعف المربّع الاوّل بالمربّع الثانى ثمّ ما اجتمع بالمربّع الثالث فيبلغ ذلك ستّة وسبعين احداً وتسعة اجزاء من ستّة عشر جزءاً من واحد فاذا نقصنا منه العدد المركّب من جملة الثلثة الاعداد وذلك واحد وخمسون وتسعة اجزاء من ستّة عشر جزءاً من واحد كان الباقى منه خمسة وعشرين احداً وهو عدد مربّع وضلعه خمسة

فقد وجدنا ثلثة اعداد على التحديد الذى حُدّ لنا وهى تسعة واربعون احداً واحد واحد وتسعة اجزاء من ستّة عشر جزءاً من واحد وذلك ما اردنا ان نجد

---

2736 (783): In uncis seclusa verba addidi.

2741 (784): Pro فنضاعف praebet codex, ut videtur, مصاهف.

ك نريد ان نجد ثلثة اعداد مربّعة اذا ضوعف الاوّل بالثانى وما اجتمع بالثالث ثمّ نقصنا ما بلغ ذلك منْ العدد المركّب من جملة الثلثة الاعداد كان الذى يبقى منه مربّعاً

فلنفرض المربّع الاوّل اربعة آحاد والمربّع الثانى اربعة اجزاء من خمسة وعشرين جزءًا من واحد والمربّع الثالث مالاً ثمّ نضاعف المربّع الاوّل بالمربّع الثانى ثمّ ما اجتمع بالمربّع الثالث فيكون ذلك ستّة عشر جزءًا من خمسة وعشرين جزءًا من مال فلننقصه من العدد /المركّب من ٢٣ جملة الثلثة الاعداد وهو مال واربعة آحاد واربعة اجزاء من خمسة وعشرين جزءًا من واحد فيكون الباقى تسعة اجزاء من خمسة وعشرين من مال واربعة آحاد واربعة اجزاء من خمسة وعشرين جزءًا من واحد فنحتاج ان يكون مربّعاً فلنفرض ضلعه ثلثة اخماس شىء وواحداً ونضربه فى مثله فيكون تسعةَ اجزاء من خمسة وعشرين جزءًا من مال وشيئاً وخُمسَ شىء وواحداً فهو يعادل تسعة اجزاء من خمسة وعشرين جزءًا من مال ومائة جزء واربعة اجزاء من خمسة وعشرين جزءًا من واحد فنلقى التسعة الاجزاء من الخمسة والعشرين جزءًا من مال والواحد المشتركة ليبقى نوع واحد يعادل نوعاً واحداً فيبقى ثلثون جزءًا من خمسة وعشرين جزءًا من شىء تعادل تسعة وسبعين جزءًا من خمسة وعشرين جزءًا من واحد فالشىء الواحد يعادل تسعة وسبعين جزءًا من ثلثين جزءًا من الواحد وقد كنّا فرضنا العدد المربّع ⟨الثالث⟩ مالاً فضلعه تسعة وسبعون جزءًا من ثلثين جزءًا من الواحد ويكون المربّع ستّة الٰف ومأتين وأحد واربعين جزءًا من تسع مائة جزء من الواحد [فهو العدد الثالث] والعدد الاوّل اربعة آحاد كما كنّا فرضناه والعدد الثانى اربعة اجزاء من خمسة وعشرين جزءًا من الواحد على ما فرضناه فاذا ضاعفنا العدد الاوّل اعنى الاربعة الآحاد بالعدد الثانى اعنى الاربعة الاجزاء من الخمسة والعشرين الجزء من الواحد ثمّ ضاعفنا ما اجتمع بالعدد الثالث اعنى ستّة الٰف جزء ومائتى جزء وأحد واربعين جزءًا من تسع مائة جزء من واحد يكون ذلك تسعة وتسعين الفاً وثمان مائة وستّة وخمسين جزءًا

---

2761 (785): وخُمسَ شىء : وخمسا in cod.

2768 (786): الثالث addidi. Vide adn. seq.

2770 (787): Verba فهو العدد الثالث a quodam lectore addita esse censeo.

2774 (788): الجزء من : الحرف in cod.

من اثنين وعشرين الفاً وخمس مائة فاذا نقصناه من العدد المركّب من الثلثة الاعداد وذلك اربعة آحاد واربعة اجزاء من خمسة وعشرين / جزءاً من واحد وستّة الٰف جزء ومائتا ⟨جزء⟩ وأحد واربعون جزءاً من ١٢٤ تسع مائة جزء من واحد اعنى مائتى ⟨الف وتسعة واربعين⟩ الفاً وستّمائة وخمسة وعشرين جزءاً من اثنين وعشرين الفاً وخمس مائة يكون الباقى مائة الف جزء وتسعة واربعين الف جزء وسبع مائة وتسعة وستّين جزءاً ⟨من اثنين وعشرين الفاً وخمس مائة⟩ وهو عدد مربّع وضلعه ثلثمائة جزء وسبعة وثمنون جزءاً من مائة وخمسين جزءاً

فقد وجدنا ثلثة اعداد على التحديد الذى حُدّ لنا وهى اربعة آحاد اربعة اجزاء من خمسة وعشرين جزءاً من الواحد ستّة الٰف جزء ومائتان وأحد واربعون جزءاً من تسع مائة جزء من الواحد وذلك ما اردنا ان نجد

---

كا نريد ان نجد عددين مربّعين اذا زيد على مربّع كلّ واحد منهما العدد المركّب من جملتهما كان المجتمع من ذلك مربّعاً

ولمّا كان كلّ عدد مربّع يزاد عليه مثل ضلعه ورُبع واحد فيكون ذلك مربّعاً وجب ان نفرض احد العددين مالاً فمربّعه مال مال واذا زيد على مال مال مثل ضلعه ورُبع واحد كان ذلك مالَ مال ومالاً ورُبعَ واحد وهو عدد مربّع ضلعه مال ونصف واحد فبيّن ان العدد المركّب من جملة العددين مال ورُبع واحد وقد كنّا فرضنا احد العددين مالاً فالعدد الآخر رُبع واحد ولكنّا اذا زدنا على مربّعه وهو نصف ثُمن واحد العددَ المركّب من جملة العددين وذلك مال ورُبع واحد كان المجتمع من ذلك مالاً وخمسة اجزاء من ستّة عشر جزءاً من واحد فنحتاج

---

2777 (789): اثنين : اىى in cod.

2779 (790): Verbum quod forma praecedentis verbi exigit inserui.

(791): واربعون : واربعىں in cod.

2780 (792): Lacunam explevi.

2782-2783 (793): Denominatorem addidi.

2786 (794): ستّة : ىسه in cod.

2792 (795): فمربّعه : فمں ىعه in cod.

2793 (796): ومالاً : ومال in cod.

2796 (797): فالعدد : والعدد in cod.

ان يكون مربّعاً فنفرض ضلعه شيئاً ونصفَ واحد ونضربه فى مثله فيكون مالاً ⟨وشيئاً وربُع واحد فذلك يعادل مالاً⟩ وخمسة اجزاء من ستّة عشر جزءاً من واحد فنلقى مالاً وربُع واحد من الناحيتين جميعاً فيبقى جزء من ستّة عشر جزءاً من واحد يعادل شيئاً فالشىء جزء من ستّة عشر جزءاً من واحد وكنّا فرضنا احد المربّعين / مالاً فضلعه جزء من ستّة عشر جزءاً ٢٥
من واحد والمربّع جزء من مأتين وستّة وخمسين من واحد والعدد الآخر رُبع واحد كما فرضناه والعدد المركّب من جملتهما خمسة وستّون جزءاً من مأتين وستّة وخمسين جزءاً من واحد فاذا زيد على مربّع احد العددين الذى هو ستّة عشر جزءاً من مأتين وستّة وخمسين من واحد كان ذلك واحداً وثمنين من مأتين وستّة وخمسين جزءاً من واحد وهو عدد مربّع وضلعه تسعة اجزاء من ستّة عشر جزءاً من واحد وايضاً فان زدناه على مربّع العدد الآخر اعنى جزءاً من خمسة وستّين الفاً وخمس مائة وستّة وثلثين كان المجتمع ستّة عشر الفاً وستّمائة وأحد واربعين جزءاً من خمسة وستّين الفاً وخمس مائة وستّة وثلثين وهو عدد مربّع وضلعه مائة وتسعة وعشرون جزءاً من مأتين وستّة وخمسين

فقد وجدنا عددين على التحديد الذى حُدّ لنا وهما رُبع واحد جزء من مأتين وستّة وخمسين جزءاً من واحد وذلك ما اردنا ان نجد

---

كب نريد ان نجد عددين مربّعين اذا جُمعا كانا عدداً مربّعاً واذا ضوعف احدهما بالآخر كان ذلك عدداً مكعّباً

ولمّا كان العدد المكعّب من تضاعيف عددٍ بمثله وايضاً ما اجتمع بمثل ذلك العدد فرضنا احد العددين المربّعين مالاً وضاعفناه بمثله فكان مالَ مال فلنفرض العدد الثانى مال مال ومن البيّن انّا اذا ضاعفنا احد العددين وهو مال بالعدد الآخر وهو مال مال يكون ذلك كعب كعب وهو عدد مكعّب لانّه من عدد ضوعف بمثله وما اجتمع بمثل ذلك العدد ولكنّا اذا جمعنا العددين المربّعين كان المجتمع منهما مال مال ومالاً فنحتاج ان يكون مربّعاً فلنفرض له ضلعاً يكون مالاً ورُبع مال / ونضربه فى مثله فيكون مالَ مال وتسعة اجزاء من ستّة عشر جزءاً من ٢٦

---

2800 (798): Per homoeoteleuton omissa verba addidi.

2808 (799): واحداً : واحد in cod.

2818 (800): Verbum بصاعف ( بصاعف , ut videtur), quod codex post بمثله و inutiliter repetivit, delevi.

2824 (801): ومالاً : ومال in cod.

مال مال فهو يعادل مال مال ومالاً فنلقى المال مال المشترك من الناحيتين فيبقى تسعة اجزاء من ستّة عشر جزءاً من مال مال تعادل مالاً فلنقسم الناحيتين على مال فيخرج تسعة اجزاء من ستّة عشر جزءاً من مال تعادل واحداً فالمال الكامل يعادل ستّة عشر جزءاً من تسعة اجزاء من واحد وقد كنّا فرضنا احد العددين مالاً فأحد العددين ستّة عشر جزءاً من تسعة اجزاء من واحد والعدد الآخر مائتا جزء وستّة وخمسون جزءاً من واحد وثمنين جزءاً من الواحد واذا ضاعفنا ستّة عشر جزءاً من تسعة اجزاء من واحد بمائتى جزء وستّة وخمسين جزءاً من واحد وثمنين جزءاً من واحد كان المجتمع اربعة الف جزء وستّة وتسعين جزءاً من سبع مائة وتسعة وعشرين جزءاً من واحد وذلك عدد مكعّب ضلعه ستّة عشر جزءاً من تسعة اجزاء من واحد وايضاً متى جمعنا العددين المربّعين كان المجتمع منهما اربع مائة جزء من واحد وثمنين وذلك عدد مربّع وضلعه عشرون جزءاً من تسعة

فقد وجدنا عددين على التحديد الذى حُدّ لنا وهما ستّة عشر جزءاً من تسعة مائتا جزء وستّة وخمسون جزءاً من واحد وثمنين وذلك ما اردنا ان نجد

ونريد ان نعمل هذه المسئلة بعمل آخر هو اسهل من العمل الاوّل ونطلب عددين مربّعين اذا جُمعا كان المجتمع منهما مربّعاً وذلك ستّة عشر مالاً وتسعة اموال فنضاعف احدهما بالآخر فيكون ذلك مائة واربعة واربعين مال مال فهو يعادل عدداً مكعّباً فليكن العدد المكعّب ثمنية كعاب فالمائة والاربعة والاربعون المال مال تعادل ثمنية كعاب فنقسم كلى الناحيتين على كعب فيكون مائة واربعة واربعون شيئاً
تعادل /ثمنية آحاد فالشىء الواحد جزء من ثمنية عشر جزءاً من واحد ٢٧
وكنّا فرضنا احد العددين المربّعين تسعة اموال فضلعه ثلثة اشياء وذلك جزء من ستّة اجزاء من واحد فنضربه فى مثله فيكون جزءاً من ستّة وثلثين جزءاً من واحد وهو احد العددين والعدد الآخر فُرض ستّة عشر مالاً وضلعه اربعة اشياء وذلك جزءان من تسعة اجزاء من واحد فنضربه فى مثله فيكون ⟨اربعة⟩ اجزاء من واحد وثمنين جزءاً من واحد

---

2834 (802): وتسعين : وحمسں in codice. Vide lineam 2833.

2842 (803): آخر : احرا in cod.

2847 (804): واربعون : واربعں in cod.

2853 (805): اربعة addidi.

وهو العدد الآخر وبيّن انّا اذا جمعنا العددين المربّعين كان جميعهما خمسة وعشرين جزءًا من ثلثمائة واربعة وعشرين وذلك عدد مربّع وضلعه خمسة اجزاء من ثمنية عشر واذا ضاعفنا احد العددين اعنى جزءًا من ستّة وثلثين جزءًا من واحد بالعدد الآخر اعنى اربعة اجزاء من واحد وثمنين يكون ذلك اربعة اجزاء من ألفين وتسع مائة وستّة عشر اعنى جزءًا واحدًا من سبع مائة وتسعة وعشرين وهو عدد مكعّب وضلعه جزء من تسعة اجزاء من واحد

فقد وجدنا عددين على التحديد الذى حُدّ لنا وهما جزء من ستّة وثلثين جزءًا من واحد اربعة اجزاء من واحد وثمنين جزءًا من واحد وذلك ما اردنا ان نجد

---

كج نريد ان نجد عددين مربّعين اذا قُسم على كلّ واحد منهما عدد مربّع مفروض وجُمع ما يخرج من القَسمين كان المجتمع عددًا مربّعًا واذا جُمعت الثلثة الاعداد اعنى العددين المطلوبين والعددَ المفروض كان المجتمع مربّعًا

فليكن العدد المربّع المفروض تسعة آحاد ونريد ان نجد عددين مربّعين اذا قُسم على كلّ واحد منهما تسعة آحاد وجُمع ما يخرج من القَسم يكون ذلك عددًا مربّعًا واذا جُمعت الثلثة الاعداد اعنى العددين المطلوبين والتسعة الآحاد / المفروضة ٢٨ كان المجتمع عددًا مربّعًا وقد علمنا انّا متى قسمنا عددًا مربّعًا بقِسمين مربّعين ثمّ قسمنا على كلّ واحد من القِسمين عددًا مربّعًا كان المجتمع ممّا يخرج من القَسم عددًا مربّعًا فلنفرض عددًا مربّعًا ونقسمه بقسمين مربّعين ويكون العدد الذى نفرضه مالًا ونقسمه بقسمين ⟨مربّعين⟩ احدهما تسعة اجزاء من خمسة وعشرين جزءًا من مال والقِسم الآخر ستّة عشر جزءًا من خمسة وعشرين جزءًا من مال فليكن هذان القسمان العددين المطلوبين ونقسم تسعة آحاد على تسعة اجزاء من خمسة وعشرين جزءًا من مال فتخرج خمسة وعشرين جزءًا من مال ونقسم ايضًا تسعة آحاد على ستّة عشر جزءًا من خمسة وعشرين جزءًا من مال فيخرج القِسمُ اربعة عشر جزءًا

---

2855 (806): وعشرين (prius): وعسرون in cod.

2857 (807): جزءًا (prius): احرا in cod.

2864-2865 (808): عدد مربّع مفروض : عددا مربعا معروصا in cod.

2870 (809): واذا : مادا in cod.

2875 (810): مربّعين addidi.

ونصفَ ثُمن جزء من مال فاذا جُمع ما خرج من القَسمين كان المجتمع تسعة وثلثين جزءاً ونصفَ ثُمن جزء من مال وهو عدد مربّع ضلعه ستّة اجزاء ورُبع جزء من شيء ولكنّا اذا جمعنا الثلثة الاعداد اعني العددين المطلوبين والتسعة المفروضة يكون المجتمع مالاً وتسعة آحاد فنحتاج ان يكون مربّعاً فلنفرض ضلعه شيئاً وواحداً ونضربه في مثله فيكون مالاً وشيئين وواحداً فهو يعادل مالاً وتسعة آحاد فأُلقْ مالاً وواحداً من الناحيتين ليبقى نوع واحد يعادل نوعاً واحداً فيبقى شيئان يعادلان ثمنية آحاد فالشيء اربعة آحاد وقد كان احد العددين المطلوبين ستّة عشر جزءاً من خمسة وعشرين جزءاً من مال وضلعه اربعة اخماس شيء فضلعه اربعة اخماس اربعةٍ وذلك ستّة عشر خُمساً فنضربها في مثلها فتكون مأتين وستّة وخمسين جزءاً من خمسة وعشرين وذلك احد / العددين ⟨المطلوبين⟩ وقد كان ايضاً العدد ٢٩ الآخر تسعة اجزاء من خمسة وعشرين جزءاً من مال وضلعه ثلثة اخماس شيء والشيء اربعة آحاد فضلعه اثنا عشر خُمساً فنضربه في مثله فيكون مائة واربعة واربعين جزءاً من خمسة وعشرين جزءاً من واحد وهو العدد الآخر المطلوب فاذا قسمنا العدد المفروض اعني التسعة الآحاد التي هي مائتا جزء وخمسة وعشرون جزءاً من خمسة وعشرين جزءاً على احد العددين وهو مائتا جزء وستّة وخمسون جزءاً من خمسة وعشرين كان الذي يخرج من القَسم مائتي جزء وخمسة وعشرين جزءاً من مائتي جزء وستّة وخمسين جزءاً واذا قسمنا ايضاً التسعة الآحاد اعني المائتي جزء والخمسة والعشرين الجزء من خمسة وعشرين على العدد الآخر اعني المائة والاربعة والاربعين الجزء من خمسة وعشرين جزءاً كان الذي يخرج من القَسم مائتي جزء وخمسة وعشرين جزءاً من مائة واربعة واربعين جزءاً اعني اربع مائة جزء من مأتين وستّة وخمسين جزءاً فاذا اضفنا ذلك الى ما خرج من قسمة التسعة على العدد الآخر وهو مائتا جزء وخمسة وعشرون جزءاً من مأتين وستّة وخمسين يكون المجتمع من ذلك ستّمائة جزء وخمسة وعشرين جزءاً من مائتين وستّة وخمسين وذلك عدد مربّع

---

2888 (811): آحاد (prius): اعاد in cod.

2892 (812): المطلوبين dubitanter addidi.

2899 (813): مائتى (prius): ماسا in cod.

2904 (814): Pro جزء من praebet codex حرو وس (sc. جزء ومن).

2906 (815): وعشرون : وعسرس in cod.

ضلعه خمسة وعشرون جزءاً من ستّة عشر جزءاً من واحد فاذا جمعنا الثلثة الاعداد اعنى المأتين والستّة والخمسين جزءاً من خمسة وعشرين جزءاً من الواحد والمائة والاربعة والاربعين الجزء من الخمسة والعشرين والتسعة الآحاد التى هى مائتا جزء وخمسة وعشرون جزءاً من خمسة وعشرين كان المجتمع من ذلك ستّمائة وخمسة وعشرين جزءاً من خمسة وعشرين وهو خمسة/وعشرون احداً وهو عدد مربّع وضلعه خمسة آحاد ١٣٠ فقد وجدنا عددين على التحديد الذى حُدّ لنا وهما مائتا جزء وستّة وخمسون جزءاً من خمسة وعشرين جزءاً من واحد مائة واربعة واربعون جزءاً من خمسة وعشرين جزءاً من الواحد وذلك ما اردنا ان نجد

تمّت المقالة السادسة من كتاب ذيوفنطس وفى هذه المقالة ثلث وعشرون مسئلةً من المسائل العدديّة

بسم الله الرحمن الرحيم

المقالة السابعة من كتاب ذيوفنطس

غرضُنا ان نتكلّم فى هذا القول على كثير من المسائل العدديّة من غير ان يكون ذلك خارجاً عن جنس ما تقدّم من المسائل فى القول الرابع والخامس وان كان مخالفاً النوعَ ليكون ذلك سبباً للتمهّر وزيادةً فى الدربة والعادة

ا نريد ان نجد ثلثة اعداد مكعّبة ويكون ضلع الاوّل من ضلع الثانى فى نسبة مفروضة وضلع الثانى من ضلع الثالث فى نسبة مفروضة واذا ضوعف العدد الاوّل بالعدد الثانى وضوعف ما بلغ بالعدد الثالث كان ذلك عدداً مربّعاً

فلتكن النسبة المفروضة نسبة المثلين ونريد ان نجد ثلثة اعداد مكعّبة يكون ضلع الاوّل منها مثلى ضلع الثانى ويكون ضلع الثانى مثلى ضلع الثالث واذا ضوعف الاوّل من الثلثة الاعداد بالعدد الثانى وما بلغ بالعدد الثالث كان ذلك عدداً مربّعاً فلنفرض ضلع العدد

2910 (816): والعشرين : والعسرون in cod.

2921 (817): Verba hujus lineae rubro colore in cod.

2924 (818): النوعَ : للنوع in cod.

2926 (819): مكعّبة : مكعب in cod.

الثالث شيئاً فيكون العدد الثالث كعباً ونفرض ضلع العدد الثانى شيئين لانّه مثلاً ضلع العدد الثالث فيكون العدد الثانى ثمانية كعاب ونفرض ضلع العدد الاوّل اربعة/اشياء لانّه مثلاً ضلع العدد الثانى ٣١

فيكون العدد الاوّل اربعة وستّين كعباً ولكن اذا ضاعفنا العدد الاوّل وهو اربعة وستّون كعباً بالعدد الثانى وهو ثمنية كعاب وما بلغ بالعدد الثالث وهو كعب يكون ذلك خمس مائة واثنى عشر كعب كعب كعب فنحتاج ان يكون مربّعاً فلنفرض له ضلعاً يكون اثنين وثلثين مال مال ونضربه فى مثله فيكون الفاً واربعة وعشرين مال مال مال فهو يعادل خمس مائة واثنى عشر كعب كعب كعب فنقسم الخمس مائة والاثنى عشر كعب كعب كعب على مال مال مال مال فتخرج خمس مائة واثنا عشر شيئاً ونقسم الفاً واربعة وعشرين مال مال مال على مال مال مال مال فتكون الفاً واربعة وعشرين احداً فهى تعادل خمس مائة واثنى عشر شيئاً فالشىء الواحد احدان وقد كنّا فرضنا ضلع العدد الثالث شيئاً والشىء اثنان فضلع العدد الثالث احدان والعدد الثالث ثمنية آحاد وكنّا فرضنا ضلع العدد الثانى شيئين لانّه مثلاً ضلع العدد الثالث والشيئان اربعة فالعدد الثانى اربعة وستّون احداً وكنّا فرضنا ضلع العدد الاوّل اربعة اشياء لانّه مثلاً ضلع العدد الثانى والشىء احدان فضلع العدد الاوّل اربعة امثال احدين وذلك ثمنية آحاد فالعدد الاوّل خمس مائة واثنا عشر احداً واذا ضاعفنا العدد الاوّل وهو خمس مائة واثنا عشر احداً بالعدد الثانى وهو اربعة وستّون احداً بلغ ذلك اثنين وثلثين الفاً وسبع مائة وثمنية وستّين احداً فنضاعفه بالعدد الثالث وهو ثمنية آحاد فيكون ذلك مائتى الف واثنين وستّين الفاً ومائة واربعة واربعين احداً وهو عدد مربّع ضلعه خمس مائة واثنا عشر

---

2940 (820): مال (posterius) ab eadem manu supra lineam additum.

2941 (821): الفاً : الف in cod.

2944 (822): الفاً : الف in cod.

2945 (823): الفاً : الف in cod.

2946 (824): احدان : احدس in cod.

2949 (825): والشيئان : فالسس in cod.

2954 (826): فنضاعفه : مصاعفه in cod.

فقد وجدنا ثلثة اعداد على التحديد الذى حُدّلنا وهى خمس مائة واثنا عشر احداً واربعة وستّون احداً وثمنية آحاد وذلك ما اردنا ان نجد

ب / نريد ان نجد ثلثة اعداد مكعّبة وهى ايضاً مربّعة واذا ضوعف ٣٢ العدد الاوّل منها بالعدد الثانى وضوعف ايضاً ما اجتمع بالعدد الثالث يكون ذلك مربّعاً من ضلع مربّع

فلنفرض العدد الاوّل جزءاً من اربعة وستّين جزءاً وهو عدد مكعّب وضلعه رُبع واحد وهو ايضاً عدد مربّع وضلعه نصف واحد ونفرض العدد الثانى اربعة وستّين احداً وهو عدد مكعّب وضلعه اربعة آحاد وهو ايضاً عدد مربّع وضلعه احدان ونفرض العدد الثالث كعب كعب وهو عدد مكعّب وضلعه مال وهو ايضاً عدد مربّع وضلعه كعب فاذا ضوعف العدد الاوّل اعنى الجزء من اربعة وستّين جزءاً من الواحد بالعدد الثانى اعنى اربعة وستّين احداً بلغ ذلك واحداً واذا ضوعف الواحد بالعدد الثالث اعنى كعب كعب بلغ ذلك كعب كعب فنحتاج ان يكون ضلعه مربّعاً [وإنما اعنى بضلعه فى هذا الموضع جذرَه] ولكن ضلع كعب كعب هو كعب فنعادل بكعب عدداً مربّعاً اعنى اربعة اموال ونقسم الناحيتين على مال فيكون شىء يعادل اربعة آحاد فهو الشىء وهو ضلع الكعب والكعب اربعة وستّون وكنّا فرضنا العدد الثالث كعب كعب وهو من ضرب كعب فى مثله والكعب اربعة وستّون احداً فنضرب اربعة وستّين احداً فى مثلها فتكون اربعة الٰف وستّة وتسعين احداً وذلك العدد الثالث واذا ضاعفنا العدد الاوّل وهو جزء من اربعة وستّين جزءاً من واحد بالعدد الثانى وهو اربعة وستّون احداً بلغ ذلك واحداً فاذا ضاعفنا بالواحد العدد الثالث وهو اربعة الٰف وستّة وتسعون كان ذلك اربعة الٰف وستّة وتسعين وهو عدد مربّع وضلعه اربعة وستّون وهو ايضاً عدد مربّع ضلعه ثمنية آحاد

---

2959 (827): و in واربعة et in وثمنية ab eadem (ut videtur) manu supra lineam in codice. Vide etiam adn. 833.

2964 (828): جزءاً (posterius): احدا in cod.

2972 (829): In uncis seclusa verba interpolamentum esse censeo.

2974 (830): شىء : سا in cod.

2977 (831): اربعة (prius): احد in cod.

2981 (832): كان: فان in codice. Vide adn. 878.

فقد وجدنا ثلثة اعداد على التحديد الذى حُدّ لنا وهى جزء من
اربعة وستّين جزءاً من واحد واربعة وستّون احداً واربعة الٰف / وستّة ١٢٣
وتسعون احداً وذلك ما اردنا ان نجد

جـ نريد ان نجد عدداً مربّعاً من ضلع مربّع اذا قسمناه بثلثة اقسام يكون كلّ قسم منها مكعّباً

فلنفرض ضلع هذا العدد مالاً فالعدد مال مال ونريد ان نقسم مال مال بثلثة اقسام يكون كلّ قسم منها مكعّباً فلنفرض القسم الاوّل كعباً والقسم الثانى ثمنية كعاب والقسم الثالث اربعة وستّين كعباً ومن البيّن ان كلّ قسم من هذه الاقسام مكعّب ولكن جملة الثلثة الاقسام ثلثة وسبعون كعباً فهى تعادل العدد المقسوم اعنى المال مال فنقسم مال مال على كعب فيكون شيئاً ونقسم ثلثة وسبعين كعباً على كعب فتكون ثلثة وسبعين احداً فهى تعادل شيئاً فالشىء ثلثة وسبعون احداً وكنّا فرضنا ضلع العدد المقسوم مالاً فهو مربّع الثلثة والسبعين احداً وذلك خمسة الٰف وثلثمائة وتسعة وعشرون احداً والعدد المقسوم مربّع خمسة الٰف وثلثمائة وتسعة وعشرين وهو ثمنية وعشرون الف الف وثلثمائة الف وثمنية وتسعون الفاً ومائتان وأحد واربعون وكنّا فرضنا احد الاقسام كعباً والكعب من ضرب ثلثة وسبعين فى ثلثة وسبعين وما بلغ فى ثلثة وسبعين وذلك ثلثمائة الف وتسعة وثمنون الفاً وسبعة عشر احداً فهو عدد مكعّب وهو احد الاقسام الثلثة والقسم الآخر ثمنية امثاله لانّا فرضناه ثمنية كعاب وذلك ثلثة الٰف الف ومائة الف واثنا عشر الفاً ومائة وستّة وثلثون والقسم الآخر اربعة وستّون كعباً فهو مثل القسم الاوّل اربعاً وستّين مرّة وذلك اربعة وعشرون الف الف وثمان مائة الف وسبعة وتسعون الفاً وثمنية وثمنون احداً وبيّن انّا اذا جمعنا هذه الاقسام الثلثة التى كلّ قسم منها مكعّب كان العدد المركّب من

---

2984 (833): و in utroque واربعة eadem manu supra lineam additum.

2988 (834): فالعدد : والعدد in cod.

2989 (835): مكعّباً: مكعب in cod.

2995 (836): فهو: وهو in cod.

3001 (837): عدد مكعّب : العدد المكعب in cod.

3004 (838): اربعاً : اربعه in cod.

3006 (839): التى: الدى in cod.

جملتها ثمنية وعشرين الف الف وثلثمائة الف وثمنية وتسعين الفاً ومأتين / وأحد واربعين وذلك العدد المقسوم وهو عدد مربّع من ضلع مربّع ١٣٤

فقد وجدنا عدداً على التحديد الذى حُدّ لنا وهو ثمنية وعشرون الف الف وثلثمائة وثمنية وتسعون الفاً ومائتان وأحد واربعون وذلك ما اردنا ان نجد

دَ نريد ان نقسم عدداً مكعّباً من ضلع مربّع بثلثة اقسام يكون كلّ قسم منها مربّعاً

فلنفرض ضلع المكعّب مالاً فيكون المكعّب كعب كعب ونريد ان نقسم كعب كعب بثلثة اقسام كلّ قسم منها مربّع فلنطلب ثلثة اعداد اذا جُمعت كان المجتمع منها مربّعاً ويكون كلّ واحد منها مربّعاً ووِجْدَان ذلك سهل على ما تقدّم فيكون احد الاعداد واحداً والثانى اربعة آحادٍ والثالث اربعة اتساع واحد فلنجعل كلّ عدد من هذه الاعداد اموالَ اموال فيكون العدد الاوّل مال مال والثانى اربعة اموال مال والثالث اربعة اتساع مال مال ومن اجل انّا نريد ان نقسم عدداً مكعّباً بثلثة اقسام مربّعة فلنفرض كلّ قسم من الاقسام الثلثة عدداً من هذه الثلثة الاعداد وجملتها ⟨ذلك العدد المكعّب⟩ فيكون العدد المركّب من جملتها تسعة واربعين جزءاً من تسعة اجزاء من مال مال فذلك يعادل العدد المكعّب اعنى الكعب كعب فنقسم كلّ ذلك على مال مال فيكون مال يعادل تسعة واربعين جزءاً من تسعة اجزاء من واحد ومن اجل انّا فرضنا ضلع العدد المكعّب مالاً والمال تسعة واربعون جزءاً من تسعة اجزاء من الواحد فهو ضلع المكعّب والعدد المكعّب من ضرب تسعة واربعين جزءاً فى مثلها وما اجتمع فى تسعة واربعين جزءاً وذلك مائة وسبعة عشر الفاً وستّمائة وتسعة واربعون

---

3007-3008 (840): وتسعين الفاً ومأتين وأحد واربعين : وسعون العا وماساں واحد واربعون in codice ( وعشرين, lin. 3007, recte scriptum).

3016 (841): Suntne verba ويكون كلّ واحد منها مربّعاً interpolata? Melius enim praebuisset textus مربّعة post اعداد in linea 3015.

3020-3021 (842): عدداً مكعّباً : عدد المكعب in cod.

3022 (843): Deficientia verba addidi.

3026 (844): والمال : فالمال in cod.

3028 (845): ضرب : صلع in cod.

جزءاً من سبع مائة وتسعة وعشرين جزءاً من الواحد ومن اجل انّا فرضنا
احد الاقسام مالاً فهو تسعة واربعون جزءاً من تسعة اجزاء من ⟨الواحد
ومن⟩ اجل انّا فرضنا القسم/الثانى اربعة اموال فهو مائة وستّة وتسعون ٣٥
جزءاً من تسعة والقسم الثالث اربعة اتساع مال وذلك مائة وستّة
وتسعون جزءاً من سبع مائة وتسعة وعشرين جزءاً من الواحد واذا جُمعت
هذه الثلثة الاقسام كانت مثل العدد المكعّب

فقد وجدنا عدداً على التحديد الذى حُدّ لنا وهو مائة الف وسبعة
عشر الفاً وستّمائة وتسعة واربعون جزءاً من سبع مائة وتسعة وعشرين
وذلك ما اردنا ان نجد

ه̄ نريد ان نجد عدداً مكعّباً من ضلع مكعّب اذا ضوعف بعددين
احدهما مكعّب والآخر مربّع وجُمع ذلك كان ⟨المجتمع⟩ عدداً مربّعاً

فلنفرض ضلع العدد المكعّب عدداً مكعّباً وليكن ثمنية آحاد فيكون
العدد المكعّب خمس مائة واثنى عشر احداً ونريد ان نجد عددين
احدهما مكعّب والآخر مربّع اذا ضوعف كلّ واحد منهما بخمس مائة
واثنى عشر احداً وجُمع ذلك كان المجتمع مربّعاً فلنفرض العدد
المكعّب كعباً والمربّع مالاً ونضرب كعباً ومالاً فى خمس مائة واثنى عشر
فيكون ⟨الجميع من⟩ ذلك خمس مائة واثنى عشر كعباً وخمس مائة
واثنى عشر مالاً فنحتاج ان يكون عدداً مربّعاً فنفرض له ضلعاً يكون
اربعة وستّين شيئاً ونضربها فى مثلها فتكون اربعة الف مال وستّة
وتسعين مالاً فهى تعادل خمس مائة كعب واثنى عشر كعباً وخمس
مائة مال واثنى عشر مالاً فنلقى خمس مائة واثنى عشر مالاً من الناحيتين
فيبقى خمس مائة واثنا عشر كعباً تعادل ثلثة الٰف وخمس مائة واربعة
وثمنين مالاً فنقسم الناحيتين على مال فيخرج ثلثة الٰف وخمس مائة
واربعة وثمنون احداً تعادل خمس مائة شئ واثنى عشر شيئاً فالشئ

---

3031-3032 (846): Omissum (forsan per homoeoteleuton) addidi.

3034 (847): وتسعون : وسعس in cod.

3040 (848): المجتمع addidi.

3046 (849): فيكون : وىكوں in cod.

(850): الجميع من addidi. Dubito enim num و in expressione كعباً ومالاً (lin. 3045) sensu additionis intellegendum sit.

3053 (851): وثمنون : وىمىں in cod.

(852): واثنى : واىسا in cod.

الواحد سبعة آحاد ومن اجل انّا فرضنا العدد المربّع مالاً وضلعه شئ والشئ سبعة آحاد والمال تسعة واربعون احداً ⟨فالعدد المربّع تسعة واربعون احداً⟩ وايضاً من اجل انّا فرضنا العدد المكعّب كعباً والكعب/من ضرب المال فى الشئ فيكون العدد المكعب ٣٦ ثلثمائة وثلثة واربعين ولكنّا اذا ضاعفنا العدد المكعّب الذى فرضنا ضلعه مكعّباً اعنى الخمس مائة والاثنى عشر بالعدد المكعّب الذى هو ثلثمائة وثلثة واربعون احداً يكون ذلك مائة الف وخمسة وسبعين الفاً وستّمائة وستّة عشر وايضاً فاذا ضاعفنا الخمس مائة والاثنى عشر بالعدد المربّع اعنى تسعة واربعين احداً يكون ذلك خمسة وعشرين الفاً وثمنية وثمنين فاذا أُضيفَ اليه المائة الف والخمسة والسبعون الفاً وستّمائة وستّة عشر يكون المجتمع من ذلك مائتى الف وسبع مائة واربعة وذلك عدد مربّع وضلعه اربع مائة وثمنية واربعون

فقد وجدنا عدداً على التحديد الذى حُدّ لنا وهو خمس مائة واثنا عشر احداً وذلك ما اردنا ان نجد

وَ نريد ان نجد عددين مربّعين يكون العدد المركّب من جميعهما مربّعاً واذا ضوعف احدهما بالآخر يكون المجتمع من العدد المركّب من جملة العددين فى نسبة مفروضة

ولكن النسبة المفروضة لا تكون الّا عدداً مربّعاً لان كلّ عددين مربّعين فان نسبة الاكبر منهما الى الاصغر لا تكون الّا عدداً مربّعاً وايضاً فمقدار الاصغر منهما الى الاكبر لا يكون الّا مربّعاً

فلتكن النسبة المفروضة نسبة التسعة الامثال فلنفرض العدد المركّب من جملة العددين مالاً ونقسم مالاً بقسمين مربّعين فليكن احدهما ستّة عشر جزءاً من خمسة وعشرين جزءاً من مال والقسم الآخر تسعة اجزاء من خمسة وعشرين جزءاً من مال ولكنّا اذا ضاعفنا احد القسمين بالآخر يكون ذلك مائة واربعة واربعين جزءاً من ستّمائة وخمسة وعشرين

---

3055-3056 (853): Uncis inclusa verba addidi.

3057 (854): والكعب : فالمكعب in cod.

3063 (855): والسبعون : والسعس in cod.

3065 (856): واربعون : واربعس in cod.

3073 (857): الّا : لا in cod.

3074 (858): الامثال : للامسال in cod.

جزءًا من مال مال فنحتاج ان يكون معادلاً لتسعة امثال العدد المركّب من العددين المربّعين اعنى تسعة اموال فنقسم مائة واربعة واربعين جزءًا من ستّمائة وخمسة/وعشرين جزءًا من مال مال على مال ٣٧ فتخرج مائة واربعة واربعين جزءًا من ستّمائة وخمسة وعشرين جزءًا من مال ونقسم تسعة اموال على مال فتخرج تسعة آحاد فهى معادلة للمائة والاربعة والاربعين جزءًا من ستّمائة وخمسة وعشرين جزءًا من مال فالمال الكامل يعادل تسعة وثلثين احداً وجزءًا من ستّة عشر جزءًا من واحد وقد كان احد العددين ستّة عشر جزءًا من خمسة وعشرين جزءًا من مال وذلك خمسة وعشرون احداً والعدد الآخر تسعة اجزاء من خمسة وعشرين جزءًا من مال وذلك اربعة عشر احداً وجزء من ستّة عشر من واحد وجملة العددين تسعة وثلثون ونصف ثُمن وهو عدد مربّع ضلعه ستّة ورُبع واذا ضاعفنا احد العددين بالآخر وذلك خمسة وعشرون فى اربعة عشر ونصف ثُمن ⟨يكون ذلك ثلثمائة وواحداً وخمسين احداً ونصفاً ونصف ثُمن⟩ وهو تسعة امثال جملة العددين اعنى التسعة والثلثين والنصف ثُمن

فقد وجدنا عددين على التحديد الذى حُدّ لنا وهما خمسة وعشرون احداً اربعة عشر احداً ونصف ثُمن واحد وذلك ما اردنا ان نجد

ز نريد ان نقسم عدداً مربّعاً من ضلع مكعّب بثلثة اقسام اىّ قسمين جُمعا كان المجتمع منهما مربّعاً

فلنفرض ضلع العدد المربّع كعباً فيكون العدد المربّع كعب كعب ونريد ان نقسم كعب كعب بثلثة اقسام اىّ قسمين جُمعا كان ⟨جميعهما⟩ مربّعاً ⟨فلنطلب ثلثة اعداد اىّ عددين منها جُمعا كان جميعهما مربّعاً⟩ ويكون العدد المركّب من الثلثة الاعداد مربّعاً ووجود ذلك سهل على ما قد بيّنّا فى المسئلة السادسة من المقالة الثالثة فيكون

---

3079 (859): لتسعة : سعه in cod.

3085 (860): يعادل : معادل in cod.

3088 (861): ستّة عشر : سه وعسر in cod.

3091-3092 (862): Per homoeoteleuton omissa verba restitui.

3099 (863): Verbum جميعهما addidi.

3100-3101 (864): Per homoeoteleuton omissa (ut opinor) addidi.

3101 (865): ووجود ذلك : ووجودلك in cod.

3102 (866): Pro من codicis فى scripsi.

العدد الاوّل ثمانين احداً والثانى ثلثمائة وعشرين احداً والثالث احد واربعين احداً وجملة الثلثة الاعداد اربع مائة وأحد واربعون احداً فلنجعل بدلاً من الآحاد اموالَ اموال فتكون جملة الثلثة/الاعداد ٤٠ اربع مائة وأحد واربعين مال مال فهى تعادل كعب كعب فنقسم الناحيتين على مال مال فيخرج من قسمة كعب كعب على مالِ مال مالٌ ويخرج من قسمة اربع مائة وأحد واربعين مال مال على مال ⟨مال⟩ اربع مائة وأحد واربعون احداً فهى تعادل مالاً فالمال اربع مائة وأحد واربعون احداً فالمال المال هو ما يجتمع من ضرب اربع مائة وأحد واربعين فى مثلها وذلك مائة الف واربعة وتسعون الفاً واربع مائة وأحد وثمنون ومن اجل انّا فرضنا احد الاقسام الثلثة ثمنين مال مال يكون خمسة عشر الف الف وخمس مائة الف وثمنية وخمسين الفاً واربع مائة وثمنين احداً وايضاً من اجل انّا فرضنا القسم الثانى ثلثمائة وعشرين مال مال يكون اثنين وستّين الف الف ومائتى الف وثلثة وثلثين الفاً وتسع مائة وعشرين وايضاً فمن اجل انّا فرضنا القسم الثالث أحد واربعين مال مال يكون سبعة الٰف الف وتسع مائة الف وثلثة وسبعين الفاً وسبع مائة وأحد وعشرين احداً ولمّا كان العدد المقسوم بهذه الثلثة الاقسام هو العدد المركّب من جملتها يكون خمسة وثمنين الف الف وسبع مائة وستّة وستّين الفاً ومائة وأحد وعشرين وذلك عدد مربّع وضلعه تسعة الٰف ومائتان وأحد وستّون وهذا الضلع عددٌ مكعّب ضلعه واحد وعشرون ومن اجل ان القسم الاوّل من الاقسام الثلثة خمسة عشر الف الف وخمس مائة الف وثمنية وخمسون الفاً واربع مائة وثمنون احداً والقسم الثانى اثنان وستّون الف الف ومائتا الف وثلثة وثلثون الفاً وتسع مائة وعشرون احداً يكون العدد المركّب من جميعهما سبعة وسبعين الف الف وسبع مائة الف واثنين وتسعين الفاً واربع مائة احد /وهو عدد مربّع وضلعه ثمنية الٰف وثمان مائة وعشرون احداً وايضاً ٤١ فمن اجل ان القسم الثانى اثنان وستّون الف الف ومائتا الف وثلثة

---

3104 (867): الثلثة : الباله in cod.

3105 (868): اموالَ اموال : الاموال اموال in cod.

3108 (869): Deficiens verbum restitui.

3116 (870): انّا : ما in cod.

3117 (871): سبعة : سعه in cod.

3121 (872): ومائتان وأحد وستّون : وماس واحد وسس in cod.

3123 (873): وخمسون... وثمنون : وحمسس... وسس in cod.

وثلثون الفاً وتسع مائة وعشرون احداً والقسم الثالث سبعة الٰف الف وتسع مائة الف وثلثة وسبعون الفاً وسبع مائة وأحد وعشرون احداً يكون العدد المركّب من جملتهما سبعين الف الف ومائتى الٰف وسبعة الٰف وستّمائة وأحد واربعين وذلك عدد مربّع وضلعه ثمنية الٰف وثلثمائة وتسعة وسبعون وايضاً فمن اجل ان القسم الثالث سبعة الٰف الف وتسع مائة الف وثلثة وسبعون الفاً وسبع مائة وأحد وعشرون احداً والقسم الاوّل خمسة عشر الف الف وخمس مائة الف وثمنية وخمسون الفاً واربع مائة وثمنون احداً فان العدد المركّب من جملتهما يكون ثلثة وعشرين الف الف وخمس مائة الف واثنين وثلثين الفاً ومائتين وواحداً وذلك عدد مربّع وضلعه اربعة الٰف وثمان مائة وأحد وخمسون احداً

فقد وجدنا عدداً على التحديد الذى حُدّ لنا وهو خمسة وثمنون الف الف وسبع مائة الف وستّة وستّون الفاً ومائة وأحد وعشرون وذلك ما اردنا ان نجد

ولمّا انتهينا الى إستتمام عمل هذه المسئلة اردنا ان نعملها عملاً ثانياً هو اسهل من العمل الاوّل فلنبدأ بالسؤال قبل العمل نريد ان نجد عدداً مربّعاً من ضلع مكعّب اذا قُسم بثلثة اقسام يكون كلّ قسمين منها ⟨اذا جُمعا⟩ عدداً مربّعاً فلنفرض العدد المربّع الذى نريد قسمتَه اربعة وستّين احداً وهو عدد مربّع وضلعه مكعّب ونريد ان نقسم اربعة وستّين احداً ثلثةَ اقسام اىّ قسمين جُمعا كان جميعهما مربّعاً فلنطلب ثلثة اعداد اذا جُمعت كان جميعها مربّعاً واىّ عددين منها جُمعا كان جميعهما / مربّعاً وقد بيّنّا ذلك فى المسئلة السادسة ٣٨ من المقالة الثالثة بياناً نستغنى به عن الاعادة فيكون احد هذه الثلثة الاعداد المطلوبة ثلثمائة وعشرين احداً والعدد الثانى ⟨ثمنين احداً⟩ والعدد الثالث واحداً واربعين احداً والعدد المركّب من جملة هذه الاعداد اربع مائة وأحد واربعون احداً وهو عدد مربّع

ولو كان هو العددَ الذى نريد ان نقسمه كنّا قد انتهينا الى ما اردنا ولكن العدد الذى نريد قسمته اربعة وستّون احداً فلنأخذ من كلّ عدد

---

3135 (874): وخمسون : وحمسس in cod.

3136 (875): وثمنون : ومسس in cod.

3137 (876): وواحداً : وواحد in cod.

3145 (877): اذا جُمعا addidi.

3149 (878): كان : فان in cod.

3151 (879): Deficientia verba restitui.

من الثلثة الاعداد التى جملتها اربع مائة وأحد واربعون احداً ⟨عدداً⟩ مقدارُه من العدد الذى نأخذه منه كمقدار الاربعة والستّين من الاربع مائة والواحد والاربعين وذلك ان نضرب كلّ عدد من الثلثة الاعداد فى اربعة وستّين فيكون المجتمع اجزاء من اربع مائة وأحد واربعين ومن اجل ان العدد الاوّل ثلثمائة وعشرون احداً يكون اذا ضاعفناه بأربعة وستّين احداً عشرين الفاً واربع مائة وثمنين فهى اجزاء من اربع مائة وأحد واربعين ومن اجل ان القسم الثانى ثمنون احداً يكون اذا ضاعفناه بأربعة وستّين خمسة الٰف ومائة وعشرين فهى اجزاء من اربع مائة وأحد واربعين جزءاً من الواحد وايضاً فمن اجل ان القسم الثالث احد واربعون احداً يكون اذا ضاعفناه بأربعة وستّين ألفين وستّمائة واربعة وعشرين وهى اجزاء من اربع مائة وأحد واربعين فقد قسمنا الاربعة والستّين بثلثة اقسام اذا جُمع الاوّل والثانى كان جميعهما خمسة وعشرين الفاً وستّمائة جزء من اربع مائة وأحد واربعين وذلك عدد مربّع وضلعه مائة وستّون جزءاً من واحد وعشرين واذا جُمع الثانى والثالث يكون جميعهما سبعة الٰف وسبع /مائة واربعة واربعين جزءاً ١٣٩ من اربع مائة وأحد واربعين وذلك عدد مربّع وضلعه ثمنية وثمنون جزءاً من واحد وعشرين واذا جُمع الثالث والاوّل كان جميعهما ثلثة وعشرين الفاً ومائة و⟨اربعة اجزاء من⟩ اربع مائة وأحد واربعين وهو عدد مربّع وضلعه مائة واثنان وخمسون جزءاً من واحد وعشرين

فقد وجدنا عدداً على التحديد الذى حُدّ لنا وهو اربعة وستّون احداً وقسمناه بثلثة اقسام وهى عشرون الفاً واربع مائة وثمنون جزءاً من اربع مائة وأحد واربعين خمسة الٰف ومائة وعشرون جزءاً من اربع مائة وأحد واربعين ألفان وستّمائة واربعة وعشرون جزءاً من اربع مائة وأحد واربعين وذلك ما اردنا ان نجد

ح نريد ان نجد عدداً مربّعاً من ضلع مكعّب اذا زيد عليه عدد ما كان مربّعاً وان زيد عليه ايضاً مثلاً ذلك العدد كان مربّعاً

فلنفرض العدد المربّع اربعة وستّين احداً وهو عدد مربّع من ضلع مكعّب ونريد ان نجد عدداً ان زيد على اربعة وستّين كان المجتمع

---

3156 (880): عدداً addidi.

3164 (881): جزءاً : احرا in cod.

3173 (882): Uncis inclusa verba addidi.

(883): Loco وأحد واربعين praebet codex واربعه .

3181 (884): مثلاً : مثلى in cod.

مربّعاً وان زيد مثلاه على اربعة وستّين كان المجتمع ايضاً مربّعاً فلنلتمس ذلك فى عدد مربّع غير الاربعة والستّين ونطلب عدداً مربّعاً اذا زدنا عليه عدداً ما كان مربّعاً وان زدنا عليه مثلى ذلك العدد كان المجتمع مربّعاً وكلّ عدد مربّع فانّا اذا زدنا عليه مثل جذريه وواحدٍ يكون المجتمع مربّعاً فانّا نفرض العدد المربّع مالاً ونزيد عليه جذريه وواحداً فيكون مالاً وشيئين وواحداً وذلك عدد مربّع وضلعه شئ وواحد ولكنّا اذا زدنا على المال مثلى شيئين وواحدٍ اعنى اربعة اشياء واثنين كان المجتمع مالاً واربعة اشياء واثنين فنريد ان يكون مربّعاً فلنفرض له ضلعاً / يكون شيئاً الّا اثنين ونضربه فى مثله فيكون مالاً واربعة ١٤٢ آحاد الّا اربعة اشياء فهو يعادل مالاً واربعة اشياء وأحدين فنزيد على كلّ ما معنا اربعة اشياء ونلقى مالاً بمال فيبقى ثمنية اشياء واثنان تعادل اربعة آحاد فنلقى احدين من الناحيتين فيبقى ثمنية اشياء تعادل اثنين فالشئ الواحد رُبع واحد والمال نصف ثُمن واحد وقد كان العدد المزاد عليه شيئين وواحداً وذلك واحد ونصف والعدد المزاد عليه ايضاً ضعف الواحد والنصف وذلك ثلثة فلنضرب ⟨كلّ⟩ ذلك فى ستّة عشر ليكون المال عدداً صحيحاً فيكون المال واحداً والعدد المزاد عليه اربعة وعشرين والعدد المزاد عليه ايضاً ضعف الاربعة والعشرين وذلك ثمنية واربعون ومن البيّن انّا اذا زدنا على الواحد اربعة وعشرين احداً يكون خمسة وعشرين احداً وذلك عدد مربّع وايضاً فانّا اذا زدنا على الواحد ضعف الاربعة والعشرين اعنى ثمنية واربعين يكون تسعة واربعين وذلك عدد مربّع

فلو كان العدد المربّع المفروض واحداً كنّا قد انتهينا الى ما اردنا ولكنّه اربعة وستّون ومن اجل ان الاربعة والستّين تعدّ الواحد اربعاً وستّين مرّة وجب ان نضاعف كلّ واحد من العددين المزيدين اعنى الاربعة والعشرين والثمنية والاربعين بأربعة وستّين ولكنّا اذا ضاعفنا

---

3187 (885): وواحدا in codice, وواحدٍ scripsi.

3190 (886): وواحدا in codice, وواحدٍ scripsi.

3198 (887): كلّ addidi.

3199 (888): عدداً : عدد in cod.

3200 (889): اربعة bis in cod.

3201 (890): اذا زدنا : اذا اردنا in codice. Vide adn. 253,314.

3206 (891): اربعاً : اربعه in cod.

3207 (892): نضاعف : نضلعف in cod.

اربعة وعشرين بأربعة وستّين يكون ذلك الفاً وخمس مائة وستّة وثلثين وهو العدد المزاد على الاربعة والستّين واذا ضاعفنا ثمنية واربعين بأربعة وستّين يكون مبلغ ذلك ثلثة الٰف واثنين وسبعين وذلك ضعف العدد الاوّل فاذا زدنا على الاربعة والستّين الفاً وخمس مائة وستّة
١٤٣ وثلثين يكون ذلك الفاً وستّمائة وذلك عدد مربّع /وضلعه اربعون
ولكنّا اذا زدنا على الاربعة والستّين ضعف الالف وخمس مائة وستّة وثلثين اعنى ثلثة الٰف واثنين وسبعين يكون مبلغ ذلك ثلثة الٰف ومائة وستّة وثلثين وهو عدد مربّع وضلعه ستّة وخمسون

فقد وجدنا عددين احدهما ضعف الآخر اذا زيد كلّ واحد منهما على عدد مربّع من ضلع مكعّب يكون المجتمع مربّعاً وهما الف وخمس مائة وستّة وثلثون احداً ثلثة الٰف واثنان وسبعون احداً وذلك ما اردنا ان نجد

ط نريد ان نجد عدداً مربّعاً من ضلع مكعّب ان نقصنا منه عدداً ما كان الباقى مربّعاً وان نقصنا ايضاً منه مثلى ذلك العدد يكون الباقى مربّعاً

فلنفرض العدد المربّع اربعة وستّين وهو عدد مربّع وضلعه مكعّب ولانّا نريد ان نجد عدداً ان نقصناه من اربعة وستّين يكون الباقى مربّعاً وان نقصنا مثليه من اربعة وستّين يكون الباقى مربّعاً فانّا نلتمس ذلك فى عدد مربّع غير الاربعة والستّين ونطلب عدداً مربّعاً ان نقصنا منه عدداً ما كان الباقى مربّعاً وان نقصنا منه ضعف ذلك العدد يكون مربّعاً وكلّ عدد مربّع ننقص منه مثل جذريه الّا واحداً فان الباقى يكون مربّعاً فلنفرض المربّع مالاً وننقص منه مثل جذريه الّا واحداً فيكون الباقى مربّعاً ولكنّا اذا نقصنا منه مثلى جذريه الّا واحداً اعنى اربعة اجذاره الّا اثنين يكون الباقى منه مالاً وأحدين الّا اربعة اشياء فنحتاج ان يكون مربّعاً فلنفرض له ضلعاً يكون شيئاً الّا ثلثة آحاد ونضربه فى مثله فيكون مالاً وتسعة آحاد الّا ستّة اشياء فهى تعادل مالاً وأحدين الّا
١٤٤ اربعة اشياء فنلقى مالاً وأحدين /الّا اربعة اشياء من الناحيتين جميعاً
فيبقى شيئان يعادلان سبعة آحاد فالشىء الواحد ثلثة آحاد ونصف

---

3222 (893): مثلى : مىل in cod.

3224 (894): وستّين : وسىوں in cod.

3226 (895): فانّا : واىا in cod.

3231 (896): مثلى : مىل in cod.

والمال اثنا عشر احداً وربع والعددان المنقوصان منه ستّة واثنا عشر فنضرب كلّ ذلك فى اربعة ليكون المال عدداً صحيحاً فيكون المال تسعة واربعين والعددان المنقوصان اربعة وعشرين وثمنية واربعين فلو كان العدد المربّع تسعة واربعين كنّا قد انتهينا الى ما اردنا ولكنّه اربعة وستّون والاربعة والستّون تعدّ التسعة والاربعين مرّةً وخمسة عشر جزءاً من تسعة واربعين ⟨جزءاً من مرّة⟩ ولمّا كان ذلك كذلك وجب ان نزيد على ⟨العددين⟩ المنقوصين اعنى الاربعة والعشرين والثمنية والاربعين خمسة عشر جزءاً من تسعة واربعين جزءاً منهما فلنضرب اربعة وعشرين فى اربعة وستّين فيكون ذلك الفاً وخمس مائة وستّة وثلثين ⟨وهى⟩ اجزاء من تسعة واربعين جزءاً من الواحد وذلك العدد المنقوص من الاربعة والستّين ونضرب ايضاً ثمنية واربعين فى اربعة وستّين فيكون ذلك ثلثة الٰف واثنين وسبعين وذلك اجزاء من تسعة واربعين وهو العدد المنقوص ايضاً من الاربعة والستّين وهو ضعف العدد الاوّل ومن اجل ان العدد الاوّل الف وخمس مائة وستّة وثلثون جزءاً من تسعة واربعين فانّا اذا نقصناه من الاربعة والستّين اعنى ثلثة الٰف ومائة وستّة وثلثين جزءاً من تسعة واربعين يكون الباقى الفاً وستّمائة جزء من تسعة واربعين وذلك عدد مربّع وضلعه اربعون سُبعاً وايضاً فمن اجل ان العدد الثانى هو ضعف العدد الاوّل اعنى ثلثة الٰف واثنين وسبعين ⟨جزءاً من تسعة واربعين⟩ فانّا اذا نقصناه/من الاربعة والستّين اعنى ثلثة الٰف ومائة وستّة وثلثين جزءاً ١٤٥

---

3237 (897): والمال : اوالمال in cod.

(898): و in والعددان ab eadem manu supra lineam scriptum.

3241 (899): وستّون : وسس in cod.

3242 (900): جزءاً : احرا in cod.

(901): Verba جزءاً من مرّة addidi.

3243 (902): العددين addidi.

3244 (903): منهما : سها in cod.

3245 (904): الفاً : الف in cod.

3246 (905): وهى addidi et ideo pro verbo حرا codicis اجزاء scripsi.

3253 (906): الفاً : الف in cod.

(907): تسعة : سعه in cod.

3254 (908): Post الثانى praebet codex الدى.

3255 (909): Denominatorem addidi.

من تسعة واربعين يكون الباقى اربعة وستّين جزءاً من تسعة واربعين وذلك عدد مربّع وضلعه ثمنية اجزاء من سبعة

فقد وجدنا عددين احدهما ضعف الآخر اذا نُقص كلّ واحد منهما من عدد مربّع من ضلع مكعّب يكون الباقى مربّعاً وهما ثلثة الف واثنان وسبعون جزءاً من تسعة واربعين الفٌ وخمس مائة وستّة وثلثون جزءاً من تسعة واربعين وذلك ما اردنا ان نجد

ى نريد ان نجد عدداً مربّعاً من ضلع مكعّب وعدداً اذا زدناه عليه كان المجتمع مربّعاً وان نقصناه من العدد المربّع كان الباقى مربّعاً

فليكن العدد المربّع اربعة وستّين ونلتمس عدداً مربّعاً غير الاربعة والستّين اذا زيد عليه عدد ما كان المجتمع مربّعاً وان نقصنا منه ذلك العدد كان الباقى منه مربّعاً وكلّ عدد مربّع ننقص منه مثل جذريه الّا واحداً فان الباقى يكون مربّعاً فلنفرض العدد المربّع مالاً والعدد المنقوص منه شيئين الّا واحداً ولكنّا اذا زدنا شيئين الّا واحداً على مال يكون المجتمع مالاً وشيئين الّا واحداً فنحتاج الى ان يكون مربّعاً فلنفرض له ضلعاً يكون شيئاً الّا ثلثة آحاد ونضربه فى مثله فيكون مالاً وتسعة آحاد الّا ستّة اشياء فهو يعادل مالاً وشيئين الّا واحداً فنلقى مالاً الّا ستّة اشياء وواحداً من الناحيتين فيبقى ثمنية اشياء تعادل عشرة آحاد فالشىء الواحد يعادل واحداً وربعاً والمال خمسة وعشرون جزءاً من ستّة عشر جزءاً من الواحد وقد كان العدد المنقوص منه شيئين الّا واحداً وذلك اربعة وعشرون جزءاً من ستّة عشر جزءاً من الواحد وكذلك العدد المزيد عليه / فلنضرب كلّ ذلك فى ستّة عشر فيكونَ المال ٦ عدداً صحيحاً فيكون المال خمسة وعشرين والعدد المزيد عليه اربعة وعشرين والعدد المنقوص اربعة وعشرين

فلو ان العدد الذى فرضناه كان خمسة وعشرين كنّا قد صرنا الى ما طلبنا ولكن العدد المفروض اربعة وستّون والاربعة والستّون تعدّ الخمسة والعشرين مرّتين واربعة عشر جزءاً من خمسة وعشرين جزءاً من مرّة فنحتاج ان نضرب العدد المزاد وهو العدد المنقوص ايضاً اعنى الاربعة والعشرين فى اربعة وستّين فيكون ذلك الفاً وخمس مائة وستّة

---

3259 (910): Pro الاصغر codicis الآخر scripsi.

3263 (911): وعدداً : وحد عددا in cod.

3284 (912): الفاً : الف in cod.

وثلثين وهى اجزاء من خمسة وعشرين وذلك العدد الذى نزيده على اربعة وستّين وننقصه من اربعة وستّين ومن البيّن انّا اذا زدنا الفاً وخمس مائة وستّة وثلثين جزءاً من خمسة وعشرين ‹على اربعة وستّين اعنى الفاً وستّمائة جزء من خمسة وعشرين› يكون ذلك ثلثة الف ومائة وستّة وثلثين ‹جزءاً من خمسة وعشرين› وهو عدد مربّع وضلعه ستّة وخمسون جزءاً من خمسة واذا نقصنا الفاً وخمس مائة وستّة وثلثين جزءاً من خمسة وعشرين جزءاً من اربعة وستّين اعنى الفاً وستّمائة جزء من خمسة وعشرين يكون الباقى اربعة وستّين جزءاً من خمسة وعشرين وذلك عدد مربّع وضلعه ثمنية اجزاء من خمسة

فقد وجدنا عدداً اذا نقصناه من عدد مربّع من ضلع مكعّب يكون الباقى مربّعاً وان زدناه عليه كان المجتمع مربّعاً وهو الف وخمس مائة وستّة وثلثون جزءاً من خمسة وعشرين جزءاً من الواحد وذلك ما اردنا ان نجد

يا نريد ان نقسم عدداً مربّعاً مفروضاً بقسمين اذا زيد عليه احدهما كان المجتمع مربّعاً وان نُقص منه الآخر كان الباقى مربّعاً

فليكن العدد المفروض خمسة وعشرين احداً ونريد ان نقسم خمسة وعشرين /بقسمين اذا زدنا على الخمسة والعشرين احدهما كان المجتمع ١٤٧ عدداً مربّعاً وان نقصنا من الخمسة والعشرين القسم الآخر كان الباقى مربّعاً ولنلتمس ان نجد مربّعاً ما نقسمه بقسمين اذا زدنا عليه احدهما ونقصنا منه الآخر كان بعد الزيادة والنقصان مربّعاً وقد علمنا انّا اذا زدنا على مال جذريه وواحداً كان المجتمع مالاً وشيئين وواحداً وذلك عدد مربّع وان نقصنا من المال جذريه الّا واحداً كان الباقى مالاً وواحداً الّا جذرين وذلك عدد مربّع ولانّا نريد ان يكون العدد المزيد والمنقوص اذا جُمعا مالاً ولكنّهما اذا جُمعا اربعة اشياء فالاربعة الاشياء

---

3286 (913): الفاً : الف in cod.

3287-3288 (914): Per homoeoteleuton omissa verba restitui.

3289 (915): Uncis inclusa verba addidi.

3290 (916): الفاً : الف in cod.

3291 (917): الفاً : الف in cod.

(918): جزء : حرا in cod.

3292 (919): وستّين : وسون in cod.

3299 (920): الآخر : احدهما in cod.

تعادل مالاً فنقسم كلّ ذلك على شیء فيكون شیء يعادل اربعة آحاد ومن اجل ان الشیء هو ضلع المال يكون المال ستّة عشر وقد كان العدد المزيد عليه شيئين وواحداً وذلك تسعة آحاد والعدد المنقوص منه شيئين الّا واحداً وذلك سبعة آحاد والتسعة والسبعة اذا جُمعا كانا ستّة عشر احداً فقد صرنا الى مطلوبنا

ولكن العدد المفروض خمسة وعشرون احداً فلنضرب التسعة الآحاد في الخمسة والعشرين فيكون ذلك مأتين وخمسة وعشرين فنقسمها على الستّة عشر فتكون مأتين وخمسة وعشرين جزءاً من ستّة عشر جزءاً وذلك احد قسمي الخمسة والعشرين اعني القسم المزيد وايضاً فلنضرب السبعة الآحاد في الخمسة والعشرين ⟨فيكون ذلك مائة وخمسة وسبعين⟩ ونقسم ذلك على ستّة عشر فيكون مائة وخمسة وسبعين جزءاً من ستّة عشر وهو القسم الآخر اعني المنقوص من الخمسة والعشرين ومن البيّن انّا اذا زدنا المأتين والخمسة والعشرين الجزء على الخمسة والعشرين الاحد اعني اربع مائة جزء من ستّة عشر كان المجتمع ستّمائة جزء وخمسة وعشرين جزءاً من ستّة عشر وذلك عدد مربّع ضلعه خمسة وعشرون جزءاً / ٤٨ من اربعة وايضاً فانّا اذا نقصنا القسم الآخر اعني المائة والخمسة والسبعين جزءاً من ستّة عشر من الاربع مائة الجزء كان الباقي مأتين وخمسة وعشرين جزءاً من ستّة عشر وذلك عدد مربّع وضلعه خمسة عشر جزءاً من اربعة واذا جُمع القسمان كانا خمسة وعشرين احداً

فقد قسمنا الخمسة والعشرين قسمين على التحديد الذی حُدّ لنا وهما مائتان وخمسة وعشرون جزءاً من ستّة عشر مائة وخمسة وسبعون جزءاً من ستّة عشر وذلك ما اردنا ان نعمل

ولمّا لم يُمكن ان نجد عدداً مربّعاً اذا قسمناه ⟨بقسمين⟩ وزدنا عليه كلّ واحد منهما كان مربّعاً اردنا ان نأتي بما يُمكن

---

3312-3313 (921): اذا جُمعا كانا : ادا حمعب كاىب in cod.

3314 (922): وعشرون : وعسرس in cod.

3315 (923): فنقسمها : ممسمها in cod.

3318-3319 (924): In uncis seclusa verba addidi, sed dubitanter.

3330 (925): نعمل : نحد (sc. نجد) in codice. Vide etiam adn. 933, 940,955 (recte autem in linea 3343).

3331 (926): بقسمين addidi.

يب فنقول نريد ان نقسم عدداً مربّعاً مفروضاً بقسمين اذا نقصنا منه كلّ واحد منهما كان الباقي مربّعاً

فليكن العدد المفروض خمسة وعشرين احداً ونريد ان نقسم خمسة وعشرين احداً بقسمين اذا نقصنا منه كلّ واحد منهما كان الباقي مربّعاً فلنلتمس هذه الشريطة في مربّع ما وكلّ مربّع يُقسم بقسمين مربّعين فانّا اذا نقصنا كلّ واحد من القسمين من المربّع فان الباقي وهو القسم الآخر هو مربّع وقد تقدّم في كتابنا هذا العمل به ويكون احد القسمين ستّة عشر والقسم الآخر تسعة

فقد قسمنا الخمسة والعشرين بقسمين اذا نقصنا كلّ واحد منهما من خمسة وعشرين كان الباقي مربّعاً وهما تسعة آحاد ستّة عشر احداً وذلك ما اردنا ان نعمل

يج نريد ان نقسم عدداً ⟨مربّعاً⟩ مفروضاً ثلثةَ اقسام اذا زيد عليه كلّ واحد منها كان المجتمع مربّعاً

وليكن العدد المفروض خمسة وعشرين احداً ونريد ان نقسم خمسة وعشرين احداً / ١٤٩ ثلثةَ اقسام اذا زيد كلّ واحد منها على الخمسة والعشرين كان المجتمع مربّعاً ومن اجل انّا اذا قسمنا عدداً مربّعاً ثلثةَ اقسام وزدنا كلّ قسم منها على العدد المقسوم يجتمع من ذلك ثلثة اعداد يكون العدد المركّب من جملتها مثل اربعة امثال العدد المقسوم فمن اجل ذلك يكون اذا قسمنا الخمسة والعشرين ثلثةَ اقسام وزدنا كلّ قسم منها على الخمسة والعشرين تكون جملة الثلثة الاعداد مائة احد فلنقسم المائة ثلثةَ اقسام مربّعة وليكن كلّ قسم منها اكثر من خمسة وعشرين وقد تقدّم في كتابنا هذا كيف نقسم العدد المربّع بأقسام مربّعة ونستغني عن اعادة العمل فيكون احد الاقسام ستّة وثلثين والقسم الآخر ثلثين احداً وثلثمائة وسبعين جزءاً من ثمان مائة وأحد واربعين جزءاً من الواحد والقسم الآخر ثلثة وثلثين احداً واربع

---

3339 (927): هو مربّع : وهو مربع in codice. Forsan sunt verba وهو القسم الآخر و interpolata; vide adn. 770.

3342 (928): احداً eadem manu supra lineam.

3344 (929): مربّعاً addidi.

3351 (930): فمن اجل : ومں احل in cod.

3355 (931): اعادة : العاده in cod.

(932): احد : احدا in cod.

مائة وأحد وسبعين جزءاً من ثمان مائة وأحد واربعين جزءاً من الواحد ولمّا كان كلّ واحد من هذه الاقسام الثلثة انّما هو مركّب من الخمسة والعشرين ومن قسم واحد من اقسام الخمسة والعشرين فانّا اذا نقصنا من كلّ واحد من هذه الاقسام الثلثة خمسة وعشرين كان الذى يبقى من كلّ قسم هو قسم من اقسام الخمسة والعشرين ولكنّا اذا نقصنا خمسة وعشرين من ستّة وثلثين يكون الباقى احد عشر وهو القسم الاوّل من اقسام الخمسة والعشرين وايضاً فلننقص الخمسة والعشرين من القسم الثانى وهو ثلثة وثلثون احداً واربع مائة وأحد وسبعون جزءاً من ثمان مائة وأحد واربعين فيكون الباقى ثمنية آحاد واربع مائة/وأحد وسبعين ١٥٠ جزءاً من ثمان مائة وأحد واربعين وهو القسم الثانى من اقسام الخمسة والعشرين وايضاً فانّا اذا نقصنا الخمسة والعشرين من القسم الثالث وهو ثلثون احداً وثلثمائة وسبعون جزءاً من ثمان مائة وأحد واربعين يكون الباقى خمسة آحاد وثلثمائة وسبعين جزءاً من ثمان مائة وأحد واربعين وهو القسم الثالث من اقسام الخمسة والعشرين وقد تجتمع هذه الثلثة الاقسام فتكون خمسة وعشرين ويزاد كلّ واحد منها على الخمسة والعشرين فيكون المجتمع عدداً مربّعاً

فقد قسمنا الخمسة والعشرين ثلثةَ اقسام اذا زيد كلّ واحد منها على الخمسة والعشرين كان المجتمع عدداً مربّعاً والاقسام هذه الاوّل احد عشر احداً الثانى ثمنية آحاد واربع مائة وأحد وسبعون جزءاً من ثمان مائة وأحد واربعين جزءاً من الواحد الثالث خمسة آحاد وثلثمائة وسبعون جزءاً من ثمان مائة وأحد واربعين وذلك ما اردنا ان نعمل

يدَ نريد ان نقسم عدداً مربّعاً مفروضاً بثلثة اقسام اذا نُقص منه كلّ واحد منها كان الباقى مربّعاً

فلنفرض العدد المربّع خمسة وعشرين احداً ونريد ان نقسم خمسة وعشرين احداً ثلثةَ اقسام اذا نقصنا كلّ واحد من الاقسام من الخمسة والعشرين يكون الباقى مربّعاً ومن اجل انّا اذا قسمنا الخمسة والعشرين ثلثةَ اقسام ونقصنا كلّ واحد من الاقسام من الخمسة والعشرين وجدت

3378 (933): نعمل : نحد in cod.

3379 (934): Numeratio propositionis XIV in margine scripta est; non enim reliquit librarius necessarium spatium neque in fine lineae in qua praecedens propositio concluditur, neque in initio lineae sequentis.

من ذلك ثلثةُ اعداد يكون العدد المركّب من جملتها خمسين
فلنقسم الخمسين بثلثة اقسام مربّعة وليكن كلّ واحد منها اقلّ من خمسة
وعشرين وقد تقدّم فى كتابنا هذا كيف نقسم عدداً بأقسام مربّعة
فلنفن بما تقدّم عن الاعادة فيكون احد الاقسام ستّة عشر احداً والقسم
الآخر اثنان وعشرون/احداً وثلثة اجزاء من مائة وتسعة وستّين جزءاً من ١٥١
الواحد والقسم الآخر احد عشر احداً ومائة وستّة وستّون جزءاً من مائة
وتسعة وستّين جزءاً من الواحد ولمّا كان كلّ قسم من هذه الاقسام
مساوياً للخمسة والعشرين اذا نُقص منها كلّ واحد من اقسامها وجب
ان نلقى كلّ قسم من هذه الاقسام الثلثة من الخمسة والعشرين فيكون
الباقى من الخمسة والعشرين هى اقسام الخمسة والعشرين ولكنّا اذا
ألقينا ستّة عشر احداً من خمسة وعشرين كان الباقى تسعة آحاد وهو
القسم الواحد من اقسام الخمسة والعشرين وايضاً فاذا ألقينا اثنين
وعشرين احداً وثلثة اجزاء من مائة وتسعة وستّين جزءاً من خمسة وعشرين
كان الباقى احدين ومائة وستّة وستّين جزءاً من مائة وتسعة وستّين وهو
القسم الآخر وايضاً فاذا ألقينا احد عشر احداً ومائة وستّة وستّين جزءاً
من مائة وتسعة وستّين من خمسة وعشرين يكون الباقى ثلثة عشر احداً
وثلثة اجزاء من مائة وتسعة وستّين جزءاً من الواحد ⟨وهو القسم الآخر⟩
ويكون العدد المركّب من هذه الاقسام الثلثة اعنى التسعة والاحدين
والمائة والستّة والستّين جزءاً من مائة وتسعة وستّين والثلثة عشر وثلثة
اجزاء من مائة وتسعة وستّين خمسةَ وعشرين احداً واذا نُقص كلّ قسم
من هذه الاقسام الثلثة من الخمسة والعشرين كان الباقى عدداً مربّعاً

فقد قسمنا الخمسة والعشرين بثلثة اقسام اذا نُقص كلّ واحد منها
من الخمسة والعشرين كان الباقى عدداً مربّعاً وهى ثلثة عشر احداً
وثلثة اجزاء من مائة وتسعة وستّين جزءاً من واحد تسعة آحاد احد
عشر احداً ومائة وستّة وستّون جزءاً من مائة وتسعة وستّين وذلك ما
اردنا/ان نعمل ١٥٢

---

3388 (935): فلنفن : فلنعى in cod.

3393 (936): Post نلقى addit codex من.

3394 (937): الباقى : melius dixisset interpres التى تبقى. Vide etiam adn. 168.

3401 (938): Verba وهو القسم الآخر addidi.

3404 (939): Ante خمسةَ habet codex و.

3410 (940): نعمل : نحد in cod.

يه نريد ان نقسم عدداً مربّعاً مفروضاً بأربعة اقسام يكون قسمان من الاربعة الاقسام اذا نُقص كلّ واحد منهما من العدد المربّع المفروض كان الباقي مربّعاً ويكون ايضاً القسمان الباقيان من الاربعة الاقسام اذا زيد كلّ واحد منهما على العدد المربّع المفروض يكون المجتمع عدداً مربّعاً

فليكن العدد المربّع المفروض خمسة وعشرين احداً ونريد ان نقسم خمسة وعشرين اربعةَ اقسام يكون قسمان من الاربعة الاقسام اذا نُقص كلّ واحد منهما من الخمسة والعشرين كان الباقي مربّعاً وقسمان اذا زيد كلّ واحد منهما على الخمسة والعشرين كان المجتمع مربّعاً فلنلتمس هذه الشريطة في عدد ما مربّع ومن اجل انّا اذا زدنا على عدد مربّع اعني مالاً مثل ضلعه مرّتين وواحدٍ كان المجتمع مربّعاً فانّا نجعل احد الاقسام شيئين وواحداً وايضاً فانّا ان زدنا على مال مثل ضلعه اربعَ مرّات واربعة آحاد فان الذي يجتمع مربّع فلنفرض القسم الآخر المزيد اربعة اشياء واربعة آحاد ويكون العدد المركّب من القسمين المزيدين ستّة اشياء وخمسة آحاد وايضاً فانّا اذا نقصنا من المال مثل ضلعه مرّتين الّا واحداً اعني شيئين الّا واحداً كان الباقي مربّعاً فانّا نفرض احد القسمين المنقوصين شيئين الّا واحداً وايضاً من اجل انّا اذا نقصنا من العدد المربّع اعني المال مثل ضلعه اربعَ مرّات الّا اربعة آحاد يكون الباقي عدداً مربّعاً فانّا نجعل القسم الآخر المنقوص اربعة اشياء الّا اربعة آحاد ويكون العدد المركّب من القسمين ⟨المنقوصين⟩ ستّة اشياء الّا خمسة آحاد وقد كان العدد المركّب من القسمين المزيدين ستّة اشياء وخمسة آحاد فالعدد

---

3417 (941): Post اقسام addit codex verbum قسمة, quod delevi. Fortasse legendum est قسمةً, accusativo scilicet specificationis, ut invenitur in pag.197-199 editionis a L. Nix curatae Mechanicorum Heronis vel in prop.II,11 Elementorum ex interpretatione Hajjāj ibn Yūsuf.

(942): قسمان : قسمين in cod.

(943): Pro اذا praebet codex, ut videtur, مادا.

3421 (944): Pro وواحدا codicis وواحدٍ (ut in lineis 3187 & 3190) scripsi.

3425 (945): القسمين : القسم in cod.

3431 (946): المنقوصين addidi.

المركّب من الاربعة الاقسام اثنا عشر/شيئاً فهى تعادل مالاً لانّه انّما ١٥٣
كان غرضُنا ان نقسم المال بأربعة اقسام فنقسم مالاً على شىء فيكون شيئاً ونقسم اثنى عشر شيئاً على شىء فتكون اثنى عشر احداً فالشىء يعادل اثنى عشر احداً والشىء ضلع المال فالمال مائة واربعة واربعون احداً وقد كان احد القسمين المزيدين عليه شيئين وواحداً فهو خمسة وعشرون احداً وكان القسم الآخر المزيد اربعة اشياء واربعة آحاد فهو اثنان وخمسون احداً وايضاً فقد كان احد القسمين المنقوصين شيئين الّا واحداً فهو ثلثة وعشرون والقسم الآخر اربعة اشياء الّا اربعة آحاد فهو اربعة واربعون احداً فقد بلغنا ما اردنا من إلتماس هذا العدد المربّع ولم نبلغ ما نريد من إستتمام المسئلة

فلو كان العدد المفروض مائة واربعة واربعين كنّا قد انتهينا الى ما نريده ولكنّه خمسة وعشرون احداً ومن اجل ذلك وجب ان نضرب كلّ قسم من اقسام المائة والاربعة والاربعين فى الخمسة والعشرين ونقسم ما اجتمع من ذلك على مائة واربعة واربعين ولكنّا اذا ضربنا الاوّل من الاقسام وهو خمسة وعشرون فى الخمسة والعشرين كان المجتمع من ذلك ستّمائة وخمسة وعشرين واذا قسمناه على مائة واربعة واربعين كان ستّمائة وخمسة وعشرين جزءاً من مائة واربعة واربعين وذلك احد القسمين المزيدين على الخمسة والعشرين وايضاً فمن اجل ان القسم الآخر المزيد اثنان وخمسون احداً فانّا نضرب اثنين وخمسين احداً فى خمسة وعشرين فيكون ذلك الفاً وثلثمائة ⟨واذا قسمناه على مائة واربعة واربعين كان الفاً وثلثمائة⟩ جزء من مائة واربعة واربعين وهو القسم الآخر المزيد وايضاً من اجل ان احد القسمين المنقوصين ثلثة وعشرون احداً فانّا نضرب خمسة وعشرين فى ثلثة وعشرين/ فتكون ١٥٤
خمس مائة وخمسة وسبعين ونقسمه على مائة واربعة واربعين فيكون خمس مائة وخمسة وسبعين جزءاً من مائة واربعة واربعين وهو العدد المنقوص من الخمسة والعشرين وايضاً من اجل ان القسم الآخر المنقوص اربعة واربعون احداً فانّا نضرب اربعة واربعين احداً فى خمسة وعشرين

---

3436 (947): والمال : فالمال in cod.

3443 (948): Pro كنّا praebet codex عنا .

3452-3453 (949): In uncis seclusa, forsan per homoeoteleuton omissa, verba addidi; sed vide adn. 952.

3455 (950): Expectandum erat نضرب ثلثة وعشرين فى خمسة وعشرين .

فتكون الفاً ومائة ⟨ونقسمه على مائة واربعة واربعين فيكون الفاً ومائة⟩ جزء من مائة واربعة واربعين وهو القسم الآخر المنقوص من الخمسة والعشرين ومن البيّن انّا اذا جمعنا هذه الاقسام الاربعة كانت خمسة وعشرين احداً وان زدنا على الخمسة والعشرين كلّ واحد من القسمين المزيدين كان المجتمع مربّعاً وان نقصنا من الخمسة والعشرين كلّ واحد من القسمين المنقوصين كان الباقى مربّعاً

فقد قسمنا خمسة وعشرين بأربعة اقسام على الشرط الذى شُرط لنا وهى الزائدان ستّمائة وخمسة وعشرون جزءاً من مائة واربعة واربعين الفٌ وثلثمائة جزء من مائة واربعة واربعين المنقوصان خمس مائة وخمسة وسبعون جزءاً من مائة واربعة واربعين الفٌ ومائة جزء من مائة واربعة واربعين وذلك ما اردنا ان نعمل

وبمثل هذا العمل نعمل مسئلة يكون سؤالها نريد ان نقسم عدداً مربّعاً مفروضاً بثمنية اقسام تكون اربعة اقسام اذا زدنا كلّ واحد منها على المربّع المفروض كان المجتمع مربّعاً والاربعة الاقسام ⟨الباقية⟩ اذا نقصنا كلّ واحد منها من العدد المفروض كان الذى يبقى عدداً مربّعاً

يو نريد ان نجد ثلثة اعداد مربّعة وتكون ايضاً متناسبة ولكن اذا نُقص الاوّل من الثانى كان الباقى مربّعاً واذا نُقص الثانى من الثالث كان الباقى مربّعاً

ومن اجل انّه اذا كانت ثلثة اعداد مربّعة/وهى ايضاً متناسبة اذا ٥٥
نُقص الاوّل من الثانى كان الباقى مربّعاً فمن طبع هذه الاعداد ان
تكون اذا نُقص الثانى من الثالث كان الباقى مربّعاً فلنفرض العدد
الاوّل واحداً والعدد الثالث مالَ مال فمن اجل ذلك يكون العدد
الثانى مالاً ولكنّا اذا نقصنا العدد الاوّل وهو واحد من العدد الثانى

---

3460 (951): الفاً:الف in cod.

(952): Uncis inclusa verba addidi.

3462 (953): *Alif* posterius verbi اذا scripsit librarius supra lineam.

3463 (954): وان : فان in cod.

3470 (955): نعمل : نحد in cod.

3473 (956): الباقية addidi.

3483 (957): و verbi واحد eadem manu supra lineam.

وهو مال يكون الباقى مالاً الّا واحداً فنحتاج ان يكون عدداً مربّعاً فلنفرض له ضلعاً من شىء الّا احدين ونضربه فى مثله فيكون مالاً واربعة آحاد الّا اربعة اشياء فهو يعادل مالاً الّا واحداً فنزيد على الناحيتين اربعة اشياء وواحداً فيكون مال واربعة اشياء تعادل مالاً وخمسة آحاد فنلقى المال المشترك فيبقى خمسة آحاد تعادل اربعة اشياء فالشىء الواحد يكون واحداً ورُبعاً ومن اجل انّا فرضنا العدد الثانى مالاً وضلعه شىء والشىء واحد ورُبع اعنى خمسة اجزاء من اربعة يكون المال خمسة وعشرين جزءاً من ستّة عشر جزءاً من الواحد ولكن العدد الثالث فُرض مال مال وهو من ضرب المال فى مثله وذلك ستّمائة وخمسة وعشرون جزءاً ⟨من مأتين وستّة وخمسين فالعدد الثالث ستّمائة وخمسة وعشرون جزءاً⟩ من مأتين وستّة وخمسين جزءاً من الواحد والعدد الاوّل على ما فرضناه اعنى انّا فرضناه واحداً ولكنّا اذا نقصنا العدد الاوّل وهو واحد من العدد الثانى وهو خمسة وعشرون جزءاً من ستّة عشر يكون الباقى تسعة اجزاء من ستّة عشر جزءاً من الواحد وهو عدد مربّع ضلعه ثلثة اجزاء من اربعة وايضاً فانّا اذا نقصنا العدد الثانى وهو خمسة وعشرون جزءاً من ستّة عشر جزءاً اعنى اربع مائة جزء من مأتين وستّة وخمسين جزءاً من العدد الثالث وهو ستّمائة وخمسة وعشرون جزءاً من مأتين وستّة وخمسين يكون الباقى مأتين وخمسة وعشرين جزءاً من مأتين وستّة وخمسين جزءاً من الواحد وهو عدد مربّع /وضلعه خمسة ١٥٦
عشر جزءاً من ستّة عشر

فقد وجدنا ثلثة اعداد على التحديد الذى حُدّ لنا وهى واحد اربع مائة جزء من مأتين وستّة وخمسين ستّمائة وخمسة وعشرون جزءاً من مأتين وستّة وخمسين وذلك ما اردنا ان نجد

**يز نريد ان نجد اربعة اعداد مربّعة وتكون ايضاً متناسبة و⟨يكون⟩ العدد المركّب من جميعها مربّعاً**

ومن اجل انّه اذا كانت اربعة اعداد متناسبة فضرب الاوّل فى الرابع مثل الثانى فى الثالث فانّا نفرض العدد المربّع الاوّل واحداً والرابع ستّة عشر مالاً والعدد الثانى اموالاً تكون اذا زدناها على ستّة

---

3487 (958): مال : مالا in cod.

3493-3494 (959): In uncis seclusa verba addidi, sed dubitanter.

3507 (960): يكون addidi.

عشر مالاً كان المجتمع اموالاً عددها مربّع وذلك تسعة اموال لانّا اذا
زدنا تسعة اموال على ستّة عشر مالاً كان ⟨المجتمع⟩ خمسة وعشرين مالاً
وهو عدد مربّع وضلعه خمسة اشياء ولكن العدد الثانى اذا ضُرب فى
الثالث كان مثل العدد الاوّل اذا ضوعف بالرابع والعدد الاوّل اذا
ضوعف بالرابع يكون ستّة عشر مالاً فنقسم ستّة عشر مالاً على تسعة اموال
فتكون واحداً وسبعة اتساع وهو العدد الثالث ومن اجل ⟨ذلك فان⟩
العدد المركّب من الاربعة الاعداد خمسة وعشرون مالاً وأحدان وسبعة
اتساع احد فانّا نحتاج ان يكون مربّعاً فلنفرض له ضلعاً يكون خمسة
اشياء وثُلثَ واحد ونضربه فى مثله فيكون خمسة وعشرين مالاً وثلثة اشياء
وثُلثَ شىء وتُسعَ واحد فهو يعادل خمسة وعشرين مالاً وأحدين
وسبعة اتساع احد فنلقى المشتركات من الناحيتين فيبقى ثلثة اشياء
وثُلث ⟨شىء⟩ تعادل احدين وثُلثى احد فالشىء الواحد ثمنية اجزاء
من عشرة اجزاء من الواحد ومن اجل ان ضلع العدد الثانى ثلثة اشياء
والعدد ⟨الثانى⟩ تسعة اموال / يكون ضلعه اربعة وعشرين جزءاً من ٥٧
عشرة ويكون العدد الثانى خمس مائة وستّة وسبعين جزءاً من مائة
وايضاً فمن اجل ان الرابع فُرض ستّة عشر مالاً وضلعه اربعة اشياء
والشىء ثمنية اجزاء من عشرة تكون الاربعة الاشياء اثنين وثلثين جزءاً
من عشرة وهو ضلع العدد الرابع والعدد الرابع الف واربعة وعشرون
جزءاً من مائة من الواحد ومن اجل انّا فرضنا العدد الاوّل واحداً
يكون واحداً كما فرضناه وكنّا فرضنا العدد الثالث واحداً وسبعة اتساع
فهو كما فرضناه واحد وسبعة اتساع وكلّ عدد من هذه الاربعة الاعداد
مربّع والعدد المركّب من جملتها ستّة عشر الفاً وتسع مائة جزء من تسع
مائة وهو عدد مربّع وضلعه مائة وثلثون جزءاً من ثلثين جزءاً من الواحد

فقد وجدنا اربعة اعداد على التحديد الذى حُدّ لنا وهى على
الولاء واحد" خمس مائة وستّة وسبعون جزءاً من مائة واحد" وسبعة
اتساع الفٌ واربعة وعشرون جزءاً من مائة جزء من واحد وذلك ما اردنا
ان نجد

---

3513 (961): المجتمع addidi.

3517 (962): Verba ذلك فان addidi.

3523 (963): شىء deest in cod.

3525 (964): الثانى addidi.

3536 (965): Pro الولاء scripsit librarius الولا ل الاول .

يح نريد ان نجد اربعة اعداد مربّعة وهى ايضاً متناسبة وتكون اذا نُقص الاوّل من الثانى كان الباقى مربّعاً وان نُقص الثانى من الثالث كان الباقى مربّعاً وان نُقص الثالث من الرابع كان الباقى مربّعاً

⟨وكنّا⟩ قد وجدنا ⟨انّه⟩ فى طبع العدد ان كلّ اربعة اعدان متناسبة وهى ايضاً مربّعة وكان العدد الاوّل منها اذا نُقص من العدد الثانى يكون الباقى مربّعاً فالعدد الثالث ايضاً اذا نُقص من العدد الرابع كان الباقى مربّعاً ومن اجل ذلك نطلب اعداداً اربعةً مربّعةً متناسبةً ⟨اذا نُقص الاوّل من الثانى كان الباقى مربّعاً وان نُقص الثانى من الثالث كان الباقى مربّعاً⟩ فلنفرض العدد الاوّل ما شئنا من الآحاد بعد ان تكون مربّعة فلنفرضه تسعة آحاد ومن اجل انّا اذا
نقصنا الاوّل من الثانى يكون الباقى / مربّعاً فلنفرض الثانى ما شئنا من ١٥٨
العدد المربّع الذى اذا نُقص منه تسعة آحاد كان الباقى مربّعاً فلنفرضه خمسة وعشرين احداً ولنفرض العدد الرابع ما اردنا من الاموال بعد ان يكون مربّعاً فلنفرضه مالاً واحداً ولمّا كان ضرب الاوّل وهو تسعة آحاد فى العدد الرابع وهو مال تسعةَ اموال وجب ان يكون ضرب العدد الثانى وهو خمسة وعشرون احداً فى العدد الثالث ايضاً تسعة اموال فالعدد الثالث تسعة اجزاء من خمسة وعشرين جزءاً من مال ولكنّا اذا نقصنا العدد الثانى وهو خمسة وعشرون احداً من العدد الثالث وهو تسعة اجزاء من خمسة وعشرين جزءاً من مال كان الباقى تسعة اجزاء من خمسة وعشرين جزءاً من مال الّا خمسة وعشرين احداً فنحتاج ان يكون مربّعاً فلنفرض له ضلعاً يكون ثلثة اخماس شىء الّا واحداً ونضربه فى مثله فيكون تسعة اجزاء من خمسة وعشرين جزءاً من مال وواحداً الّا ستّة اخماس شىء فذلك يعادل تسعة اجزاء من خمسة وعشرين جزءاً من مال الّا خمسة وعشرين احداً فلنزد على الناحيتين جميعاً ستّة اخماس شىء وخمسة وعشرين احداً ونلقى التسعة الاجزاء المشتركة فيبقى ستّة اخماس شىء تعادل ستّة وعشرين احداً فالشىء الواحد مائة وثلثون جزءاً من ستّة ومن اجل ان العدد الرابع فُرض مالاً وضلعه شىء والشىء مائة وثلثون جزءاً من ستّة فالعدد الرابع

---

3542 (966): Verba وكنّا et انّه addidi.

3544 (967): فالعدد : والعدد in cod.

3546-3547 (968): In uncis seclusa verba, quae necessaria esse existimo, addidi.

ستّة عشر الفاً وتسع مائة جزء من ستّة وثلثين جزءاً من الواحد وايضاً فمن اجل ان العدد الثالث تسعة اجزاء من خمسة وعشرين جزءاً من مال يكون ستّة الٰف واربعة وثمنين جزءاً من ستّة وثلثين جزءاً من الواحد فاذا نقصنا العدد الاوّل وهو تسعة آحاد من العدد الثانى وهو خمسة وعشرون احداً يكون الباقى /ستّة عشر احداً وهو عدد مربّع وضلعه اربعة ٥٩ آحاد وان نقصنا العدد الثانى وهو خمسة وعشرون احداً اعنى تسع مائة جزء من ستّة وثلثين جزءاً من الواحد من العدد الثالث وهو ستّة الٰف واربعة وثمنون جزءاً من ستّة وثلثين جزءاً من الواحد يكون الباقى خمسة الٰف ومائة واربعة وثمنين جزءاً من ستّة وثلثين جزءاً من الواحد وهو عدد مربّع وضلعه اثنان وسبعون جزءاً من ستّة اجزاء من الواحد وايضاً فاذا نقصنا العدد الثالث وهو ستّة الٰف واربعة وثمنون جزءاً من ستّة وثلثين من العدد الرابع وهو ستّة عشر الفاً وتسع مائة جزء من ستّة وثلثين كان الباقى عشرة الٰف جزء وثمان مائة جزء وستّة عشر جزءاً من ستّة وثلثين جزءاً من الواحد وهو عدد مربّع ضلعه مائة واربعة اجزاء من ستّة

فقد وجدنا اربعة اعداد على التحديد الذى حُدّ لنا وهى تسعة آحاد وخمسة وعشرون احداً وستّة الٰف واربعة وثمنون جزءاً من ستّة وثلثين وستّة عشر الفاً وتسع مائة جزء من ستّة وثلثين جزءاً من الواحد وذلك ما اردنا ان نجد

تمّت المقالة السابعة من كتاب ذيوفنطس فى الجبر والمقابلة وهى ثمانى عشرة مسئلةً

وتمّ الكتاب والحمد لله ربّ العالمين ووقع الفراغ من نسخه بتاريخ يوم الجمعة الثالث من صفر سنة خمس وتسعين وخمس مائة حامداً لله تعالى ومصلّياً على نبيّه محمّد وآله اجمعين

---

3567 (969): الفاً : الف in cod.

3569 (970): واربعة : واربع in cod.

3573 (971): Pro جزء من scripsit librarius حروں, copulatione ut videtur verborum حرو (sc. جزء) et من .

3579 (972): جزء وستّة : حرو من سه in cod.

(973): جزءاً : احرا in cod.

3587 (974): ثمانى عشرة : ىمىه عسر in cod.

3589 (975): حامداً : حامدں in cod.

Part Five

# Arabic Index

This index contains all words of any pertinence to the text of Books IV–VII (excluding words which occur only in the *incipit*, lines 1–5, and in the *explicit*, lines 3588–90). The Greek equivalents, where listed, are of course given only *à titre d'indication*; in most cases they have been arrived at after comparison with similar passages in the Greek Diophantus.

The basic reference dictionary used has been Wehr's (original German edition); for words or meanings not found in Wehr, we have referred to some other dictionaries, or to Arabic original texts or translations from the Greek.

*atà* (I): 8 (+*ʿalà*); 3332 (+*bi*).

*taʾattin*: (nomen verbi *atà*, V) 439 (but see p. 99, n. 48).

*min ajl*: 1°. +*anna*: 67, 68, 601, 905–6, 977, 983, 1725, 3020 passim. Gr. ἐπεί, ἐπεὶ γάρ, as in Hypsicles (e.g., lines 25, 76). See also *wajaba*, 2°.
2°. +*ḏālika*: 170, 637, 692, 991, 1750, 3444, 3482, (3517), 3545.

*aḥad*: 1°. M̊, sc. μονάς: 51, 52 (bis), 66, 67, 81 (bis) passim.
2°. (prior): 134, 136, 202, 275, 374 passim;
(primus): 3001, 3017, 3031, 3112, 3150, 3355, 3388. Cf. *āḫar*, 2°.
3°. in the expression of a fraction $m/n$, $1 \leq m < n \leq 10$: 482, 483, 485, 506, 618 passim; less frequent than *wāḥid* (q.v.).
N.B. On *aḥad* in association with tens in the expression of numerals, see p. 37.

*aḫaḏa* (I): 1°. =λαμβάνειν (to take, e.g., the difference, the root, the half; cf. D.G., I,92,20; 134,25; 330,9): 960, 1348, 1352, 1354, 1808, 1812, 1918, 1992, 1997, 2054, 2102, 3155, 3157.
2°. =ζητεῖν(?): 1174 (and app.). Cf. D.G., I,120,14 (where, however, the meaning is rather that of ἐκτιθέναι).
3°. + *fī*: 1240, (1389).

*maʾḫaḏ*: 379 (bis), 388, 412. Gr. ἀγωγή?

*āḫar*: 1°. ἕτερος = other (of two): 48, 88, 134, 136, 173 passim.
2°. repeated in enumeration of more than two objects: 3001–3, 3356–57, 3389–90. Gr. (ὁ μέν...) ὁ δέ...ὁ δέ (D.G., I,374,3–4 and 17–18), (εἷς μέν)...ὁ δὲ ἕτερος...ὁ δέ (370,11). The same use of *āḫar* is found in other mathematical texts, cf. Kutsch 69, lines 8–9 (=(εἷς μέν)...ἕτερος δέ...τρίτος δέ); Abū Kāmil, *Alg.*, 87v,7–8; 105v,14–15.
3°. ἕτερος = other, different: 38, 200, 262, 343, 1030, 1034, 1066 passim.

*adà* (II): 299, 2542.

*iḏ*: 8 (*ammā iḏ*), 17, 205, 2505.

*iḏā*: 1°. 35, 48, 50, 106, 107 (bis) passim.
2°. +*mā*: 1323, 1327, 1449, 1455, 1542.
N.B. 1°. In the statement of two parallel conditions, *iḏā* commonly introduces the first and *in* the second (cf. Reckendorf, *A.S.*, p. 484; *S.V.*, p. 685): 377, 973–75, 1065–66, 1069–70, 1412–13 passim; one finds *iḏā*... *wa-iḏā* as well (277–78, 442–43, 519–20, 852, 956–57 passim), while *in*... *wa-in* is rarely used (cf. 2661–63 (cf. 2637–38), 3221–22 seqq. (cf. 3180–81)).
2°. The verb of the apodosis can be in the imperfect as well as in the perfect; both tenses are found in the formulation of VII,15.

*iḏan*: Gr. οὖν, ἄρα.
1°. written *iḏan*: 46, 51, 64, 77, 104 passim.
2°. written *iḏan*: 1705, 1717, 1753, 1778, 1826 passim.
The second spelling does not supersede the former one, as is seen, e.g., in lines 1706, 1724.

*aṣl*: 1361. Cf. p. 120, n. 89.

*mu'allaf*: 1033, 1264. Gr. συγκείμενος, but only in the sense of *Arithmetica* II,9 (a number being *the sum* of two squares). Otherwise, συγκείμενος is translated in our text by *murakkab* or by *mujtama*ᶜ.

Other occurrences: Klamroth, 298 (*ullifa*); Apoll.-Nix, 14; Ṭūsī, e.g., VIII,5; Heron, *Mech.* (Nix), 199,6; Samaw'al, *Bāhir*, 150,16–17 (allegedly quoting Diophantus: cf. p. 12).

*ilà*: Besides its use after various verbs, *ilà* is found in the expression of a ratio; see, e.g., 411, 1626, 1629 (bis), 1632, 1666. Cf. *min*, 2°.

*ammā*: 1°. *ammā* ... *fa* = (μέν) ... δέ: (50), 128, 270, 626 (bis), 640 passim.
2°. *ammā iḏ*: 8.

*illā*: Gr. ⋀. Cf. *ġair*.
228, 254, 257, 713, 717 passim.

In those cases in which two terms are subtracted, we find either *illā* ... *wa* (1827, 2620, 2621–22, 2626, 2628 passim) or *illā* ... *wa-illā* (1867, 1942, 1971, 2063, 2065 passim). The same in, e.g., Abū Kāmil's *Algebra* (cf. 93ᵛ,4 and 7), al-Karajī's *Badī*ᶜ, 124ʳ and 125ᵛ.

*innamā*: ἤτοι, or used for emphasis: 80 (interp. ?; cf. p. 31, no. 13), 205, 559, 811, 2972 (interp.), 3359, 3433.

*ānif*$^{an}$: 1414, 1701.

*awwal*: 15, 262, 416, 422, 428 passim.
Feminine: *ūlà* in 318, 1027; cf. app. crit., note 347 (and p. 33, no. 23). Otherwise, *awwala*$^{h}$ (forma vitiosa), as in 988, 989, 1026, 1123, 1124, 1257 passim.
Adv. *awwal*$^{an}$: 411, 584, 712, 1341.

*al-ān*: 136, 1159, 1594.

*ayy*: 282, 285, 1179, 3096, 3099 passim. Note the plural *ayyat* in 283, 285.

*aiḍ*$^{an}$: ἔτι, πάλιν.

1°. = also, again: 10, 12–14, 76, 92, 118 passim. Used to point out the second of two considered quantities, e.g., in 1025, 3198, 3200, 3249.

2°. at the beginning of a sentence, *wa-aiḍ*$^{an}$ (*fa-*) introduces an alternative reasoning (e.g., 296, 965), a new aspect of a problem (e.g., 730, 1549), a subsequent step in the analysis (e.g., 983, 1020) or in the synthesis (e.g., 1151, 1188).

N.B. The occurrence of an initial *wa-aiḍ*$^{an}$ (*fa-*) is frequent in (and characteristic of) translations of Greek works, and it may be understood both as a Grecism (ἔτι, ἔτι δέ) and a Syriacism (*tūḇ*, *tūḇ dēn*).

Other examples of *wa-aiḍ*$^{an}$ (*fa-*) are Georr, 71 and Endreß, 66 (ἔτι, ἔτι δέ); Hypsicles, line 76 and Ḥajjāj, prop. I,1 (πάλιν).

*bada'a* (I): 3143.

*badal*$^{an}$ *min*: 3105. Cf. *maqām*.

*baʿd*: 1°. *baʿda*: 9, 35, 221, 226, 233, 737 passim.
2°. *min baʿdi* (cf. Reckendorf, *A.S.*, p. 475): 745, 838, 876, 939, 2073.
3°. *baʿda an* (not "after that", but "provided that"): 283, 285, 1398, 2356, 3548, 3552. Found with the same meaning in other mathematical works, e.g., al-Karajī's *Badīʿ*, $95^{r}$–$95^{v}$ (Anbouba, 62,10 and 15); Abū Kāmil's *Algebra*, $80^{r}$,12 and 20. Gr. μόνον ἵνα? (as in D.G., I,94,15). Cf. Kutsch, 293.

*baʿḍ*: 1°. = fraction, Gr. μόριον (μέρος?): 255. Used also in al-Karajī's *Faḫrī*: see Woepcke, *Extrait*, 22 or supra, p. 188, n. 24. Compare the use of *baʿḍ* in al-Ḫwārizmī's *Alg.*, 119,10 seqq.
2°. repeated, expresses the reciprocity (ἀλλήλων): 38 (bis). Cf., e.g., Klamroth, 295; Georr, 62–63 and 208–9.

*baġà* (VII): *yanbaġī* = δεῖ (cf. *iḥtāja*): 35, 438, 468, 516, 538 passim. Associated with a verb other than *kāna*, *yanbaġī* may well render the verbal adjective in -τέος (cf. Georr, 94); as in 740 and 1034 (διαιρετέον), 603 and 993 (ζητητέον), 2047 (ἰστέον). See also under *arāda*, 2°, α.

*baqiya* (I): (κατα-)λείπεσθαι; λοιπός (as, e.g., in D.G., I,16,19).
1°. to remain, to result (after a subtraction): 224, 233, 236, 252, 254 passim.
2°. to remain, to result (other operations involved); cf. *kāna*, *ḫaraja*, etc.:
—division: (737), 876.
—restoration and reduction: 257, 1208, 2073, 2115 (cf. D.G., I,226,14; 254,18).
—restoration, reduction and division: 745.
3°. auxiliary to *ʿādala* (Gr. λοιπός . . . ἴσος): 215, 243, 257 (cf. 2°), 263, 686 passim.

*baqīya*$^{h}$: remainder of: 1209, (1210), 2033, 2122, 2159.

*bāq*$^{in}$: λοιπός (adj. or subst.).
1°. adj.: 983, 987, 1019, 1025 (bis), 1026 passim.
2°. subst. 853, 1505, 1607, 1869, 2099 passim.

*bal*: 2061. Gr. (οὐκ . . .) ἀλλά, as in D.G., I,218,20; 246,6.

*balaġa* (I): 2686, 2719, 2742, 2751, 2928, 2933 passim; 3441, 3442 (= *intahà ilà*).

*mablaġ*: 3211, 3215; cf. app. crit., note 208.

*bāna* (I): 171 (or form II? Cf. app. crit.).

*bāna* (II): δεικνύναι.
1°. to show = to expound: 1034, 2181, 3102, 3149.
2°. to find; syn. *tabayyana*, *wajada*: 1538, 2086. Also in the concluding words of problems IV,1–6 (afterwards replaced by *wajada*); Gr. ὅπερ ἔδει δεῖξαι/εὑρεῖν (perhaps another translation of this expression in 581).

*bāna* (V): to find (cf. *bayyana*): 271, 461, 489, 508, 555, 841 passim. Gr. δεικνύναι? (*wa-qad tabayyana* = ἐδείχθη δέ).

*bāna* (X): 851, 1100; in these two places, *istabāna* introduces corollaries, and such is its use in Ṭūsī's Euclid also (see the corollaries in I,10; I,15 etc. and in III,1).

*bayyin*: *bayyin anna*, or *min al-bayyin anna*, is used at the beginning of sentences indicating, generally, that one of the requirements of a problem has been fulfilled; thus it can be found in the analyseis as well as in the apodeixeis of problems. See 2488, 2581, 2620 (see p. 69), 2648, 2794, 2820, 2990–91 and 399 (and 401), 2635, 2854, 3005, 3201, 3286.

Both *bayyin* and *min al-bayyin* probably stand for φανερόν or δῆλον.

*bayān*: 2209, 2438, 3150.

*talā* (I): 13.

*tamma* (I): 1615, 2168, 2918, 3586, 3588 (end of the Books and of the ms).

*tamām*: 597 (*ʿalà 'l-tamām*).

*istitmām*: 3142, 3442.

*ṯabata* (I): (+ *ʿalà*) 1361, 1409, (1494). See app., n. 450.

*ṯabata* (IV): 2541 (cf. app.).

*ṯumma*: 165, 279, 412, 449, 560, 737, 1116 passim.

*muṯanna*ⁿ: see under *musāwā*ʰ, 2°.

*mustaṯna*ⁿ: 1030, 1812.

*jabara* (I):

1°. alone: to restore, i.e., to make an expression (the side of an equation) consist of positive terms only, by adding to it its subtracted terms (taken positively). Compare with that the (non-mathematical) meaning given by Blachère et al. in their *Dictionnaire*, p. 1297: "the girl was sold in order that, with her price, the sum might be completed" (*ḥattà yujbar al-māl min ṯamanihā*). The added terms by means of which the deficiency is removed are introduced by the (instrumental) *bi*.
229, 2557, 2592.

No Greek correspondent is known: the restoration of one side *and* the increasing of the other side by the same quantity are conceived as simultaneous operations in D.G. (which is the case for *jab*(*a*)*r*(*a*) in our text only when it is associated with (*mu*)*qābala*⁽ʰ⁾).

2°. with *qābala*; the *two* terms then mean:

(α) to restore and reduce (an equation).
257, 1208, 2114, 2149, 2379.
Cf. Greek κοινὴ προσκείσθω ἡ λεῖψις καὶ (ἀφῃρήσθω) ἀπὸ ὁμοίων ὅμοια, e.g., D.G., I,26,27–28; 90,17–18.

(β) = to restore (no common term to reduce; cf. p. 65, n. 36). 2246, 2400; 2513 (where the suppression of common magnitudes is indicated by a following *alqà al-muštarakāt*).

There is the same usage in al-Ḫwārizmī's *Algebra* (p. 31,14–15), but its genuineness is made dubious by the Latin translation, which has only "restaurabis" (Libri, *Hist.*, I,280,8).

($\gamma$) = to restore and solve (or: and divide by the power of lesser degree). 2342.

*qābala* alone in the sense of "to solve" is found in al-Ḫwārizmī's *Algebra* (37,18; 41,8; 114,1 and 19—in Libri's text, *qābala* is rendered in the first instance by "operare" and in the second by "facere"), and also in al-Karajī's *Badī*ʿ (see my study on it, p. 303).

*jabr*: 1°. alone:
35 (post.; (interp.) def. of the term).
2°. with *muqābala*ʰ (cf. "def. XI" of D.G.).
9,35 (prius), 1040, 2073, 2115, 2150; 3586.
3°. with *muqābala*ʰ and *qisma*ʰ (cf. "def. XIII").
745, 1493.

There are no known Greek equivalents for the Arabic words *jabr* and *muqābala*ʰ, although the appearance of two words to denote the common addition and the common removal would have been an expected development. This need was apparently felt by Planudes, who, in his commentary, simply uses the words πρόσθεσις and ἀφαίρεσις (cf. D.G., II,171 seqq.).

*jiḏr*: 21–34, 49, 51 (all interp.), 2506, 2659, 2972 (interp.), 3187–88, 3229–31, 3267, 3305–7.

Gr. πλευρά, used both in the sense of *ḍila*ʿ, latus, and of *jiḏr*, radix (the latter quite clearly in D.G., I,310,9). Except in the (interpolated) passage in 2972, *jiḏr* and *ḍila*ʿ are used synonymously (cf. 3187 with 3421).

N.B. The plural *juḏūr* is found in two (interpolated) places, in lines 49 (nine roots) and 51, while *ajḏār* appears in 3231 (four roots). Arabic mathematicians do not seem to make a distinction between the regular plural and the plural of paucity, at least not according to the number (see Luckey, *Richtigkeitsnachweis*, 98–100).

*jirmī*: 14. Used for στερεός (see p. 67, n. 42).

*juz*ʾ: μέρος, μόριον. Cf. *ba*ʿ*ḍ*.
1°. used in the expression of a general fraction *m*/*n* (see p. 39):
256 (bis), 324 (bis), 325 (bis), 327, 331 (bis) passim.
2°. = aliquot fraction: 282–85, 346 (post.), 386, 393. Cf. p. 95, n. 33. The *ayy juz*ʾ *au ayyat ajzā*ʾ in IV,14 is clearly the μέρος τι ἢ μέρη found in the Greek "IV",33. Comp. also lines 386 (aliquot) with 393 (non-aliquot).

*ja*ʿ*ala* (I): 1°. syn. *faraḍa*, i.e., τάσσειν: 81, 109, 111, 262, 263 passim.
= τάσσειν ἐν (cf. D.G., I,120,18): 1077, 1082, 1385, 1386, 1398 passim.
2°. various senses of "to make": 11; 15 (+imperfect); 763 (+impf. of ʿ*ādala*) and 2068, 2111 (+*mu*ʿ*ādil*) = ποιεῖν ... ἴσον; 873 (+*muštarak*; cf. Apoll.-Nix, 14), 1484; passim.
3°. *ja*ʿ*ala* ... *ajzā*ʾ *min* = ἀναλύειν εἰς μόριον?
2459, 2464.

*jama*ʿ*a* (I): συντιθέναι.
1998, 2173, 2175, 2667, 2679, 2816, 2823, 2836 passim.

*jamaʿa* (VIII): γίνεσθαι, ποιεῖν, etc. See p. 66.

1°. to result (from: *min*).

(α) after an addition: 208, 218, 238, 247, 899, 957, 974, 1003 passim.

N.B. The verb can be used alone (2259, 2410, 3423), or be followed by a *min* referring to the operation (e.g., 1511, 1517 (post.)), to the two addends (e.g., 974, 1003), or to one addend (e.g., 899, 1092, 2125).

See also *mujtamaʿ*, 2°, α, in fine.

(β) after a multiplication: (121), 194, 201, 303, 305, 315, 316, 325, 347 passim.

N.B. The verb can be used alone (e.g., 361, 370), or be followed by a *min* referring to the operation (e.g., 201, 303) or to the multipliers (see 371, 413, 1107).

2°. to be added.

2672, 3371. Syn. *jumiʿa* (comp. 2672 with 2679).

3°. auxiliary to *ʿādala.*

985, 1129, 1215, 1263, 1300, 1308, 1374, 1484.

*jamʿ*: 84. Cf. app. crit.

*jamīʿ*:

1°. the whole (e.g., we divide "the whole", i.e., the two sides of the equation): 39, 48, 65, 78, 92; 1362 passim.

2°. the sum: 41, 57, 73, 627, 682 passim; constructed with genitive or *min* (as in 41,820). Cf. *jumla^h^*.

3°. *jamīʿ^an^*: 1487, 2631, 2656, 2731, 2801 passim; *ʿalà/min al-nāḥiyatain jamīʿ^an^* = *ʿalà/min kil(t)à al-nāḥiyatain.*

*majmūʿ*:

1°. (adj.) 623, 677, 706, 797, 824, 2001 passim. Cf. app., note 243.

2°. (subst.; cf. *jumla^h^*) 1955, 2059, 2068, 2069, 2112, 2113, 2147, 2148; *majmūʿ* does not supersede *jumla^h^* (cf. 1958, 2008, 2061).

*mujtamaʿ*:

1°. result (from a multiplication).

(α) (subst.) alone (e.g., 443 (bis), 2834), or with *min* referring to the operation (e.g., 201, 812) or (apparently) to the multiplicands (e.g., 436, 466).

(β) (adj.; +*min ḍarb*) 866, 868, 869, 870, 1280 passim.

2°. result (from an addition).

(α) (subst.) alone (e.g., 2243, 2428, 3264), or with *min* referring to the operation (1170, 1172), or to the addends (918, 2823; 2667, 3016). Also synonymous with *jamīʿ*, *jumla^h^* as in 2016, 2178, 2873; cf. the particular use of *ijtamaʿa* in 1817.

(β) (adj.)

—(with *min*) syn. *murakkab min*: 991.

—(alone) = resulting from an addition (ant. *bāq^in^*): 987, 988, 990, 1303, 1305, 1380, (1382); 2028.

*jumlaʰ*: 44, 55, 75, 82, 642, 846 passim. Syn. *jamīʿ*, *majmūʿ*.

N.B. The word "sum" is often omitted in Greek when the sum's constituents are mentioned (cf., e.g., D.G., I,40,15; 42,26; 146,2 and 5; 152,8–9; 190,7; 354,17; Euclid, *Elem.*, VII, def. 22). This omission is frequent in the second part of our Arabic translation (as in lines 1990, 1994–95, 1998, 2044; see also 2816, 3035—cf. 2843). Such an omission occurs also in some texts Arabic in origin (e.g., *Badīʿ*, fol. 99ᵛ and 106ʳ⁻ᵛ (titles), 126ʳ⁻ᵛ).

*jins*: 205, 2923. Gr. γένος; cf. pp. 261 and 263.

*jawāb*: 17.

*ḥattà*: 39, 46, 61, 63, 77, 88 passim; chiefly consecutive (not in line 39: final (=ἕως?)).

*ḥadda* (I): to impose (a condition); occurs (in association with *taḥdīd*) in the final statements (συμπεράσματα) of problems V,14; VI,1–23; VII,1–7, 11, 16–18. Analogously used is *šaraṭa*.

*ḥadda* (II): to determine; used as a synonym of *wajada* (cf. 1613 with 1593) in the final statements of problems IV,40, 41, 42.b, 43, 44.c; V,2, 4, 13. Otherwise used in lines 1594; 1702 (interp. ?; see p. 128, n. 9).

*taḥdīd*: see *ḥadda* (I).

*ḥaṣala* (I): 1355.

*taḥṣīl*: 12.

*ḥifẓ*: 11.

*taḥlīl*: ἀνάλυσις; see p. 48.
1122, 1409, 1494.

*ḥāja* (VIII): δεῖν. Cf. *baġà* (VII), *wajaba*.
+ *an*: 44, 75, 122, 142, 163 passim.
+ *illà an*: 145, (168), 2060, (2355), 2726–27, 3270.

*ḥājaʰ*: 2271 (*laisa bi-nā ḥājaʰ*).

*ḥāṭa* (IV): 100, 102, 112, 114, 116 passim. Gr. περιέχειν. See p. 90, n. 15.

*ḥāla* (X): 2418.

*ḥāza* (II): 1944, with the sense attributed to it by Dozy, *Suppl. dict. ar.*, i.e., "faire disparaître"; syn *aḏhaba*.

N.B. The reading *jabara* naturally comes to mind in 1944, since the word is written without diacritical points. But the phrase can hardly refer to a restoration, whereas it makes perfect sense with the meaning given by Dozy. Thus our interpretation.

We are also inclined to read as *ḥayyaza* the *jabara* of a similar passage in al-Ḫwārizmī's *Algebra* (24,9), even though the Latin translator also read *jabara* (cf. Libri, *Hist.* I,274,7: "restaurant").

*ḥīnaʾiḏⁱⁿ*: 122, 143, 146, (169).

*ḫaraja* (I): γίνεσθαι, etc. (except 4°).

1°. to result (from: *min*)

—after a division: 21, 22 ($1^{um}$), 24 (bis), 25; 564 passim.

—after a multiplication: 22 ($2^{um}$ and $3^{um}$), 26, 30; 1186 passim.

N.B.: The verb can be used alone (e.g., 331, 1186), or with *min*, variously used (see, e.g., 149, 1378; 50, 108; 124, 127).

2°. to come out as (sister of *kāna*; see Caspari–Wright, II,103, n.). 80, 406, 2547, 2879, 2880, 3082.

Other examples: Hypsicles, lines 74, 79, 104; Abū Kāmil, *Algebra* $80^r$,3; $84^r$,18; $87^v$,8–9.

3°. Auxiliary to *ᶜādala*: 65, 142, 146, 166, 215, 230 passim.

4°. to be soluble (of a problem): 2047, 2417.

*ḫaraja* (X): 600.

*ḫārij*: (+*ᶜan*) 2923.

*ḫāṣṣa*$^h$: 584, 597, 609. Gr. ἰδιότης?

*ḫaṭṭī*: 10. Gr. γραμμικός (cf. Nicomachus, *Introd. arithm.* II,7,3; Ṯābit translates by *ḫuṭūṭī*). See p. 175.

*muḫālif*: 2924.

*muḫtalif*: 1265 (=*āḫar*, 1034), 1420, 1808, 1809, 1811 passim. Ant. *mutasāwi*$^n$. Gr. ἕτερος, ἄνισος.

*tadbīr*: 609.

*durba*$^h$: 16, 2925. Both times associated with *ᶜāda*$^h$.

*daraja*$^h$: 16 (bis).

*daᶜā* (I): 439.

*ḏā*: *min ḏā*: 591 (several *ᶜalà ḏā* in Endreß, e.g., 69–70).

*hā-ka-ḏā*: 1410.

*ḏālika*: The difference between *hāḏa* and *ḏālika* is not strictly observed in our text (nor is it in others; cf. Georr, 63), except in the particular case of line 1562 seqq.

*ka-ḏālika*: see *ka-*.

*li-ḏālika*: 269, 398, 427, 456, 481 passim.

*wa-ḏālika* + conj.:

1°. = nam: 1828, 1869, 1943. Gr. γάρ, δῆλον γάρ; cf. Endreß, 63 and 83 seqq.; common in Menelaus' *Sphärik* (cf. 39,5; 42,11; 43,11 and 19 passim). Syn. *wa-ḏālika li-anna* (1442).

2°. = igitur: 2461, 3158. Syn. *aᶜnī an*.

*ḏakara* (I): 123, 608 (interp.), 1158 (interp.), 1344, 1701, 2017 (interp. ?; cf. p. 32, no. 21), 2159.

*ḏikr*: (associated with *taqaddama*) 172, 714, 1410.

*maḏkūr*: 610.

*ḏahaba* (I): 1872. Also in al-Ḫwārizmī's *Algebra* 18,17; 24,12.

*ḏahaba* (IV): (1829). Cf. *ḥayyaza.*

*ḏū*: 1410 (*masāʾil ḏawāt al-ṭarafain*).

*Ḏiyūfanṭus* 1, 7, 1615, 1618, 2168, 2171, 2918, 2921, 3586. In 1 and 1618 with the epithet *al-iskandarānī*. See. p. 4, n. 4.

*raʾà* (I): 12.

*murabbaʿ*: (subst. and adj.) Gr. τετράγωνος (τετραγωνικός).
2, 7, 20, 41, 45 (bis) passim (τετραγωνικός, e.g., 3548; cf. D.G., I,300,1).
N.B.: *murabbaʿ murabbaʿ al-ḍilaʿa*: 453, 454, (538–39), 542, 580. Gr. τετράγωνος πλευρὰν ἔχων τετράγωνον (cf. D.G., I,296,3 and 5–6; 362,6). Syn. *murabbaʿ min ḍilaʿ murabbaʿ*, as in 2963, 2986, 3008.

*martaba*$^{h}$: 11.

*rajaʿa* (I): 1361, 1389. Gr. ἀνατρέχειν. Syn. *ʿāda.*

*rasama* (I): 17 (bis), 18.

*raqiya* (I): 15.

*rakiba* (II): 1361, 1409, 1493. Gr. συντιθέναι (not in the sense of "to add" but "to make the synthesis (*tarkīb*) of").

*tarkīb*: 1390. Gr. σύνθεσις, with the meaning explained on p. 48.

*murakkab*: συγκείμενος.
1°. *murakkab min* = συγκείμενος ἐκ = compositus per additionem: 1807, 1810, 1999, 2709, 2777, 3080, 3101, 3359, 3402, 3424 passim.
Also found in Hypsicles, lines 4, 8 passim, as a substantive (= ὁ συγκείμενος). See under *muʾallaf.*
2°. *murakkab min jumla*$^{h}$*/jamīʿ*: 2687, 2692, 2719, 2724, 2744, 2751 passim/3068, 3125, 3508.
3°. + *maʿa*: 14 (συντεθείς?).

*rāda* (IV): A. Used as an auxiliary—presumably in a periphrastic translation.
1°. Repeated use.
(α) In the formulations of problems:
*nurīd an najid* (*naqsim* in V,9, 12; VII,4, 7, 11–15), with which all problems begin, corresponds to the Greek aorist II infinitive εὑρεῖν (διελεῖν) of D.G.
Isḥāq uses the very same expression (Klamroth, 286), while Ḥajjāj commonly inserts a *nubayyin kaifa* (found exceptionally in Ṭūsī, as in VIII,2 and 4: usual is *li-nā an*, cf. Klamroth, ibid.).
The expression *nurīd an* etc. may occur again in the restatement of problems involving given numbers once the values of these numbers have been settled in the ἔκθεσις; see 277, 376, 419, 441, 470 passim.
(β) In the conclusions of problems:
*wa-ḏālika mā aradnā an najid* (*nubayyin*, IV,1–6; *nafʿal*, V,9 and 12; *naʿmal*, VII,12) corresponds to the Greek ὅπερ ἔδει εὑρεῖν (δεῖξαι, ποιῆσαι). The same expressions are found in the translations of the *Elements* (see Klamroth, 286).
2°. Sporadic use.

*arāda* seems to have been used as an auxiliary for rendering various formulations such as "let us", "we shall (now)", "we must". Thus, it may have served with an appropriate word (most often a verb in the subjunctive preceded by *an*) to translate:
(α) the verbal adjective in -τέος (cf. Endreß, 75).
(β) again, the idea of necessity, *arāda* playing the same rôle as *iḥtāja* (cf. 3191 with 3232); thus in Hypsicles, line 112, *fa-nurīd an naʿlam* stands for δεῖ δὴ εὑρεῖν.
(γ) this idea of necessity merely amounts in some cases to a simple future; an example is found in the Arabic version of Galen's *Anatomica* (with a change of person and voice): cf. Simon, I, pp. L–LI (first lines);
(δ) the adhortative subjunctive (cf. Endreß, 75);
(ε) a simple participle with a future sense, as may be the case in 3145–46 (*al-ʿadad al-murabbaʿ allaḏī nurīd qismatahū* reminding one of ὁ διαιρούμενος τετράγωνος in the sense of "partiendus quadratus", as in D.G., I,92,5–6);
(ζ) finally, *arāda* may simply belong to Arabic phraseology, and have no correspondent in the Greek: cf. Endreß, 67–68; see also Ḥajjāj II,6 and 9, in which *nurīd an nubayyin anna* stands for λέγω ὅτι. Note, however, that our text does not seem to be characterized by this kind of verbose phraseology (see p. 67).

To one or the other of the above categories belong the following instances of the Arabic Diophantus: 226, 280, 711, 1344, 2002, 2242, 2493, 2842, 3020, 3142 (apod. to *lammā*), 3146 (prius: cf. supra, (ε)), 3155, 3191 (cf. supra, (β)), 3307, 3332 (apod. to *lammā*).

**Remark.** Such uses of *arāda*, attested in many translations from the Greek, do not necessarily apply to every *arāda*. This is particularly true, as far as our text is concerned, for point (β), since D.G. uses θέλειν much in the sense of obligation (so in 192,19; 196,11 and following pages). As to point (γ), one should keep in mind the use in koine-Greek of θέλειν for expressing the future from hellenistic times onwards (cf., e.g., Dieterich, *Untersuchungen*, 245–46).

B. Used alone.
See, e.g., 202, 240; 1461, 1906; 3240, 3444. In particular, synonymous with *šāʾa* in expressing arbitrariness of choice (61, 118, 3551).

*raib*: 2541: *lā raiba fīhi* (but see app.).

*muzdawij*: 11. Gr. συζυγής? Cf. Apollonius, *Con.* I, def. (p. 6,1 of Nix's excerpt of the Arabic text); Hypsicles, line 34.

*zāda* (I):
1°. + *ʿalà*, and Acc. = προστιθέναι τινί τι. Syn. *aḍāfa*.
207, 220, 229, 237, 760, 770 passim.
—with *muštarak(at)$^{an}$*
717, 735, 835, 964, 985 passim. Cf. pp. 65–66.

2°. + ʿalà, and Acc. = ὑπερέχειν τινός τινι.
272, 691, 712, 752, 861 passim.
N.B. In the sense of ὑπερέχειν, *zāda* is constructed in the modern usage with *bi* instead of the Acc.; note that while Ṯābit's Apollonius has the first construction (Nix, 13), his Nicomachus apparently has the second (Kutsch, 102,3).

*ziyādaʰ*:
1°. πρόσθεσις.
36 (interp.), 221, 1469, 1484, 1511, 1518 passim. Ant. *nuqṣān*.
2°. ὑπεροχή. Cf. *faḍl*.
690, 728, 730, 738, 826, 844 passim.
3°. (non-mathematical) 2924.

*mazīd*:
1°. (substantive) 873. Cf. Freytag, *Lexicon*: "accessio, augmentum, incrementum". Syn. *ziyādaʰ*.
2°. (participle) προστιθέμενος.
3207, 3277, 3278, 3307, 3311 passim. Ant. *manqūṣ*; see lines 3307–8: *al-ʿadad al-mazīd wa'l-manqūṣ* = ὁ προστιθέμενος καὶ ἀφαιρούμενος (ἀριθμός), D.G., I,28,13 and 22–23.

*muzād*: Syn. of *mazīd*. 3197, 3198, 3200 (bis), 3210, 3283. This form is found in Johnson's *Dictionary*, p. 1168 (under: *mazād* (A), in fine).

*zāʾid*:
1°. added, positive = ὑπάρχων. Ant. *nāqiṣ*.
1068, 1100, 1622, 1714, 1830, (1871), 1945 passim.
2°. (+ *ʿalà*): added (to). Syn. *mazīd*.
2007, 2009; 3467.

*suʾāl*: formulation (of a problem). 3143, 3471.

*masʾalaʰ*: 8, 9, 13, 17, 18, 160, 273, 318, 328, 447 passim.
—*masāʾil ʿadadīyaʰ* = προβλήματα ἀριθμητικά (D.G., I,4,10), προβλήματα ἐν τοῖς ἀριθμοῖς (ibid., 2,3): 8, 1618, 2168, 2919, 2922.
—*masāʾil muhayyaʾaʰ*: προβλήματα πλασματικά: 496, 1801 (cf. 439).
Explicit reference to a problem (namely III,6): 3102, 3149–50.

*sabab*: 1870 (*bi-sababi*), 2924.

*saṭḥī*: ἐπίπεδος (cf. Nicomachus, *Introd. arith.*, II,7,3; Ṯābit: *musaṭṭaḥ*).
10. Cf. p. 175.

*musaṭṭaḥ*: ἐπίπεδος. Cf. Klamroth, 297; Euclid-Ṭūsī, VII, deff.
541. See *saṭḥī*.

*salaka* (I): 15.

*maslak*: 15, 18.

*samà* (II): 14, 23, 27, 31, 799, 802, 829 passim. Gr. καλεῖν.

*musammaⁿ*: 858, 859.

*sahl*: ῥᾴδιος. 2180, 2208, 2437, 3017, 3102; (*ashal*:) 2842, 3143.

*suhūla*[h]: 171 (*min suhūla*[h]; interpolated—if from Greek times: διὰ τὴν εὐχρηστίαν?).

*sawiya* (III):
1°. ἴσος εἶναι, ἰσοῦν (= ʿ*adala*, I and III): 335.
2°. to make the same, identical (cf. *musāwā*[h], 1°): 1240.

*sawiya* (VI): to be the same, identical. 969.

*sawiya* (VIII): to be equal one to another. 993, 1127, 1175.

*musāwā*[h]: 1°. identicalness. 1363, 1417. Cf. line 969 (*sawiya*, VI).
2°. *al-musāwā*[h] *al-muṯannā*[h] = διπλοισότης, διπλῆ ἰσότης, διπλῆ ἴσωσις. 960, 977 (cf. app. crit.), 1348.

*musāw*[in]: ἴσος τινί (*li*). 52, 164, 181, 385, 524, 571 passim.

*mutasāw*[in]: ἴσος (absolute). 36 (interp.), 190, 454, 455, 1033, 1264 passim.

*šabaha* (IV): 955 (*mā ašbahahū*).

*mutašābih*: see *muštarak*.

*šaraṭa* (I): Found, associated with *šarṭ*, in the συμπεράσματα of problems IV,7, 12, 13, 16–18, 21–24, 32, 33, 37–39; V,6; VII,15. Cf. *ḥadda*.
Otherwise: 447.

*šaraṭa* (VIII): used as *šaraṭa* (I) in the conclusions of problems IV,40, 41; V,4, 15, 16.

*šarṭ*: see the two previous words. Similarly constructed with *arāda* in IV,31; V,1 and 2.
Otherwise: 559.

*šarīṭa*[h]: 3337, 3420.

*mušārak*: 283; (cf. notes 422, 676 of the app. crit.). Gr. σύμμετρος.

*muštarak*: κοινός.
1°. (adj.) 214, 243, 686, 696, 718 passim. More examples under *jaʿala*, *zāda*, *alqà*, *naqaṣa*. On *muštarak*, see also pp. 65–66.
2°. (subst.)
—*al-muštarak*: 2631 (single power with coefficient unity);
—*al-muštarakāt*: 2183 (cf. app.), 2212, 2513, 2594, 3522.
—*al-mutašābihāt al-muštarakāt*: 2559, 2675–76 (cf. app.), 2700.

*šāʾa* (I): 43, 45, 76, 87, 90 passim. Only in expressions such as *kam* (*mā*) *šiʾnā* (*min*) = ὅσος δήποτε. See also *arāda*, B.

*šaiʾ*: 1°. sense of τι: 12, 1240. See also p. 67, n. 40. Cf. *mā*, 2°.
2°. = ς (ἀριθμός), sc. *x* (unknown): 21, 22, 24–27; 42, 43 (bis) passim. Cf. under ʿ*adad*, N.B.

*ṣiḥḥa*[h]: 2288.

*ṣaḥīḥ*: 1°. correct (of treatment, resolution): 2541.
2°. integral (of number): 3199, 3238, 3278. Gr. ὅλος, ὁλόκληρος.

*ṣāḥib*: 2289 = "related" (of problems).

*ṣaġīr*: 1°. 1176–77, 1215–16, 1261 (bis), 1263, 1487, 1488 (bis). See p. 115, n. 75.
  2°. *aṣġar* = ἐλάττων: 42, (53), 54, 57, 60 passim. Ant. *aʿẓam*, *akbar*.

*ṣināʿa$^h$*: 13. Gr. τέχνη(?), i.e., art, science.

*ṣāba* (IV): 1999. Syn. *ṣāra* (+*ilà*), *intahà* (+*ilà*).

*ṣāra* (I): γίνεσθαι.
  1°. —*naṣīr ilà an naṭlub*: 170, 692, 1665. Cf. D.G., I,214,7
  —*qad ṣirnā ilà mā ṭalabnā/maṭlūbinā*: 2505, 3280–81, 3313.
  2°. syn. εἶναι, ποιεῖν: 221, 1203, 1536, 1757, 1760, 1816, 1991 passim.
  3°. auxiliary to *ʿādala*: 735, 1022, (1205), 1251.

*ḍaraba* (I) πολλαπλασιάζειν (ἐπί: *fī*). Syn. *ḍāʿafa*.
  20, 22, 25, 30, 154, 194, 200, 275 passim.

*ḍarb*: πολλαπλασιασμός.
  23, 26, 30, 31, 80, 110 passim.

*maḍrūb*: 1°. (subst.) product of multiplication; *al-maḍrūb* (*min*) = ὁ ὑπό.
  471, 472, 546 (bis), 561–63, 568 passim.
  2°. (adj.) multiplied by (*fī*); Gr. πολλαπλασιασθείς (ἐπί).
  801, 2265, 2268, 2274 (interp.), 2277, 2280, 2299 passim.

*ḍaʿafa* (III): πολλαπλασιάζειν (ἐπί: *bi*). Syn. *ḍaraba*.
  2685, 2689, 2690, 2706, 2718, 2722, 2741 passim. Does not supersede *ḍaraba*; cf., e.g., 2696, 2728, 2760; 3045, 3158, 3198.

*ḍiʿf*: διπλασίων.
  1384, 1419, 1420, 3198, 3200, 3203 passim.

*taḍʿīf*: πολλαπλασιασμός.
  38.
  This word is repeatedly used in the sense of "product" in Hypsicles (see lines 29, 31, 33, 38, 43 passim); Ṭūsī employs it both in the senses of "product" and "duplication" (cf. p. 215,4 (IX,16 = *Elem.* IX,15) with the formulations in IX,32 and 34).

*taḍāʿīf*: πολλαπλασιασμός.
  2818. Cf. note 800 of the app. crit.

*ḍilaʿ*, *ḍilʿ*: πλευρά (see also under *jiḏr*).
  20, 21, 42, 43, 45, 46 passim.

*ḍāfa* (IV): προστιθέναι (τινί: *ilà*). Syn. *zāda*.
  210, 217, 240, 246, 1809, 1812, (2188), 2904, 3063.

*ṭabʿ*: 3480 (*min ṭabʿ*), 3542 (*fī ṭabʿ*). Gr. (τῇ) φύσει, φυσικῶς; see, e.g., Georr, 231.

*ṭaraf*: equation (of a proposed system; see p. 111, n. 64).
  969, 1241, 1363, 1410 (*masāʾil ḏawāt al-ṭarafain*), 1418.
  N.B. The word *ṭaraf* is employed in other Arabic texts dealing with algebra to mean the *side* of an equation (=*nāḥiya$^h$*, *jiha$^h$*): cf. al-Ḫwārizmī,

*Alg.*, 184–85 (Rosen's excerpts from an Arabic text and a Persian one); Ḥājjī Ḫalīfaʰ (Flügel), II,583,8–9 (art. *jabr*).

In mathematical texts translated from the Greek, *ṭaraf* renders ἄκρος (sc. ὅρος: in a progression; cf. Klamroth, 301–2 (301: rather "ἄκρος")), or πέρας (in geometry; cf. Nix's Apollonius, 13; Klamroth, 297).

*ṭarīq*: 204 (*min ṭarīq*), 1343 (*ᶜalà ṭarīq*).

*ṭalaba* (I): ζητεῖν.

1°. =to look for: 170, 412, 603, 609, 692, 714, 733 passim. Syn. *iltamasa*.

2°. =to require that (*an*); cf., e.g., D.G., I,158,5; 244,4; 256,1. 712, 730, 1341, 1412.

3°. =to examine, to solve: 1522, 1549, 1594.

N.B. This verb occasionally takes the sense of ἐκτιθέναι, in the Greek text (cf. in D.G., p. 96,10 with 214,9; 272,11 etc.) as well as in the Arabic version (961, 1350).

*ṭalab*: 1363, 1417, 1774.

*maṭlūb*: ζητούμενος.

279, 344, 345, 354, 356; 3313 (τὸ ζητούμενον) passim.

*ẓahara* (I): 249.

*ẓāhir*: (+*anna*) 541, 1117, 1669, 1755. Gr. φανερόν, δῆλον; cf. (*min al-*) *bayyin*.

*ᶜadda* (I):

1°. =to measure (if $A/B = k$, $B$ "measures" $A$ by $k$): 516. Gr. μετρεῖν τι (Acc.) κατά τι (*bi*). Cf. D.G., I,134,16–18 and 22–23; 136,14 seqq.; passim. See also *Elem.*, VII, def. 3 ($k$ integral).

2°. =to number (if $A/B = k$, $A$ "numbers" $B$ by $k$): 3206, 3241, 3281.

In Greek, one normally uses μετρεῖσθαι, as is done in D.G. One passage, however, is ambiguous: Tannery's 220,19 (Vat. gr. 191: μετροῦσιν ἀριθμοὺς β κατὰ μ° β). (Another passage shows some confusion: 242,3; Vat. gr. 191 has the same reading as B, including the τουτέστι κατὰ πλευρὰς β̄ τῆς $\Delta^Y$, which looks very much like a later addition.)

*ᶜadad*: 1°. =ἀριθμός.

10, 14, 41 (bis), 51, 57 passim.

2°. =πλῆθος.

180 (post.), 571 ($1^{um}$), 594 ($1^{um}$), 691 (prius), 739, 982 (post.) passim.

N.B. The word is frequently omitted in the latter case; see, e.g., 572, 595, 690, 744, 989–90, 1025 passim; comp. also 1025 with 1026. This omission also occurs in the Greek text: as Tannery pointed out (D.G., II,264; cf. also p. 267 (δύναμις)), "interdum οἱ ἀριθμοί dicitur pro coefficiente $x$". Note that this use is found in original Arabic mathematical texts as well; see, e.g., Luckey, *Richtigkeitsnachweis*, 98 seqq.

*ᶜadadī*: ἀριθμητικός. Vide under *masʾalaʰ*.

*ᶜadala* (I): 1°. ἴσος (εἶναι), ἰσοῦν.

62, 64, 66, 79, 120, 214 passim.

2°. equate, ἰσοῦν (τινι: *bi*): 280.

Much more common (for both meanings) is the third form.

*ʿadala* (III): 1°. 9, 38, 40, 45, 47, 52 passim.

2°. (+*bi*) 142, 145, 165, (167), 306, 307 passim. See also p. 47.

N.B. Instead of the expression that an aggregate "is equal to a square", which is a common way of saying that it must be made a square, we sometimes find the ambiguous expression that an aggregate "is a square" (see, e.g., 761–62, 1077, 1405, 1528–29; cf. the ἔσται κύβος in D.G., I,438,11). The same occurs in al-Karajī's *Badīʿ*: see fol. 122$^{r-v}$.

*ʿadala* (VIII): 1074. Syn. *istawà*.

*muʿādala*$^{h}$: ἰσότης, ἴσωσις. Equation (i.e., resulting one in a problem; cf. p. 111, n. 64).

969, 988–990, 997, 1026–28, 1040, 1086, 1123–25, 1240, (1357), 1417, 1426, 1492.

*muʿādil*: ἴσος (τινί: *li*): 180, 281, 282, 285–87 passim. Cf. *musāw*$^{in}$.

*ʿarafa* (I): 17, 1408.

*maʿrifa*$^{h}$: 1361, 1389.

*ʿaẓīm*: Syn. *kabīr*; ant. *ṣaġīr*.

1°. 1177, 1215–16, 1260, 1262–63, 1308, 1488, 1490. See p. 115, n. 75.

2°. *aʿẓam* = μείζων: 43, 44, 54, 55, 58, 61 passim.

*ʿaks*: inverse. 204, 206, 328, 378, 955.

A problem or a treatment is the "inverse" of another one if it is formed from the latter by exchanging the names of the powers or the values of the given numbers. Gr. ἐναντίον? ἐναλλάξ? (cf. D.G., I, e.g., 194,18—cf. 194,7).

*ʿalima* (I):

1°. γινώσκειν.

609, 615, 1360, 1361, 1389.

2°. *ʿalimnā anna*: does *not* refer to anything previous, but rather states s.th. obvious (Aristotle, *Gen. An.* (arab.), 254: φανερὸν ὅτι) or involving some simple computations.

2440, 2872, 3304; 2458 (introduces the apodosis to *lammā*).

This use of *ʿalima* is not peculiar to translators; it also occurs in other (mathematical) works: see, e.g., al-Ḫwārizmī, *Algebra*, 6,3 or 34,6–7 (+*matà*), al-Karajī, *Faḫrī* V,12.

*maʿlūm*: δοθείς, δεδομένος.

1658, 1660, 1663, 1669, 1696 passim. See also p. 228.

That *maʿlūm* and *mafrūḍ* are synonymous in our text appears when one compares lines 1632 with 1669, 1691 with 1693, 1740 with 1742. This is also true for other treatises: in Ḥajjaj II,14, too, the two words render δοθείς.

*mutaʿallim*: 11.
Gr. ἀρχόμενος? (see D.G., I,2,10; 16,5).
The meaning "student", "beginner" (not: "educated person") is clear in, e.g., Heron's *Mechanics* (Nix: 63,13–14; 71,10). In Bergsträsser's *Ḥunain*, it translates the Greek εἰσαγόμενος (see text, p. 6,7 and (Register) p. 48, no. 116; its antonym there is *mustakmil*: see text, p. 7,21).

*ʿamada* (V): 2436.

*ʿamila* (I):
1°. πλάσσειν (ἀπό: *min*), κατασκευάζειν (cf. D.G., I,314,4).
45 (bis), 76, 241, 349, 457 (bis), 628; 1764, 1766 passim.
Cf. *farada*, 1°.
2°. (trans.) to treat, solve (problem): 204, 262, 977, 2541, 2842, 3142, 3471.
(abs. or with *fī*) 343, 614 (bis), 960, 1241, 1348, 1363, 1417.
3°. *wa-ḏālika mā aradnā an naʿmal* = ὅπερ ἔδει ποιῆσαι: 3343. (3330, 3378, 3410, 3470 by correction). Syn. *faʿala*.

*ʿamal*: treatment, resolution. 12, 35, 172 (cf. app.), 960, 977, 1363, 1410, 1480, 1559, 1601, 2047, 2288, 2418 passim.

*ʿinda*: πρός (in: λόγος τινὸς πρός τι). Cf. *ilà*, *min*.
1562–64, 1566–70, 1909, 1934, 1939.

*aʿnī*: τουτέστιν, λέγω δή.
1°. 35, 120, 132, 284 (+*an*), 316, 317 passim. Introducing (supposed) glosses (as τουτέστιν often does in Greek) in 35, 2391
2°. =ἔστω: 2973, 3421. Cf. D.G., I,144,15 (but: II,xlvi, 5).

*maʿna*$^{n}$: 12.

*ʿāda* (I): 1409. Syn. *rajaʿa*.

*ʿāda*$^{h}$: 16, 2925. Cf. *durba*$^{h}$.

*iʿāda*$^{h}$: 3150, (3355), 3388.

*ʿain*: (*wāḥid*) *bi-ʿainihī* = ὁ αὐτός.
1358, 1388.

*ġaraḍ*: 2539, 2922, 3434.

*ġaliṭa* (II): 1808.

*ġaniya* (I): 3388.

*ġaniya* (X): 3150, 3355. Syn. with the previous.

*ġair*: 1°. =without: 977; (*min ġair an*:) 1343, 2922–23.
2°. =other than: 3185, 3227, 3265.
3°. Synonymous with *illā*, but much less frequently used (as παρά compared to ∧). 1204, 1806, 1822, 1827, 1838.

*fa*: (+subj., =so that—if not copyist's mistake for *li*): 2495, 3277.

*farada* (I):
1°. to put; Gr. τάσσειν, πλάσσειν.

(α) type: *faraḍa al-murabbaʿ/ḍilaʿ al-murabbaʿ* (+assigned value).
42, 43 (bis), 60 (post.), 63, 74 (bis) passim.

(β) type: *faraḍa al-murabbaʿ min ḍilaʿ* (+assigned value).
53, 54, 60 (prius), 61, 67 passim. Cf. Greek πλάσσειν ἀπὸ πλευρᾶς (e.g., D.G., I,126,4 and 12). Syn. *ʿamila.*

(γ) type: *faraḍa ḍilaʿ al-murabbaʿ min* (+assigned value). Cf. e.g., D.G., I,244,5.
94 (post.), 124 (post.), 146, (182; cf. app.), 270.

(δ) same expression, but with the coefficient of the power not specified (Gr. τάσσειν ἐν, as in D.G., I,120,18; 136,3): 124 (prius), 141, 144, 164, (166), 180 passim.
Further on, the *min* is suppressed: (121; cf. app.), 684, 733, 978, 984, 1016, 1020 passim.

(ε) (Later) variant of (α)–(γ): *faraḍa li'l-murabbaʿ ḍilaʿan yakūn* + Acc./*min*: 2340, 2589, (2673), 2695, 2824 passim/3485. Cf. also 528, 715.

2°. to take, choose (τάσσειν, as in D.G., I,136,14): 411 (post.), 2874, 2875.

3°. +conj. = to suppose, stipulate: 318, 328, 591.

*mafrūḍ*: 1°. δοθείς, δεδομένος. Syn. *maʿlūm* (q.v.).
275, 302, 373, 376, 406, 410 passim.
2°. chosen, put.
413 (cf. 411, *faraḍnā* posterius); 1024, 1118, 1121, 1170, 1173 passim.

*faḍl*: ὑπεροχή.
1°. = difference (between: *baina*). Syn. *tafāḍul.*
263 (interp.), 960, 1349, 1354, 1816, 1891 passim.
2°. = excess (over: *ʿalà*). Syn. *ziyādah.*
2616.

*tafāḍul*: ὑπεροχή.
59, 62, 69, 71, 86, 89, 97, 98, 651, 654, 670 passim.

*faʿala* (I): *wa-ḏālika mā aradnā an nafʿal* = ὅπερ ἔδει ποιῆσαι: 1907, 1984.
Syn. *ʿamila.*

*fann*: 13, 16 (bis), 951.

*fāta* (I): 12.

*qabila* (III):
1°. alone.

(α) equate = ἰσοῦν (τι: Acc., τινι: *bi*).
1352, 1355, 2182, 2417.
Commonly used by, e.g., al-Karajī in this sense (see Woepcke, *Extrait*, p. 64, or my study on the *Badīʿ*, p. 303).

(β) restore, i.e., = *jabara*(?). Perhaps a mistake (cf. p. 65, n. 36).
2212.

2°. with *jabara*: see under *jabara.*

*qabla*: 3143.

*qibal*: (*min qibal ḏālika*) 1360, 1389.

*muqābala*$^{h}$:

1°. alone: reduction.

36 ((interp.) def. of the term), 212, (737).

N.B. *muqābala*$^{h}$ in line 212 might also be translated as "equation", in accordance with one of the meanings of *qābala* (1°,α); this usage occurs in al-Karajī's works (see my study on the *Badī*$^{c}$, p. 303, and compare *Faḫrī* V,28 with V,6: *ḥattà yumkin al-muqābala*$^{h}$ and *ḥattà yumkin al-mu*$^{c}$*ādala*$^{h}$ are used in the same sense). The phrase in our manuscript could also be an Arabic addition.

2°. in association with *jabr*: see thereunder.

*qad*: 1°. +perfect: 8, 18, 37, 57, 71, 84 passim. See also *kāna*, 3°,α.

2°. +imperfect: 3371. In this case, *qad* does not have the usual sense of "perhaps". The use by Qusṭā of a *qad* "in konstatierendem Sinne" was noted by Nix in the preface to his edition of Heron's *Mechanica* (*Opera*, II,xliii–xliv) and, after him, by Daiber (*Placita*, pp. 10 and 447). In this connection, note that

(1) the Greek construction underlying *wa-qad* (...) *al-aqsām* (3371–72)—and similar passages without *qad* (see, e.g., 2672, 2679, 3034–35)—may well have been a genitivus absolutus;

(2) in any case—and this is also true for the instances mentioned by Nix (loc. cit.)—there is no need for a specific Greek equivalent to *qad*, its function being most probably that defined by Brockelmann (after Nöldeke): "Endlich aber kann *qad* vor dem Impf. wie vor dem Perf. einfach als Bekräftigung dienen" (*Grundriss*, II, p. 508); examples of this *qad* ("einfach bekräftigend") in Reckendorf, *A.S.*, pp. 302–3. See also Samawʾal, *Bāhir*, p. 231,4, where *qad* precedes *yanbaġī* at the beginning of the diorism belonging to prop. I,16 of Diophantus (see above, p. 12).

*miqdār*: quantity, amount.

1°. μέγεθος. 1389, 1626, 1631, 1663, 1669, 1696 ,1702, 1772, 1774.

2°. =measure: 245; *bi'l miqdār allaḏī huwa* = τοιούτων (μονάδων) οἵων ἐστὶν ἡ μία, or perhaps with *miqdār* rendering μέτρον.

3°. μέτρησις(?). 3073, 3157 (bis).

*qadama* (V): 8, 172, 204, 205, 240, 412, 714, 1241, 1342, 1410, 1418 passim.

*mutaqaddim*: (379), 610, 615, 714 (+*li*), 846, 1364 passim.

*aqrab*: 2543. Syn. *ashal*. ἁπλούστερος?

*qasama* (I):

1°. =dividere (+$^{c}$*alà*). Gr. παραβάλλειν, μερίζειν (παρά/εἴς τι).

20, 21, 23–25, 27 passim.

2°. =partiri. Gr. διαιρεῖν τι εἰς.

(α) in = *bi*: 740, 1034, 1265 (bis), 1419, 1885, 1890, 1906, 1961, 1966 passim.

(*β*) in = Acc.: 3147, 3328, 3344, 3346, 3348, 3351, 3353, 3374, 3381, 3383, 3416.

The latter does not supersede the former; cf. 3166, 3176, 3354, 3379 passim.

N.B. *qasama*, "ut mos est mathematicorum, constr. cum *bi*, non cum accus. partium in quas res dividitur" (Nallino, *Albatenii Opus*, II,349). This is not true for all mathematical texts (see al-Ḫwārizmī, *Alg.* 25,8; 26,18 passim; Abū Kāmil, *Alg.* 85r,8; 85v,1–2 passim), but for many (e.g., Heron, *Mech.* 77,13; 189,14; 191,3–4; 193, 8 and 16 passim; Apoll.–Nix, 14; Ṯābit: Luckey, *Richtigkeitsnachw.*, 114, and Kutsch, 331).

*qasm*: (nomen verbi *qasama*) 149, 150, 204, 1394, 2491, 2531, 2550, 2577, 2583, 2865 passim.

*qism*: 1°. παραβολή (=quotiens), μερισμός: 322, 549, 570, 573, 575, 576 passim.
2°. διῃρημένος: 742, 744, 1035, 1885, 1890, 1891 passim.
3°. =type, class: 951.

*qismah*: 1°. =divisio (παραβολή): 38, 205, 291, 564, 566, 570, 589, 590 passim.
2°. =quotiens (παραβολή, μερισμός): 2427, 2430, 2463 (cf. 2478), 2542.
3°. =partitio (διαίρεσις): 3146, 3155.

*maqsūm*:
1°. 571, 576, 577, 579, 594, 599 passim; + *ʿalaihi*: divisor (sim. *maḍrūb fīhi*); without *ʿalaihi*, as in 2539 (cf. 2540): dividendus (sim. *maḍrūb*).
2°. (α) =partitus (cf., e.g., D.G., I,138,12): 3349, 3351.
(*β*) =partiendus (cf., e.g., D.G., I,92,5–6): 2992 seqq., 3118; cf. 3145–46, 3155.

*qaṣada* (I): 1343 (+*li*; cf. Freytag, *Lexicon*); 2542 (+*ilà*).

*aqʿad*: lower in degree; a corresponding Greek word is not known.
39, 47, 65, 78, 126, 149, 165, 185, 451, 480, 661, 688, 809, 837, 1359, 1375, 2184.
The word occurs in other mathematical works; see the *Badīʿ*, 98v (Anbouba, 64,19): *ʿalà'l-wāḥid min aqʿad al-marātib* (proprie: *al-martabatain*); similarly in the *Faḫrī* (V,16, 17, 18, 21, 28, 30 seqq.).

*aqall*: ἐλάττων; ant. *akṯar*.
1218, 2002, 2005, 2010, 2061, 2377, 2436, 3386.

*qāla* (I):
1°. 20, 560, 584, 684, 733, 977, 1241, 1364, 2143, 2146, 3333.
2°. used in stating an equation: 713, 1029, 1127, 1175, 1291, 1344, 1391, 1414 (bis), 1594. Similarly used in al-Karajī's *Badīʿ*, e.g., 101r–v (*qulta*).

*qaul*: 1°. βιβλίον; syn. *maqālah*. 1615, 2922, 2923.
2°. =treatise: 8.

*maqāla*$^h$: βιβλίον. See the previous word.
1, 7, 1618, 2168, 2171, 2918 (bis), 2921, 3586.
*al-maqāla*$^h$ *al-t̲āniya*$^h$: 741–42, 1035, 2438.
*al-maqāla*$^h$ *al-t̲ālat̲a*$^h$: 3102, 3150.

*qāma* (IV): (+*maqām*) 1997.

*qāma* (X): 164.

*maqām*: vide *qāma* (IV).

*qiyās*: 379. Gr. ἀναλογία? Cf. Toomer, *Diocles*, 147. The usual meaning in Arabic mathematical texts is "reasoning", hence "resolution".

*ka-*: e.g., 123, 162, 614, 1034, 1414, 2356; indicates the equality only in expressions like *ka-nisba*$^h$ (e.g., 1667, 1700), *ka-miqdār* (3157).

*ka-d̲ālika*: ὁμοίως.
203, 204, 453, 568 (τοιοῦτοι(?)), 1164 passim.
In phraseology: *lammā kāna d̲ālika ka-d̲ālika*, 3242.

*akbar*: Syn. *a*ᶜ*ẓam* (μείζων). 2540 (bis), 3072, 3073.
N.B. Our reading (*akbar* instead of *akt̲ar*) is arbitrary inasmuch as the words are written without diacritical points (cf. p. 22). But certain translators seem to distinguish between *akbar*,*a*ᶜ*ẓam*/*aṣġar* and *akt̲ar*/*aqall* (cf. Klamroth, 291).
Note that D.G. sometimes uses μείζων where the Arabic would probably have had *akt̲ar*: e.g., 246,26 or 364,10 (cf. 36,6).

*kataba* (I): 13.

*kitāb*: treatise (the *Arithmetica*).
—at the beginnings or ends of Books: 1, 7, 1615, 1618, 2168, 2171, 2918, 2921, 3586, 3588.
—within Books: 2181, 3339, 3354, 3387. Cf. *qaul*, 2°.

*kat̲ra*$^h$: 1808.

*kat̲īr*: 1°. 8, 13, 17, 2922.
2°. *akt̲ar*; Gr. πλείων (μείζων: see under *akbar*, in fine). Ant. *aqall*.
18, 212, 1310, 1410, 1887, 1910, 2356, 3353.

*ka*ᶜ*b*: $K^Y$, sc. $x^3$.
20 (bis; def.), 21, (22), 23, 25; 42, 44 (bis) passim.
N.B.: The plural can be *ki*ᶜ*āb*, *ku*ᶜ*ūb* or *ak*ᶜ*ub* (Freytag, *Lex.*).
Our text has only the first form. Al-Karajī uses the last two forms (see *Badī*ᶜ, 105$^{r-v}$, *Faḫrī* V,6/*Badī*ᶜ, 78$^r$, *Faḫrī* V,1).

*ka*ᶜ*b ka*ᶜ*b*: $K^YK$.
31 (bis; def), 179, 184, 186, 449–50, 450, 478 passim.
*Plural*: (α) *ki*ᶜ*āb ki*ᶜ*āb*: 144–45, 147, 147–48, 163, 165 passim.
(β) *ki*ᶜ*āb ka*ᶜ*b*: 149, 897, 2417, 2419, 2420.

*kaᶜb kaᶜb kaᶜb*: (not in the extant Greek text).
799 (def.), 802, 806, 808, 811 passim. Occurs only in problems IV,29–33, 42–44; V,4–6; VII,1.
*Plural*: (α) *kiᶜāb kaᶜb kaᶜb*: 898, 1400, 1717–18, 1718–19, 1768, 1768–69.
(β) *kiᶜāb kiᶜāb kiᶜāb*: 1747–48.

*kaᶜb kaᶜb māl*: (not in the extant Greek text). Syn. *māl māl māl māl.*
802 (def.), 803, 805–6, 806, 807 (bis) passim, but only in problems IV,29–33, 42–44; V,4–6.
*Plural*: (α) *kiᶜāb kiᶜāb māl*: 833, (834–35), 1365, 1366, 1368, 1374 passim.
(β) *kiᶜāb kaᶜb māl*: 1358, 1370, 1372, 1373, 1479.

*kaᶜb māl*: $\Delta K^{Y}$. Syn. *māl kaᶜb, kaᶜb maḍrūb fī māl* (cf. p. 45).
140, 146, 148 (bis), 149.
*Plural*: (α) *kiᶜāb amwāl*: 123, 142, 145.
(β) *kiᶜāb māl*: 125–26, 128.

*mukaᶜᶜab*: κύβος.
2, 7, 41, 42 (bis), 43 passim.

*kafà* (VIII): 2288.

*kull*: 1°. ἕκαστος, ἑκάτερος.
19, 3353; mostly with *wāḥid*: 341, 413, 436, 466 passim.
2°. ὅλος, adj. or subst.
283, 381, 385, 393, 939 (= *kāmil*), 2032, 2304, 2451, 3194.
3°. introducing a theorem or a general statement: 20, 415, 524, 547, 655, 782, 862 passim. Greek initial πᾶς (in D.G. 260,1 (cf. D.A. 2791); 296,12).

*kullamā*: 1807. See note 731 of the app.

*kilà*: —with a masculine noun: 545, 969, 1417.
—with a pronoun: 1484 (see app. crit.).
—with a feminine noun (see the orthographical remark, p. 29): occurs regularly when a common addition or subtraction is made: 36 and 37 (interp.), 686, 696, 735 passim (see synonymous expression under *jamīᶜ*, 3°); it can be omitted: 718 and 835 (see note 247 of the crit. app.), 2278, 2323, 2358 passim. For common division: 2560, 2847.

*kalama* (V): 2922.

*kam*: 43, 45, 61, 76, 87 passim. Always with *šiʾnā* (*aradnā*), used to render ὅσος δήποτε.

*kāmil*: whole. 2702, 2829, 3085. See *kull*, 2°.

*kāna* (I):
1°. *kāna* "complete".
610, 1344.
Complete *kāna* with *min* (as, e.g., in 525, 553): see under *min.*
2°. *kāna* "incomplete".
(α) to be, to become (εἶναι, ποιεῖν, γίνεσθαι).
43, 46, 53 (bis), 54, 55 passim;

after an addition: 211, 240, 761, 771, 1076 (post.) passim;
after a multiplication: 278 (bis), 279, 280, 301, 326 passim;
after a division: 784, 811, 981 (bis), 1404, 2359 passim;
(after a subtraction: *baqiya*).

N.B. *kāna* agrees naturally with the logical subject, i.e., the increased number, the multiplicand or the dividend. Note, however, the agreement with the multiplier in line 477, and with the added numbers in line 1527.

(*β*) single *kāna* for several subjects/predicates (unlike, e.g., 666 and 3388 seqq.), as attested by the external (oblique) form.
two: 275, 545, 663, 1597, 1754, 1766, 1804, 1853, 1940, 1941, 2138 (prius), 2668;
three: 3102, 3150, 3199 (post.), 3238 (post.), 3355 (comp. with 3388);
four: 1968.

N.B. The *li-yakun* appearing (like the Greek ἔστω) at the beginning of an ἔκθεσις is never repeated (see, e.g., (497), 518, 540, 1265).

3°. auxiliary *kāna*.

(*α*) to express the pluperfect: either with no *qad* (e.g., 1361, 1409), or with *qad* interposed (e.g., 18, 3154), or with *qad* preceding the two verbs (e.g., 1342, 2120).

(*β*) to express the Latin/Greek imperfect: 581 (ἔδει).

(*γ*) to express the future-perfect: 414 (but see app.).

(*δ*) in different constructions and with various verbs, see, e.g., 153, 929, 2416, 3351; (subj.:) 495, 516, 538, 572, 574, 595, 691, 712.

(*ε*) (most commonly) in connection with *ʿadala* (I/III): 44, 75, 79, 91, 93 (prius), passim. It is sometimes omitted, however: 46, 51, 64, 104, 108 passim. Also, it may be written only once for two equations (cf. 2°,*β*): see 1014–15, 1238–39, 1770.

**Remark.** Asyndetic connection of *kāna* incomplete with *ʿādala*: 979, 1017, 1021, 1242–43, 1250 passim (cf. 1471–72 with 1473–74). Cf. 1203 (*ṣāra*), 936 (*baqiya*). See D.G., I,200,12–13; 230,2–3; 250,5 (cf. 250,19).

*kāʾin*: 606, 659, 1555, 1570, 1811, 1816 (bis), 1820 passim.
*al-kāʾin min* = ἀπό: *al-murabbaʿ al-kāʾin min* (*ḍilaʿ fulānī*) = ὁ ἀπό (τινος πλευρᾶς) τετράγωνος.

*makān*: *li-makān*: Dozy, *Suppl.*, II,503: “à cause de”.
386, 393. *li-makān al-juzʾ* (*al-ajzāʾ*) = Gr. διὰ τὸ μόριον? (Cf. D.G., I,254,13).

*kaifa*: 18, 2061, 3354, 3387.

*li-*: 1°. conjunction + subjunctive.
Meaning consecutive (e.g., 42, 225, 1072) or final (e.g., 566, 573, 590).
2°. conjunction + jussive.
61, 63, 88, 318, 440 passim; *fal-nafriḍ* = τετάχθω, *fal-yakun* = ἔστω, etc.
3°. preposition.
(*α*) (mostly) with personal pronoun: 13, 18 (bis), 39, 65, 142 passim.

(β) *al-ʿadad allaḏī li'l-nisbaʰ al-mafrūḍaʰ*: 438, 1910, 1914, 1935, 1937, 1963, 1965 (*al-amṯāl allatī...*). Periphrastic for πηλικότης? (cf. p. 99, n. 47).

(γ) For simple genitive (see Georr, 103): 49 (interp.). We have corrected the other places: see notes 503; 391, 435, 614, 858 (cf. p. 29); of these, all except the first seem merely to be scribal errors.
Cf. also lines 528, 715, and *faraḍa*, 1°, ε.

*li-allā*: 12, 1808.

*li-ḏālika*: see *ḏālika*.

*li-kai*: 212, 2002.

*li-makān*: see *makān*.

*lā*: 3072, 3073.

*lākin(na)*: (*lākin* clearly in 1264). Gr. ἀλλά (also δέ: Klamroth, 309; cf. Aristotle, *Gen. An.*, 6).

1°. *lākin(na)* restrictive: "but", "however", with various intensities.
—3314; in particular after an hypothetical clause: 3155, 3206, 3444;
—in the formulation of a problem (=such (however) that): 2372, 3476;
—introducing a diorism: 3071;
—synonymous with *baʿd an*: 2377 (cf. 2356).

2°. "weak" *lākin(na)*, marking some kind of transition in the resolution, such as: introducing steps in the analysis, of various kinds (see 144, 1264, 1999, 2338), among which the transition from some consideration to its (numerical) application (e.g., 312, 347, 503, 689) or, after one of the problem's conditions/equations has been satisfied, the passage to the next one (e.g., 2493, 2623, 2650, 2823); in the synthesis, *lākin(na)* sometimes introduces verifications of conditions (e.g., 2085, 2405, 3058, 3214, 3495).

*laqiya* (IV): ἀφαιρεῖν (αἴρειν) τι ἀπό (*min*) τινος. Syn. *naqaṣa*.

1°. =to remove (a common, *muštarak*, quantity from the two sides of an equation): 214, 242, 696, 807, 875, 967 passim.
Cf. Klamroth, 310; Ḥajjāj, e.g., II,11. Cf. also p. 66.
Note the expression *alqà mālᵃⁿ bi-māl* in 3194, which also appears in al-Ḫwārizmī's *Algebra* (e.g., 47, 17) and in Abū Kāmil's *Algebra* (see fol. 79ᵛ, 20).

2°. =to subtract: 1814, 1831, 1868, 2055, 2079, 2102 passim.
(The former meaning still occurs: cf. 1874, 1896 passim).

*ilqā*ʾ: 36 (interp.).

*lam*: 17, 2418, 3331, 3442.

*lammā*: ἐπεί.

1°. =cum: 569, 592 (with impf. in the apod.), 2336, 2344, 2383, 2457 passim.
See also *ʿalima*, *wajaba*.

2°. =postquam: 3142.

*lamasa* (VIII): ζητεῖν. Syn. *ṭalaba*.
200, 201, 205, 3185, 3226, 3265, 3303 (+*an*), 3337, 3420.

*iltimās*: 3441.

*lau*: εἰ (... ἄν).
1100, 3154, 3205, 3240, 3280, 3443.

*laisa*: 2060, 2271.

*mā*: τι.
1°. =id quod: 8, 10 (bis), 12, 13 (bis), 14 passim.
2°. =aliquid: 2481 (post.), 2577. Cf. *šaiʾ*, 1°.
3°. (in apposition) = quidam: 1624, 1626, 3180, 3186, 3221 passim.

*matà*: 1°. =when. Syn. *iḏā*: 16, 20, 2440, 2836.
2°. =whenever (general statement): 2047, 2872. Gr. ὅταν.

*miṯl*: Expressing likeness.
(α) "times" (for a multiple); Gr. -πλάσιον. Cf. *marra*ʰ.
210, (217), 220, 232, 235 (post.) passim. See also under *nisba*ʰ.
(β) (not multiplied) various renderings, such as "the same as", "equal to" (see, e.g., 22, 26, 30 (prius); 1500, 1629, 1795). *miṯl* can frequently remain untranslated (see, e.g., 207, 223, 235 (prius); 1502 (2ᵘᵐ, 3ᵘᵐ), 2297, 2334).
(γ) expressions:
—*fī miṯlihī* (-*hā*), after *ḍaraba*; Gr. ἐφ' ἑαυτόν.
23, 30, 80, 111, 381, 449 (bis) passim. Syn. *fī nafsihī*.
—*miṯla an*: 952.
—*miṯla mā*: 172.
—*ʿalà miṯl mā ... fa-li* (+jussive): 1402, 2057–58.

*miṯāl*: 19 (=παράδειγμα); 951, 1291 (*ʿalà miṯāl mā* (*qad*) *waṣafnā*).

*marra*ʰ: 208, 209, 224, 225, 238, 240 passim. In particular, *al-marrāt* means "the multiplicative factor" or "the multiple" (see, e.g., 209, 253; 240).

*maʿa*: 14, 393, 438, 732, 755, 773 passim.

*makuna* (IV): 11, 12 (or II?), 17, 212, 1264 (+*an*), 1410, 3331 (+*an*), 3332.
Gr. δύνασθαι, δυνατὸς εἶναι.

*min*: 1°. Various uses in nominal or verbal sentences corresponding to ἐκ (e.g., 26 (post.), 30 (post.), 80), ἐν (e.g., 141, 144, 1350), ἀπό (e.g., 92, 308 (prius), 2822) or replacing a simple genitive (e.g., 282 and 284 (cf. D.G., I,274,5), 1358).
See also *faraḍa*, *ʿamila*.
2°. =to (in the expression of a ratio). Cf. *ilà*, *ʿinda*.
319 (prius), 409, 441, 470, 1663, 1670 (=*ilà*, cf. 1632), 1696, 2236, 2258, 2926 passim.

*tamahhur*: 2924. The word is given in Zenker's *Dictionnaire*, p. 310, with the meaning "habileté, finesse d'esprit".

*māl*: $\Delta^{Y}$, sc. $x^2$. See also p. 30, no. 1.
21 seqq., 46, 47 (bis) passim.

*māl kaᶜb*: $\Delta K^{Y}$. Syn. *kaᶜb māl, kaᶜb maḍrūb fī māl* (cf. p. 45).
27 (bis; def.), 30, 32, 34, 150 (bis), 870, 871, 872, 875, 876.
*Plural*: *māl kiᶜāb*: 120.

*māl māl*: $\Delta^{Y}\Delta$.
23 (bis; def.), 25, 28, 29, 31 passim.
*Plural*: (α) *amwāl amwāl*: 103, 104, 108, (167), 685, 687, 716, 760 passim.
(β) *amwāl māl*: 697, 717, 736, 759, 833, 859, 868 passim.

*māl māl māl māl*: No correspondent in D.G. Syn. *kaᶜb kaᶜb māl* (see p. 45).
2671, 2672, 2674, 2675, 2941 passim (only in problems VI,17 and VII,1).

*naḥnu*: 978, 984, 1016, 1364, 1371.

*naḥw*: *ᶜalà naḥw mā*: 1241, 1409, 1494, 1893; 1774 (*ᶜalà'l-naḥw allaḏī*).

*nāḥiyaʰ*: μέρος (cf. D.G., I,14,13; 98, 16). Syn. *jihaʰ*.
36 and 37 (interp.), 39, 47, 48, 64, 65 passim.

*nazala* (IV): (+*anna*) to suppose.
1400.
Commonly used in this sense by translators. See, e.g., Galen, *De diaeta*, 78,20; 80,2; 82,3; Klamroth, 310–11; Ḥajjāj I,6; I,7 passim; see also Pappus, *Comment. Euclid X*, 289.

*nisbaʰ*: λόγος (πρός: *ilà, min*).
320, (329), (409), 411 (bis), 414, 416, 435, 438, 440 (bis) passim.
N.B. The two ways of expressing a numerical ratio, with *ilà* and with *miṯl*, appear together in lines 1562–64, 1667–68.

*mutanāsib*:
1°. = in continuous proportion: 3476, 3479. Gr. ἐν τῇ γεωμετρικῇ ἀναλογίᾳ (e.g., D.G., I,310,4; 312,6) or simply ἀνάλογον (see ibid. 234, 14; *Elem.*, V, deff. 9–10).
2°. = in proportion: 3507, 3509, 3539, 3542. Gr. ἀνάλογον (see, e.g., *Elem.*, V, prop. 16).

*nafs*: *fī nafsihī* = ἐφ'ἑαυτόν. Syn. *fī miṯlihī*.
1076, 1090, 1556.

*naqaṣa* (I): ἀφαιρεῖν (αἴρειν) τι ἀπό (*min*) τινος. Syn. *alqà*.
1°. to remove (a common, *muštarak*, quantity from the two sides of an equation).
686, 764, 785, 979, 1017, 1023 passim.
2°. to subtract.
223, 235, 251, 853, 936, 946 passim.

*nuqṣān*: ἀφαίρεσις.
226, (233), 990, 1469, 1520, 1700, 1870, 3304.

*nāqiṣ*:

1°. subtracted, negative. Ant. *zāʾid*.

36 (interp.), 1068, 1101, 1251, 1307, 1622, 1714 passim.

2°. +*min*: subtracted from. Ant. *zāʾid*, *mazīd*.

1084, 1262, 1483, 1863.

3°. +Acc.: deficient by, less.

1169, 1172, 1180, 1181, 1203, 1208 passim.

Greek: forms of λείπειν; cf. D.G., I,14,5 seqq.; 138,16; see also Tannery, *Symbole de soustr.* = *Mém. sc.*, III,208–12.

*manqūs*: ἀφαιρούμενος (ἀπό: *min*). Ant. *mazīd* (q.v.).
2156, 2341, 3237, 3239, 3243, 3247, 3249, 3269, 3275 passim.

*nahà* (VI): 35. See the next word.

*nahà* (VIII): 9, 969, 997, 1040, 1086, 1136, 1182, 1221, 1270, 1314, 1357, 1387, 1418 (+*ilà an*), 1426, 1493, 2002, 2150, 2540 (+*ilà an*), 3142 passim.

N.B. *intahà bi-nā al-ʿamal ilà* (2540) = *tanāhà bi-nā al-ʿamal ilà* (35–37).

*nauʿ*: εἶδος.

1°. Non-mathematical sense.

10 (post.), 14, 15; clearly in 19, 1418 (syn. *fann*: cf. 951), 2924.

2°. In mathematical usage, εἶδος can mean (algebraical) *term* as well as, specifically, *power* (comp. D.G., I,210,1–2 with 14,2). See ambiguous use in line 37.

9, 10 (prius), 37 (bis), 38, 39, 1808–16, 2631 (bis), 2765 (bis), 2887 (bis).
*nauʿ wāḥid yuʿādil nauʿan wāḥidan* (cf. most of the above ref.) = ἓν εἶδος ἑνὶ (εἴδει) ἴσον (D.G., def. XI and, e.g., probl. II,10–12).

*hāhunā*: 1302, 1344.

*muhayyaʾ*: πλασματικός. See p. 192.
439, 496, 1801.

*wa*: A few *wa*'s may have the sense of *fa*, e.g., those in 2677 (post.) and 2678 (post.). Cf. Georr, 76–77: (*wa*-) . . . *wa*- = (μὲν) . . . δέ, and Kutsch, 347–48.

*wajaba* (I): δεῖν. Cf. *iḥtāja*, *inbaġà*, *arāda* (2°,β).

1°. independent: 574, 1210.

2°. introducing the apodosis to a causal clause

—beginning with *lammā*: 2337, 2792, 3243, 3392, 3553.

—beginning with *min ajl anna*: 3207.

We have rendered the *wajaba an nafriḍ* in line 2792, where there is no inherent obligation, by a simple future.

*wajada* (I): εὑρίσκειν.

1°. (repeated use)

(α) in the formulations of problems: 41, 59, 73, 86, 100 passim.
Cf. *arāda*, 1°,α.

(β) in the conclusions of problems: 57, 71, 84; 157 and 158, 197 and 198, 220 and 222 passim. Cf. *arāda*, 1°,β.
2°. (other occurrences)
(α) = to find: e.g., 172, 201, 414, 584, 693, 716; 3384.
(β) = to solve (problem): 951.

*wujūd*: εὕρεσις (το εὑρεῖν).
1°. Finding, discovery: 610, 1343, 1344; 3101–2 (*wa-wujūd ḏālika sahl* corresponding to the τοῦτο δὲ ῥᾴδιον in D.G.). Cf. *wijdān*.
2°. Resolution (of problems): 18. Cf. D.G., I,2,3; also found with this meaning in al-Karajī's *Badī*ᶜ (fol. 122ᵛ).

*wijdān*: nomen verbi of *wajada* (together with *wujūd*; given in Freytag, *Lexicon*, and Lane, *Dictionary*).
2180, 2208, 2437, 3016 (*wa-wijdān ḏālika sahl*; cf. *wujūd*).

*jihaʰ*:
1°. 343, 388, 1344, 1348, (1411). Gr. τρόπος (343: in a periphrase for ἄλλως).
The meaning "modus" (Freytag, *Lexicon*), synonymous to that of *wajh*, is extremely common in translations from the Greek.
2°. side (of an equation); Gr. μέρος. Syn. *nāḥiyaʰ*.
1307, 1374.

*wajh*: 1°. 262 (in a periphrase for ἄλλως; cf. *jihaʰ*, 1°).
2°. = aspect: 1594.

*wāḥid*: εἷς.
(α) = one: 9, 10, 37, 39 (post.), 42 (bis) passim.
(β) = unit ($x^n$ is a "unit" of $mx^n$): 39 (prius), 451, 809 (post.), 837, 1359, 1375.
(γ) in the expression of a fraction $m/n$: "*m* parts of *n* parts *of one* (*of the unit*)"; cf. p. 39.
*min wāḥid*: 331–32, 631, 634, 637, 639 passim.
*min al-wāḥid*: 2187–89, 2193, 2195, 2196, 2216 passim.
This last formulation does not supersede the previous one; see lines 2217, 2370, 2518, 2525, 2537 (cf. 2538), 2915 (cf. 2916).
Particular case: *wāḥidⁱⁿ* (cf. *aḥad*, 3°): 315, 323, 324, 326, 327 (bis), 332 (post.), 340, 508, 512 passim; comp. 2039 with 2042.
(δ) = first: 3396. Cf. Kutsch, 69, lines 8–9.
(ε) = same: 205, 379; 1388 (+ *bi-ᶜainihī*).
(ζ) + *kull*: see thereunder.

*ausaṭ*: μέσος. 1629 (bis), 1666 (bis), 1699, 1700.

*waṣafa* (I): 18, 37, 560, 741, 951, 1291, 1364 passim.

*ṣifaʰ*: 1343, 1507.

*waḍaᶜa* (I): 79. Syn. *faraḍa*.

*mauḍiᶜ*: 2972 (interp.).

*wafiqa* (VIII): 1179 (*ayy ʿadadain ittafaqā* = δύο ἀριθμοὶ τυχόντες), 2061 (*kaif ittafaqa* = ὡς ἔτυχεν).

*ittifāq*: 1343.

*waqaʿa* (I): 2139.

*wilāʾ*: 3536 (*ʿalàʾl-wilāʾ* = κατὰ τὸ ἑξῆς). Cf. app.

*yasīr*: 2270.

# Appendix

# Conspectus of the Problems of the *Arithmetica*

In this conspectus of the problems of the extant Books of the *Arithmetica*, we have adopted the following conventional symbols:

$a, b, c, d$: required magnitudes.
$k, l, j, h, m, n, p, q, r$: given quantities (i.e., given numbers, including multipliers and ratios; supposed to be positive).
* indicates that a problem is interpolated. There is no asterisk in dubious cases (see, for I,26, p. 195; for III,1–4, p. 52; for "IV",3, p. 198).
A indicates that there is an alternative resolution in the text (whether genuine or interpolated).
D marks the presence of a diorism.
$D^p$ is used instead of D when the diorism leads to a constructible problem.

**I,1:** $\begin{cases} a + b = k, \\ a - b = l. \end{cases}$

**I,2:** $\begin{cases} a + b = k, \\ a = mb. \end{cases}$

**I,3:** $\begin{cases} a + b = k, \\ a = mb + l. \end{cases}$

**I,4:** $\begin{cases} a - b = k, \\ a = mb. \end{cases}$

D **I,5:** $\begin{cases} a + b = k, \\ \dfrac{1}{m} a + \dfrac{1}{n} b = l. \end{cases}$

D **I,6:** $\begin{cases} a + b = k, \\ \frac{1}{m}a - \frac{1}{n}b = l. \end{cases}$

**I,7:** $a - k = m(a - l)$.

D **I,8:** $a + k = m(a + l)$.

D **I,9:** $k - a = m(l - a)$.

**I,10:** $a + k = m(l - a)$.

**I,11:** $a + k = m(a - l)$.

**I,12:** $\begin{cases} a_1 + b_1 = a_2 + b_2 = k, \\ a_1 = mb_2, \\ a_2 = nb_1. \end{cases}$

**I,13:** $\begin{cases} a_1 + b_1 = a_2 + b_2 = a_3 + b_3 = k, \\ a_1 = mb_2, \\ a_2 = nb_3, \\ a_3 = pb_1. \end{cases}$

D **I,14:** $a \cdot b = m(a + b)$.

**I,15:** $\begin{cases} a + k = m(b - k), \\ b + l = n(a - l). \end{cases}$

D **I,16:** $\begin{cases} a + b = k, \\ b + c = l, \\ c + a = j. \end{cases}$

D **I,17:** $\begin{cases} a + b + c = k, \\ b + c + d = l, \\ c + d + a = j, \\ d + a + b = h. \end{cases}$

A **I,18:** $\begin{cases} a + b - c = k, \\ b + c - a = l, \\ c + a - b = j. \end{cases}$

D, A **I,19:** $\begin{cases} a + b + c - d = k, \\ b + c + d - a = l, \\ c + d + a - b = j, \\ d + a + b - c = h. \end{cases}$

**I,20:** $\begin{cases} a + b + c = k, \\ a + b = mc, \\ b + c = na. \end{cases}$

D, A **I,21:** $$\begin{cases} a - b = \frac{1}{m} c, \\ b - c = \frac{1}{n} a, \\ c - k = \frac{1}{p} b.^{1} \end{cases}$$

**I,22:** $$\left(a - \frac{1}{m} a\right) + \frac{1}{p} c = \left(b - \frac{1}{n} b\right) + \frac{1}{m} a = \left(c - \frac{1}{p} c\right) + \frac{1}{n} b.$$

**I,23:** $$\left(a - \frac{1}{m} a\right) + \frac{1}{q} d = \left(b - \frac{1}{n} b\right) + \frac{1}{m} a$$
$$= \left(c - \frac{1}{p} c\right) + \frac{1}{n} b = \left(d - \frac{1}{q} d\right) + \frac{1}{p} c.$$

**I,24:** $$a + \frac{1}{m}(b + c) = b + \frac{1}{n}(c + a) = c + \frac{1}{p}(a + b).$$

**I,25:** $$a + \frac{1}{m}(b + c + d) = b + \frac{1}{n}(c + d + a)$$
$$= c + \frac{1}{p}(d + a + b) = d + \frac{1}{q}(a + b + c).$$

**I,26:** $$\begin{cases} k \cdot a = \square, \\ l \cdot a = \sqrt{\square}. \end{cases}$$

$D^p$ **I,27:** $$\begin{cases} a + b = k, \\ a \cdot b = l. \end{cases}$$

$D^p$ **I,28:** $$\begin{cases} a + b = k, \\ a^2 + b^2 = l. \end{cases}$$

**I,29:** $$\begin{cases} a + b = k, \\ a^2 - b^2 = l. \end{cases}$$

$D^p$ **I,30:** $$\begin{cases} a - b = k, \\ a \cdot b = l. \end{cases}$$

**I,31:** $$\begin{cases} a^2 + b^2 = n(a + b), \\ a = mb. \end{cases}$$

**I,32:** $$\begin{cases} a^2 + b^2 = n(a - b), \\ a = mb. \end{cases}$$

**I,33:** $$\begin{cases} a^2 - b^2 = n(a + b), \\ a = mb. \end{cases}$$

---

[1] Diophantus takes $1/m = 1/n = 1/p$.

**I,34:** $\begin{cases} a^2 - b^2 = n(a - b), \\ a = mb. \end{cases}$

**Corollaries:** (a) $\begin{cases} a \cdot b = n(a + b), \\ a = mb. \end{cases}$

(b) $\begin{cases} a \cdot b = n(a - b), \\ a = mb. \end{cases}$

**I,35:** $\begin{cases} b^2 = na, \\ a = mb. \end{cases}$

**I,36:** $\begin{cases} b^2 = nb, \\ a = mb. \end{cases}$

**I,37:** $\begin{cases} b^2 = n(a + b), \\ a = mb. \end{cases}$

**I,38:** $\begin{cases} b^2 = n(a - b), \\ a = mb. \end{cases}$

**Corollaries:** (a) $\begin{cases} a^2 = nb, \\ a = mb. \end{cases}$

(b) $\begin{cases} a^2 = na, \\ a = mb. \end{cases}$

(c) $\begin{cases} a^2 = n(a + b), \\ a = mb. \end{cases}$

(d) $\begin{cases} a^2 = n(a - b), \\ a = mb. \end{cases}$

**I,39:** $\left.\begin{array}{ll} \text{(a)} & (a + k)l - (k + l)a = (k + l)a - (l + a)k, \\ \text{(b)} & (a + k)l - (l + a)k = (l + a)k - (k + l)a, \\ \text{(c)} & (k + l)a - (a + k)l = (a + k)l - (l + a)k, \end{array}\right\} k > l.$

* **II,1:** $a + b = m(a^2 + b^2)$.

* **II,2:** $a - b = m(a^2 - b^2)$.

* **II,3:** (a) $a \cdot b = m(a + b)$.
(b) $a \cdot b = m(a - b)$.

* **II,4:** $a^2 + b^2 = m(a - b)$.

* **II,5:** $a^2 - b^2 = m(a + b)$.

D * **II,6:** $\begin{cases} a - b = k, \\ (a^2 - b^2) - (a - b) = l. \end{cases}$

D * **II,7:** $(a^2 - b^2) - m(a - b) = l$.

A **II,8:** $a^2 + b^2 = k^2$.

**II,9:** $a^2 + b^2 = k = k_1^2 + k_2^2$.

**II,10:** $a^2 - b^2 = k$.

A II,11: $\begin{cases} a + k = \square, \\ a + l = \square'. \end{cases}$

II,12: $\begin{cases} k - a = \square, \\ l - a = \square'. \end{cases}$

A II,13: $\begin{cases} a - l = \square, \\ a - k = \square'. \end{cases}$

II,14: $\begin{cases} a + b = k, \\ c^2 + a = \square, \\ c^2 + b = \square'. \end{cases}$

II,15: $\begin{cases} a + b = k, \\ c^2 - a = \square, \\ c^2 - b = \square'. \end{cases}$

II,16: $\begin{cases} a + k^2 = \square, \\ b + k^2 = \square', \\ a = mb. \end{cases}$

A * II,17: $$\left(a - \left(\frac{1}{m}a + k\right)\right) + \left(\frac{1}{p}c + j\right) = \left(b - \left(\frac{1}{n}b + l\right)\right) + \left(\frac{1}{m}a + k\right) = \left(c - \left(\frac{1}{p}c + j\right)\right) + \left(\frac{1}{n}b + l\right).$$

* II,18: $$\begin{cases} \left(a - \left(\frac{1}{m}a + k\right)\right) + \left(\frac{1}{p}c + j\right) = \left(b - \left(\frac{1}{n}b + l\right)\right) + \left(\frac{1}{m}a + k\right) = \left(c - \left(\frac{1}{p}c + j\right)\right) + \left(\frac{1}{n}b + l\right), \\ a + b + c = h. \end{cases}$$

II,19: $a^2 - b^2 = r(b^2 - c^2)$.

II,20: $\begin{cases} a^2 + b = \square, \\ b^2 + a = \square'. \end{cases}$

II,21: $\begin{cases} a^2 - b = \square, \\ b^2 - a = \square'. \end{cases}$

**II,22:** $\begin{cases} a^2 + (a+b) = \square, \\ b^2 + (a+b) = \square'. \end{cases}$

**II,23:** $\begin{cases} a^2 - (a+b) = \square, \\ b^2 - (a+b) = \square'. \end{cases}$

**II,24:** $\begin{cases} (a+b)^2 + a = \square, \\ (a+b)^2 + b = \square'. \end{cases}$

**II,25:** $\begin{cases} (a+b)^2 - a = \square, \\ (a+b)^2 - b = \square'. \end{cases}$

**II,26:** $\begin{cases} a \cdot b + a = \square, \\ a \cdot b + b = \square', \\ \sqrt{\square} + \sqrt{\square'} = k. \end{cases}$

**II,27:** $\begin{cases} a \cdot b - a = \square, \\ a \cdot b - b = \square', \\ \sqrt{\square} + \sqrt{\square'} = k. \end{cases}$

**II,28:** $\begin{cases} a^2 \cdot b^2 + a^2 = \square, \\ a^2 \cdot b^2 + b^2 = \square'. \end{cases}$

**II,29:** $\begin{cases} a^2 \cdot b^2 - a^2 = \square, \\ a^2 \cdot b^2 - b^2 = \square'. \end{cases}$

**II,30:** $\begin{cases} a \cdot b + (a+b) = \square, \\ a \cdot b - (a+b) = \square'. \end{cases}$

**II,31:** $\begin{cases} a \cdot b + (a+b) = \square, \\ a \cdot b - (a+b) = \square', \\ a + b = \square''. \end{cases}$

**II,32:** $\begin{cases} a^2 + b = \square, \\ b^2 + c = \square', \\ c^2 + a = \square''. \end{cases}$

**II,33:** $\begin{cases} a^2 - b = \square, \\ b^2 - c = \square', \\ c^2 - a = \square''. \end{cases}$

**II,34:** $\begin{cases} a^2 + (a+b+c) = \square, \\ b^2 + (a+b+c) = \square', \\ c^2 + (a+b+c) = \square''. \end{cases}$

**II,35:** $\begin{cases} a^2 - (a+b+c) = \square, \\ b^2 - (a+b+c) = \square', \\ c^2 - (a+b+c) = \square''. \end{cases}$

**III,1:** $\begin{cases}(a+b+c)-a^2=\square,\\(a+b+c)-b^2=\square',\\(a+b+c)-c^2=\square''.\end{cases}$

**III,2:** $\begin{cases}(a+b+c)^2+a=\square,\\(a+b+c)^2+b=\square',\\(a+b+c)^2+c=\square''.\end{cases}$

**III,3:** $\begin{cases}(a+b+c)^2-a=\square,\\(a+b+c)^2-b=\square',\\(a+b+c)^2-c=\square''.\end{cases}$

**III,4:** $\begin{cases}a-(a+b+c)^2=\square,\\b-(a+b+c)^2=\square',\\c-(a+b+c)^2=\square''.\end{cases}$

A **III,5:** $\begin{cases}a+b+c=\square,\\a+b-c=\square',\\b+c-a=\square'',\\c+a-b=\square'''.\end{cases}$

A **III,6:** $\begin{cases}a+b+c=\square,\\a+b=\square',\\b+c=\square'',\\c+a=\square'''.\end{cases}$

**III,7:** $\begin{cases}a-b=b-c,\\a+b=\square,\\b+c=\square',\\c+a=\square''.\end{cases}$

**III,8:** $\begin{cases}a+b+c+k=\square,\\a+b+k=\square',\\b+c+k=\square'',\\c+a+k=\square'''.\end{cases}$

**III,9:** $\begin{cases}a+b+c-k=\square,\\a+b-k=\square',\\b+c-k=\square'',\\c+a-k=\square'''.\end{cases}$

**III,10:** $\begin{cases}a\cdot b+k=\square,\\b\cdot c+k=\square',\\c\cdot a+k=\square''.\end{cases}$

**III,11:** $\begin{cases}a\cdot b-k=\square,\\b\cdot c-k=\square',\\c\cdot a-k=\square''.\end{cases}$

**III,12:** $\begin{cases} a \cdot b + c = \square, \\ b \cdot c + a = \square', \\ c \cdot a + b = \square''. \end{cases}$

**III,13:** $\begin{cases} a \cdot b - c = \square, \\ b \cdot c - a = \square', \\ c \cdot a - b = \square''. \end{cases}$

**III,14:** $\begin{cases} a \cdot b + c^2 = \square, \\ b \cdot c + a^2 = \square', \\ c \cdot a + b^2 = \square''. \end{cases}$

A **III,15:** $\begin{cases} a \cdot b + (a + b) = \square, \\ b \cdot c + (b + c) = \square', \\ c \cdot a + (c + a) = \square''. \end{cases}$

**III,16:** $\begin{cases} a \cdot b - (a + b) = \square, \\ b \cdot c - (b + c) = \square', \\ c \cdot a - (c + a) = \square''. \end{cases}$

**III,17:** $\begin{cases} a \cdot b + (a + b) = \square, \\ a \cdot b + a = \square', \\ a \cdot b + b = \square''. \end{cases}$

**III,18:** $\begin{cases} a \cdot b - (a + b) = \square, \\ a \cdot b - a = \square', \\ a \cdot b - b = \square''. \end{cases}$

**III,19:** $\begin{cases} (a + b + c + d)^2 + a = \square, \\ (a + b + c + d)^2 - a = \square', \\ (a + b + c + d)^2 + b = \square'', \\ (a + b + c + d)^2 - b = \square''', \\ (a + b + c + d)^2 + c = \square^{\mathrm{IV}}, \\ (a + b + c + d)^2 - c = \square^{\mathrm{V}}, \\ (a + b + c + d)^2 + d = \square^{\mathrm{VI}}, \\ (a + b + c + d)^2 - d = \square^{\mathrm{VII}}. \end{cases}$

* **III,20:** $\begin{cases} a + b = k, \\ c^2 - a = \square, \\ c^2 - b = \square'. \end{cases}$

* **III,21:** $\begin{cases} a + b = k, \\ c^2 + a = \square, \\ c^2 + b = \square'. \end{cases}$

**IV,1:** $b^3 + a^3 = \square.$

**IV,2:** $b^3 - a^3 = \square.$

**IV,3:** $b^2 + a^2 = \text{❒}.$

**IV,4:** $b^2 - a^2 = \boxdot$.

**IV,5:** $b^2 \cdot a^2 = \boxdot$.

**IV,6:** $b^3 \cdot a^2 = \square$.

**IV,7:** $b^3 \cdot a^2 = \boxdot$.

**IV,8–9:** $b^3 \cdot a^3 = \square$.

**Corollaries:** (a) $\dfrac{b^3}{a^3} = \square$,

(b) $\dfrac{b^2}{a^2} = \boxdot$, etc.

**IV,10:** $a^3 + k \cdot a^2 = \square$.

**IV,11:** $a^3 - k \cdot a^2 = \square$.

**IV,12:** $a^3 + k \cdot a^2 = \boxdot$.

A **IV,13:** $a^3 - k \cdot a^2 = \boxdot$.

A **IV,14:** $\begin{cases} k \cdot a = \boxdot, \\ l \cdot a = \square'. \end{cases}$

A **IV,15:** $\begin{cases} k \cdot a = \boxdot, \\ l \cdot a = \square. \end{cases}$

**Corollary:** $b^3 = m \cdot a^2$.

**IV,16:** $\begin{cases} k \cdot b = \boxdot, \\ k \cdot a = \sqrt[3]{\boxdot}. \end{cases}$

D$^p$ **IV,17:** $\begin{cases} k \cdot b^2 = \boxdot, \\ k \cdot a^2 = \sqrt[3]{\boxdot}, \\ b = ma. \end{cases}$

D **IV,18:** $\begin{cases} k \cdot b^3 = \square, \\ k \cdot a^3 = \sqrt{\square}, \\ b = ma. \end{cases}$

D$^p$ **IV,19:** $\begin{cases} k \cdot a = \boxdot, \\ l \cdot a = \sqrt[3]{\boxdot}. \end{cases}$

D$^p$ **IV,20:** $\begin{cases} k \cdot a^3 = \square, \\ l \cdot a^3 = \sqrt{\square}. \end{cases}$

D$^p$ **IV,21:** $\begin{cases} k \cdot a^2 = \boxdot, \\ l \cdot a^2 = \sqrt[3]{\boxdot}. \end{cases}$

D$^p$ **IV,22:** $\begin{cases} k \cdot a^3 = \boxdot, \\ l \cdot a^3 = \sqrt[3]{\boxdot}. \end{cases}$

**IV,23:** $(b^2)^2 + (a^2)^2 = \boxdot$.

**IV,24:** $(b^2)^2 - (a^2)^2 = \boxdot$.

**IV,25:** $(a^3)^2 + (b^2)^2 = \square$.

**IV,26:** (a) $(a^3)^2 - (b^2)^2 = \square.$

(b) $(b^2)^2 - (a^3)^2 = \square.$

**IV,27:** $(a^3)^2 + k \cdot b^2 = \square.$

**IV,28:** $(b^2)^2 + k \cdot a^3 = \square.$

**IV,29:** $(a^3)^3 + (b^2)^2 = \square.$

**IV,30:** $(a^3)^3 - (b^2)^2 = \square.$

**IV,31:** $(b^2)^2 - (a^3)^3 = \square.$

**IV,32:** $(a^3)^3 + k \cdot a^3 \cdot b^2 = \square.$

**IV,33:** $(a^3)^3 - k \cdot a^3 \cdot b^2 = \square.$

**Corollaries:** (a) $(b^2)^2 + k \cdot a^3 \cdot b^2 = \square$, etc.

(b) $(b^2)^3 + k \cdot a^3 \cdot b^2 = \square$, etc.

A **IV,34:** $\begin{cases} a^3 + b^2 = \square, \\ a^3 - b^2 = \square'. \end{cases}$

**IV,35:** $\begin{cases} b^2 + a^3 = \square, \\ b^2 - a^3 = \square'. \end{cases}$

**IV,36:** $\begin{cases} a^3 + k \cdot a^2 = \square, \\ a^3 - l \cdot a^2 = \square'. \end{cases}$

**IV,37:** $\begin{cases} a^3 + k \cdot a^2 = \square, \\ a^3 + l \cdot a^2 = \square'. \end{cases}$

**IV,38:** $\begin{cases} a^3 - l \cdot a^2 = \square, \\ a^3 - k \cdot a^2 = \square'. \end{cases}$

**IV,39:** $\begin{cases} k \cdot a^2 - a^3 = \square, \\ l \cdot a^2 - a^3 = \square'. \end{cases}$

**IV,40:** $\begin{cases} (b^2)^2 + a^3 = \square, \\ (b^2)^2 - a^3 = \square'. \end{cases}$

**IV,41:** $\begin{cases} a^3 + (b^2)^2 = \square, \\ a^3 - (b^2)^2 = \square'. \end{cases}$

A[2] **IV,42:** (a) $\begin{cases} (a^3)^3 + (b^2)^2 = \square, \\ (a^3)^3 - (b^2)^2 = \square'. \end{cases}$

(b) $\begin{cases} (b^2)^2 + (a^3)^3 = \square, \\ (b^2)^2 - (a^3)^3 = \square'. \end{cases}$

**IV,43:** $\begin{cases} (a^3)^3 + k \cdot (b^2)^2 = \square, \\ (a^3)^3 - l \cdot (b^2)^2 = \square'. \end{cases}$

**IV,44:** (a) $\begin{cases} (a^3)^3 + k \cdot (b^2)^2 = \square, \\ (a^3)^3 + l \cdot (b^2)^2 = \square'. \end{cases}$

[2] IV,42,a only.

(b) $\begin{cases}(a^3)^3 - l\cdot(b^2)^2 = \square,\\ (a^3)^3 - k\cdot(b^2)^2 = \square'.\end{cases}$

(c) $\begin{cases}k\cdot(b^2)^2 - (a^3)^3 = \square,\\ l\cdot(b^2)^2 - (a^3)^3 = \square'.\end{cases}$

**V,1:** $\begin{cases}(b^2)^2 + k\cdot a^3 = \square,\\ (b^2)^2 - l\cdot a^3 = \square'.\end{cases}$

**V,2:** $\begin{cases}(b^2)^2 + k\cdot a^3 = \square,\\ (b^2)^2 + l\cdot a^3 = \square'.\end{cases}$

**V,3:** $\begin{cases}(b^2)^2 - l\cdot a^3 = \square,\\ (b^2)^2 - k\cdot a^3 = \square'.\end{cases}$

**V,4:** $\begin{cases}(b^2)^2 + k\cdot(a^3)^3 = \square,\\ (b^2)^2 - l\cdot(a^3)^3 = \square'.\end{cases}$

**V,5:** $\begin{cases}(b^2)^2 + k\cdot(a^3)^3 = \square,\\ (b^2)^2 + l\cdot(a^3)^3 = \square'.\end{cases}$

**V,6:** $\begin{cases}(b^2)^2 - l\cdot(a^3)^3 = \square,\\ (b^2)^2 - k\cdot(a^3)^3 = \square'.\end{cases}$

Dᵖ **V,7:** $\begin{cases}a + b = k,\\ a^3 + b^3 = l.\end{cases}$

Dᵖ **V,8:** $\begin{cases}a - b = k,\\ a^3 - b^3 = l.\end{cases}$

Dᵖ **V,9:** $\begin{cases}a + b = k,\\ a^3 + b^3 = l(a - b)^2.\end{cases}$

Dᵖ **V,10:** $\begin{cases}a - b = k,\\ a^3 - b^3 = l(a + b)^2.\end{cases}$

Dᵖ **V,11:** $\begin{cases}a - b = k,\\ a^3 + b^3 = l(a + b).\end{cases}$

Dᵖ **V,12:** $\begin{cases}a + b = k,\\ a^3 - b^3 = l(a - b).\end{cases}$

**V,13:** $\begin{cases}k\cdot a^2 + l = u + v,\\ u + a^3 = ⧉,\\ v + a^3 = ⧉'.\end{cases}$

**V,14:** $\begin{cases}k\cdot a^2 - l = u + v,\\ a^3 - u = ⧉,\\ a^3 - v = ⧉'.\end{cases}$

**V,15:** $\begin{cases}k\cdot a^2 - l = u + v,\\ a^3 + u = ⧉,\\ a^3 - v = ⧉'.\end{cases}$

**V,16:** $\begin{cases} k \cdot a^2 - l = u + v, \\ a^3 - u = \cube, \\ v - a^3 = \cube'. \end{cases}$

* **VI,1:** $\begin{cases} (a^3)^2 + (b^2)^2 = \square, \\ a = mb. \end{cases}$

* **VI,2:** $\begin{cases} (a^3)^2 - (b^2)^2 = \square, \\ a = mb. \end{cases}$

* **VI,3:** $\begin{cases} (b^2)^2 - (a^3)^2 = \square, \\ a = mb. \end{cases}$

* **VI,4:** $\begin{cases} (a^3)^2 + a^3 \cdot b^2 = \square, \\ a = mb. \end{cases}$

* **VI,5:** $\begin{cases} (b^2)^2 + a^3 \cdot b^2 = \square, \\ a = b. \end{cases}$

* **VI,6:** $\begin{cases} a^3 \cdot b^2 - (a^3)^2 = \square, \\ a = b. \end{cases}$

* **VI,7:** $\begin{cases} a^3 \cdot b^2 - (b^2)^2 = \square, \\ a = b. \end{cases}$

* **VI,8:** $a^3 \cdot b^2 + \sqrt{a^3 \cdot b^2} = \square.$

* **VI,9:** $a^3 \cdot b^2 - \sqrt{a^3 \cdot b^2} = \square.$

* **VI,10:** $\sqrt{a^3 \cdot b^2} - a^3 \cdot b^2 = \square.$

* **VI,11:** $(a^3)^2 + a^3 = \square.$

**VI,12:** $\begin{cases} a^2 + \dfrac{a^2}{b^2} = \square, \\ b^2 + \dfrac{a^2}{b^2} = \square', \end{cases} \quad a > b.$

**VI,13:** $\begin{cases} a^2 - \dfrac{a^2}{b^2} = \square, \\ b^2 - \dfrac{a^2}{b^2} = \square', \end{cases} \quad a > b.$

**VI,14:** $\begin{cases} \dfrac{a^2}{b^2} - b^2 = \square, \\ \dfrac{a^2}{b^2} - a^2 = \square', \end{cases} \quad a > b.$

**VI,15:** $\begin{cases} a^2 + (a^2 - b^2) = \square, \\ b^2 + (a^2 - b^2) = \square', \end{cases} \quad a > b.$

**VI,16:** $\begin{cases} a^2 - (a^2 - b^2) = \square, \\ b^2 - (a^2 - b^2) = \square', \end{cases} \quad a > b.$

**VI,17:** $\begin{cases} a^2 + b^2 + c^2 = \square, \\ a^2 = b, \\ b^2 = c. \end{cases}$

**VI,18:** $a^2 \cdot b^2 \cdot c^2 + (a^2 + b^2 + c^2) = \square.$

**VI,19:** $a^2 \cdot b^2 \cdot c^2 - (a^2 + b^2 + c^2) = \square.$

**VI,20:** $(a^2 + b^2 + c^2) - a^2 \cdot b^2 \cdot c^2 = \square.$

**VI,21:** $\begin{cases} (a^2)^2 + (a^2 + b^2) = \square, \\ (b^2)^2 + (a^2 + b^2) = \square'. \end{cases}$

A **VI,22:** $\begin{cases} a^2 + b^2 = \square, \\ a^2 \cdot b^2 = \text{(cube)}'. \end{cases}$

**VI,23:** $\begin{cases} \dfrac{k^2}{a^2} + \dfrac{k^2}{b^2} = \square, \\ a^2 + b^2 + k^2 = \square'. \end{cases}$

* **VII,1:** $\begin{cases} a^3 \cdot b^3 \cdot c^3 = \square, \\ a = mb, \\ b = mc. \end{cases}$

* **VII,2:** $(a^2)^3 \cdot (b^2)^3 \cdot (c^2)^3 = \square^2.$

* **VII,3:** $(a^2)^2 = a_1^3 + a_2^3 + a_3^3.$

* **VII,4:** $(a^2)^3 = a_1^2 + a_2^2 + a_3^2.$

* **VII,5:** $(a^3)^3 \cdot b^3 + (a^3)^3 \cdot c^2 = \square.$

D * **VII,6:** $\begin{cases} a^2 \cdot b^2 = r(a^2 + b^2), \\ a^2 + b^2 = \square. \end{cases}$

A **VII,7:** $\begin{cases} (a^3)^2 = a_1 + a_2 + a_3, \\ a_1 + a_2 = \square, \\ a_2 + a_3 = \square', \\ a_3 + a_1 = \square''. \end{cases}$

**VII,8:** $\begin{cases} (a^3)^2 + 2b = \square, \\ (a^3)^2 + b = \square'. \end{cases}$

**VII,9:** $\begin{cases} (a^3)^2 - b = \square, \\ (a^3)^2 - 2b = \square'. \end{cases}$

**VII,10:** $\begin{cases} (a^3)^2 + b = \square, \\ (a^3)^2 - b = \square'. \end{cases}$

**VII,11:** $\begin{cases} a^2 = a_1 + a_2, \\ a^2 + a_1 = \square, \\ a^2 - a_2 = \square'. \end{cases}$ $a^2$ given

**Remark:** $\begin{cases} a^2 = a_1 + a_2, \\ a^2 + a_1 = \square, \\ a^2 + a_2 = \square', \end{cases}$ not soluble.

**VII,12:** $\begin{cases} a^2 = a_1 + a_2, & a^2 \text{ given} \\ a^2 - a_1 = \square, \\ a^2 - a_2 = \square'. \end{cases}$

**VII,13:** $\begin{cases} a^2 = a_1 + a_2 + a_3, & a^2 \text{ given} \\ a^2 + a_1 = \square, \\ a^2 + a_2 = \square', \\ a^2 + a_3 = \square''. \end{cases}$

**VII,14:** $\begin{cases} a^2 = a_1 + a_2 + a_3, & a^2 \text{ given} \\ a^2 - a_1 = \square, \\ a^2 - a_2 = \square', \\ a^2 - a_3 = \square''. \end{cases}$

**VII,15:** $\begin{cases} a^2 = a_1 + a_2 + a_3 + a_4, & a^2 \text{ given} \\ a^2 + a_1 = \square, \\ a^2 + a_2 = \square', \\ a^2 - a_3 = \square'', \\ a^2 - a_4 = \square'''. \end{cases}$

**Corollary:** $\begin{cases} a^2 = \sum_{k=1}^{8} a_k, & a^2 \text{ given} \\ a^2 + a_i = \square_i, & i = 1, \ldots, 4, \\ a^2 - a_j = \square_j, & j = 5, \ldots, 8. \end{cases}$

**VII,16:** $\begin{cases} a^2 - b^2 = \square, \\ b^2 - c^2 = \square', \\ a^2 : b^2 = b^2 : c^2. \end{cases}$

**VII,17:** $\begin{cases} a^2 + b^2 + c^2 + d^2 = \square, \\ a^2 : b^2 = c^2 : d^2. \end{cases}$

**VII,18:** $\begin{cases} a^2 - b^2 = \square, \\ b^2 - c^2 = \square', \\ c^2 - d^2 = \square'', \\ a^2 : b^2 = c^2 : d^2. \end{cases}$

* **"IV",1:** $\begin{cases} a + b = k, \\ a^3 + b^3 = l. \end{cases}$

* **"IV",2:** $\begin{cases} a - b = k, \\ a^3 - b^3 = l. \end{cases}$

**"IV",3:** $\begin{cases} a \cdot b = \text{cube}, \\ a^2 \cdot b = \sqrt[3]{\text{cube}}. \end{cases}$

"IV",4: $\begin{cases} a^2 + b = \square, \\ a + b = \sqrt{\square}. \end{cases}$

"IV",5: $\begin{cases} a + b = \square, \\ a^2 + b = \sqrt{\square}. \end{cases}$

"IV",6: $\begin{cases} a^3 + c^2 = \text{❐}, \\ b^2 + c^2 = \square'. \end{cases}$

A "IV",7: $\begin{cases} b^2 + c^2 = \text{❐}, \\ a^3 + c^2 = \square'. \end{cases}$

"IV",8: $\begin{cases} a^3 + b = \text{❐}, \\ a + b = \sqrt[3]{\text{❐}}. \end{cases}$

"IV",9: $\begin{cases} a + b = \text{❐}, \\ a^3 + b = \sqrt[3]{\text{❐}}. \end{cases}$

"IV",10: $a^3 + b^3 = a + b.$

"IV",11: $a^3 - b^3 = a - b.$

"IV",12: $a^3 + b = a + b^3.$[3]

"IV",13: $\begin{cases} a + 1 = \square, \\ b + 1 = \square', \\ (a + b) + 1 = \square'', \\ (a - b) + 1 = \square'''. \end{cases}$

"IV",14: $a^2 + b^2 + c^2 = (a^2 - b^2) + (b^2 - c^2) + (a^2 - c^2).$

"IV",15: $\begin{cases} (a + b)c = k, \\ (b + c)a = l, \\ (c + a)b = j. \end{cases}$

"IV",16: $\begin{cases} a + b + c = \square, \\ a^2 + b = \square', \\ b^2 + c = \square'', \\ c^2 + a = \square'''. \end{cases}$

"IV",17: $\begin{cases} a + b + c = \square, \\ a^2 - b = \square', \\ b^2 - c = \square'', \\ c^2 - a = \square'''. \end{cases}$

"IV",18: $\begin{cases} a^3 + b = \text{❐}, \\ b^2 + a = \square'. \end{cases}$

---

[3] This problem is similar to the preceding one.

**"IV",19:** $\begin{cases} a \cdot b + 1 = \square, \\ b \cdot c + 1 = \square', \\ c \cdot a + 1 = \square''. \end{cases}$ *in indeterminato.*

**"IV",20:** $\begin{cases} a \cdot b + 1 = \square, \\ b \cdot c + 1 = \square', \\ c \cdot d + 1 = \square'', \\ d \cdot a + 1 = \square''', \\ c \cdot a + 1 = \square^{\text{IV}}, \\ d \cdot b + 1 = \square^{\text{V}}. \end{cases}$

**"IV",21:** $\begin{cases} a - b = \square, \\ b - c = \square', \\ a - c = \square'', \\ a:b = b:c. \end{cases}$

**"IV",22:** $\begin{cases} a \cdot b \cdot c + a = \square, \\ a \cdot b \cdot c + b = \square', \\ a \cdot b \cdot c + c = \square''. \end{cases}$

**"IV",23:** $\begin{cases} a \cdot b \cdot c - a = \square, \\ a \cdot b \cdot c - b = \square', \\ a \cdot b \cdot c - c = \square''. \end{cases}$

**"IV",24:** $\begin{cases} a + b = k, \\ a \cdot b = \text{❒} - \sqrt[3]{\text{❒}}. \end{cases}$

**"IV",25:** $\begin{cases} a + b + c = k, \\ a \cdot b \cdot c = \text{❒}, \\ (a - b) + (b - c) + (a - c) = \sqrt[3]{\text{❒}}. \end{cases}$

**"IV",26:** $\begin{cases} a \cdot b + a = \text{❒}, \\ a \cdot b + b = \text{❒}'. \end{cases}$

**"IV",27:** $\begin{cases} a \cdot b - a = \text{❒}, \\ a \cdot b - b = \text{❒}'. \end{cases}$

A **"IV",28:** $\begin{cases} a \cdot b + (a + b) = \text{❒}, \\ a \cdot b - (a + b) = \text{❒}'. \end{cases}$

**"IV",29:** $a^2 + b^2 + c^2 + d^2 + (a + b + c + d) = k.$

**"IV",30:** $a^2 + b^2 + c^2 + d^2 - (a + b + c + d) = k.$

A **"IV",31:** $\begin{cases} a + b = 1, \\ (a + k)(b + l) = \square. \end{cases}$

**"IV",32:** $\begin{cases} a + b + c = k, \\ a \cdot b + c = \square, \\ a \cdot b - c = \square'. \end{cases}$

**"IV",33:** $\begin{cases} a + \frac{m}{n} b = r\left(b - \frac{m}{n} b\right), \\ b + \frac{m}{n} a = p\left(a - \frac{m}{n} a\right). \end{cases}$

**Lemma:** $a \cdot b + (a + b) = k,$ *in indeterminato.*

D **"IV",34:** $\begin{cases} a \cdot b + (a + b) = k, \\ b \cdot c + (b + c) = l, \\ c \cdot a + (c + a) = j. \end{cases}$

**Lemma:** $a \cdot b - (a + b) = k,$ *in indeterminato.*

D **"IV",35:** $\begin{cases} a \cdot b - (a + b) = k, \\ b \cdot c - (b + c) = l, \\ c \cdot a - (c + a) = j. \end{cases}$

**Lemma:** $a \cdot b = r(a + b),$ *in indeterminato.*

**"IV",36:** $\begin{cases} a \cdot b = r(a + b), \\ b \cdot c = p(b + c), \\ c \cdot a = q(c + a). \end{cases}$

**"IV",37:** $\begin{cases} a \cdot b = r(a + b + c), \\ b \cdot c = p(a + b + c), \\ c \cdot a = q(a + b + c). \end{cases}$

**"IV",38:** $\begin{cases} a(a + b + c) = \triangle, \\ b(a + b + c) = \square', \\ c(a + b + c) = ⧈''. \end{cases}$

**"IV",39:** $\begin{cases} a - b = r(b - c), \\ a + b = \square, \\ b + c = \square', \\ c + a = \square''. \end{cases}$

**"IV",40:** $\begin{cases} a^2 - b^2 = r(b - c), \\ a + b = \square, \\ b + c = \square', \\ c + a = \square''. \end{cases}$

**"V",1:** $\begin{cases} a - k = \square, \\ b - k = \square', \\ c - k = \square'', \\ a:b = b:c. \end{cases}$

**"V",2:** $\begin{cases} a + k = \square, \\ b + k = \square', \\ c + k = \square'', \\ a:b = b:c. \end{cases}$

"V",3: $\begin{cases} a + k = \square, \\ b + k = \square', \\ c + k = \square'', \\ a \cdot b + k = \square''', \\ b \cdot c + k = \square^{\mathrm{IV}}, \\ c \cdot a + k = \square^{\mathrm{V}}. \end{cases}$

"V",4: $\begin{cases} a - k = \square, \\ b - k = \square', \\ c - k = \square'', \\ a \cdot b - k = \square''', \\ b \cdot c - k = \square^{\mathrm{IV}}, \\ c \cdot a - k = \square^{\mathrm{V}}. \end{cases}$

"V",5: $\begin{cases} a^2 \cdot b^2 + c^2 = \square, \\ b^2 \cdot c^2 + a^2 = \square', \\ c^2 \cdot a^2 + b^2 = \square'', \\ a^2 \cdot b^2 + (a^2 + b^2) = \square''', \\ b^2 \cdot c^2 + (b^2 + c^2) = \square^{\mathrm{IV}}, \\ c^2 \cdot a^2 + (c^2 + a^2) = \square^{\mathrm{V}}. \end{cases}$

"V",6: $\begin{cases} a - 2 = \square, \\ b - 2 = \square', \\ c - 2 = \square'', \\ a \cdot b - c = \square''', \\ b \cdot c - a = \square^{\mathrm{IV}}, \\ c \cdot a - b = \square^{\mathrm{V}}, \\ a \cdot b - (a + b) = \square^{\mathrm{VI}}, \\ b \cdot c - (b + c) = \square^{\mathrm{VII}}, \\ c \cdot a - (c + a) = \square^{\mathrm{VIII}}. \end{cases}$

**Lemma 1:** $a \cdot b + (a^2 + b^2) = \square.$

**Lemma 2:** $\begin{cases} a_1^2 + b_1^2 = c_1^2, \\ a_2^2 + b_2^2 = c_2^2, \\ a_3^2 + b_3^2 = c_3^2, \\ a_1 \cdot b_1 = a_2 \cdot b_2 = a_3 \cdot b_3. \end{cases}$

"V",7: $\begin{cases} a^2 + (a + b + c) = \square, \\ a^2 - (a + b + c) = \square', \\ b^2 + (a + b + c) = \square'', \\ b^2 - (a + b + c) = \square''', \\ c^2 + (a + b + c) = \square^{\mathrm{IV}}, \\ c^2 - (a + b + c) = \square^{\mathrm{V}}. \end{cases}$

**Lemma:** $\begin{cases} a \cdot b = k^2, \\ b \cdot c = l^2, \\ c \cdot a = j^2. \end{cases}$

**"V",8:** $\begin{cases} a \cdot b + (a + b + c) = \square, \\ a \cdot b - (a + b + c) = \square', \\ b \cdot c + (a + b + c) = \square'', \\ b \cdot c - (a + b + c) = \square''', \\ c \cdot a + (a + b + c) = \square^{\text{IV}}, \\ c \cdot a - (a + b + c) = \square^{\text{V}}. \end{cases}$

D **"V",9:** $\begin{cases} a + b = 1, \\ a + k = \square, \\ b + k = \square'. \end{cases}$

**"V",10:** $\begin{cases} a + b = 1, \\ a + k = \square, \\ b + l = \square'. \end{cases}$

D **"V",11:** $\begin{cases} a + b + c = 1, \\ a + k = \square, \\ b + k = \square', \\ c + k = \square''. \end{cases}$

**"V",12:** $\begin{cases} a + b + c = 1, \\ a + k = \square, \\ b + l = \square', \\ c + j = \square''. \end{cases}$

**"V",13:** $\begin{cases} a + b + c = k, \\ a + b = \square, \\ b + c = \square', \\ c + a = \square''. \end{cases}$

**"V",14:** $\begin{cases} a + b + c + d = k, \\ a + b + c = \square, \\ b + c + d = \square', \\ c + d + a = \square'', \\ d + a + b = \square'''. \end{cases}$

**"V",15:** $\begin{cases} (a + b + c)^3 + a = \text{cube}, \\ (a + b + c)^3 + b = \text{cube}', \\ (a + b + c)^3 + c = \text{cube}''. \end{cases}$

**"V",16:** $\begin{cases} (a + b + c)^3 - a = \text{cube}, \\ (a + b + c)^3 - b = \text{cube}', \\ (a + b + c)^3 - c = \text{cube}''. \end{cases}$

**"V",17:**
$$\begin{cases} a - (a+b+c)^3 = \text{❒}, \\ b - (a+b+c)^3 = \text{❒}', \\ c - (a+b+c)^3 = \text{❒}''. \end{cases}$$

**"V",18:**
$$\begin{cases} a+b+c = \square, \\ (a+b+c)^3 + a = \square', \\ (a+b+c)^3 + b = \square'', \\ (a+b+c)^3 + c = \square'''. \end{cases}$$

**"V",19:**[4] (a)
$$\begin{cases} a+b+c = \square, \\ (a+b+c)^3 - a = \square', \\ (a+b+c)^3 - b = \square'', \\ (a+b+c)^3 - c = \square'''. \end{cases}$$

(b)
$$\begin{cases} a+b+c = \square, \\ a - (a+b+c)^3 = \square', \\ b - (a+b+c)^3 = \square'', \\ c - (a+b+c)^3 = \square'''. \end{cases}$$

(c)
$$\begin{cases} a+b+c = k, \\ (a+b+c)^3 + a = \square, \\ (a+b+c)^3 + b = \square', \\ (a+b+c)^3 + c = \square''. \end{cases}$$

(d)
$$\begin{cases} a+b+c = k, \\ (a+b+c)^3 - a = \square, \\ (a+b+c)^3 - b = \square', \\ (a+b+c)^3 - c = \square''. \end{cases}$$

**"V",20:**
$$\begin{cases} a+b+c = \dfrac{1}{m}, \\ a - (a+b+c)^3 = \square, \\ b - (a+b+c)^3 = \square', \\ c - (a+b+c)^3 = \square''. \end{cases}$$

**"V",21:**
$$\begin{cases} a^2 \cdot b^2 \cdot c^2 + a^2 = \square, \\ a^2 \cdot b^2 \cdot c^2 + b^2 = \square', \\ a^2 \cdot b^2 \cdot c^2 + c^2 = \square''. \end{cases}$$

**"V",22:**
$$\begin{cases} a^2 \cdot b^2 \cdot c^2 - a^2 = \square, \\ a^2 \cdot b^2 \cdot c^2 - b^2 = \square', \\ a^2 \cdot b^2 \cdot c^2 - c^2 = \square''. \end{cases}$$

[4] The extant text has only the beginning of problem (a) and the end of problem (d).

**"V",23:** $\begin{cases} a^2 - a^2 \cdot b^2 \cdot c^2 = \square, \\ b^2 - a^2 \cdot b^2 \cdot c^2 = \square', \\ c^2 - a^2 \cdot b^2 \cdot c^2 = \square''. \end{cases}$

**"V",24:** $\begin{cases} a^2 \cdot b^2 + 1 = \square, \\ b^2 \cdot c^2 + 1 = \square', \\ c^2 \cdot a^2 + 1 = \square''. \end{cases}$

**"V",25:** $\begin{cases} a^2 \cdot b^2 - 1 = \square, \\ b^2 \cdot c^2 - 1 = \square', \\ c^2 \cdot a^2 - 1 = \square''. \end{cases}$

**"V",26:** $\begin{cases} 1 - a^2 \cdot b^2 = \square, \\ 1 - b^2 \cdot c^2 = \square', \\ 1 - c^2 \cdot a^2 = \square''. \end{cases}$

**"V",27:** $\begin{cases} a^2 + b^2 + k = \square, \\ b^2 + c^2 + k = \square', \\ c^2 + a^2 + k = \square''. \end{cases}$

**"V",28:** $\begin{cases} a^2 + b^2 - k = \square, \\ b^2 + c^2 - k = \square', \\ c^2 + a^2 - k = \square''. \end{cases}$

**"V",29:** $(a^2)^2 + (b^2)^2 + (c^2)^2 = \square.$

**"V",30:** $\begin{cases} k \cdot a + l \cdot b = \square, \\ (a + b)^2 = \square + j. \end{cases}$[5]

**"VI",1:** $\begin{cases} a^2 = b^2 + c^2, \\ a - b = \cube, \\ a - c = \cube'. \end{cases}$

**"VI",2:** $\begin{cases} a^2 = b^2 + c^2, \\ a + b = \cube, \\ a + c = \cube'. \end{cases}$

**"VI",3:** $\begin{cases} a^2 = b^2 + c^2, \\ \frac{1}{2}b \cdot c + k = \square. \end{cases}$

**"VI",4:** $\begin{cases} a^2 = b^2 + c^2, \\ \frac{1}{2}b \cdot c - k = \square. \end{cases}$

**"VI",5:** $\begin{cases} a^2 = b^2 + c^2, \\ k - \frac{1}{2}b \cdot c = \square. \end{cases}$

**"VI",6:** $\begin{cases} a^2 = b^2 + c^2, \\ \frac{1}{2}b \cdot c + b = k. \end{cases}$

[5] The enunciation is in the form of an epigram.

**"VI",7:** $\begin{cases} a^2 = b^2 + c^2, \\ \frac{1}{2}b \cdot c - b = k. \end{cases}$

**"VI",8:** $\begin{cases} a^2 = b^2 + c^2, \\ \frac{1}{2}b \cdot c + (b + c) = k. \end{cases}$

**"VI",9:** $\begin{cases} a^2 = b^2 + c^2, \\ \frac{1}{2}b \cdot c - (b + c) = k. \end{cases}$

**"VI",10:** $\begin{cases} a^2 = b^2 + c^2, \\ \frac{1}{2}b \cdot c + (a + b) = k. \end{cases}$

**"VI",11:** $\begin{cases} a^2 = b^2 + c^2, \\ \frac{1}{2}b \cdot c - (a + b) = k. \end{cases}$

**Lemma 1:** $\begin{cases} a^2 = b^2 + c^2, \\ b - c = \square, \\ b = \square', \\ \frac{1}{2}b \cdot c + c = \square''. \end{cases}$

**Lemma 2:** $\begin{cases} k \cdot a^2 + l = \square, \\ k + l = p^2. \end{cases}$

**"VI",12:** $\begin{cases} a^2 = b^2 + c^2, \\ \frac{1}{2}b \cdot c + b = \square, \\ \frac{1}{2}b \cdot c + c = \square'. \end{cases}$

**"VI",13:** $\begin{cases} a^2 = b^2 + c^2, \\ \frac{1}{2}b \cdot c - b = \square, \\ \frac{1}{2}b \cdot c - c = \square'. \end{cases}$

**"VI",14:** $\begin{cases} a^2 = b^2 + c^2, \\ \frac{1}{2}b \cdot c - a = \square, \\ \frac{1}{2}b \cdot c - b = \square'. \end{cases}$

**Lemma:** $\begin{cases} k \cdot a^2 - l = \square, \\ k \cdot n^2 - l = p^2, \end{cases} \qquad a^2 > n^2.$

**"VI",15:** $\begin{cases} a^2 = b^2 + c^2, \\ \frac{1}{2}b \cdot c + a = \square, \\ \frac{1}{2}b \cdot c + b = \square'. \end{cases}$

**"VI",16:** "To find a right-angled triangle such that, one of the acute angles being bisected, the number of the (measure of the) bisectrix is rational."

**"VI",17:** $\begin{cases} a^2 = b^2 + c^2, \\ \frac{1}{2}b \cdot c + a = \square, \\ a + b + c = \text{⧉}'. \end{cases}$

**"VI",18:** $\begin{cases} a^2 = b^2 + c^2, \\ \tfrac{1}{2}b \cdot c + a = \cube, \\ a + b + c = \square'. \end{cases}$

**"VI",19:** $\begin{cases} a^2 = b^2 + c^2, \\ \tfrac{1}{2}b \cdot c + b = \square, \\ a + b + c = \cube'. \end{cases}$

**"VI",20:** $\begin{cases} a^2 = b^2 + c^2, \\ \tfrac{1}{2}b \cdot c + b = \cube, \\ a + b + c = \square'. \end{cases}$

**"VI",21:** $\begin{cases} a^2 = b^2 + c^2, \\ a + b + c = \square, \\ (a + b + c) + \tfrac{1}{2}b \cdot c = \cube'. \end{cases}$

**"VI",22:** $\begin{cases} a^2 = b^2 + c^2, \\ a + b + c = \cube, \\ (a + b + c) + \tfrac{1}{2}b \cdot c = \square'. \end{cases}$

**"VI",23:** $\begin{cases} a^2 = b^2 + c^2, \\ a^2 = \square + \sqrt{\square}, \\ \dfrac{a^2}{b} = \cube' + \sqrt[3]{\cube'}. \end{cases}$

**"VI",24:** $\begin{cases} a^2 = b^2 + c^2, \\ a = \cube + \sqrt[3]{\cube}, \\ b = \cube', \\ c = \cube'' - \sqrt[3]{\cube''}. \end{cases}$

# Bibliography

Abenbeder (Ibn Badr): *Compendio de álgebra*, ed., trad. y estudio por J. Sánchez Pérez. Madrid, 1916.

Abū Kāmil: *Algebra*, i.e., *Kitāb fī'l-jabr wa'l-muqābala*$^{h}$. Ms Beyazıt 19046 (*olim* Kara Mustafa Paşa 379).
*See also:* Anbouba, A.; Levey, M.; Sesiano, J.

Abū'l-Faraj (bar Hebraeus): *Historia orientalis*, arab. ed. & lat. vert. E. Pococke. Oxoniae, 1672.

Anbouba, A.: *Un Algébriste arabe, Abu Kamil Šuǧāʿ ibn Aslam*, in: *Horizons techniques du Moyen-Orient*, 3 (1963), pp. 6–15.
*L'Algèbre al-Badīʿ d'al-Karaǧī*, éd., avec introd. et notes (=Publications de l'Université libanaise, Sect. des études mathém., II). Beyrouth, 1964.
*Un Traité d'Abū Jaʿfar [al-Khazin] sur les triangles rectangles numériques*, in: *Journal for the Hist. of Arabic Science*, 3 (1979), pp. 134–178.

Apollonius: *Apollonii Pergaei quae graece exstant cum commentariis antiquis*, ed. & lat. vert. J. L. Heiberg (2 vol.). Lipsiae, 1891–93.
*See also:* Nix, L.

Archimedes: *Archimedis Opera omnia cum commentariis Eutocii*, ed. & lat. vert. J. L. Heiberg (3 vol.). Lipsiae, 1910–15$^{2}$.

Aristotle: *Aristotelis Opera*, ed. Academia regia borussica (5 vol.). Berolini, 1831–70.
*Generation of animals. The Arabic translation commonly ascribed to Yaḥyâ ibn al-Biṭrîq*, ed., with introd. and gloss., by J. Brugman and H. J. Drossaart Lulofs. Leiden, 1971.
*See also:* Heath, Th.

Bachet de Méziriac, C. G.: *Diophanti Alexandrini Arithmeticorum libri sex et De numeris multangulis liber unus*, gr. et lat. cum comment. Lutetiae Paris., 1621.

Beeston, A.: *The Marsh manuscript of Apollonius's Conica*, in: *The Bodleian Library Record*, IV (1952–53), pp. 76–77.

Bergsträsser, G.: *Ḥunain ibn Isḥāq über die syrischen und arabischen Galen-Übersetzungen*, hrsg. und übers., in: *Abh. für die Kunde des Morgenlandes*, XVII, 2 (1925).

al-Bīrūnī: *The book of instruction in the elements of the art of astrology*, text, with transl. by R. R. Wright. London, 1934.

Blachère, R., *et al.*: *Dictionnaire arabe-français-anglais*. Paris, 1967–

Blemmydes, Nicephorus: *See* Nicephorus Blemmydes.

Bombelli, R.: *L'algebra*. Bologna, 1572.

Brockelmann, C.: *Grundriss der vergleichenden Grammatik der semitischen Sprachen* (2 Bde). Berlin, 1908–13.

Casiri, M.: *Bibliotheca arabico-hispana Escurialensis* (2 vol.). Matriti, 1760–70.

Caspari, C., Wright, W., *et al.*: *A grammar of the Arabic language* (2 vol.). Cambridge, 1896–98[3].

*Catalogus codicum manuscriptorum orientalium qui in Museo Britannico asservantur*, II (codd. arab.). Londini, 1846–71.

Chuquet, N.: *See* Marre, A.

Curtze, M.: *Eine Studienreise*, in: *Centralblatt für Bibliothekswesen*, 16 (1899), pp. 257–306.

Cydones, Demetrius: *See* Demetrius Cydones.

Daiber, H.: *Die arabische Übersetzung der Placita Philosophorum* (Diss.). Saarbrücken, 1968.

Demetrius Cydones: *Correspondance*, publ. par R.-J. Loenertz (2 vol., = Studi e Testi, 186 & 208). Città del Vaticano, 1956–60.

D. G. = Diophantus graecus: *See* Diophantus, ed. Tannery.

Diels, H.: *Doxographi graeci*. Berolini, 1879.

Dieterich, K.: *Untersuchungen zur Geschichte der griechischen Sprache* (=Byzant. Archiv, I). Leipzig, 1898.

Diophantus: *Diophanti Alexandrini Opera omnia cum graecis commentariis*, ed. & lat. vert. P. Tannery (2 vol.). Lipsiae, 1893–95.
*See also*: Bachet de Méziriac, C. G.; Heath, Th.; Sesiano, J.; Tannery, P.; Vogel, K.

Dozy, R.: *Supplément aux dictionnaires arabes* (2 vol.). Leyde, 1881.

*DSB* = *Dictionary of scientific biography* (16 vol.). New York, 1970–80.

Endreß, G.: *Die arabischen Übersetzungen von Aristoteles' Schrift De Caelo* (Diss.). Frankfurt am Main, 1966.

Euclid: *Euclidis Opera omnia*, edd. J. L. Heiberg et H. Menge (8 vol.). Lipsiae, 1883–1916.
*See also*: Heath, Th.; Klamroth, M.; Peyrard, F.

Euclid-Ḥajjāj: *Codex Leidensis 399,1*, i.e., *Euclidis Elementa ex interpretatione al-Hadschdschadschii cum commentariis al-Narizii*, edd. & lat. vertt. R. Besthorn et J. L. Heiberg (libri I–IV), G. Junge, J. Raeder et W. Thomson (libri V-VI), (6 fasc.). Hauniae, 1893–1932.

Euclid-Ṭūsī: *Kitāb taḥrīr uṣūl li-Uqlīdis min taʾlīf ḫōgaʰ* (ut fertur) *Naṣīr al-Dīn al-Ṭūsī*. Romae, 1594.

Euler, L. (Lagrange, J. L.): *Vollständige Anleitung zur Algebra, mit den Zusätzen von Joseph Louis Lagrange* (=Opera omnia I,1). Leipzig/Berlin, 1911.

Eutocius: *See* Apollonius, ed. Heiberg (vol. II); Archimedes.

Fleischer, H. L.: *Kleinere Schriften* (3 Bde). Leipzig, 1885–88.

Flügel, G.: *Ueber arabische und persische (...) Wortabkürzungen und die geheime Bedeutung der Buchstaben insbesondere*, in: *Zeitschr. d. deutsch. morgenl. Ges.*, 7 (1853), pp. 87–92.

della Francesca, Piero: *See* Piero della Francesca.

Freytag, G. W.: *Lexicon arabico-latinum* (4 vol.). Halis Sax., 1830–37.

Gabrieli, G.: *Nota biobibliografica su Qusṭā ibn Lūqā*, in: *Rendiconti della R. Acc. dei Lincei*, Cl. di sc. mor., stor. e filolog., s. V, XXI (1912), pp. 341–382.

Galen: *De partibus artis medicativae*, (...), *De diaeta in morbis acutis* (...), arab. ed. & angl. vert. M. Lyons (=Corpus medicorum graec., suppl. orient., II). Berolini, 1969.

Georgius Pachymeres: *See* Tannery, P.

Georr, Kh.: *Les Catégories d'Aristote dans leurs versions syro-arabes*. Beyrouth, 1948.

Ginzel, F. K.: *Handbuch der mathematischen und technischen Chronologie* (3 Bde). Leipzig, 1906–14.

Gollob, E.: *Ein wiedergefundener Diophantuscodex*, in: *Zeitschr. f. Math. u. Phys.*, hist.-liter. Abt., 44 (1899), pp. 137–140.

Graf, G.: *Der Sprachgebrauch der ältesten christlich-arabischen Literatur*. Leipzig, 1905.

Gulchīn-i Maʿānī, A.: *Fihrist-i kutub-i ḫaṭṭī-yi kitābḫāna-i Āstān-i quds-i Riḍawī*, VIII. Mashhad, 1350 (1971/72).

(al-)Ḥajjāj: See "Euclid-Ḥajjāj".

Ḥājjī Ḫalīfa[h]: *Lexicon bibliographicum et encyclopaedicum*, ed. & lat. vert. G. Flügel (8 vol.). Lipsiae, 1835–58.

Heath, Th.: *The thirteen Books of Euclid's Elements* (3 vol.). Cambridge, $1926^2$.
*Diophantus of Alexandria. A study in the history of Greek algebra*. Cambridge, $1910^2$.
*A history of Greek mathematics* (2 vol.). Oxford, 1921.
*Mathematics in Aristotle*. Oxford, 1949.

Heiberg, J. L.: *Anonymi logica et quadriuium, cum scholiis antiquis*, in: *D. Kgl. Danske Vidensk. Selsk.*, hist.-filol. Medd., XV,1. København, 1929.
*See also*: Apollonius; Archimedes; Euclid; Euclid-Ḥajjāj; Heron.

Heron: *Heronis Alexandrini Opera quae supersunt omnia* (praeter Belopoiica), edd. W. Schmidt, L. Nix, H. Schöne, J. L. Heiberg (5 vol. in 6). Lipsiae, 1899–1914.

Howell, M. S.: *A grammar of the classical Arabic language* (4 vol. in 7 pts). Allahabad, 1883–1911.

Hunger, H.: *Die hochsprachliche profane Literatur der Byzantiner* (2 Bde, = Hdbuch d. Altertumswiss., XII,5,1 & 2). München, 1978.

al-Ḫwārizmī: *The Algebra of Mohammed ben Musa*, ed. and transl. by F. Rosen. London, 1831.

Hypsicles: *Die Aufgangszeiten der Gestirne*, hrsg. (gr. u. arab.) u. übers. von V. de Falco und M. Krause, mit einer Einf. von O. Neugebauer, in: *Abh. d. Akad. d. Wiss. in Göttingen*, philol.-hist. Kl., 3. F., 62 (1966).

Ibn abī Uṣaibiʿa[h]: *Kitāb ʿuyūn al-anbāʾ fī ṭabaqāt al-aṭibbāʾ*, ed. A. Müller (2 vol.). Cairo, 1882.

Ibn al-Nadīm: *Kitâb al-Fihrist*, hrsg. mit Anm. von G. Flügel, J. Roediger und A. Müller (2 Bde). Leipzig, 1871–72.

Ibn al-Qifṭī: *See* Casiri, M.

Impellizeri, S.: *La letteratura bizantina*. Firenze/Milano, 1975.

Iriarte, J.: *Regiae Bibliothecae Matritensis codices graeci mss*, I. Matriti, 1769.

Johnson, F.: *A Dictionary, Persian, Arabic, and English*. London, 1852.

al-Karajī: *See* Anbouba, A.; Sesiano, J.; Woepcke, F.

Klamroth, M.: *Über den arabischen Euklid*, in: *Zeitschr. d. deutsch. morgenl. Ges.*, 35 (1881), pp. 270–326 & 788.

Kühnel, E.: *Islamische Schriftkunst* (=Monographien künstler. Schrift, IX). Berlin/ Leipzig, 1942.

Kutsch, W.: *Ṯābit b. Qurra's arabische Übersetzung der 'Αριθμητικὴ Εἰσαγωγή des Nikomachos von Gerasa*, hrsg. mit Wörterverz. (=Recherches publ. sous la dir. de l'Inst. de Lettres orient. de Beyrouth, IX). Beyrouth, 1959.

Lagrange, J. L.: *Additions à l'Algèbre d'Euler*. See Euler.

Lane, E. W.: *An Arabic-English Lexicon* (8 pts). London, 1863–93.

Lemerle, P.: *Le premier humanisme byzantin* (=Bibliothèque byzant., Etudes, VI). Paris, 1971.

Leonard of Pisa: *Scritti*, pubbl. da B. Boncompagni (2 vol.: (I) *Liber abbaci;* (II) *Practica geometriae*, *Flos*, *Epistola ad Theodorum*, *Liber quadratorum*). Roma, 1857–62.

Levey, M.: *The Algebra of Abū Kāmil*. Madison, 1966.

Libri, G.: *Histoire des sciences mathématiques en Italie* (4 vol.). Paris, 1838–41.

Luca Pacioli: *See* Pacioli, L.

Luckey, P.: *Ṯābit b. Qurra über den geometrischen Richtigkeitsnachweis der Auflösung der quadratischen Gleichungen*, in: *Sitz.-Ber. d. Akad. d. Wiss. zu Leipzig*, math.-phys. Kl., 93 (1941), pp. 93–114.

*Die Rechenkunst bei Ǧamšīd b. Masʿūd al-Kāšī, mit Rückblicken auf die ältere Geschichte des Rechnens*, (1944, posthumously edited) in: *Abh. für die Kunde des Morgenlandes*, XXXI, 1 (1951).

Marre, A.: *Notice sur Nicolas Chuquet et son Triparty en la science des nombres*, in: *Bullettino di bibliografia e di storia delle scienze matem. e fis.*, 13 (1880), pp. 555–659, 693–814.

Maximus Planudes: *Maximi monachi Planudis epistulae*, ed. M. Treu. (Programm d. Kgl. Friedrichs-Gymn. zu Breslau) Breslau, 1886–90.
*See also:* Diophantus, ed. Tannery (vol. II); Wendel, C.

Menelaus: *Die Sphärik von Menelaos aus Alexandrien in der Verbesserung von Abū Naṣr Manṣūr b. ʿAlī b. ʿIrāq*, von M. Krause, in: *Abh. d. Ges. d. Wiss. zu Göttingen*, philol.-hist. Kl., 3. F., 17 (1936).

Nallino, C.: *Al-Battānī sive Albatenii Opus astronomicum*, ed. et vers. lat. cum adnot. (3 vol., = Pubbl. del R. Osservatorio di Brera in Milano, XL, 1–3). Mediolani, 1899–1907.

Nesselmann, G.: *Beha-eddin's Essenz der Rechenkunst*, hrsg. und übers. Berlin, 1843.

Neugebauer, O.: *Mathematische Keilschrift-Texte* (3 Bde, = Quellen u. Studien z. Gesch. d. Math., Astr. u. Phys., Abt. A, 3, I–III). Berlin, 1935–37.

Nicephorus Blemmydes: *Curriculum vitae et carmina*, ed. A. Heisenberg. Lipsiae, 1896.

Nicolaus Rhabdas: *See* Tannery, P.

Nicomachus: *Nicomachi Geraseni Pythagorei Introductionis arithmeticae libri II*, ed. R. Hoche. Lipsiae, 1866.
*See also:* Kutsch, W.

Nix, L.: *Das fünfte Buch der Conica des Apollonius von Perga in der arabischen Übersetzung des Thabit ibn Corrah*, (teilweise) hrsg. und übers. Leipzig, 1889.
*See also:* Heron, *Opera* (vol. II: *Mechanica*).

Ostrogorsky, G.: *Geschichte des byzantinischen Staates* (=Hdbuch d. Altertumswiss., XII,1,2). München, 1940.

Pachymeres, Georgius: *See* Tannery, P.

Pacioli, L.: *Summa de Arithmetica, Geometria, Proportioni e Proportionalita.* Venetiis, 1494.

Pappus: *Pappi Alexandrini Collectionis quae* (graece) *supersunt*, ed. & lat. vert. cum comment. F. Hultsch ( 4 t. in 3 vol.). Berolini, 1875–78.
*The commentary of Pappus on Book X of Euclid's Elements*, Arabic text, transl. and comment. by W. Thomson and G. Junge (=Harvard Semitic Series, VIII). Cambridge (Mass.), 1930.
*See also:* Rome, A.

Peyrard, F.: *Les Oeuvres d'Euclide, en grec, en latin et en français* (3 vol.). Paris, 1814–18.

Piero della Francesca: *Trattato d'abaco*, a cura e con introd. di G. Arrighi. Pisa, 1970.

Planudes, Maximus: *See* Maximus Planudes.

Praechter, K.: *Hypatia*, in: Pauly-Wissowa, *Real-Encyclop. d. class. Altertumswiss.*, IX,1 (Stuttgart, 1914), pp. 242–249.

Proclus: *Procli Diadochi in primum Euclidis Elementorum librum commentarii*, ed. G. Friedlein. Lipsiae, 1873.
*See also*: Steck, M.

Ptolemy: *Syntaxis mathematica* (sc. Almagest), ed. J. L. Heiberg (2 vol.). Lipsiae, 1898–1903.

Reckendorf, H.: *Die syntaktischen Verhältnisse des Arabischen* (2 Tle). Leiden, 1895–98.
*Arabische Syntax.* Heidelberg, 1921.

Rhabdas, Nicolaus: *See* Tannery, P.

Rome, A.: *Le Troisième Livre des Commentaires sur l'Almageste par Théon et Hypatie*, in: *Annales de la Soc. scient. de Bruxelles*, 46 (1926), pp. 1–14.
*Commentaires de Pappus et de Théon d'Alexandrie sur l'Almageste* (3 t., = Studi e Testi, 54, 72, 106). Città del Vaticano, 1931–43.

de Sacy, S.: *Grammaire arabe* (2 vol.). Paris, $1831^2$.

Samaw'al: *Al-Bahir en Algèbre*, éd. avec notes par S. Ahmad et R. Rashed. Damas, 1972.

Sesiano, J.: *The Arabic text of Books IV to VII of Diophantus' 'Αριθμητικά in the translation of Qusṭā ibn Lūqā*, ed. with transl. and comm. (Diss.). Providence, 1975.
*Diophantus of Alexandria*, in: *DSB*, XV, pp. 118–122.
*Les Méthodes d'analyse indéterminée chez Abū Kāmil*, in: *Centaurus*, 21 (1977), pp. 89–105.
*Le Traitement des équations indéterminées dans le Badīʿ fī'l-ḥisāb d'Abū Bakr al-Karajī*, in: *Archive for Hist. of Ex. Science*, 17 (1977), pp. 297–379.

Simon, M.: *Sieben Bücher Anatomie des Galen*, hrsg. mit Übers. u. Komm. (2 Bde). Leipzig, 1906.

Steck, M., Schönberger, L.: *Proklus Diadochus: Kommentar zum ersten Buch von Euklids "Elementen"*. Halle (Saale), 1945.

Suter, H.: *Die Mathematiker und Astronomen der Araber und ihre Werke* (=*Abh. z. Gesch. d. math. Wiss. mit Einschluss ihrer Anw.*, X. Heft). Leipzig, 1900.

Ṯābit ibn Qurra[h]: *See* Kutsch, W.: Luckey, P.; Nix, L.

Tannery, P.: *A quelle époque vivait Diophante?*, in: *Bulletin des sciences mathém.*, 2e s., 3 (1879), pp. 261–269.
*L'article de Suidas sur Hypatia*, in: *Annales de la Fac. des Lettres de Bordeaux*, 2 (1880), pp. 197–201.
*La perte de sept livres de Diophante*, in: *Bulletin des sciences mathém.*, 2e s., 8 (1884), pp. 192–206.
*Notice sur les deux Lettres arithmétiques de Nicolas Rhabdas*, éd. et trad., in: *Notices et extraits des mss de la Bibliothèque Nationale*, XXXII,1 (1886), pp. 121–252.
*Rapport sur une mission en Italie*, in: *Archives des missions scient. et litt.*, 3e s., 14 (1888), pp. 409–455.
*Psellus sur Diophante*, in: *Zeitschr. f. Math. u. Phys.*, hist.-liter. Abt., 37 (1892), pp. 41–45.
*Notions historiques*, in: *Notions de Mathématiques* par J. Tannery (Paris, 1903), pp. 324–348.
*Sur le symbole de soustraction chez les Grecs*, in: *Bibliotheca mathematica*, 3. F., 5 (1904), pp. 5–8.
*Quadrivium de Georges Pachymère* (=Studi e Testi, 94). Città del Vaticano, 1940.
*Mémoires scientifiques* (17 vol.). Toulouse/Paris, 1912–50.
*See also*: Diophantus.

Theon of Alexandria: *See* Rome, A.; Toomer, G.

Thureau-Dangin, F.: *Textes mathématiques babyloniens*. Leyde, 1938.

Toomer, G.: *Theon of Alexandria*, in: *DSB*, XIII, pp. 321–325.
*Diocles: On burning mirrors* (=Sources in the Hist. of mathem. and phys. Sciences, I). Berlin/Heidelberg, 1976.

Turyn, A.: *Codices graeci vaticani saeculis XIII et XIV scripti annorumque notis instructi*. Città del Vaticano, 1964.
*Dated Greek manuscripts of the thirteenth and fourteenth centuries in the libraries of Italy* (2 vol.). Urbana, 1972.

(al-)Ṭūsī: *See* "Euclid-Ṭūsī".

Vogel, K.: *Zur Berechnung der quadratischen Gleichungen bei den Babyloniern*, in: *Unterrichtsbl. f. Math. u. Naturwiss.*, 39 (1933), pp. 76–81.

*Byzantine Science*, in: *Cambridge Mediaeval History*, IV,2 (Cambridge, 1967), pp. 264–305, 452–470.

*Diophantus of Alexandria*, in: *DSB*, IV, pp. 110–119.

Wehr, H.: *Arabisches Wörterbuch für die Schriftsprache der Gegenwart* (2 Tle). Leipzig, 1952.

Wendel, C.: *Planudea*, in: *Byzant. Zeitschrift*, 40 (1940), pp. 406–445.

Wilson, N. G.: *Three Byzantine scribes*, in: *Greek, Roman, and Byzantine Studies*, 14 (1973), pp. 223–228.

Woepcke, F.: *Extrait du Fakhrî*. Paris, 1853.

*Recherches sur plusieurs ouvrages de Léonard de Pise*. (1,III,A:) *Traduction d'un fragment anonyme sur la formation des triangles rectangles en nombres entiers, et* (1,III,B) *d'un traité sur le même sujet par Aboû Dja'far Mohammed Ben Alhoçaïn*, in: *Atti dell'Acc. Pont. de' Nuovi Lincei*, 14 (1861), pp. 211–227, 241–269; 301–324, 343–356.

Zenker, J.: *Dictionnaire turc-arabe-persan* (2 vol.). Leipzig, 1866–76.

# General Index

This index consists of six parts:

1. Index adnotationum.
   Index of the critical notes discussed or alluded to elsewhere in the book (observe that some references are given, not to the notes' numbers, but to the corresponding lines of the Arabic text).
2. Index codicum.
   Index of manuscripts used or mentioned.
3. Index graecitatis.
   Index of Greek scientific words defined or discussed.
4. Index auctorum (veterum ac recentiorum).
   Index of authors (a few historical personnages are also included).
5. Index propositionum librorum sex qui Graece supersunt.
   Index to the problems of the Greek *Arithmetica* referred to in this edition.
6. Index rerum ad Diophanti *Arithmetica* spectantium.
   Index to the *Arithmetica*, by subject.

## 1. Index adnotationum

**1:** 22*n*. **3:** 28. **4:** 87*n*. **6:** 22*n*. **7:** 30. **9:** 30. **11:** 66*n*. **15:** 28. **17:** 66. **19:** 36. **21:** 66. **27:** 66. **28–29:** 31, 66. **30:** 87*n*. **32:** 22, 296. **35:** 22, 33*n*. **36:** 31. **39:** 66. **41:** 449. **43:** 66, 66*n*. **44:** 87*n*, 286. **50:** 286. **52:** 66. **54:** 23. **57:** 32, 35*n*. **58:** 66. **68:** 29, 316. **74:** 47. **75:** 22, 330. **84:** 66. **87:** 66. **88:** 22*n*. **90:** 21. **91–92, 95:** 30, 185. **97:** 32. **100:** 22. **108:** 42. **109:** 30, 95*n*. **124:** 22, 28, 34. **125:** 42. **126–127:** 36*n*. **132:** 33 (B.26), 298. **134:** 22, 33. **136:** 28. **139:** 66. **142:** 21. **143:** 40. **147:** 66. **150:** 66. **151:** 66*n*. **157:** 36. **161:** 32, 35*n*, 98*n*. **163:** 66. **164:** 22, 36. **167:** 22*n*. **168:** 66, 422. **171:** 66*n*. **172:** 22, 22*n*, 28, 99*n*. **173:** 22*n*. **174:** 37*n*. **175:** 66. **177:** 36. **178:** 87*n*, 286. **181:** 66*n*. **191:** 329. **198:** 23. **208:** 36. **212:** 36. **215:** 36, 301. **216:** 30, 103*n*. **218:** 22*n*. **219:** 23. **224:** 287. **226:** 66. **228:** 36. **229:** 28, 34. **230:** 36. **232:** 36. **238:** 66. **241:** 23. **243:** 319. **244:** 62. **245:** 66*n*. **246:** 313. **247:** 319. **248:** 23. **249:** 36. **250:** 62. **251–252:** 34. **253:** 23, 414. **255–256:** 66. **264:** 320. **265:** 36. **268:** 66. **270:** 67*n*. **271:** 313. **272:** 42. **274:** 42. **282:** 47. **283:** 313. **286:** 87*n*, 286. **292:** 314. **294:** 316. **303:** 36. **305:** 36. **312:** 36. **313:** 36. **314:** 414. **317:** 36. **323:** 36. **326:** 29. **328:** 29. **329:** 29*n*,

362. **330**: 28, 293. **331**: 21 (two refs.). **333**: 29. **334**: 325. **338**: 29. **340**: 37*n*. **344**: 29. **345**: 323. **346**: 315. **347**: 33. **348**: 29. **349**: 29. **350**: 313. **351**: 66. **353**: 29. **354**: 37*n*. **355**: 29. **357**: 36. **360**: 40. **363**: 66. **365**: 66*n*. **366**: 292. **374**: 36. **376**: 36. **379**: 29. **380**: 66. **381**: 66. **382**: 22*n*. **383**: 328. **388**: 36. **389**: 22, 31, 115*n*. **391**: 455. **393**: 433. **394**: 328. **396**: 23. **400–403**: 34, 33 (no. 25). **404**: 29. **405**: 66, 328. **406**: 42. **407**: 42. **410**: 29. **413**: 29. **414**: 29. **415**: 66. **416**: 63. **417**: 66. **418**: 63. **419**: 63. **420**: 66. **421**: 63. **422**: 375. **424**: 66, 328. **425**: 66. **429**: 29. **431**: 36. **432**: 23. **433**: 36. **434**: 66, 287. **435**: 29, 455. **436**: 66. **437**: 63, 337. **439**: 29. **441**: 66. **442**: 36. **444**: 66. **450**: 22. **452**: 22*n*. **453**: 66. **455**: 66. **468**: 31. **469–470**: 33. **473**: 37*n*. **476**: 36. **478**: 40. **479**: 40. **480**: 36. **495**: 29. **497**: 66, 328. **498**: 36, 341. **502**: 66. **503**: 455. **505–506**: 32. **507, 511**: 69*n*. **516**: 67*n*. **519**: 36. **520**: 67*n*, 349. **521–522**: 36. **525**: 36. **528**: 36. **529**: 36. **531**: 33*n*. **532**: 23. **535**: 38. **541**: 66. **550**: 36. **551**: 38. **552**: 353. **553**: 32. **554**: 32, 128*n*. **565**: 37*n*. **566**: 29. **570**: 36. **575**: 22. **579**: 28. **583**: 47. **589**: 32. **593**: 37*n*. **594**: 47. **598**: 47. **599**: 29, 325. **612**: 36. **614**: 29, 455. **615**: 47. **616**: 42. **619**: 36. **621**: 22. **623**: 66. **625**: 66, 365. **629**: 134*n*. **630**: 361. **632**: 66. **637**: 66*n*. **638**: 32. **639**: 22*n*. **644**: 36. **649**: 29. **650**: 47. **652**: 47. **653**: 352. **658–659**: 329. **660**: 23. **661**: 47. **662**: 36. **665**: 47. **667**: 38. **676**: 337. **685**: 23. **686**: 22*n*. **688**: 32. **692**: 31, 45*n*. **703**: 37. **708**: 22. **709**: 36. **712**: 66*n*. **716**: 31, 63–64. **721**: 23, 338. **723**: 249. **727**: 251. **729**: 47. **730**: 313. **732**: 37*n*. **735**: 36. **737**: 28. **739**: 23. **745**: 37*n*. **746**: 34. **751**: 22, 36. **752**: 40. **765**: 66. **770**: 28, 32, 420. **771**: 28, 31. **772**: 31. **776**: 66. **777**: 23. **778**: 42. **781**: 36. **782**: 23. **785**: 42. **786–787**: 32, 35*n*. **788**: 22, 36. **793**: 40. **795**: 22. **799**: 37*n*. **819**: 323. **820**: 23. **821–823**: 36. **827**: 37. **829**: 31. **831**: 67*n*. **833**: 37, 405. **837**: 66. **838**: 38. **841**: 33, 35*n*. **858**: 29, 455. **871**: 36. **878**: 405. **882–883**: 33. **891**: 38. **898**: 23. **903**: 66*n*. **904**: 36. **906**: 36. **907**: 36. **909**: 40. **912**: 36. **913**: 36. **915**: 40. **916–917**: 36. **920**: 66. **921**: 66. **923**: 66*n*. **925**: 165*n*. **927**: 32, 393. **928**: 23. **933**: 165*n*, 419. **934**: 21. **935**: 47. **937**: 66. **940**: 165*n*, 419. **948**: 314. **951**: 36. **952**: 424. **953**: 23. **955**: 165*n*, 419. **957**: 23. **963**: 42. **965**: 23*n*. **968**: 36. **969**: 36. **970**: 36. **971**: 23. **974**: 38.

## 2. Index codicum

### (1) Arabic

[Cairo] *Dār al-kutub riyāḍ. 6 m*: 8*n*.
[Istanbul] *Aya Sofya 4830*: 23*n*.
*Beyazıt 19046*: See under Abū Kāmil (Index 4).
*Esat 3157*: 60, 60*n*, 188*n*, 197.
*Laleli 2714*: 60, 60*n*, 188*n*, 197.
*Köprülü 950*: 60, 60*n*, 188*n*, 197.
[London] *British Museum 407*: 8*n*.
[Mashhad] *Āstān-i quds-i Riḍawī 295*: 21–29 (description), 29–35 (interpolations), 35–37 (progenitor), 75 (genealogical tree), 283–429 (critical edition).
[Oxford] *Bodleian Marsh 667*: 22*n*.
[Paris] *Bibliothèque Nationale 2459*: 5, 23*n*, 60, 60*nn*, 188*n*, 194n, 197.

### (2) Greek

[Istanbul] *Mon. S. Sepulchri* (Μετόχιον Παναγίου Τάφου) *355*[1] : 16*n*.
*Palatii veteris* (Topkapı Sarayı) *1*: 16*n*.
[Madrid] *Matrit.* (B.N.) *4678*: 18–19, 40, 62*n*, 74(–75).

[1] Reportedly now held by a dealer in Paris.

[Milan] *Ambros. Et 157 sup.*: 20, 74–75.
[Oxford] *Bodleian. Dorvillian. 301*: 16*n*.
[Vatican] *Vat. 190*: 16*n*.
*Vat. 191*: 18–19, 62*n*, 446.
*Vat. 1594*: 15*n*.
[Venice] *Marc. 308*: 19–20, 74–75, 446.

(3) Latin

[Paris] *Parisin.* (B.N.) *7377A*: 45*n*.

(4) (Mesopotamian tablets)

[London] *British Museum 13901*: 237.
*British Museum 85194*: 237.

3. Index graecitatis

ἄλογος: 43–44.
ἀνάλυσις: 48–49.
ἀόριστος: 56*n*.
ἀπόδειξις: 48*n*; cf. 18.
ἀριθμός: 43–44; cf. 56*n*, 444.
ἀριθμοστόν: 45.
γραμμικός: 175; cf. 440.
δεδομένος (λόγος, λόγῳ): 228; cf. 230*n*.
διορισμός: 49.
δύναμις: 43; cf. 87*n*.
τετραπλῆ δύναμις: 43; cf. 45.
δυναμοδύναμις: 43.
δυναμόκυβος: 43.
δυναμοστόν: 45.
εἶδος: 458; cf. 261–262.
ἔκθεσις: 49; cf. 106*n*.
ἐπίπεδος: 175; cf. 443.
εὐχερής: 80*n*.
κύβος: 43.
κυβόκυβος: 43.
μετρεῖν: 446.
παρισότητος ἀγωγή: *See* "method of approximation to limits", index 6, C, c.
περιέχειν: 90*n*.
πηλικότης: 99*n*; cf. 65*n*.
πλασματικός: 192–193; cf. 99*n*.
πλευρά: 437; cf. 31.
πρότασις: 49.
στερεός: 67*n*.
συμπέρασμα: 49.
σύνθεσις: 48–49.
τετραγωνίζειν: 195–196.

4. Index auctorum (veterum ac recentiorum)

Abenbeder (Ibn Badr): 45*n*.
Abū Kāmil: 9–10, 13*n*, 41–42, 45, 58, 60, 81–82, 433–435, 440, 451, 455.
Abū'l-Faraj (bar Hebraeus): 4*n*, 10, 13.
Abū'l-Wafā᾿: 10.
Anbouba, A.: 9*n*, 10, 12, 83, 435, 451.
Anthemius: 14.
Apollonius: 15, 22*n*, 67*n*, 222*n*, 275, 314*n*, 434, 437, 442–443, 446, 451.
Archimedes: 15, 16*n*, 222*n*.
Arethas: 16*n*.
Aristotle: 246*n*, 447, 455.

Bachet de Méziriac, C.: 10*n*, 19*n*, 59*n*, 87*n*.
Bahā᾿al-Dīn: 279.
Bashmakova, I.: 3*n*.
Beeston, A.: 22*n*.
Bergsträsser, G.: 448.
al-Bīrūnī: 13*n*.
Blachère, R. (*et al.*): 436.
Blemmydes: *See* Nicephorus Blemmydes.
Bombelli, R.: 44.
Brockelmann, C.: 46, 450.
Bryennius: *See* Manuel Bryennius.

Casiri, M.: 10, 13.
Caspari, C. (Wright, W.): 23*n*, 28–29, 38–39, 41, 440.
Chuquet, N.: 44.
Cossali, P.: 18.
Curtze, M.: 14*n*.
Cydones: *See* Demetrius Cydones.
Cyrinus: 15.

Daiber, H.: 8*n*, 450.
Demetrius Cydones: 18.
Diels, H.: 43*n*, 56*n*.
Dieterich, K.: 442.
Dionysius: 87*n*.
Diophantus: (life and works:) 3, 13*n*, (*De polygonis numeris*:) 3, 17; (name, in Arabic:) 4, 441.
Dozy, R.: 439, 454.

Endreß, G.: 434, 440, 442.
Euclid: 275; (*Elements*:) 12, 15, 16*n*, 18, 43, 66, 90*n*, 91*n*, 95*nn*, 98*n*, 99*n*, 168*n*, 169*n*, 224–225, 236, 277, 439, 446, 457; (*Data*:) 228*nn*.
Euclid–Ḥajjāj: 165*n*, 423*n*, 434, 441–442, 447, 455, 457.
Euclid–Isḥāq: 441.
Euclid–Ṭūsī: 62*n*, 67*n*, 434, 436, 441, 443, 445.
Euler, L.: 199.
Eutocius: 14, 67*n*, 70, 99*n*.

Fleischer, H.: 28*n*, 41.
Flügel, G.: 4*n*, 13, 23*n*, 446.
della Francesca, P.: *See* Piero della Francesca.
Freytag, G.: 443, 451–452, 459.
Friedlein, G.: 228*n*.

Gabrieli, G.: 8*n*.
Galen: 442, 457.
Georgius Pachymeres: 17.
Georr, Kh.: 434–435, 440, 445, 455, 458.
Ginzel, F.: 22.
Gollob, E.: 14*n*.
Graf, G.: 28.
Gulchīn-i Maʿānī, A.: 21, 22*n*.

al-Ḥajjāj: *See* Euclid–Ḥajjāj.
Ḥājjī Ḫalīfaʰ: 446.
al-Ḫāzin: 10, 12–13, 83.
Heath, Th.: 3*n*, 49*n*, 67*n*, 76*nn*, 80*n*, 99*n*, 168*n*, 227*n*, 228*n*, 246*nn*, 275*n*.
Heiberg, J.: 17*n*, 18, 67*n*, 99*n*, 168*n*.
Heisenberg, A.: 17.
Heron Alex.: 8*n*, 16*n*, 41, 43, 49*n*, 76*n*, 175*n*, 423*n*, 434, 448, 450–451.
St. Hippolytus: 43, 56*n*.
Hoche, R.: 16*n*, 18*n*.
Howell, M.: 38.
Hultsch, F.: 48*n*, 247*n*.
Hunger, H.: 14*n*.
al-Ḫwārizmī: 30*n*, 39–40, 42, 435–437, 439–440, 447, 451, 455.
Hypatia: 71–75.
Hypsicles: 3, 433–434, 440–442, 445.

Ibn abī Uṣaiʿbiaʰ: 8–9, 11, 13.
Ibn al-Haiṯam: 11.
Ibn al-Nadīm: 4*n*, 8–10, 13.
Ibn al-Qifṭī: 8, 10, 13.
Ibn al-Sirāj: 5.
Ibn Yūnus: 11.
Impellizeri, S.: 15, 67.
Iriarte, J.: 18.
Isḥāq b. Ḥunain: *See* Euclid-Isḥāq.

Johnson, F.: 443.
Julian Emp.: 13*n*.
Justinian Emp.: 14.

al-Karajī: 5, 9–11, 13*n*, 40–42, 44*n*, (45*n*), 57–60, 73, 81–82, 180–185, 187, 189–192, 194–195, 197, 199, 201–205, 207–209, 211–214, 434–435, 437, (439), 447, 449–452, 459.
Klamroth, M.: 67*n*, 165*n*, 434–435, 441, 443, 446, 452, 455, 457.
Krause, M.: 62*n*.
Kühnel, E.: 22*n*.
Kutsch, W.: 67*n*, 433, 435, 443, 451, 458–459.

Lagrange, J.: 227.
Lane, E.: 459.
Lemerle, P.: 15*n*.
Leon VI Emp.: 16*n*.
Leon Math.: 14–16, 67.
Leonard of Pisa: 16, 83.
Levey, M.: 45*n*.
Libri, G.: 436–437, 439.
Loenertz, R.: 18.
Luckey, P.: 44*n*, 45, 47, 437, 446, 451.

al-Maʾmūn: 15.
Manuel Bryennius: 17.
Marcellus: 15.
Marinus Neap.: 14, 228*n*.
Maximus Planudes: 14, 16–20, 71, 74–75, 437.
Menelaus: 62*n*, 440.
Michael Psellus: 16, 43–44, 56*n*.
Mirzā Rezā Khān: 21.
Muḥammad b. abī Bakr: 22, 86.
Müller, A.: 11, 13.
Muscus (Μόσχος?): 16*n*.
al-Muʿtaṣim: 15*n*.
Muzalon: *See* Theodorus Muzalon.

Nallino, C.: 41–42, 451.
Nesselmann, G.: 279.
Neugebauer, O.: 236*n*.
Nicephorus Blemmydes: 16–17.
Nicolaus Rhabdas: 17.
Nicomachus: 13*n*, 17, 67*n*, 440, 443.
Nix, L.: 314*n*, 423*n*, 434, 437, 442–443, 446, 448, 450–451.
Nöldeke, Th.: 450.

Ostrogorsky, G.: 14*n*.

Pachymeres: *See* Georgius Pachymeres.
Pacioli, L.: 44.
Pappus: 48, 70, 175*n*, 247*n*, 457.
Paulus Alex.: 15.
Peyrard, F.: 16*n*.
Piero della Francesca: 44*n*.
Planudes: *See* Maximus Planudes.
Pococke, E.: 4*n*, 10.
Praechter, K.: 72.
Proclus: 14–15, 49*n*, 228.
Psellus: *See* Michael Psellus.
Ptolemy: 15, 72, 175*n*.
Pythagoras (Pythagoreans): 43; (Pythagorean equation and triplets:) 259, 277, 279.

Qusṭā b. Lūqā: 3, 8–9, (37), 46, 60, 67, 86, 450.

Reckendorf, H.: 29, 38, 433, 435, 450.
Rhabdas: *See* Nicolaus Rhabdas.
Rome, A.: 72, 99*n*, 247*n*.
Rosen, F.: (30*n*), 39, 42, 446.

de Sacy, S.: 28*n*.
Samawʾal: 11–13, 44*n*, 57*n*, 58, 73–74, 434, 450.
Simon, M.: 28, 328*n*, 442.
Simplicius: 14.
Stamatis, E.: 3*n*.
Steck, M.: 228*n*.
"Suidas": 71–72.
Suter, H.: 11*n*, 37, 45*n*.

Ṯābit b. Qurraʰ: 67*n*, 440, 443, 451.
Tannery, P.: 3*n*, 10, 13*n*, 14*n*, 16*n*, 17–19, 40, 43*n*, 46*n*, 48*n*, 51–54, 56–59, 66, 69*n*, 71, 74, 85*n*, 87*n*, 178*n*, 195*n*, 222, 446, 458.
Theodorus Muzalon: 17, 19.
Theon Alex.: 3, 15, 70–72, 99*n*, 175*n*.
Theophilus Emp.: 15.
Thureau-Dangin, F.: 236*n*.
Toomer, G.: 8*n*, 70, 72, 452.
Treu, M.: 17.
Turyn, A.: 17*n*, 18, 20.
al-Ṭūsī: *See* Euclid-Ṭūsī.

Veselovsky, I.: 3*n*.
Vogel, K.: 3*n*, 14*n*, 16, 236.

Wehr, H.: 432.
Wendel, C.: 19.
Wilson, N.: 15*n*.
Woepcke, F.: 5, 10*n*, 30*n*, 58, 60*n*, 83, 181, 187, 194, 435, 449.
Wright, R.: 13*n*.
Wright, W.: *See* Caspari, C.

Zenker, J.: 456.

5. Index propositionum librorum sex qui Graece supersunt.

*Book I:* **3:** 56. **5:** 69*n*. **8–9:** 57. **16:** 12–13, 57, 73. **17:** 57. **18:** 54. **19:** 54, 57. **21:** 54, 57, 267*n*. **26:** 12, 51, 52*n*, 58, 73–74, 195–196, 198. **27:** 192–193, 233, 236. **28:** 192–193, 236, 238. **30:** 192–193, 236. **31–34:** 51–52, 55. **35–38:** 55. **39:** 57.
*Book II:* **1–7:** 10, 51–53, 58. **8:** 6, 54, 69*n*; see also refs. p. 6*n*. **9:** 6, 69*n*; see also refs. p. 6*n*. **10:** 6, 57; see also refs. p. 6*n*. **11:** 6–7, 54, 58, 59*n*, 124*n*, 212, 221. **12:** 57, 125*n*, 212, 221, 278*n*. **13:** 54, 58, 124*n*, 212, 221. **14–15:** 53. **16:** 227*n*, 231*n*, 270. **17:** 10, 52, 54, 58. **18:** 52, 58. **19:** 7, 57; see also refs. p. 7*n*. **22:** 60, 278*n*. **24:** 56, 178*n*. **28:** 57, 175*n*, 259*n*. **29:** 175*n*, 259*n*. **30:** 209*n*. **31:** 57. **34:** 57, 278*n*.
*Book III:* **1–4:** 52 (N.B.), 222; (**1:** also 278*n*; **4:** also 10, 58). **5:** 54, 58, 158*n*, 255, 264. **6:** 5, 52, 54–55, 57, 266–267, 275. **10:** 57, 78, 80*n*, 265*n*. **11:** 56–57, 80*n*. **13:** 232,

278*n*. **15:** 54, 78, 227*n*, 232, 255, 267*n*. **16:** 259*n*. **17:** 232. **18:** 232. **19:** 10, 59. **20–21:** 53, 58.
*Book "IV":* **1–2:** 53–54, 233(–34). **3:** 53, 198. **7:** 54, 56. **8:** 78. **19:** 56. **22:** 77*n*. **28:** 54, 56, 267*n*. **31:** 54, 77*n*, 79, 259*n*, 267*n*. **32:** 271–272. **33:** 77*n*. **34:** 49, (Lemma:) 56. **35:** 49. **36:** 77*n*. **39:** 77*n*, 232.
*Book "V":* **8:** 56. **9:** 49, 257, 271. **10:** 77*n*. **11:** 49. **12–14:** 56*n*. **19:** 56*n*. **30:** 77*n*.
*Book "VI":* **6–11:** 77*n*. **12** (Lemma II): 79. **14:** 79, 271. **15** (Lemma): 80. **22:** 77*n*.

## 6. Index rerum ad Diophanti *Arithmetica* spectantium

### A. Historical rôle and influence of the *Arithmetica*

#### a. Pre-Diophantine elements in the *Arithmetica*

Link with Mesopotamian mathematics: 76, 236–237.
Power-system: 43.
Traces of excerpts from other works?: 275.

#### b. The *Arithmetica* in Greek times

Early recension of the Diophantine text: 14, 58, 68, 75.
Interpolated problems and their placement: 4, 8, 51–54, 68, 201, 205–206, 233–234, 244–249, 263–266; (problems possibly interpolated:) 52, 195–196, 198.
Alternative resolutions (genuine or not): 54–55, 68, 185, 187, 189–190, 207, 215, 258, 267–268.
Complements of various kinds (genuine or not): 55–57, 183–184, 189–191, 197, 203, 205, 209–210, 215–216, 239–240, 271, 276.
Major commentary: 48–50, 61, 63–65, 68–72, 75, 92*n*, 157*n*, 163*n* (cf. 164*n*), 178–179, 210, 246–247, 265; (extent of:) 72–75; (probably Hypatia's:) 71–72.
Addition of the final statements: 49–50, 61–65, 67, 72–73, 75, 246.
Minor interpolations (by Greek readers?) in the text: (of the Greek Books:) 69*n*, 178*n*, 198; (of the Arabic Books:) 35*n*, 232.
Corruptions (Greek?) in the text: (of the Greek Books:) 52 (2°); (of the Arabic Books:) 30 (nos. 3–4), 63 (no. 5), 219*n*, 263*n*, 265; see also p. 67.

#### c. The *Arithmetica* in Arabic times

Designation in Arabic: 13.
Arabic translation of the *Arithmetica*: 9, 58–62, 65–68, 73–75; (extent of:) 9–11, 73–74. On individual propositions of Books I–III, see index 5.
Arabic references to the *Arithmetica*: 5, 9–13.
Arabic commentaries to the *Arithmetica*: (by Qusṭā:) 9; (by Abū'l-Wafā᾽:) 10; (by Samaw᾽al:) 11, 13, 74.
Arabic extracts from the *Arithmetica*: (by al-Karajī:) 5, 10–11, 57–60; (by Samaw᾽al:) 12–13, 73–74.
Arabic complements to the *Arithmetica*: (by al-Ḫazin:) 10; (by Abū'l-Wafā᾽:) 10; (by Ibn al-Haiṯam / Ibn Yūnus:) 11; (by (?) al-Karajī:) 11, 57–59, 181, 190, 191, 194–195; (by Samaw᾽al:) 12–13, 73.
Arabic readers' glosses incorporated into the manuscript of Books IV–VII (cf. p. 23): 29–37, 159*n*, 249 (cf. 144, *n*. 16), 450; (partially reproduced:) 33 (no. 25), 34 (no. 27), 128*n*, 134*n*.

d. The *Arithmetica* in Byzantine times

Manuscript sources of the Greek *Arithmetica*: 14, 74–75.
 Manuscripts available to Planudes: 14, 17.
 Oldest extant today: 18–20.
Byzantine references to the *Arithmetica*: 16–18.
Byzantine commentaries to the *Arithmetica*: (by Pachymeres:) 17; (by Planudes:) 14, 17.
Byzantine extracts from the *Arithmetica*: (by Psellus:) 16; (by Rhabdas:) 17–18.
Byzantine complements to the *Arithmetica*: (by Cydones:) 18.
Byzantine readers' glosses: 18, 56*n*.

B. The thirteen Books of the *Arithmetica*, in particular the Books extant in Arabic

a. Placement of the Books

Arabic extant Books: 4–5, 7–8.
Later three Greek Books: 4, 8, 54, 68*n*, 263.
(Hence) missing Books: 80, 83–84.

b. Prefaces in the *Arithmetica*'s Books

to Book I (to the work): 6, 76–78, 175, 178–179.
to Book IV: 4, 46, 175–179.
to Book VII: 8, 176, 261–263.
to Book "IV", missing?: 77.

c. Character of the Arabic Books

Link between Books IV–VII and Books I–III: 5–8, 175–176, 179, 260, 263.
Purpose of Books IV–VII: 8, 76, 176, 260–263.
 Character of Book IV: 176.
  of Book V: 221–222, 226–228, 231, 233, 237–238, 242–243.
  of Book VI: 259–260, 268*n*.
  of Book VII: 261–263, 267, 268*n*.
Prolix form of Books IV–VII: 48–50.

d. Missing Books

Supposed content: 76–78, 80, 83–84.

C. The problems of the Arabic Books

a. Didactics and presentation

Arrangement of problems: 175–176, 184, 222. See also above, B, c.
Choice of convenient solutions: 200*n*, 203, 212 (*n*. 54), 218*n*, 230*n*.
Numbering of the problems: 59, 61–62, 62*n*, 118*n* (cf. 213), 189*n*, 191, 194–195.

b. Pattern of resolution

General pattern: 49, 61.
 Formulation: 49, 62*n*, 69; (defective or badly expressed:) 62, 68, 156*n*, 158*n*, 159*n*, 161*n*, 267; (variations in:) 134*n*.
 Condition of resolution (diorism): 49, 78; (one given number:) *IV,18*; *VII,6*; (two given numbers, whence constructible problem—see pp. 192–193:) *IV,17*; *IV,19–22*; *V,7–12* (see also p. 238); (condition established:) *IV,21–22*: pp. 196–197.
 Setting of the given magnitudes: 49, 106*n*.

Analysis: 48–50, 61, 69; (abbreviated:) 70, 106*n*, 226*n*; (incomplete:) *IV,22*: 197; *IV,41*: 118*n*, 213; *IV,42,a*: 214–215; *IV,44,b*: 218–219; (defective:) *IV,8–9*: 92*n*; *IV,13,2°*: 185; *IV,40–41*: 63; *VI,13*: 251–253; (in particular, condition not mentioned or defective:) *IV,28, 29, 31, 33*: 201–204; *IV,34,1°*: 206*n*; *VI,11*: 249; *VI,12*: 250–251; *VI,22,1°*: 257; *VII,5*: 265. See also above, "Major commentary."

Synthesis (apodeixis): 48–50, 61, 69; (abbreviated or incomplete:) *IV,7*: 70–71, 91*n*; *IV,14,e*: 70, 96*n*; *IV,15,1°*: 70, 97*n*; *IV,43*: 70–71, 217; *IV,44,a*: 70–71, 217; *IV,44,c*: 49–50, 70–71, 220; *V,3*: 70; *V,5*: 70, 230; *V,9*: 70, 234; *VI,4*: 49–50, 63, 247; *VII,15*: 275; (defective:) *IV,27(–28)*: 63; *VII,4*: 64, 265.

Final statements: [see also above, A, b;] (missing:) 72–73; (incomplete:) 72, 73*n*, 112*n*; (defective:) *IV,26*: 62; *IV,37*: 115*n*; *VI,4*: 63, 246; *VI,9*: 63–64; *VI,13,1°*: 72; *VII,14*: 64; (badly formulated:) 159*n*, 161*n*, 165*n*, 185–186; (badly placed:) 108*n*.

c. Techniques of resolution

Introductory rules ("definitions"): 87–88, 109*n*, 176–179.

Final equation aimed at: 8*n*, 76–78, 175–176, 179.

Algebraic reduction of an equation: 6, 87–88, 175–176, 179.

Reduction to problems of the first Greek Books: 5–7 (on the references: 5*n*, 69).

Method of the double-equation: *IV,34*: 206–208; *IV,42,a*: 214–215. Cf. 212, 221, 231–232.

Identification of a pair of equations: *IV,34,2°*; *IV,35–41*; *IV,42,a,2°*; *IV,42,b*; *IV,43* (;*IV,44*); 207–208, 221. Cf. the references on pp. 120 (*n.* 90), 121 (*n.* 93).

Reduction of systems of equations to linear ones: *IV,42,a,3°*; *IV,44,a–c*; *V,4–6*. Cf. pp. 215–216, 227, 228.

Problems soluble by the method of approximation to limits: 274*n*, 277.

d. Varia

Problems of Books IV–VII which are determinate: *IV,15*; *IV,17–22*; *V,7–12*.

Problems of Books IV–VII constructed from identities: *V,7–12*: 237–238; *V,13–16*: 242–243.

Diophantus and determinate second-degree equations: 76–78. Cf. p. 11.

Diophantus and indeterminate second-degree equations: 7, 78–80, 260.

Some non-soluble problems: 79, 181, 199, 257, 271–272, 279.

Conspectus of the problems of the whole *Arithmetica*: 461–483.

CPSIA information can be obtained
at www.ICGtesting.com
Printed in the USA
LVOW13s2116020118
561538LV00007B/29/P

9 781461 381761